Mass Spectrometry of Peptides

Editor

Dominic M. Desiderio, M.S., Ph.D.
Departments of Neurology and Biochemistry
University of Tennessee
Memphis, Tennessee

CRC Press
Boca Raton   Ann Arbor   Boston

Library of Congress Cataloging-in-Publication Data

Mass spectrometry of peptides / editor, Dominic M. Desiderio.
       p.   cm.
    Includes bibliographical references.
    ISBN 0-8493-6293-8
    1. Proteins — Analysis — Handbooks, manuals, etc. 2. Peptides —
Analysis — Handbooks, manuals, etc. I. Desiderio, Dominic M.
    [DNLM: 1. Peptides — analysis. 2. Spectrum Analysis, Mass. QU 68M414]
QP551.M32 1991
574.19′2456 — dc20
DNLM/DLC
for Library of Congress   90-1772
                                                 CIP

Direct all inquiries to CRC Press, Inc., 2000 Corporate Blvd., N.W., Boca Raton, Florida, 33431.

© 1991 by CRC Press, Inc.

International Standard Book Number 0-8493-6293-8

Library of Congress Card Number 90-1772
Printed in the United States

PREFACE

Va pensiero, sull' ali dorate.
G. Verdi, *Nabucco*

The purpose of this book is to collect into one volume a description of the wide range of the state-of-the-art research using mass spectrometry to analyze peptides. The leaders in their respective fields describe their up-to-date research, which spans a wide range of different research topics. The book includes three main areas: ionization methods, instrumental developments, and analysis of endogenous and synthetic peptides. Ionization methods include the use of plasma desorption and laser-induced multiphoton ionization. The instrumental developments discussed include magnetic sector, quadrupole, and time-of-flight instrumentation. The section on analysis of peptides describes sample preparation, continuous-flow fast atom bombardment, analysis of various proteins and cross-linkages, interactions with metal ions, permethylated peptides, and neuropeptides.

It has been gratifying to edit such a book because of the cooperation of each author, their commitment to deadlines, and their detailed descriptions of the state-of-the-art techniques, instrumentation, and methods used for the important area of peptide analysis. We feel that, collectively, we have attained the depth of this topic that is demanded by the research currently being conducted in the areas of biotechnology, neurosciences, mutation, and clinical and biochemical analyses of peptides. We hope that this book will be found useful by researchers, teachers, and students.

Dominic Desiderio
September 1990

INTRODUCTION

Progress during the first two decades of the involvement of mass spectrometry with amino acid and peptide characterization was severely constrained by the zwitterionic character of amino acids and peptides, and most tellingly by the need for vaporization of the sample into the ion source of the mass spectrometer, a process that limited molecular size to around 1000 Daltons. By far the largest fraction of research effort during this period was devoted to development of increasingly general strategies for carrying out suitable quantitative derivitization procedures for amino acids and peptides and for methods that would enable the separation and identification of mixtures of peptides obtained from partial acid hydrolysis of proteins.

Milestones in this history include the elimination of the zwitterionic property and suppression of diketopiperazine formation during volatilization of amino acids by use of the ethyl ester derivative,[1] reduction of small polypeptides with lithium aluminum hydride to form the corresponding polyamino alcohols,[2] N,O-acetylation, and permethylation of peptides[3] illustrated by determination of the sequence and structure of the nonapeptidolipid fortuitine,[4] computer-aided interpretation of peptide sequences using high resolution mass spectra,[5] and finally the use of GCMS to separate and sequence the trimethylsilylated derivatives of polyamino alcohols.[6] Using accurate mass measurement from photoplate recording, Desiderio and co-workers identified the thyroid-stimulating hormone releasing factor, TRF.[7] The acetylation-permethylation procedure permitted the discovery of the sequence of the neuropeptide enkephalin by Morris and co-workers[8] in the mid-70s. Further work resulted in sequencing of dihydrofolate reductase, a 180 amino acid protein.[9] The polyamino alcohol/GCMS methodology culminated in the late 70s with Biemann's studies of the primary sequence of bacteriorhodopsin[10] and the development of a strategy for verification of the fidelity of cDNA sequences by mapping selected portions of the transcribed protein sequence.[11]

The chronicle of advances (including FDMS and PDMS) that occurred during the first two decades has been described in considerable detail in chapters by Vetter[12] and Biemann.[13] This review is not to say that mass spectrometry did not find an almost unique niche in protein biochemistry despite the constraints on volatility and molecular weight suffered when the primary ion sources available were limited to electron impact and chemical ionization. During this period, it was recognized gradually that mass spectrometry was becoming the method of choice for determination of posttranslational modification of proteins as well as in the identification of new amino acids.[14]

Most dramatically, this last decade has changed all that, bringing the field to its present threshold of becoming the method of choice for sequencing and structural elucidation of new peptides, proteins and glycoproteins, and recombinant proteins. The progress has been unusually rapid, due in large part to the national emphasis on the upgrading of instrumentation during this period,[15] which created an atmosphere in which funding of the development of entirely new mass spectrometric strategies could take place. However, the discovery that underivatized peptides and small proteins could be splashed off the surfaces of viscous liquids as a continuous beam of ions upon irradiation of these surface with energetic beams of atoms and ions provided the necessary initiating breakthrough in 1980.[16] The present suite of instrumental capabilities is now very powerful and will certainly be able to continue solving new problems in structural biology in the immediate future.[17,18] However, continuing capital investment in another generation of instrument innovations based on what has been learned and recently discovered[19] will permit realization of well-founded promise as well as sustained progress toward unraveling the complexities of peptides and proteins of undiscovered function, which are the essential machinery of our cells.[20]

Presently the costs of this capability are rather high and the number of laboratories involved rather low, but the power, versatility, and speed of these methods are well-suited to tackle the complexities of research in peptide and protein biochemistry. The future is bright, and the next decade is set for explosive growth of the new methods of mass spectrometry in this field.

The present arsenal permits the ready measurement of molecular weight of peptides and the components of mixtures of peptides both qualitatively and quantitatively. In the field of neuropeptides, the work of Williams and co-workers on *Xenopus laevis* is a case in point.[21] This strategy speeds the identification and cloning of the genes. Other types of mass spectrometric investigation are used to carry out the quantitation necessary for physiological experimentation.[22] Proteins themselves may be digested by specific proteolytic enzymes to generate once again mixtures readily sufficiently separated on narrow bore and microbore liquid chromatography for mass mapping,[23] most recently by LC flow-probe MS.[24] Following mass mapping, sequences of these components may be determined by the new methods of tandem mass spectrometry[25] even at the picomole level using recently available multichannel array detection systems.[23,26,27] Glycoproteins are amenable to similar strategies, as is the identification of their posttranslational modifications.[18,20]

Finally, the molecular weights of these intact proteins may be measured with considerably enhanced accuracy (± 1 to 2 Daltons) compared with SDS gels by the new method of electrospray mass spectrometry[29] such that the purity, confirmation of the complete sequence,[30] and the discovery of posttranslation modifications may be speeded.[31] These developments are readily applicable to recombinant proteins and will be used extensively in the biotechnology industry to authenticate recombinant substances and to elucidate the nature of cloning and expression problems and posttranslational modifications that may differ depending upon expression system and growth conditions, causing potential concern for changes in immunogenicity or pharmacodynamics in the case of proteins under development as therapeutic agents.[28,31]

Al Burlingame
San Francisco

REFERENCES

1. **Biemann, K., Seibl, J., and Gapp, F.,** *Biochem. Biophys. Res. Commun.,* 1, 307, 1959; *J. Am. Chem. Soc.,* 83, 3795, 1961.
2. **Biemann, K., Gapp, F., and Siebl, J.,** *J. Am. Chem. Soc.,* 81, 2274, 1959.
3. **Das, B. C., Géro, S. D., and Lederer, E.,** *Biochem. Biophys. Res. Commun.,* 29, 211, 1964.
4. **Barber, M., Jollés, P., Vilkas, E., and Lederer, E.,** *Biochem. Biophys. Res. Commun.,* 18, 469, 1965.
5a. **Senn, M., Vankataraghavan, R., and McLafferty, F. W.,** *J. Am. Chem. Soc.,* 88, 5593, 1966.
5b. **Biemann, K., Cone, C., Webster, B., and Arsenault, G. P.,** *J. Am Chem. Soc.,* 88, 5598, 1966.
6. **Nau, H.,** *Biochem. Biophys. Res. Commun.,* 59, 1088, 1974.
7. **Burgus, R., Dunn, T. F., Desiderio, D. M., Ward, D. N., Vale, W., and Guillemin, R.,** *Nature,* 269, 1870, 1969.
8. **Hughes, J., Smith, T. W., Kosterlitz, H. W., Fothergill, L., Morgan, B. A., and Morris, H. R.,** *Nature,* 258, 577, 1975.
9. **Morris, H. R., Batley, K. E., Dell, A., and Inglis, A.,** 26th Ann. Conf. Mass Spectrometry and Allied Topics, paper FB6, St. Louis, 1978.
10. **Herlihy, W. C., et al.,** *Biomed. Mass Spectrom.,* 8, 62, 1981.
11. **Gibson, B. W. and Biemann, K.,** *Proc. Natl. Acad. Sci. U.S.A.,* 81, p. 1956, 1984.
12. **Vetter, W.,** in *Biochemical Applications of Mass Spectrometry,* 1st suppl. vol., Waller, G. R. and Dermer, O. S., Eds., Wiley-Interscience, New York, 1980, 439.
13. **Biemann, K.,** *ibid.,* 469.
14. **Carr, S. A. and Biemann, K.,** *Meth. Enzymol.,* 106, 29, 1984.
15. **Handler, P.,** *Science,* 204, 474, 1979.

16. **Barber, M., Bordoli, R. S., and Sedgwick, R. D.,** in *Soft Ionization Biological Mass Spectrometry,* Morris, H. R., Ed., Heyden & Son, London, 1981, 137.

17. **Burlingame, A. L., Maltby, D., Russell, D. H., and Holland, P. T.,** *Anal. Chem.,* 60, 294R, 1988.

18. **Burlingame, A. L. and McCloskey, J. A., Eds.,** *Biological Mass Spectrometry,* Elsevier, Amsterdam, 1990.

19. **Burlingame, A. L., Millington, D., Norwood, D., and Russell, D. H.,** *Anal. Chem., Fund. Revs.,* 62, 268R, 1990.

20. **Kornberg, A.,** *Biochemistry,* 26, 6888, 1987.

21. **Gibson, B. W., Poulter, L., Williams, D. H., and Maggio, J.,** *J. Biol. Chem.,* 261, 5341, 1986.

22. **Desiderio, D. M.,** *Analysis of Neuropeptides by Liquid Chromatography and Mass Spectrometry,* Elsevier, Amsterdam, 1984.

23. **Gibson, B. W., Yu, Z., Gillece-Castro, B., Aberth, W., Walls, F. C., and Burlingame, A. L.,** in *Techniques in Protein Chemistry,* Hugli, T., Ed., Academic Press, San Diego, 1989, 135.

24. **Caprioli, R. M., Fan, T., and Cottrell, J. S.,** *Anal. Chem.,* 58, 2949, 1986.

25. **Biemann, K. and Scoble, H. A.,** *Science,* 237, 992, 1987.

26. **Burlingame, A. L.,** in Proc. 36th Ann. Conf. Mass Spectrometry and Allied Topics,1988, 727.

27. **Hill, J. A., Martin, S. A., Biller, J. E., and Biemann, K.,** 36th Ann. Conf. Mass Spectrometry and Allied Topics, San Francisco, 1988, 849.

28. **Carr, S. A., Roberts, G. D., and Hemling, M. E.,** in *Mass Spectrometry of Biological Materials,* McEwen, C. N. and Larsen, B. S., Eds., Marcel Dekker, New York, 1990, 87.

29. **Fenn, J., Mann, M., Meng, C. K., Wong, S. F., and Whitehouse, C. M.,** *Science,* 246, 64, 1989.

30. **Kaur, S., Medzihradszky, K., Baldwin, M. A., Gillece-Castro, B. L., Walls, F. C., Gibson, B. W., and Burlingame, A. L.,** in *Biological Mass Spectrometry,* Burlingame, A. L. and McCloskey, J. A., Eds., Elsevier, Amsterdam, 1990, chap. 17.

31. **Poulter, L., Gibson, B. W., Kaur, S., and Burlingame, A. L.,** in *Biological Mass Spectrometry,* Burlingame, A. L. and McCloskey, J. A., Eds., Elsevier, Amsterdam, 1990, chap. 7.

EDITOR

Dominic M. Desiderio, M.S., Ph.D., is a Professor of Neurology (Chemistry) in the Department of Neurology, a Professor of Biochemistry in the Department of Biochemistry, and the Director of the Charles B. Stout Neuroscience Mass Spectrometry Laboratory at the University of Tennessee, Memphis, Tennessee.

Dr. Desiderio received his education at the University of Pittsburgh, where he received a B.A. in 1961, and at the Massachusetts Institute of Technology, where he received M.S. (1964) and Ph. D. (1965) degrees. He was an Assistant Professor (1967 to 1971) and an Associate Professor of Chemistry (1971 to 1978) in the Institute for Lipid Research, and Associate Professor of Biochemistry (1971 to 1978) in the Department of Biochemistry, Baylor College of Medicine.

Dr. Desiderio is a member of the American Society of Biological Chemistry, American Chemical Society, American Society of Mass Spectrometry, Society for Neuroscience, and the Protein Society. He is on the Editorial Board of *The Journal of Chromatography-Biomedical Applications.* He received the first annual International Award in Mass Spectrometry and Biochemistry in Medicine, Alghero, Italy (1975), is listed in *Who's Who in America,* and was a member of the National Institutes of Health Metallobiochemistry Study Section (1985 to 1989). Dr. Desiderio is the author of more than 165 scientific papers, two books, and several book chapters. He has been an invited speaker at meetings in the U.S. and abroad. His major research interests include the use of mass spectrometry to analyze biologically important molecules, quantification of neuropeptides by mass spectrometry, and the effects of stress and pain on human neuropeptidergic systems.

CONTRIBUTORS

Alison E. Ashcroft, M.Sc., Ph.D.
Safety of Medicines Department
ICI Pharmaceuticals
Alderley Park, Macclesfield
Cheshire, U.K.

Douglas F. Barofsky, M.S., Ph.D.
Professor
Department of Agricultural Chemistry
Oregon State University
Corvallis, Oregon

R. S. Bordoli, Ph.D.
Managing Director
VG Analytical Ltd.
Manchester, U.K.

David M. Bunk, B.A.
Research/Teaching Assistant
Department of Chemistry
Texas A&M University
College Station, Texas

Kenneth L. Busch, Ph.D.
Associate Professor of Chemistry
School of Chemistry and Biochemistry
Georgia Institute of Technology
Atlanta, Georgia

Richard Caprioli, Ph.D.
Analytical Chemistry Center
University of Texas Medical School
Houston, Texas

Ronald L. Cerny, Ph.D.
Assistant Director
Department of Chemistry
University of Nebraska
Lincoln, Nebraska

Brian T. Chait, D.Phil.
Associate Professor
Laboratory of Mass Spectrometry
 and Gaseous Ion Chemistry
The Rockefeller University
New York, New York

Ling Chen
Applied BioSystems, Inc.
Foster City, California

Robert J. Cotter, Ph.D.
Director
Middle Atlantic Mass Spectrometry Facility
Johns Hopkins University
Baltimore, Maryland

Chhabil Dass, M.S., Ph.D.
Assistant Professor
Department of Neurology
University of Tennessee
Memphis, Tennessee

Peter J. Derrick, M.Sc., Ph.D.
Professor
Department of Chemistry
University of Warwick
Coventry, U.K.

Dominic M. Desiderio, M.S., Ph.D.
Departments of Neurology and
 Biochemistry
University of Tennessee
Memphis, Tennessee

W. Ens, Ph.D.
Assistant Professor
Department of Physics
University of Manitoba
Winnipeg, Manitoba, Canada

Kym F. Faull, Ph.D.
Department of Psychiatry and Biobehavioral
 Sciences and the Neuropsychiatric
 Institute
UCLA School of Medicine
Los Angeles, California

Gottfried J. Feistner, Dr. rer. nat.
Division of Immunology
Beckman Research Institute of the City
 of Hope
Duarte, California

Brian N. Green, B.Sc.
VG Analytical Ltd.
Manchester, U.K.

P. R. Griffin, Ph.D.
Department of Biology
California Institute of Technology
Pasadena, California

Michael L. Gross, Ph.D.
Professor
Department of Chemistry
University of Nebraska
Lincoln, Nebraska

C. R. Hauer, Ph.D.
Department of Chemistry
University of Virginia
Charlottesville, Virginia

D. F. Hunt, Ph.D.
Professor
Department of Chemistry
University of Virginia
Charlottesville, Virginia

Itsuo Katakuse, Ph.D.
Department of Physics
Faculty of Science
Osaka University
Toyonaka, Osaka, Japan

T. Krishnamurthy, Ph.D.
Research Directorate
U.S. Army Chemical Research,
  Development, and Engineering Center
Aberdeen Proving Ground
Aberdeen, Maryland

Terry D. Lee, Ph.D.
Department of Immunology
City of Hope
Duarte, California

David M. Lubman, Ph.D.
Associate Professor
Department of Chemistry
University of Michigan
Ann Arbor, Michigan

Ronald D. Macfarlane, M.S., Ph.D.
Professor
Department of Chemistry
Texas A&M University
College Station, Texas

P. A. Martino, Ph.D.
Department of Chemistry
University of Virginia
Charlottesville, Virginia

Hisashi Matsuda, Ph.D.
Institute of Physics
College of General Education
Osaka University
Toyonaka, Osaka, Japan

Takekiyo Matsuo, Ph.D.
Institute of Physics
College of General Education
Osaka University
Toyonaka, Osaka, Japan

A. L. McCormack, Ph.D.
Department of Chemistry
University of Virginia
Charlottesville, Virginia

William T. Moore, Ph.D.
Analytical Chemistry Center
University of Texas Medical School
Houston, Texas

Paul D. Mudgett, B.S.
Research/Teaching Assistant
Department of Chemistry
Texas A&M University
College Station, Texas

Samuel Rahbar, M.D.
Department of Hematology and BMT
City of Hope
Duarte, California

Peter Roepstorff, Ph.D.
Professor
Department of Molecular Biology
Odense University
Odense, Denmark

Keith Rose, M.A., D.Phil.
Department of Biomedical Chemistry
University Medical Center
Geneva, Switzerland

William E. Seifert, Jr., Ph.D.
Analytical Chemistry Center
University of Texas Medical School
Houston, Texas

J. Shabanowitz, Ph.D.
Department of Chemistry
University of Virginia
Charlottesville, Virginia

David L. Smith, Ph.D.
Associate Professor
Department of Medicinal Chemistry
  and Pharmacognosy
Purdue University
Lafayette, Indiana

K. G. Standing, A.M., Ph.D.
Professor
Department of Physics
University of Manitoba
Winnipeg, Manitoba, Canada

Yiping Sun, B.S.
Department of Medicinal Chemistry
  and Pharmacognosy
Purdue University
Lafayette, Indiana

Xuejun Tang, Ph.D.
Department of Physics
University of Manitoba
Winnipeg, Manitoba, Canada

Yoshinao Wada, M.D.
Osaka Medical Center and Research
  Institute for Maternal and Child Health
Izumi, Osaka, Japan

Rong Wang, Ph.D.
Middle Atlantic Mass Spectrometry Facility
Johns Hopkins University
Baltimore, Maryland

J. B. Westmore, Ph.D.
Department of Chemistry
University of Manitoba
Winnipeg, Manitoba, Canada

Barbara Wolf, B.S.
Department of Chemistry
Texas A&M University
College Station, Texas

John R. Yates, III, Ph.D.
Department of Biology
California Institute of Technology
Pasadena, California

TABLE OF CONTENTS

SECTION I. Ionization Methods

Part A: ^{252}Cf Plasma Desorption Methods

Chapter 1

Fundamental Aspects of Protein Mass Spectrometry Using 252-Californium
Plasma Desorption 3
Ronald D. Macfarlane, David Bunk, Paul Mudgett, and Barbara Wolf

Chapter 2

Plasma Desorption Mass Spectrometry of Peptides and Peptide Conjugates 17
Robert J. Cotter, Ling Chen, and Rong Wang

Chapter 3

The Analysis of Synthetic Peptides and Proteins by ^{252}Cf-Plasma Desorption
Mass Spectrometry 41
Brian T. Chait

Chapter 4

Analysis of Peptides and Proteins by Plasma Desorption Mass Spectrometry 65
Peter Roepstorff

Part B: Laser-Induced Multiphoton Ionization

Chapter 5

Laser-Induced Multiphoton Ionization of Peptides in Supersonic
Beam/Mass Spectrometry 87
David M. Lubman

SECTION II. Instrumental Developments

Part A: Magnetic Sector Instruments

Chapter 6

The Molecular Weight Determination of Large Peptides by Magnetic Sector
Mass Spectrometry 109
B. N. Green and R. S. Bordoli

Chapter 7

Four-Sector Tandem Mass Spectrometry of Peptides 121
Alison E. Ashcroft and Peter J. Derrick

Part B: Quadrupole Fourier Transform

Chapter 8

Peptide Sequence Analysis by Triple Quadrupole and Quadrupole Fourier
Transform Mass Spectrometry 139
**D. F. Hunt, T. Krishnamurthy, J. Shabanowitz, P. R. Griffin, J. R. Yates, III,
P. A. Martino, A. L. McCormack, and C. R. Hauer**

Part C: Time-of-Flight Instruments

Chapter 9

Correlation Measurements in a Reflecting Time-of-Flight Mass Spectrometer 159
K. G. Standing, W. Ens, and J. B. Westmore

SECTION III. Analysis of Peptides

Part A: Sample Preparation
Chapter 10
Sample Preparation and Matrix Selection for Analysis of Peptides by FAB
and Liquid SIMS 173
Kenneth L. Busch

Part B: LC-MS Analysis
Chapter 11
On-Line Methods for Peptide Analysis by Continuous-Flow FABMS 201
William E. Seifert, Jr., William T. Moore, and Richard M. Caprioli

Part C: Analysis of Protein Products
Chapter 12
Investigation of Amino Acid Mutations by High Resolution Mass Spectrometry 223
Hisashi Matsuda, Takekiyo Matsuo, Itsuo Katakuse, and Yoshinao Wada

Chapter 13
The Mass Spectral Analysis of Hemoglobin Variants 257
Terry D. Lee and Samuel Rahbar

Part D: Protein Cross-Linkages
Chapter 14
Detection and Location of Disulfide Bonds in Proteins by Mass Spectrometry 275
David L. Smith and Yiping Sun

Part E: Peptide Interactions with Metal Ions
Chapter 15
Tandem Mass Spectrometry for Determining the Amino Acid Sequence of Cyclic
Peptides and for Assessing Interactions of Peptides and Metal Ions 289
Ronald L. Cerny and Michael L. Gross

Part F: Permethylated Peptides
Chapter 16
Analysis of Permethylated Peptides by Mass Spectrometry 315
Keith Rose

Part G: Neuropeptides
Chapter 17
Applications of Mass Spectrometry for Characterization of Neuropeptides 327
Chhabil Dass

Chapter 18
Peptide-Charting Applied to Studies of Precursor Processing in Endocrine Tissues 347
Douglas F. Barofsky, Gottfried J. Feistner, Kym F. Faull, and Peter Roepstorff

Part H: Quantification of Neuropeptides
Chapter 19
Mass Spectrometry of Biologically Important Neuropeptides 367
Dominic M. Desiderio

Index 403

Section I: Ionization Methods

Part A: ^{252}Cf Plasma Desorption Methods

Chapter 1 3

Chapter 2 17

Chapter 3 41

Chapter 4 65

Part B: Laser-Induced Multiphoton Ionization

Chapter 5 87

Chapter 1

FUNDAMENTAL ASPECTS OF PROTEIN MASS SPECTROMETRY USING 252-CALIFORNIUM PLASMA DESORPTION

Ronald D. Macfarlane, David Bunk, Paul Mudgett, and Barbara Wolf

TABLE OF CONTENTS

I.   Introduction 4

II.   Role of the Matrix 5
        A.   Historical Development...5
        B.   The Nitrocellulose Matrix 6
        C.   Stoichiometry of Adsorption of Proteins on Nitrocellulose 6
        D. ^{252}Cf-PDMS of High Molecular Weight Proteins
                in Buffer Solutions 7
        E.   New Substrates for ^{252}Cf-PDMS 8

III.   Formation of Gas Phase Ions of Proteins 10

IV.   Mass Determination of a Protein by ^{252}Cf-PDMS 11

Acknowledgments 14

References 14

I. INTRODUCTION

Mass spectrometry of peptides and proteins in the mass range up to 35,000 U is now a routine method, but this technique has been feasible only in the past 5 years. Although mass spectrometry had been used extensively in other applications in the life sciences, its utility for determining the molecular weight and structure of proteins was blocked by a fundamental obstacle — the inability to form gas phase molecular ions of the protein. The fundamental difficulty was connected to two properties of peptides and proteins: involatility and thermal instability. It was not possible to increase the vapor pressure of the peptide by heating, because it decomposed. As the field of mass spectrometry developed in the 1960s, the ability to record spectra of high molecular weight species presented no problem, as was demonstrated clearly by Fales[1] when he detected molecular ions of a perfluorocarbon at m/z 3628. The first indication that the volatility/thermal instability bottleneck could be circumvented came with the discovery of field desorption by Beckey in 1970.[2] By subjecting peptides applied to surfaces to high electric field gradients, it was possible to reduce the surface binding energy of the molecule to the point where it could be thermally desorbed as an ion with minimal internal excitation. In 1974, Friedman et al.[3] proposed another route to circumvent the problem, using arguments from classical chemical kinetics to propose that rapid heating of a substrate could lead to preferential desorption of surface molecules with minimal internal excitation. At about the same time, we had been puzzling over some observations we had made when high-energy fission fragments from a [252]Californium (Cf)-source irradiated thin films of arginine and cysteine, two molecules that exhibit the involatility/thermal instability problem. We observed that, in the spectrum of ions ejected from the surface, were intact molecular ions of these species.[4] We postulated that the mechanism for the ejection process was associated with the suggestions of Friedman et al. This mechanism was proposed 15 years ago and although theoretical microscopic models have been developed that have more detail, the basic idea is still relevant: it is possible to desorb intact a thermally unstable species from a surface if the ejection process is so fast that little energy goes into the internal motions of the desorbed species. There are now many ways for excited matrices with short bursts of energy to eject the molecular ion of a protein. Most of the chapters in this book discuss techniques and applications that incorporate this "simple" concept. The energy pulse can come from a high-energy heavy ion (e.g., nuclear fission fragments from [252]Cf-decay), low-energy ion or atom bombardment (organic secondary ion mass spectroscopy [SIMS] and fast atom bombardment [FAB]), or from a laser (laser desorption). The focus of this chapter is on the technique that uses [252]Cf-fission fragments to produce gas phase molecular ions of proteins. The method is referred to as [252]Cf-plasma desorption mass spectrometry ([252]Cf-PDMS). Two other chapters in this book (Roepstroff and Cotter) review recent developments in applications of [252]Cf-PDMS to peptide/protein research. In this chapter, we discuss some of the basic principles that influence its use in these applications.

The [252]Cf-PDMS method has been used primarily to determine the molecular weight of proteins. The fragmentation pattern is relatively weak and gives little structural information. The highest molecular weight reported for a protein is currently at about 45,000 U, the molecular weight of bovine serum albumin.[5] The sensitivity is at the picomole level, and mass accuracies have been reported to be on the order of 0.1%.[6] Mass accuracy and sensitivity are among the critical issues discussed in this chapter. The method is essentially nondestructive, which means that the sample is available for further measurements after the primary analysis is complete. One of the most widely used applications of this feature is in peptide mapping by *in situ* enzymatic degradation.[7,8]

The application of [252]Cf-PDMS to protein research and development was accelerated with the introduction, in 1983, of a commercial version of our original design by a Swedish company, Bio-Ion, Nordic. It is rapidly becoming a standard instrument in the protein laboratory. Part of the reason for this usefulness, besides its ability to analyze moderately high molecular weight proteins, is its simplicity. The commercial version is computer-controlled, automated, and can be used as a routine spectroscopic instrument much like a spectrophotometer. Some new features have been added by Bio-Ion to our original design to facilitate ease of use and improve overall performance. However, based on the quality of some of the published data, it appears that the mass resolution, mass accuracy, and sensitivity are not as good as what has been achieved in other laboratories where they have constructed their own instruments.

In this chapter, we review some of the features of ^{252}Cf-PDMS that have particular relevance to applications involving proteins: the role of the matrix; current ideas about the ion desorption mechanism; and factors that influence mass resolution, mass accuracy, and sensitivity. We conclude the chapter with some projection of future developments to improve performance.

II. ROLE OF THE MATRIX

A. HISTORICAL DEVELOPMENT

The matrix is defined as the sample that is exposed to the fission fragments from the ^{252}Cf-source. In the most common geometry, the ^{252}Cf-source is located behind a thin foil, with the sample to be analyzed deposited on the surface. The fission fragment passes through the foil, depositing energy in the form of a dense cylinder of ionization that is referred to as the fission track. The matrix serves two essential roles. It is the medium for the transformation of the primary energy into a form that triggers the ejection of molecules and ions. It also is the source of protein molecules that are transformed to gas phase protein molecular ions. In our first studies of peptides,[9] we used thick samples of proteins that served both functions.

Although many successful analyses were carried out with this arrangement, this type of matrix had some disadvantages. The most critical one was the quenching of molecular ion formation by certain kinds of impurities in the matrix. A general rule of thumb of that era was that if the protein sample was pure enough that it could be crystallized, it would be possible to obtain a ^{252}Cf-PD mass spectrum for that sample. We know now that crystallinity of the matrix plays an important role in the dissipation of the high-energy density around the fission track.[10] In order to circumvent this problem, we introduced a new approach to the substrate structure by separating the energy dissipation part of the substrate from the supply of protein molecules. We accomplished this separation by adsorbing the protein onto the surface of a film. With the protein firmly attached, the film and adsorbate could be washed to remove unwanted impurities. The most common impurities were the alkali metal ions that attach to the protein through ion-dipole interactions. These ion-dipole complexes produce metastable gas phase ions that dissociate before they are detected in the mass spectrometer. Washing the film with water prior to analysis removes the alkali metal ion contaminants, resulting in much sharper mass spectra.

For our first studies using the protein adsorption method, we chose a perfluorocarbon sulfate cation exchange polymer, Nafion,[11] and demonstrated that insulin could be adsorbed onto the surface from aqueous solution as a monolayer. The molecules were tightly adsorbed to a degree that the adsorbed layer could be washed repeatedly to remove molecules that were not adsorbed to the Nafion. A ^{252}Cf-PDMS analysis of these films showed that an enhanced molecular ion yield was obtained. In addition, doubly protonated molecular ions of insulin were observed for the first time.[12] This feature was to become a signature of protein spectra in later studies. Shortly after this study came the introduction of a new matrix, nitrocellulose, which utilized the same adsorption/wash/desorb concept to produce clean monolayer films of proteins.[13] The molecular ion yields of peptides and proteins were enhanced considerably over those from the Nafion matrix. It is with this matrix that the Uppsala group, studying a series of proteins in the molecular weight range from 5000 to 45,000 U, was able to demonstrate the high molecular weight capabilities of the ^{252}Cf-PDMS method.[14] In the past few years, nitrocellulose has been used almost exclusively for protein studies using ^{252}Cf-PDMS. Some ideas have been put forth to explain why this polymer has been such an effective matrix; it is a clue to the ion formation/

desorption mechanism, which is still not well understood. A summary of what has been learned about this important matrix follows.

B. THE NITROCELLULOSE MATRIX

The preferential binding of proteins to nitrocellulose is the basis of many of the "blot-based" assay methods in protein analysis. The nitro group on the nitrocellulose plays a key role in its function. To verify the necessity for the nitro group we have transformed it to a carboxyl moiety by ultraviolet (UV)-induced photo-oxidation and have found that the binding capacity was diminished drastically. We have also acetylated the hydroxyl groups on the nitrocellulose. Except that the surface was more hydrophobic, the binding capacity and molecular ion intensity of the model peptide (insulin) used in this study, was little affected by this modification.

Chait et al.[15] measured the lifetime of insulin molecular ions desorbed from a nitrocellulose surface and found that it was much longer than when the matrix was a multilayer of insulin. This important finding demonstrated that the energy transferred to a desorbed insulin molecular ion from a nitrocellulose substrate was considerably less than when the substrate was insulin itself. The important connection between the substrate and desorbed protein was established clearly and was to become the basis for continued development of new, even more effective substrates to enhance molecular ion yields of proteins.

We have recently completed a study of the dynamics of peptide and protein adsorption on nitrocellulose and have devised a model that is consistent with the observations. The adsorption process is driven by the hydrophobic interaction.[16] Ordered water clusters form around the protein solute as a consequence of exclusion of the aqueous medium from nonpolar domains within the protein structure. Protein molecules are then repelled from the solution onto any available surface, resulting in a reordering of the ordered water clusters to a higher entropy state. The adsorption process is entropy-driven. How tightly bound the protein is to the nitrocellulose depends on the magnitude of the polar and nonpolar forces between the protein and nitrocellulose. When the peptide is small (mol wt <500 U), the nonpolar domains are also small. Little adsorption on nitrocellulose occurs from aqueous solutions because the hydrophobic interaction is not very well developed. In addition, the interaction of the peptide with the nitrocellulose is so weak that the adsorbed layer can be readily removed by exposing the surface to a layer of pure solvent (water). The adsorbed molecules (proteins and impurities) partition between the surface layer and solvent to re-establish a new thermodynamic equilibrium. If the protein is bound tightly to the surface, only a very small amount of protein will re-enter the solvent phase. However, a large fraction of the common impurities (salts, small molecules) favor partitioning into the pure water phase, so they are essentially removed from the adsorbed layer, leaving behind the more tightly-bound protein free from impurities. This process occurs when rinsing steps follow the adsorption of a high molecular weight peptide. With lower molecular weight peptides, the binding to nitrocellulose is not as great. By changing the substrate to a more nonpolar polymer, e.g., poly(ethylene terephthalate), the surface concentration of adsorbed peptide is increased.[17]

A novel application of the nitrocellulose/protein matrix is the *in situ* enzymatic degradation of an adsorbed protein followed by [252]Cf-PDMS analysis of the cleavage products, which are a collection of peptides with a distribution of sizes and hydrophobicity.[18] It was learned from these studies that, when a mixture of peptides is adsorbed on nitrocellulose, the more nonpolar peptide fragments are adsorbed preferentially and are more tightly bound, an effect consistent with the principles of our model.[18]

C. STOICHIOMETRY OF ADSORPTION OF PROTEINS ON NITROCELLULOSE

One of the factors that influences the sensitivity of [252]Cf-PDMS for the detection of proteins is the equilibrium surface concentration of adsorbed protein. We have recently developed a method for determining the surface concentration of protein adsorbed on nitrocellulose, using a variation of a procedure developed several years ago for studying the thermodynamics of the adsorption of proteins on polymer surfaces.[19] Ribonuclease was used as the model protein for our studies. The objectives of the study were to determine the minimum solution concentration of protein that would give saturation coverage of protein on the surface of nitrocellulose and to assay how much protein remained on the surface of the polymer following the rinsing steps. The minimum concentration was found to be 10^{-4} M. After adsorption took place and the aqueous layer was removed by centrifugation, the surface concentration of adsorbed ribonuclease was determined to be 10^{14} molecules per cm^2. Approximately 60% of the protein in solution was adsorbed onto the nitrocellulose. This surface concentration was an order of magnitude larger than would be expected for monolayer coverage on a smooth surface. From electron micrographs of the electrosprayed nitrocellulose, it was known that nitrocellulose has a fibrous structure, which can increase the effective area of an adsorbing surface. A ^{252}Cf-PDMS analysis of the film showed a high Na^+ content in the adsorbed layer and a broad peak from the ribonuclease due to contributions from metastable ions. When the sample was rinsed with water, only 2% of the protein was removed, but the Na^+ contamination was eliminated completely. The mass peak from the ribonuclease was considerably sharper because the metastable component was reduced, an effect first noted by the Uppsala group.[13]

D. ^{252}Cf-PDMS OF HIGH MOLECULAR WEIGHT PROTEINS IN BUFFER SOLUTIONS

As the use of ^{252}Cf-PDMS for protein analysis expands and more complex, fragile proteins are analyzed, a need has developed to standardize the procedures used to transfer these proteins to the nitrocellulose matrix with minimal perturbation of the structure of the protein. Even dissolution with 0.1% trifluoroacetic acid can result in a degradation of some proteins. Most proteins are stable in buffered solutions of near neutral pH. Phosphate-buffered saline (PBS) solution is an example. In most mass spectrometric analyses, however, the use of any salt buffers is generally avoided because its presence interferes with molecular ion formation in the mass spectrometer. However, using the techniques described previously, the buffer can be removed from the adsorbed layer by rinsing the surface with water. It was not clear whether this method would leave the protein in a state in which the basic amino acid residues in the protein would be protonated so that gas phase molecular ions would form. We carried out a study for a set of small proteins with molecular weights ranging from 13,000 to 29,000 U; we dissolved the proteins in PBS and adsorbed them onto nitrocellulose, a process followed by water rinses of the surface to remove occluded buffer salts. In all cases, protonated and multiprotonated molecular ions were observed to have a normal molecular ion yield. Examples for ribonuclease A and α-2 interferon are shown in Figure 1. The pattern of protonated and multiprotonated molecular ions was the same in both cases. This trend continued for the higher molecular weight proteins in the series, as shown in Figure 2 for papain and proteinase K. It is clear from these data that a source of protons for protonation is still available for the larger protein molecules when they are adsorbed on the nitrocellulose, even when the adsorbing and rinsing solutions are at neutral pH.

The study also revealed another feature of the ^{252}Cf-PDMS of proteins. Above 20,000 U mol wt, the protein molecular ion intensities were observed to decrease markedly with increasing molecular weight. Although this trend had been suggested in earlier studies, the use of PBS as a solvent system for our studies had an added advantage in that it minimized some of the pH-dependent variables that might be attributed to differences in protein structure within the series of proteins studied. These differences could have influenced molecular ion yields. We now feel that the loss of intensity with increasing molecular weight is inherent in the ^{252}Cf-PDMS methodology in its present form. With the matrices currently being used, the energy density at the site of protein desorption is too high for the larger protein molecules to be desorbed intact; the desorbed protein molecular ions have too much internal energy. The only variable available to solve this problem lies in the substrate: how it dissipates the energy deposited in the fission

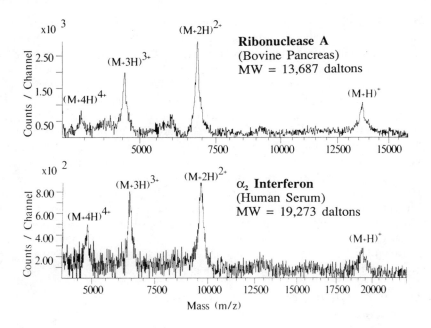

FIGURE 1. ^{252}Cf-PDMS of two proteins adsorbed onto nitrocellulose from a PBS solution. The buffer was removed by elution of the surface with water.

fragment track and how it is involved in moderating the internal energy that flows into the desorbed species. The next section outlines some of the current paths that are being followed to solve this problem, along with some encouraging new results from our laboratory.

E. NEW SUBSTRATES FOR ^{252}Cf-PDMS

The impetus for discovering new substrates for ^{252}Cf-PDMS is the possibility of finding one that will give higher protein molecular ion yields, because this improvement would increase the limit of detection for proteins. In addition, molecular ions that have lower internal excitation give sharper peaks in the mass spectrum, a development that translates into improved mass accuracy. Nitrocellulose is currently the substrate of choice for protein studies. Any new substrate that is introduced must give better results, and nitrocellulose is a good starting point for identifying the key ingredients of a viable substrate. The nitro group is essential for the adsorption because it provides relatively nonpolar sites for attachment to the nonpolar domains of the protein. This link between the protein and substrate is responsible for retention of the protein on the substrate when it is rinsed with water. In order to carry out the crucial rinsing step, the substrate must be water-insoluble. Solubility and hydrophilic/hydrophobic character are not the only factors, however, because numerous substrates have been studied that adsorb proteins and have these properties, yet they are not nearly as effective as nitrocellulose. Other factors must operate. The nitro group on the nitrocellulose is thermally labile. Some of the most intense ions in the ^{252}Cf-PD mass spectrum of nitrocellulose are derived from the nitro group. If a nitro group that is attached to an adsorbed protein via dispersion forces dissociates from the nitrocellulose during excitation by a fission fragment, the protein molecule is released from the substrate and enters the gas phase. The effect is a lower activation energy for desorption of the protein molecular ion.

We have used the model described above to identify matrices that might be suitable for the ^{252}Cf-PDMS of proteins. The criteria were: low water solubility, a mix of hydrophilic and nonpolar sites, and ease of dissociation of the nonpolar sites that effectively bind the protein to

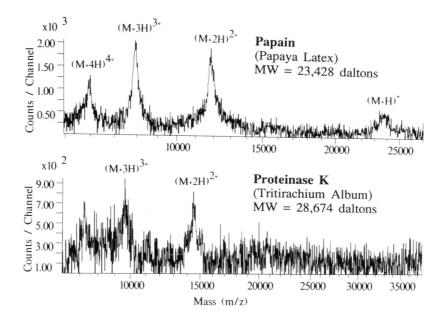

FIGURE 2. [252]Cf-PDMS of two proteins in the mass range m/z 20,000 to 30,000. The intensities of these ions are decreasing markedly with increasing molecular weight.

the substrate. A significant new development in our approach to the problem was the realization that to meet these criteria we were not limited to polymeric materials. A substrate composed of a collection of small molecules having these properties might also be a suitable substrate. We began a search for candidates, and found several that met all of the criteria.

We give as an example of the performance of this new class of substrates results obtained for one of them — anthroic acid. Anthroic acid is a derivative of anthracene with a single carboxyl group and a molecular weight of 220 U. It is water-insoluble, has both hydrophilic and nonpolar components in its structure, readily forms microcrystalline deposits, and is easily sublimable. (We have mimicked the thermal instability of the nitro group in nitrocellulose by selecting small molecules with a low heat of sublimation. The effect is the same — the energy of activation for desorption is reduced.) For this particular substrate, the molecular ion intensity for two of the model proteins used in the study, insulin and lysozyme, were comparable to what we observed using nitrocellulose as the substrate. This equivalency by itself was an achievement! However, an important new feature appeared in the mass spectra: the peak widths were extremely narrow. This observation is an indication that the internal excitation of the protein molecular ion desorbed from anthroic acid is lower than when the substrate is nitrocellulose.

To demonstrate the influence of the substrate on the quality of the mass spectrum, we have compared the mass spectrum of insulin obtained using three different adsorbing surfaces: photo-oxidized poly(ethylene terephthalate), nitrocellulose, and anthroic acid. The surface of the poly(ethylene terephthalate) was oxidized to provide additional hydrophilic sites to enhance peptide adsorption. All three matrices are effective in adsorbing insulin from aqueous solutions resulting in an adsorbed layer with a high surface coverage of protein. The [252]Cf-PD mass spectra are shown in Figure 3. The sharpness and intensities of the molecular ion peaks are directly related to the amount of internal excitation imparted to the desorbed molecular ion from the substrate. Although the photo-oxidized poly(ethylene terephthalate) is an excellent substrate for small peptides (mol wt <1500 U), the nature of the surface interaction with an adsorbed protein

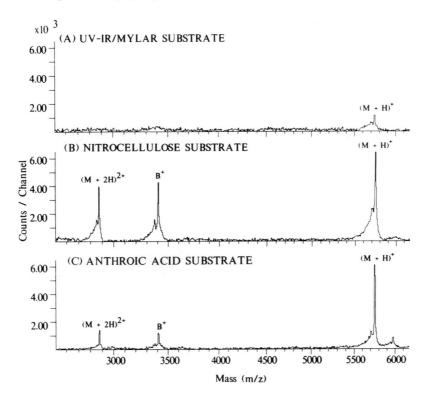

FIGURE 3.   [252]Cf-PDMS of insulin showing the influence of the substrate on molecular ion yields. The surface concentration of peptide was the same in all three cases.

is such that too much energy is transferred to the desorbed molecular ion. Most of the molecular ions formed are so short-lived that they do not survive longer than a few hundred nanoseconds, a lifetime that means that they do not contribute to the intensity of the intact molecular ion. This factor is the reason why the intensity is so much lower than for the other two substrates. Note the differences in line shapes for insulin desorbed from anthroic acid vs. that obtained using a nitrocellulose substrate. The tail on the left-hand side of the peak is due to ions that have lived long enough to survive the acceleration step, but have decayed in flight, resulting in a slight shift in flight times. This component is reduced considerably with the anthroic acid substrate.

III. FORMATION OF GAS PHASE IONS OF PROTEINS

The desorption of a protein molecule from the surface of a substrate is an essential part of the process of forming a gas phase molecular ion. However, at some point while the molecule is still within a few nanometers from the surface, an ionizing event must take place. After recording hundreds of mass spectra of proteins, the pattern is clear that protonated and multi-protonated protein molecular ions are formed most readily. The protein spectra shown in Figures 1 to 3 are good examples. Recent results from the electrospray ionization of proteins shows that the protonation sites are the basic residues in the protein.[20] In the solid state, each protonation site pairs with an anion to form an ion-pair. The dissociation of these ion-pairs into a separated ion-pair with the H^+ attached to the basic residue is a likely mechanism for the ionization. The probability of this separation occurring depends on the size of the counterion (bulky anion groups dissociate more readily) and the level of internal excitation of the protein. When the excitation energy is high, the appearance of doubly protonated molecular ions is diminished, as shown in insulin desorbed from poly(ethylene terephthalate) (Figure 3A).

The probability of the formation of a gas phase molecular ion of peptide the size of insulin by excitation with a [252]Cf-fission fragment is rather low. In the best case, 1 fission fragment track in 20 produces a stable insulin molecular ion, and the efficiency decreases significantly with the increasing size of the protein. Part of the reason for this relationship is that, because of its larger size, the desorbed protein molecular ion absorbs a higher amount of the energy that is deposited by the fission fragment. Fragment and metastable ions form to give rise to a continuum of ions on the low mass side of the molecular ion peak. Most of the effect of using different substrates to reduce the energy density is in transforming some of that continuum of ions to the peaks representing the intact molecular ion of the protein.

The most intense ion that is derived from the adsorbed protein is the CN[-] ion, the result of the pyrolysis of a protein molecule located very close to the center of the fission track. The intensity of this ion is directly proportional to the amount of protein adsorbed on the surface of the substrate, so that it can be used as an internal standard for measuring the surface concentration of a protein.[17] There is evidence that a large fraction of desorbed protein molecules are in the form of neutral molecules, and some efforts are underway to enhance molecular ion yields using postionization methods.[21] The neutral molecule ejection studies showed that as many as 1500 molecules of a small peptide are ejected per incident fission fragment from a multilayer sample. It is not known whether this neutral molecule yield is the same for a monolayer deposit of a protein.

To summarize what we know about the energetics of ion formation in [252]Cf-PDMS: it is clear that the energy density developed by the fission track is too high and that most of the protein molecules that are excited either pyrolyze to small fragment ions or desorb with too much internal excitation. Only a very small fraction of the desorbed protein molecules survives as intact molecular ions. The fact that the molecular ion intensity can be enhanced by selecting matrices that dissipate the energy more effectively, as demonstrated in the set of spectra shown in Figure 3, is an indication that the key to improved molecular ion yields and to extending the mass range is linked to control of the energy density near the site of molecular ion desorption.

IV. MASS DETERMINATION OF A PROTEIN BY [252]Cf-PDMS

Most of the past applications of [252]Cf-PDMS to protein research have focused on the determination of the molecular weight of a protein or the component parts that have been cleaved from the parent structure by an enzymatically induced fragmentation. We have discussed some of the problems encountered in producing a gas phase molecular ion of a protein. Once a gas phase molecular ion has been formed, the rest of the analysis involves an accurate determination of the mass of the ion, the final topic discussed in this chapter. The majority of [252]Cf-PDMS analyses have been carried out using the time-of-flight (TOF) method to generate mass spectra. Ions that are desorbed from the surface of a sample are accelerated in an electric field to an energy that is dependent on the electric potential and the charge of the ion. For example, the spectra shown in Figure 3 were recorded with an acceleration voltage of 15 kV. The molecular ion of insulin $(M + H)^+$ has an energy of 15 keV, and the doubly protonated molecular ion, shown on the left side of the spectra, has an energy of 30 keV. With a constant acceleration potential, an ion accelerated through that 15 kV potential will attain a velocity that is proportional to its charge/mass ratio. The ion velocity can be measured by the TOF of the ion through a fixed distance in the mass spectrometer. Through the relationships that exist between kinetic energy, velocity, TOF, and mass/charge, the TOF scale can be converted readily to a mass/charge scale. The first step in measuring the mass of an ion is determining the centroid of the mass peak in terms of TOF. The TOF scale is converted to a mass scale by measuring accurately the TOF of ions of known mass in the TOF spectrum.[22] How accurately we need to establish the mass scale depends on the accuracy desired for the mass determination of the protein.

To develop an appreciation for the problem, let us assume that we wish to know the mass of a protein with a mol wt 20,000 U to an accuracy of 0.5 U. From the standpoint of the information desired by the protein chemist, this accuracy is a reasonable objective, and it is also an attainable experimental goal. The calibration ions that are selected are in the low mass region (< m/z 100) because the intensities of many of the ions in this mass range are among the highest in the entire mass spectrum. If all of the ions had precisely the same kinetic energy after acceleration in the electric field, the error in the measurement of the TOF of the calibration ions and the molecular ions associated with the protein would be so low that the 0.5 U accuracy would be achieved easily because the electronic time resolution for modern time-interval digitizers is very high, exceeding 1 part in 200,000.[23] However, several contributions to the TOF of an ion make it difficult to achieve this resolution. The most important factor is the initial kinetic energy that is imparted to the ion during the desorption process and is associated with the mechanism of ion formation and ejection. Each ion type has a kinetic energy distribution characterized by a mean kinetic energy and distribution width.[24] The width of the distribution is large enough so that it broadens the TOF spectrum of a protein molecular ion to an extent that the isotopic pattern cannot be resolved. This phenomenon means that the TOF of the centroid of the isotopic distribution is measured and the mass value derived from the measurement is the isotopically averaged value. Another effect that changes the kinetic energy of the ions is a local electrical charging of the surface of the substrate surrounding the fission track that repels positive ions from the surface and decelerates negative ions.[25] This effect is due to the presence of a dielectric layer on the surface of the substrate, and varies from one sample to the next. To complicate the issue further, not all ion types are influenced in the same manner, probably because they are not ejected from the same part of the fission track or in the same time frame relative to the dissipation of the energy of the fission track. The influences of initial kinetic energy fluctuations on mass accuracy can be minimized by using high acceleration voltages, and this possibility is one of the reasons why voltages in the 15 to 20 kV regime are used.

Another contribution to perturbations in the TOF of an ion arises from the stability of the desorbed ion. If the ion decays in flight while it is being accelerated in the electric field above the surface of the sample, then the products of the fragmentation will not contribute to the TOF distribution of the protein molecular ion. If decay occurs after the ion passes through the acceleration field, then the products of the decay will maintain the TOF of the parent ion, except for the kinetic energy release associated with the fragmentation process. However, the few electron volt energy release superimposed on the 15 keV ion energy is sufficient to introduce a perturbation in the line-shape of the protein mass peak in the TOF spectrum.[15] This effect induces a small shift in the centroid of the TOF distribution that influences the mass determination of the ion. The fragment ion and neutral species can be resolved from the stable component of the distribution using electrostatic retardation fields at the end of the flight tube, with voltage values approaching 90% of the acceleration voltage. Fragment ions with energies less than 90% of the energy of the parent ion are repelled, and those with energies between 90 to 98% of the primary energy have new TOF values that are sufficiently different from those of the primary ion that they are resolved from the primary distribution. The neutral component is also resolved from the main distribution because it is unaffected by the retardation field. These effects are shown in Figure 4 for a series of spectra recorded for insulin. Figure 4A is the spectrum for the molecular ion region recorded with an acceleration voltage of 10 kV. Three peaks are present in the spectrum due to the protonated and doubly protonated molecular ion, and a fragment ion from the beta chain of the peptide. The metastable component forms a shoulder on the low mass side of the TOF distribution, shifting the centroid to a slightly lower mass. Figure 4B is the spectrum obtained with a 9 kV retardation potential in front of the ion detector at the end of the flight tube. The shoulder is now gone, and the peak is considerably sharper. The mass of insulin derived from this spectrum was within 0.5 U of the known average mass for this peptide. Figure 4C shows the spectrum with a retardation potential larger than the

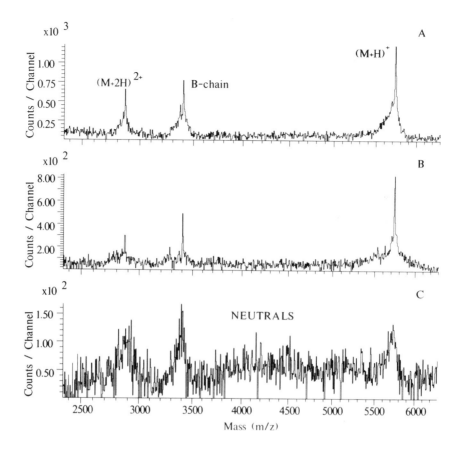

FIGURE 4.   [252]Cf-PDMS of insulin showing the influence of metastable species on the line-shape of the mass peak using a 10 kV acceleration voltage. (A) no retardation voltage; (B) 9 kV retardation voltage; (C) 10.25 kV retardation voltage.

acceleration potential. Only neutral species are detected under these conditions. These species are the neutral fragments formed from primary insulin molecular ions in the field-free region of the spectrometer.

The presence of a metastable component to the line shape of the primary molecular ion is a reflection of the excess energy that is introduced to the desorbed species from the substrate. We showed earlier in this chapter that this component can be controlled to some extent by the substrate. The spectrum obtained for insulin using the anthroic acid substrate (Figure 3C) has a resolution comparable to the one obtained using the nitrocellulose substrate and incorporating the ion retardation filter. Thus, the ultimate mass accuracy that can be achieved for a protein using [252]Cf-PDMS is, like the sensitivity question, linked to our quest for the "ultimate" substrate — a chemical problem that embodies the contributions of many different disciplines from the chemical sciences.

The interdisciplinary feature of the subject is the theme of this chapter. Although the final goal is the development of a mass spectrometry-based analysis method for the life sciences, the attainment of this goal demands an appreciation of the science of complex surfaces, the chemical physics of energy relaxation in solids, and the organic chemistry that occurs under extreme conditions of high energy density and short time scales. The interdependence of these diverse features is perhaps the fascination of this field; something of interest exists for everyone and the final product of the fundamental studies is even useful!

ACKNOWLEDGMENTS

We wish to acknowledge the scientific contributions of John Hill, David Jacobs, and Z.H. Huang to the work described here, and to Dennis Shelton and Mark Miller for providing technical support. The results originating from Texas A&M were from ongoing projects supported by the National Institutes of Health (GM - 26096), the National Science Foundation (CHE - 8604609), and the Welch Foundation (A - 258). We express our gratitude to these sponsors.

REFERENCES

1. **Fales, H. M.,** The mass spectrum of a compound of formula C72H24O8F128N4P4 and molecular weight 3628., *Anal. Chem.,* 38, 1058, 1966.
2. **Beckey, H. D.,** Principles of field ionization and field desorption mass spectrometry, *International Series in Analytical Chemistry,* Vol. 61, Belcher, R. and Freiser, H., Eds., Pergamon Press, Oxford, 1977.
3. **Beuhler, R. J., Flanigan, E., Greene, L. J., and Friedman, L.,** Proton transfer mass spectrometry of peptides. A rapid heating technique for underivatized peptides containing arginine, *J. Am. Chem. Soc.,* 96, 3990, 1974.
4. **Torgerson, D. F., Skowronski, R. P., and Macfarlane, R. D.,** New approach to the mass spectrometry of non-volatile compounds, *Biochem. Biophys. Res. Commun.,* 60, 616, 1974.
5. **Jonsson, G., Hedin, A., Hakansson, P., Sundqvist, B. U. M., Bennich, H., and Roepstorff, P.,** Compensation for non-normal ejection of large molecular ions in plasma desorption mass spectrometry, *Rapid Commun. Mass Spectrom.,* 3, 190, 1989.
6. **Sundqvist, B., Hedin, A., Hakansson, P., Kamensky, I., Salehpour, M., and Sawe, G.,** Plasma desorption mass spectrometry (PDMS). Limitations and possibilities, *Int. J. Mass Spectrom. Ion Processes,* 65, 69, 1985.
7. **Roepstorff, P., Nielsen, P. F., Klarskov, K., and Hojrup, P.,** Optimization and use of nitrocellulose as a matrix for peptide and protein analysis by plasma desorption mass spectrometry, in *The Analysis of Peptides and Proteins by Mass Spectrometry,* McNeal, C. J., ed., Wiley, Chichester, U.K., 1988, 55.
8. **Tsarbopoulos, A., Becker, G. W., Occolowitz, J. L., and Jardine, I.,** Peptide and protein mapping by 252Cf-plasma desorption mass spectrometry, *Anal. Biochem.,* 171, 113, 1988.
9. **Macfarlane, R. D. and Torgerson, D. F.,** Californium-252 plasma desorption mass spectroscopy, *Science,* 191, 920, 1976.
10. **Macfarlane, R. D., McNeal, C. J., and Jacobs, D. L.,** The role of ordered surface aggregates in ion desorption by 252Cf-PDMS, *J. Phys. (Paris),* 50(C2), 21, 1989.
11. **Jordan, E. A., Macfarlane, R. D., Martin, C. A., and McNeal, C. J.,** Ion-containing polymers: a new matrix for 252Cf-plasma desorption mass spectrometry, *Int. J. Mass Spectrom. Ion Phys.,* 53, 345, 1983.
12. **Macfarlane, R. D.,** A mass spectrometric method for studying liquid-surface partitioning of biomolecules, in *Mass Spectrometry in the Health and Life Sciences,* Burlingame, A. L. and Castignoli, N., Jr., Eds., Elsevier, Amsterdam, 1985, 85.
13. **Jonsson, G. P., Hedin, A. B., Hakansson, P. L., Sundqvist, B. U. M., Save, B. G., Nielsen, P. F., Roepstorff, P., Johansson, K. E., Kamensky, I., and Lindberg, M. S. L.,** *Anal. Chem.,* 58, 1084, 1986.
14. **Roepstorff, P. and Sundqvist, B.,** Plasma desorption mass spectrometry of high molecular weight biomolecules, in *Mass Spectrometry in Biomedical Research,* Gaskell, S. J., Ed., Wiley, Chichester, U.K., 1986, chap.15.
15. **Chait, B. T.,** A study of the fragmentation of singly and multiply charged ions produced by 252Cf fission fragment bombardment of polypeptides bound to nitrocellulose, *Int. J. Mass Spectrom. Ion Processes,* 78, 237, 1987.
16. **Norde, W.,** Adsorption of proteins at solid surfaces, in *Adhesion and Adsorption of Polymers,* Vol. 2, Lee, L. H. Y., Ed., Plenum Press, New York, 1980, 801.
17. **Macfarlane, R. D., Hill, J. C., Geno, P. W., and Mudgett, P. D.,** Improving sensitivity for 252Cf-PDMS of peptides/proteins, in *The Analysis of Peptides and Proteins by Mass Spectrometry,* McNeal, C. J., Ed., Wiley, Chichester, U.K., 1988, 239.
18. **Roepstorff, P., Nielsen, P. F., Klarskov, K., and Jojrup, P.,** Applications of plasma desorption mass spectrometry in peptide and protein chemistry, *Biomed. Environ. Mass Spectrom.,* 16, 9, 1988.
19. **Dillman, W. J. and Miller, I. F.,** On the adsorption of serum proteins on polymer surfaces, *J. Coll. Interface Sci.,* 44, 221, 1973.

20. **Meng, C. K., Mann, M., and Fenn, J. B.,** Interpreting mass spectra of multiply charged ions, *Anal. Chem.,* 61, 1702, 1989.
21. **Salehpour, M., Hakansson, P., Sundqvist, B., and Widdiyasekera,** Total molecular yields for fast heavy ion induced desorption of biomolecules, *Nucl. Instrum. Methods,* 813, 752, 1984.
22. **Macfarlane, R. D.,** Californium-252 plasma desorption mass spectrometry. Large molecules, software, and the essence of time, *Anal. Chem.,* 55, 1247A, 1983.
23. **Turko, B. T., Macfarlane, R. D., and McNeal, C. J.,** 252Cf-plasma desorption mass spectrometry multistop time digitizer, *Int. J. Mass Spectrom. Ion Phys.,* 53, 353, 1983.
24. **Furstenau, N., Knippelberg, W., Krueger, F. R., Weisz, G., and Wien, K.,** Experimental investigation about the mechanism of fission-fragment induced desorption, *Z. Naturforsch.,* 32a, 711, 1977.
25. **Macfarlane, R. D., Hill, J. C., and Jacobs, D. L.,** Dynamics of ion emission from heavy-ion tracks, *J. Trace Microprob. Tech.,* 4, 281, 1986-7.

Chapter 2

PLASMA DESORPTION MASS SPECTROMETRY OF PEPTIDES AND PEPTIDE CONJUGATES

Robert J. Cotter, Ling Chen, and Rong Wang

TABLE OF CONTENTS

I.   Introduction 18

II.   Confirmation of Peptide Synthesis: Human Mutant Insulin 18

III. Peptide Mixtures 19
    A.   Surveying Pancreatic Peptides 20
    B.   Tryptic Mapping by PDMS 21

IV. Glycosylation Sites and Heterogeneity 22
    A.   Ovomucoid Third Domains 22
    B.   Gylcopeptides from Bovine Fetuin 24
    C.   The High-Mannose Glycopeptide from Trypanosomes 25

V.   Phosphorylated Peptides: Heterogeneity of Riboflavin-Binding Protein 31

VI. Post-Translational Modifications: Murine Elongation Factor 31

VII. Dimers and Adducts 31
    A.   Insulin Dimerization in Long-Term Preparations 32
    B.   Nonenzymatic Formation of Insulin-Glutathione Mixed Disulfides 36
    C.   Metal-Linked Dimers of the *Tat III* Cysteine-Rich Region 36

VIII. Conclusions 36

Acknowledgments 37

References 39

I. INTRODUCTION

Mass spectrometry traditionally has been employed for the elucidation of molecular structure from spectra that exhibit both molecular and fragment ions. The mass-to-charge ratio of the parent radical-ions ($M^{+\cdot}$) observed in electron impact (EI) mass spectra, or the even-electron protonated species (MH^+) observed in chemical ionization (CI) mass spectra are used to determine the molecular weight (mol wt) of the compound being analyzed. Many of the ions formed by these methods undergo fragmentation, producing ions whose mass-to-charge ratios can be used to determine the presence and location of functional groups and other details of the chemical structure.[1] In general, CI mass spectra exhibit less fragmentation (and hence less structural information) than EI mass spectra. However, the CI method has been useful for compounds that fragment so extensively by EI that a molecular ion cannot be observed. Because establishment of the molecular weight is a crucial first step in identifying the chemical structure, analyses by chemical ionization and EI are often both carried out to provide complementary information.

In recent years, the capabilities for determining molecular structure by mass spectrometry have increased greatly beyond those available from these two established techniques with the introduction of several new ionization methods, known generally as *desorption* techniques. These methods include *field desorption,*[2] *fast atom bombardment,*[3] *laser desorption*[4] and *plasma desorption*[5] mass spectrometry (PDMS). Like chemical ionization, they are considered *soft* ionization techniques, producing stable, even-electron molecular ions (MH^+ and MNa^+ or $MK^{+)}$ in the presence of alkali ion salts. An added advantage is that the compounds to be analyzed do not need to be volatile, and therefore these methods have been remarkably successful in the analysis of much larger structures, such as peptides, glycolipids, oligosaccharides, and oligonucleotides. Fragmentation is observed in desorption mass spectra, but (fortuitously) occurs primarily at glycosidic or amide bonds, resulting in a series of peaks whose mass *differences* correspond to the basic building blocks of the biomolecule. Thus, the plasma desorption mass spectrum of Asp-Arg-Val-Tyr-Ile-His-Pro-Phe shown in Figure 1 exhibits a series of peaks, which reveals the primary amino acid sequence of this octapeptide.

For much larger molecules, fragmentation is observed rarely in plasma desorption mass spectra. Thus, the PDMS method, used in conjunction with a TOF mass analyzer, provides an accurate, but low resolution, measurement of the molecular weight. Such information can be combined with the more detailed structural information available from higher resolution methods; however, molecular weight measurements themselves can often answer essential structural questions.

In this chapter, several examples of the application of the PDMS technique to the primary structural analysis of peptides and their conjugates are summarized. The mass spectral measurements were all obtained on a Bio-Ion, Nordic (Uppsala, Sweden) BIN 10K plasma desorption mass spectrometer located at the Middle Atlantic Mass Spectrometry Facility.

II. CONFIRMATION OF PEPTIDE SYNTHESIS: HUMAN MUTANT INSULIN

A simple but accurate determination of the molecular weight of a peptide is an important measurement for comparing peptides with a high degree of homology, revealing deletions or insertions of amino acids, or verifying the structures of chemically synthesized or recombinant DNA-derived products. In a recent review,[6] it was noted that PDMS was used to verify chemical modifications to the A- and B-chain of *relaxins*, peptides that have tertiary but not sequence homology to insulins,[7] as well as to provide a rapid interspecies comparison. Jardine et al.[8] have used the PDMS technique to confirm the molecular weight of intact recombinant human interleukin-2 (hIL-2).

Familial hyperinsulinemia is an autosomal dominant disorder characterized by the synthesis

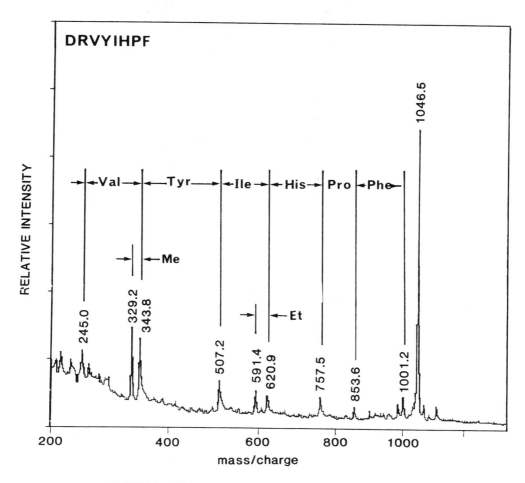

FIGURE 1.   PDMS of the peptide: Asp-Arg-Val-Tyr-Ile-His-Pro-Phe.

of structurally abnormal proinsulin and a marked increase in immunoreactive proinsulin-like material in the serum. For two families with this condition, it has been shown that mutations affect the dibasic amino acids linking the C-peptide to the insulin A-chain.[9] A third instance of familial hyperinsulinemia was reported by Gruppuso and co-workers.[10] However, cloning and sequencing both alleles of the insulin genes from two affected members of this family by Chan et al.[11] revealed that a point mutation in the B-chain coding region of the insulin gene was associated with hyperinsulinemia in this family. A single nucleotide substitution in the codon for residue 10 of proinsulin (CAC→GAC) exchanges an aspartic acid for histidine in the insulin B-chain region. A human insulin analog, B-10 Asp-insulin, corresponding to the mutant proinsulin, has been synthesized recently and evaluated for its biological activity.[12] This analogue exhibits fivefold greater potency than natural insulin in binding to the insulin receptor. PDMS was performed on 10 μg of the synthetic *superactive* insulin (Figure 2). The molecular weight was determined by averaging the measurements of both singly and doubly charged ions, and confirmed the substitution of aspartic acid for histidine.[13]

III. PEPTIDE MIXTURES

For the analysis of mixtures, the observation of molecular ions and the absence of fragmentation can be considerable advantages. Several examples follow in which the assumption is made that all of the peaks correspond to intact molecular species. These examples include a survey of

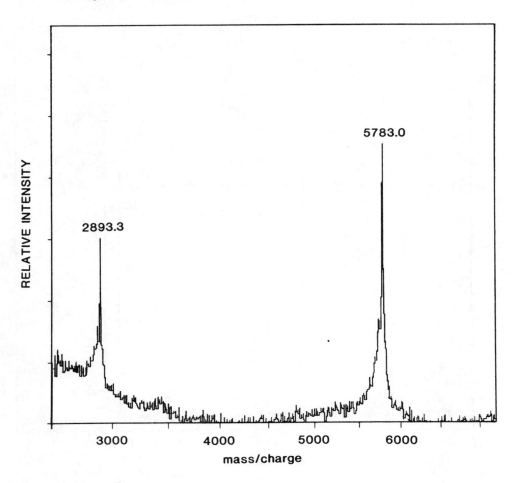

FIGURE 2.   PDMS of B-10 Asp mutant human insulin. (From Alai, M., Ph.D. thesis, Johns Hopkins University, Baltimore, 1988.)

the pre-, and prohormones, hormones, and connecting peptides in the catfish pancreas, mapping of the tryptic fragment mixture from recombinant human growth hormone, and assessment of glycosylation heterogeneity in ovomucoid third domains and in the high mannose glycopeptide from trypanosomes.

A. SURVEYING PANCREATIC PEPTIDES

The catfish pancreas is a well-characterized system whose major peptide components are known. An extract of catfish pancreas contains peptides differing greatly in size, composition, and degree of post-translational modification. The cDNA for two different somatostatins has been cloned from catfish.[14,15] One of the somatostatins (SST) is a 14-residue peptide[16] identical to that from several other species; the other is a 22-residue somatostatin[17,18] that is *O*-glycosylated at Thr-5. Because the structures of the precursors to both SST-14 and SST-22 have been deduced from their cDNA sequences, peptides derived from nonhormonal portions of prosomatostatin can be identified from their masses. The major biologically active fragments of pro-glucagon have also been purified,[19] as have insulin and the C-peptide of proinsulin.[20] Recently, therefore, Andrews et al.[21] obtained plasma desorption mass spectra of this well-characterized system in order to evaluate its potential for surveying peptides in a crude mixture.

PDMS of a crude, desalted extract of piscine endocrine pancreas provides mass information for the major biologically active peptide hormones present in this tissue (Figure 3 and Table 1). Abundant molecular ions are observed in the positive ion mode for SST-14 (1638 amu), O-glycosylated SST-22 (2944 amu), glucagon (3512 amu), glucagon-like peptide (3785 amu), insulin (5550 amu), and other prohormone-derived peptides. Protonated species and sodium adducts are both apparent in the mass spectrum, as shown in Figure 4 for the molecular ion region for SST-14. A number of other molecular ions are observed, including somatostatin-26$_{1-10}$ (1014 amu) and the entire portion of prosomatostatin-22 remaining after removal of SST-22 (6465 amu). The data obtained by this method have also resulted in the identification of the third major product of proglucagon processing in catfish pancreas: glicentin-related polypeptide. Subtractive Edman degradation of the entire extract, followed by PDMS analysis, was also explored as a method for confirming mass assignments. Figure 5 shows the expanded region containing the molecular ions of des-Ala somatostatin-26$_{1-10}$.[21]

B. TRYPTIC MAPPING BY PDMS

Fast atom bombardment (FAB) and PDMS have both been used to provide tryptic maps, i.e., mass spectra that contain (primarily) molecular ions corresponding to the peptide fragments resulting from a tryptic digest. FAB- or PD-mapping does not require physical separation of the fragments in a tryptic digest, and can be less ambiguous than tryptic maps provided by high performance liquid chromatography (HPLC), when the measured masses can be compared directly to masses calculated from a known or expected sequence. However, in many cases, not all of the tryptic fragments are observed in the mass spectrum. For the simple case of glucagon, which produces four fragments upon treatment with trypsin, selective desorption of some components observed in FAB mass spectra has been related to the relative hydrophobicities of each peptide fragment.[22] Because hydrophobicity is generally thought to affect the competition among different components in a mixture for the surface sites on the liquid FAB matrix, it would seem that this property is less relevant in PDMS, where samples are desorbed from the solid phase. Nielsen and Roepstorff[23] have investigated the PDMS spectra of tryptic digests of glucagon, porcine insulin, and chicken lysozyme. They concluded that there were, in fact, suppression effects in PDMS, but that they were due to the net charge at pH 7.4 rather than to the relative hydrophobicities of the tryptic fragments.

Recently, researchers at Genentech[24] have investigated the potential of FAB mass spectrometry for mapping the unfractionated tryptic digest of the 23 kDa recombinant DNA-derived human growth hormone (rhGH). The digest contains 21 tryptic fragments (Table 2), with the largest (T6-T16) at 3762 amu. They were able to observe all but the T14 peptide (which is very hydrophilic) and the T6-T16 disulfide-linked fragment, which is suppressed by the presence of fragments of lower molecular weight. This 23 kDa peptide was then analyzed by PDMS. The analysis was carried out in two steps. First, the digest was absorbed to a nitrocellulose-coated foil, and the mass spectrum shown in Figure 6 was obtained. The sample was then removed, washed with deionized water, and its PDMS spectrum was obtained again as shown in Figure 7.

In the mass spectrum of the unwashed digest (Figure 6), the higher mass fragments are clearly supressed. T4 and T9 are observed, but their signal-to-noise ratios are considerably lower than those for the fragments below 1500 amu. The negative ion PDMS spectrum (shown in the insert of Figure 6) reveals only three tryptic fragments: T8, T2, and T11, which are also observed as intense ions in the positive ion spectrum. Washing the sample foil removes cations, such as sodium, which interfere with the desorption of peptide ions. In this case, when the sample is washed and the mass spectrum re-recorded, the high mass fragments, T9, T10, T4 and T6-T16 are observed with a greater signal-to-noise ratio than the fragments below 1500 amu (Figure 7). Several of the fragments observed in the first spectrum disappear altogether. To some extent, this result can be correlated with the greater hydrophobicities of these peptides, which enable them

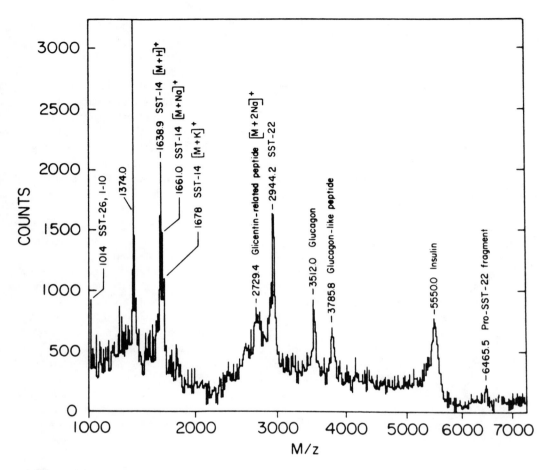

FIGURE 3.   Plasma desorption TOF mass spectrum of a crude pancreatic extract. The sample was prepared by acidic ethanol extraction of catfish endocrine pancreas followed by ether precipitation. (From Andrews, P. C., Alai, M., and Cotter, R. J., *Anal. Biochem.,* 174, 23, 1988. With permission.)

to survive the washing step.[25] This experiment, in turn, suggests that an effective protocol for surveying tryptic fragments by PDMS should include obtaining mass spectra before and after removal by washing of the smallest and most hydrophilic tryptic fragments.

IV. GLYCOSYLATION SITES AND HETEROGENEITY

The extent of glycosylation and heterogeneity can be assessed easily by comparing the molecular weights of glycosylated and unglycosylated peptides. Alternatively, when the amino acid sequence is known, more detailed information can be gained from comparison of the PD-maps of the tryptic fragments from glycosylated and unglycosylated peptides, or by PDMS analysis of the purified glycopeptides. Some examples of these approaches follow.

A. OVOMUCOID THIRD DOMAINS

Ovomucoids, glycoproteins present in avian egg white, are strong inhibitors of serine proteinases. Ovomucoids consist of three tandem domains, each of which is homologous to single-domain pancreatic secretory trypsin inhibitors.[26] Laskowski and co-workers[27] have determined the amino acid sequence and hypervariability of enzyme inhibitor contact residues of carbohydrate-free ovomucoid third domain from more than 100 avian species. The third

TABLE 1
Peptides Found in Catfish Pancreatic Extract

Peptide	MH+ calc	MH+ measured
SST-14	1638.9	1639
SST-26$_{1-10}$	1013.2	1014
Prosomatostatin-14$_{25-87}$ connecting peptide	6331.0	N.O.
SST 22-c (sialylated)	3234.5	N.O.
SST 22-a	2944.0	2944
Prosomatostatin-22$_{25-81}$ connecting peptide	6491.3	6466
Glucagon	3511.8	3512
Glucagon-like peptide	3786.1	3786
Glicentin-related polypeptide	2729[a]	2729[a]
Insulin	— [b]	5550[c]

[a] Mass indicated is for the $(M+2Na-H)^+$ ion.

[b] Not known.

[c] Estimated value.

Adapted from Andrews, P. C., Alai, M., and Cotter, R. J., *Anal. Biochem.*, 174, 23, 1988.

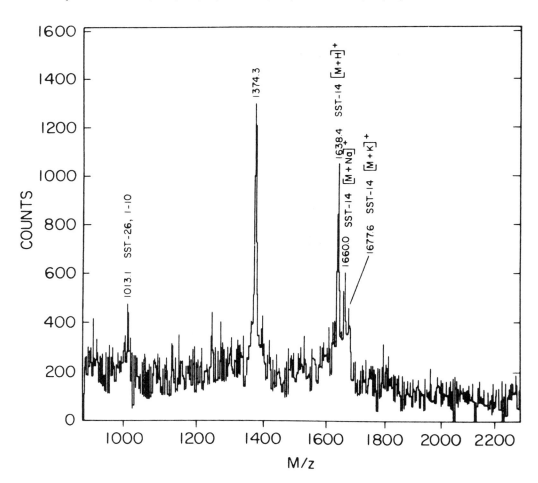

FIGURE 4. Expansion of the PDMS of crude pancreatic extract around the region of SST-14. (From Andrews, P. C., Alai, M., and Cotter, R. J., *Anal. Biochem.*, 174, 23, 1988. With permission.)

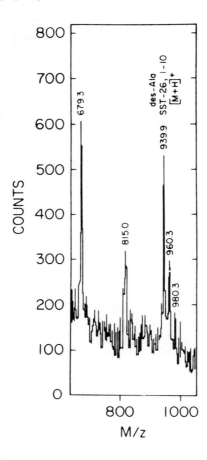

FIGURE 5.   PDMS of pancreatic extract after one cycle of manual Edman degradation. The region surrounding SST-26$_{1-10}$ is shown. The des-Ala mass provided confirmation of the mass assignment for SST-26$_{1-10}$. (From Andrews, P. C., Alai, M., and Cotter, R. J., *Anal. Biochem.,* 174, 23, 1988. With permission.)

domain of the ovomucoid from Japanese quail (OV3JPQ) exhibits serine/glycine polymorphism at position 162.[26] The allelic forms of the intact, carbohydrate-free third domain have been isolated and separated, and their secondary, tertiary, and crystallographic structures have been determined by X-ray studies and crystallographic refinement.[28]

Recently, crystals from the intact, fully glycosylated, polymorphic ovomucoid third domain have been obtained for crystallographic studies. PDMS was then employed to determine the number and possible heterogeneity of the *N*-acetylglucosamine and mannose residues attached to asparagine at position 175. The PDMS spectrum of the carbohydrate-free OV3JPQ-containing serine at position 162 is shown in Figure 8; the spectrum of the glycosylated, polymorphic form is shown in Figure 9. The molecular ion at 6945 amu corresponds to the average weight of a 70% serine/30% glycine ovomucoid with two *N*-acetylglucosamines and three mannoses (see Table 3). The peak at 6783 amu corresponds to the loss of a single mannose, and the presence of corresponding doubly charged ions in the same ratio of the singly charged ions suggests that this peak represents a minor component rather than a fragment ion.[29]

B. GLYCOPEPTIDES FROM BOVINE FETUIN

It has been shown by Spiro and co-workers[30,31] that bovine fetuin has three *O*-linked and three *N*-linked oligosaccharides per molecule, and the structure of the *N*-linked triantennary oligosaccharides has been reported by Nilsson et al.[32] In this study, PDMS was used to assess the

TABLE 2
rhGH Tryptic Peptides[a]

Peptides	Residues	Sequence	MH+ calc
T1	1—9	MFPTIPLSR	1062.3
T2	10—17	LFDNAMLR	980.2
T3	18—20	AHR	383.4
T4	21—39	LDQLAFDTYQEFEEAIPK	2343.6
T5	40—42	EQK	404.5
T6	43—65	YSFLQNPQTSLCFSESIPTPSNR	2617.9
T6-T16			3764.3
T7	66—71	EETQQK	762.9
T8	72—78	SNLELLR	845.0
T9	79—95	ISLLLIQSWLEPVQFLR	2056.5
T10	96—116	SVFANSLVYGASDSNVYDLLK	2263.5
T11	117—128	DLEEGIQTLMGR	1362.6
T12	129—135	LEDGSPR	773.8
T13	136—141	TGQIFK	693.8
T14	142—146	QTYSK	626.7
T15	147—159	FDTNSHNDDALLK	1490.6
T16	160—168	NYGLLYCFR	1149.4
T17	169—169	K	
T17-T18-T19			1382.7
T18	170—173	DMDK	
T18-T19			1254.5
T19	174—179	VETFLR	
T20	180—184	IVQCR	
T20-T21			1401.6
T21	185—192	SVEGSCGF	

Adapted from Chen, L., Cotter, R. J., and Stults, J. T., *Anal. Biochem.*, 183, 190, 1989.

molecular weight heterogeneity of glycopeptides (6 to 12 amino acids) from each of the three N-linked glycosylation sites of bovine fetuin.[33] The glycopeptides were separated by anion exchange (DEAE Sephacel) chromatography, yielding five fractions containing N-linked glycopeptides, which were purified further by reversed-phase HPLC. The amino acid composition of each fraction was determined and compared with known amino acid sequences[34] in the vicinity of the three glycosylation sites. PDMS were obtained for each of the purified fractions.

Figure 10 shows the PDMS of fraction 1. The peaks at 2691, 2713, and 2734 amu correspond to the MH+, MNa+, and (M+2Na-H)+ ions for the triantennary oligosaccharide N-linked to the Asn-Cys site of the tryptic peptide ANCSVR (deduced from the amino acid composition). The smaller peaks at 2326 and 2351 amu are the MH+ and MNa+ ions for a biantennary oligosaccharide linked at the same site. Table 4 shows the results for all five of the fractions. For example, the molecular ions observed in fraction 2 correspond to attachment of a triantennary (R2) oligosaccharide to the same Asn-Cys site for the tryptic peptide VLDPTLANCSVR. Fraction 3 also exhibits biantennary (R1) and triantennary (R2) heterogeneity at the Asn-Asp glycosylation site, whereas fractions 4 and 5 reveal attachment of only triantennary oligosaccharide at the Asn-Gly site.[33]

C. THE HIGH-MANNOSE GLYCOPEPTIDE FROM TRYPANOSOMES

African trypanosomes are parasitic protozoa that are transmitted by the tsetse fly and live exclusively in the bloodstream of their mammalian host. Trypanosomes are covered with a surface coat composed of approximately 10^7 molecules of a variant surface glycoprotein (VSG), a polypeptide of around 500 amino acids, which is approximately 10% carbohydrate by weight.[35,36] The VSG of the IlTat 1.3 variant of *Trypanosoma brucei* has two Asn-linked glycan

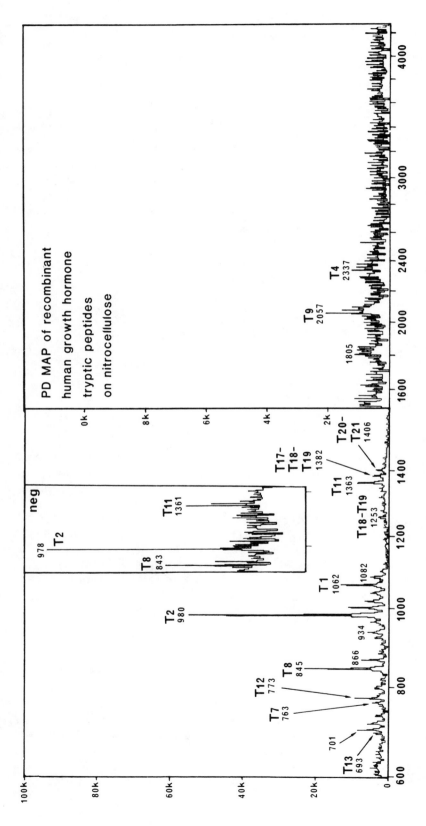

FIGURE 6.   Positive ion PDMS of the tryptic digest of rhGH deposited on a nitrocellulose surface. (Insert: a portion of the negative ion mass spectrum.) (From Chen, L., Cotter, R. J., and Stults, J. T., *Anal. Biochem.*, in press. With permission.)

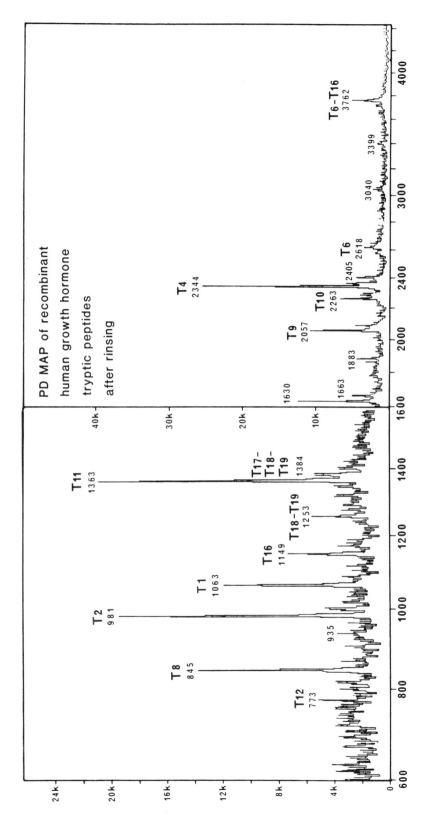

FIGURE 7. Positive ion PDMS of the same tryptic digest after washing with deionized water. (From Chen, L., Cotter, R. J., and Stults, J. T., *Anal. Biochem.*, in press. With permission.)

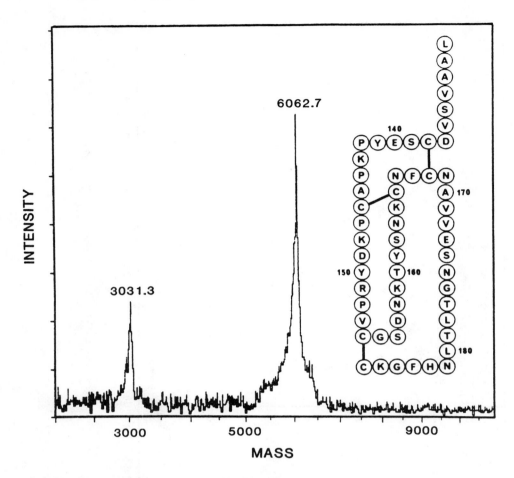

FIGURE 8.   Positive ion PDMS of carbohydrate-free Ser[162] OV3JPQ. From Wang, R., Cotter, R. J., Lin, T.-Y., and Laskowski, M., Jr., *Rapid Comun. Mass Spectrom.,* 2, 71, 1988. With permission.)

moieties, as well as a glycosylphosphatidylinositol membrane anchor.[37] The complete structure of this anchor has been reported recently by Ferguson et al.[38] The two N-linked glycosylation sites of this protein contain remarkably different types of oligosaccharides. Those sugars at Asn-419 are almost exclusively biantennary, whereas those at Asn-432 are high mannose species. The general structure of gp432 glycopeptide (which contains the latter site):

$$\text{Phe–Asn(GlcNAc}_2\text{–Man}_n\text{)–Glu–Thr–Lys}$$

has been reported.[37] The gp432 glycopeptides were radiolabeled *in vivo* with [³H]mannose, and size fractionation of the free oligosaccharides released by EndoH-cleavage between the two GlcNAc residues suggested that gp432 contained four to seven mannose units.[37] A PDMS spectrum was then obtained for the intact, nonradiolabeled glycopeptide mixture (Figure 11). The peaks observed correspond to the structures summarized in Table 5, and indicate that the gp432 glycopeptide contains five to nine mannose units.[39] The latter determination by PDMS is expected to be more accurate, because the altered metabolic conditions present in the low-glucose culture medium used for incorporation of the [³H]mannose radiolabel may result in a population differing from that of the bulk VSG. In addition, plasma desorption provides a direct measurement that obviates the need for radiolabeling and processing by EndoH prior to analysis.[39]

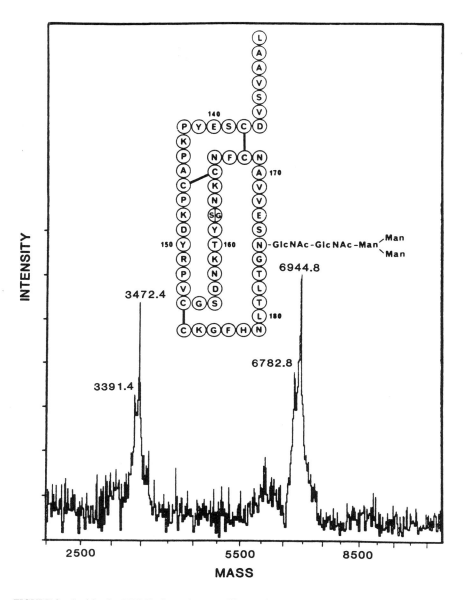

FIGURE 9. Positive ion PDMS glycosylated Ser[162] / Gly[162] OV3JPQ. (From Wang, R., Cotter, R. J., Lin, T.-Y., and Laskowski, M., Jr., *Rapid Commun. Mass Spectrom.*, 2, 71, 1988. With permission.)

TABLE 3

Calculated Average Molecular Weights of Carbohydrate-free and Glycosylated Ovo-mucoid Third Domains from Japanese Quail

Ser[162]**OV3JPQ**:	6061.7
Gly[162]**OV3JPQ**:	6031.7
(Man)$_3$(GlcNAc)$_2$-Ser[162]**OV3JPQ**:	6953.8
(Man)$_3$(GlcNAc)$_2$-Gly[162]**OV3JPQ**:	6923.8

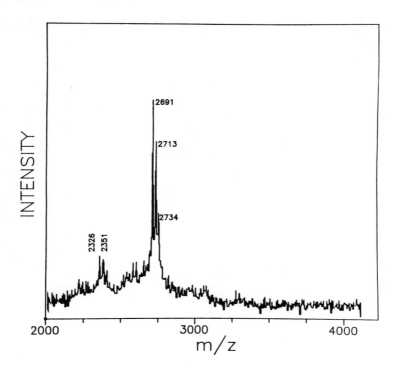

FIGURE 10. PDMS of the glycopeptide from bovine fetuin corresponding to the Asn-Cys glycosylation site. (From Townsend, R. R., Alai, M., Hardy, M. R., and Fenselau, C., *Anal. Biochem.*, 171, 180, 1988. With permission.)

TABLE 4
Molecular Ions of Site-specific *N*-Linked Glycopeptides from Bovine Fetuin

Glycosylation site	Fraction	Glycopeptides	MH⁺ calc	MH⁺ obs	MNa⁺ obs	M+2Na-H⁺ obs
Asn-Cys	1	AN(R$_1$)CSVR	2329	2326	2351	
		AN(R$_2$)CSVR	2694	2691	2713	2735
	2	VLDPTLAN(R$_2$)CSVR	3343	3430	3451	3472
Asn-Asp	3	LAPLN(R$_1$)DSR	2510	2507		
		LAPLN(R$_2$)DSR	2875	2870		
Asn-Gly	4	NAESN(R$_2$)GSYLQ	3072		3091	
	5	NAESN(R$_2$)GSYL	2944		2960	

Galβ(1,4 or 3)GlcNAcβ(1,2)Manα(1,6)
$\quad\quad\quad\quad\quad\quad\quad\quad\quad\quad\quad$ \
$\quad\quad\quad\quad\quad\quad$ Manβ(1,4)GlcNAcβ(1,4)GlcNAc→ = R_1
$\quad\quad\quad\quad\quad\quad\quad\quad\quad$ /
Galβ(1,4 or 3)GlcNAcβ(1,2)Manα(1,3)

Galβ(1,4 or 3)GlcNAcβ(1,2)Manα(1,6)
$\quad\quad\quad\quad\quad\quad\quad\quad\quad\quad\quad$ \
$\quad\quad\quad\quad\quad\quad$ Manβ(1,4)GlcNAcβ(1,4)GlcNAc→ = R_2
$\quad\quad\quad\quad\quad\quad\quad\quad\quad$ /
Galβ(1,4 or 3)GlcNAcβ(1,2)Manα(1,3)
$\quad\quad\quad\quad\quad\quad\quad\quad$ /
Galβ(1,4 or 3)GlcNAcβ(1,4)

Adapted from Townsend, R. P., Alai, M., Hardy, M. R., and Fenselau, C., *Anal. Biochem.*, 171, 180, 1988.

V. PHOSPHORYLATED PEPTIDES: HETEROGENEITY OF RIBOFLAVIN-BINDING PROTEIN

Riboflavin-binding protein from hen egg white is a phosphoglycoprotein that possesses two Asn-linked complex oligosaccharides and eight phosphoserine residues. Glycosyl and phosphoryl groups have both been implicated in the protein's function,[40] which is to deliver riboflavin for use by the chick embryo. The phosphoryl groups are confined to a single short region of the protein for which the amino acid sequence and locations of the phosphoryl groups have been determined.[41,42] Positive and negative ion FAB mass spectra were both obtained for the tryptic fragment containing the phosphate groups, from which a mol wt of 3206 D, corresponding to an octaphosphoryl structure, was determined.[42] A portion of the negative-ion mass spectrum used to establish the positions of the phosphoryl groups is shown in Figure 12. The eight phosphates are attached to serine residues in a portion of that fragment:[42]

Ser–Glu–pSer–pSer–Glu–Glu–pSer–pSer–pSer–Met–pSer–pSer–pSer–Glu–Glu–

Riboflavin-binding protein is heterogeneous with respect to both glycosyl and phosphoryl groups. Thus, in addition to the octaphosphoryl peptide, there are also forms with fewer phosphoryl groups. Anion-exchange chromatography was used to separate homologs of the tryptic phosphopeptide, and four peptide peaks were obtained and analyzed by PDMS.[43] The spectra shown in Figure 13 correspond to the A, B, and D chromatographic fractions. Amino acid analysis of fractions C and D indicated that these fractions were pure and had an amino acid composition:

$Ser_9,Glu_5,His_2,Leu_2,Met,Ala,Cys,Gln,Lys$

Comparison of the protonated molecular ions observed in Figure 13 with the calculated masses indicated that the mixture contained peptides carrying five, seven, and eight phosphoryl groups (Table 6).

VI. POST-TRANSLATIONAL MODIFICATIONS: MURINE ELONGATION FACTOR

Strategies for revealing other post-translational modifications are similar to those used for glycopeptides and phosphopeptides. Knowledge of the amino acid sequence is an advantage when only the molecular ion (and a few fragment ions) are observed.

Elongation Factor 1α (EF-1α) is an important eukaryotic translation factor that transports charged aminoacyl-tRNA from the cytosol to the ribosomes during polypeptide synthesis. Recently, Whiteheart et al.[44] used metabolic radiolabeling of this 53 kDa protein with [3H]ethanolamine to reveal a novel amide-linked ethanolamine-phosphoglycerol post-translational modification. Tryptic digestion of radiolabeled murine EF-1α yielded two major [3H]ethanolamine-labeled peptides. Microsequencing of peptide I indicated that this peptide had an amino acid sequence: F-A-"blank"-L-K, with the "blank" coinciding with the [3H]ethanolamine-labeled amino acid derivative. Compositional analysis of this peptide showed that it contained Phe, Ala, Leu, Lys, Glx, ethanolamine, glycerol, and phosphate. Dansylation analysis demonstrated that the amine group of the ethanolamine is blocked. The plasma desorption mass spectrum and the final structure of this post-translationally modified peptide are shown in Figure 14.[44]

VII. DIMERS AND ADDUCTS

Proton-bound dimers, heterodimers, and adducts are observed often in plasma desorption

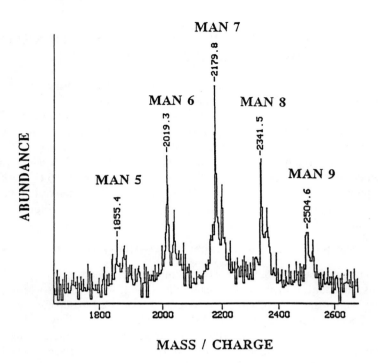

FIGURE 11. PDMS of the high mannose gp432 glycopeptide obtained from *Trypanosoma brucei*. (From Bean, M. F., Brangs, J. D., Doering, T. L., Englund, P. T., Hart, G. W., Fenselau, C., and Cotter, R. J., *Anal. Chem.*, 61, 2686, 1989. With permission.)

TABLE 5
Molecular Weights of gp432 Glycopeptide

Species	Monoisotopic mass (MH⁺)	Average mass (MH⁺)
FNETK GlcNAc$_2$Man$_4$	1692.68	1693.80
FNETK GlcNAc$_2$Man$_5$	1854.73	1855.96
FNETK GlcNAc$_2$Man$_6$	2016.78	2018.12
FNETK GlcNAc$_2$Man$_7$	2178.84	2180.27
FNETK GlcNAc$_2$Man$_8$	2340.89	2342.43
FNETK GlcNAc$_2$Man$_9$	2502.94	2504.59

mass spectra as artifacts of the technique. The formation of covalent dimers can, however, be distinguished from these artifacts by their masses (i.e., peptide bonds will result in molecular ions 18 amu lower, and disulfide bonds will result in molecular ions 2 amu lower than the proton-bound dimers) and by the presence of doubly- and triply-charged ions.

A. INSULIN DIMERIZATION IN LONG-TERM PREPARATIONS

Insulin preparations degrade by deamination of Asn at the 21-position of the A-chain or the 3-position of the B-chain, yielding Asp at the A21 and B3 sites.[45] The two resulting mon-odesamido isomers possess full biological potency, and their immunogenicity in rabbits is not significantly different from that of insulin. The structures of other major degradation products of insulin are not known in detail. Based upon chromatographic and electrophoretic behavior of

33

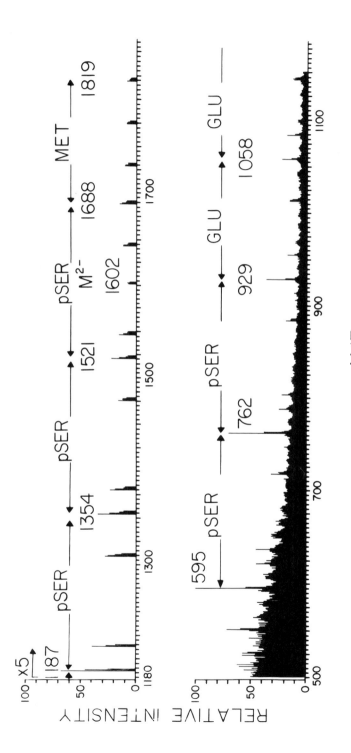

FIGURE 12. Negative ion FAB mass spectrum of the phosphopeptide from riboflavin binding protein. (From Fenselau, C., Heller, D. N., Miller, M. S., and White, H. B., III, *Anal. Biochem.*, 150, 309, 1985. With permission.)

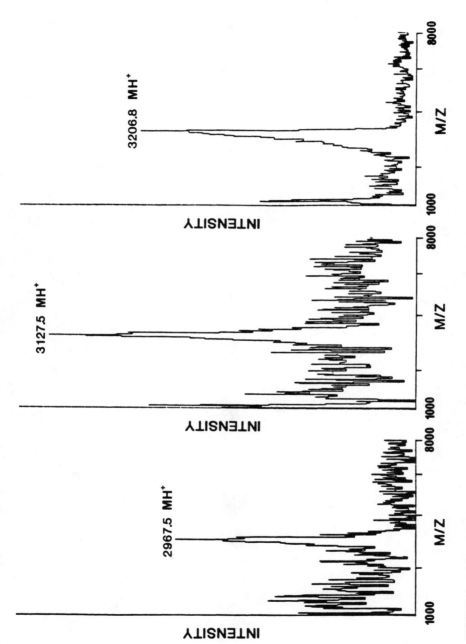

FIGURE 13. PDMS of chromatographic peaks A, B, and D (see Table 6) used to determine the number of phosphorylated serine residues on wRBP phosphopeptide. (From Vaughn, V. L., Wang, R., Fenselau, C., and White, H. B., III, *Biochem. Biophys. Res. Commun.*, 147, 115, 1987. With permission.)

TABLE 6
PDMS Analysis of wRBP Phosphopeptides

Chromatography peak	$(M+H)^+$	No. of phosphorylated serine residues	Calc mol wt of phosphopeptide
A	2967.5	5	2565.8
B	3127.5	7	3125.8
C	3125.4	7	3125.8
	3210.7	8	3205.7
D	3206.8	8	3205.7

From Vaughn, V. L., Wang, R., Fenselau, C., and White, H. B., III, *Biochem. Biophys. Res. Commun.*, 147, 115, 1987. With permission.

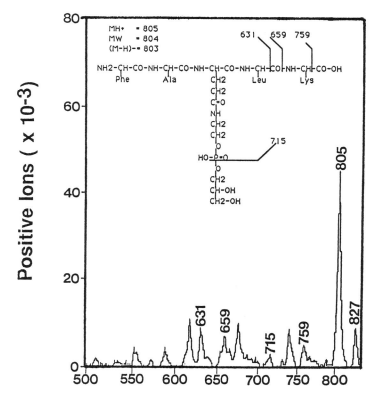

FIGURE 14. Structure and PDMS of *peptide I* from murine EF-1α. (From Whitheart, S. W., Shenbagamurthi, P., Chen, L., Cotter, R. J., and Hart, G. W., *J. Biol. Chem.*, 264, 14334, 1989. With permission.)

the products and their A- and B-chains, these degradants are thought to be isomeric dimers of insulin with a covalent link between the N-terminal amino group in one molecule with a C-terminal group in the A-chain of another insulin molecule. Although the immunogenicity of the dimers in rabbits was not significantly different from monomeric insulin, biological potency is reduced significantly.[46] PDMS was used to verify the formation after 12 months of two isomeric insulin dimers, isolated from a zinc suspension by gel chromatography on Biogel P30 in 1 M acetic acid.[47] The PDMS spectrum of one of the dimers is shown in Figure 15. Singly, doubly, and triply charged ions are observed for the dimer. At the same time, the absence of a doubly charged ion for the monomer (normally observed in PDMS spectra of insulins) confirms the existence of a covalent dimer.

B. NONENZYMATIC FORMATION OF INSULIN-GLUTATHIONE MIXED DISULFIDES

The thiol-disulfide interchange of glutathione and disulfide of peptides and proteins have been the subject of numerous studies. A number of enzymes has been proposed to catalyze this reaction.[48] Varandani[49] has advocated that insulin is first split at the disulfide bonds by glutathione-insulin transhydrogenase into A- and B-chains, which are then hydrolyzed by specific peptidases. On the other hand, much evidence supports the idea that the degradation of insulin requires receptor binding, internalization, and digestion in the lysozymes.[50] Until recently, no laboratory has been successful in isolating and demonstrating the transient insulin-glutathione mixed disulfide. Recently, however, Alai and Fenselau[51] bound bovine insulin to a glutathione agarose column, which was eluted with glutathione-ammonium acetate, and obtained the PDMS (Figure 16), which showed the insulin-glutathione mixed disulfide (both singly and doubly charged ions) and an increase in the amount of B-chain along with its mixed disulfide.

C. METAL-LINKED DIMERS OF THE *TAT III* CYSTEINE-RICH REGION

The tat protein from human immunodeficiency virus (HIV) is a viral transactivator,[52] and is essential for viral replication.[53,54] Tat is 86 amino acids long, and contains a highly basic region (2 lysines and 6 arginines within 9 residues) and a cysteine-rich region (7 cysteines within 16 residues) as shown in Figure 17.[55,56] Frankel et al.[57] have shown that purified tat can form metal-linked dimers. The dimer structure contains four metal ions, and it appears that two or more of them bridge the cysteine-rich regions from each monomer. From proteolytic digestion studies and circular dichroism spectra, it appears that the structural effects of metal binding are localized primarily, if not exclusively, to the cysteine-rich region.

For this reason, the 18-amino acid peptide, tat_{21-38}, containing the cysteine region, was synthesized (Figure 17) and PDMS measurements of the purified peptide confirmed the molecular weight. Then 7 nmol of peptide was incubated with 3 equivalents of $CdCl_2$ in Tris-HCl buffer (pH 7.0). The complex was lyophilized and redissolved in water, deposited on a nitrocellulose foil, and washed with 0.01% acetic acid to remove excess Cd^{+2} ions. The PDMS spectrum of the complex in Figure 18 was recorded, and shows two broad peaks due to the monomer and the Cd^{2+}-linked dimer. The interpretation of the fine structure of these peaks is also shown in Figure 18, and indicates that the Cd^{2+}-tat_{21-38} complex includes four Cd^{2+} ions.[58] PDMS also confirmed the formation of Zn^{2+}-tat_{21-38} dimers with four Zn^{2+} ions. The tat_{21-38} peptide can also combine with the intact protein to form metal-linked heterodimers. It has been suggested that, if these heterodimers are unable to transactivate viral transcription, then tat_{21-38} could be an important compound for designing drugs to treat acquired immune deficiency syndrome (AIDS).

VIII. CONCLUSIONS

When combined with other techniques, molecular weight measurements by PDMS can be of critical importance. When the amino acid sequence of a protein is known, it is of course possible to predict the products from tryptic or other enzymic digests. In a PD-map of an unfractionated or partially fractionated digest, differences between the measured and predicted masses for the peptides can reveal modifications, blocking groups, or errors in synthesis. PD-mapping can also be employed to identify disulfide linkages in proteins carrying several cysteines. For example, cysteine residues are found on positions 54, 166, 183, and 190 in the rhGH discussed previously.

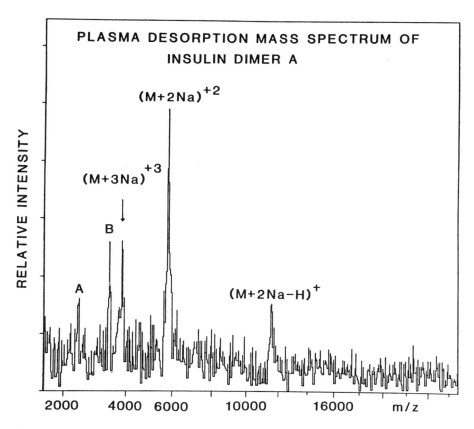

FIGURE 15. PDMS of a porcine insulin dimer. (From Unger, S. W., Brange, J., Lauritano, A., Demirev, P., Wang, R., and Cotter, R. J., *Rapid Commun. Mass Spectrom.*, 2, 109, 1988. With permission.)

The presence of peaks with masses corresponding to T6-T16 and T20-T21 (less 2 amu) in the spectra of Figures 6 and 7 reflect the disulfide bond pairs: Cys^{54}-Cys^{166} and Cys^{183}-Cys^{190}. At the same time, the masses correlated with the T18-T19 and T17-T18-T19 peptides correspond to fragments that are covalently bound and reflect incomplete cleavage at these sites. Thus, although mass spectrometry in general has become a powerful technique for the primary amino acid sequencing of peptides, it may be most valuable for the elucidation of glycopeptide, phosphopeptide, and other structural details of the fully-processed protein.

ACKNOWLEDGMENTS

The plasma desorption mass spectrometer was purchased by a grant (RR 02727) from the Division of Research Resources of the National Institutes of Health, and supported by a grant (DMB 85-15390) from the National Science Foundation. The authors acknowledge the contributions of a number of collaborators (C. Schwabe, E. Bullesbuch, D.F. Steiner, P.C. Andrews, J.E. Dixon, J.T. Stults, M. Laskowski, T.-Y. Lin, Y.C. Lee, R.R. Townsend, M. Hardy, P.T. Englund, G.W. Hart, J.D. Bangs, T.L. Doering, H. B. White, V.L. Vaughn, S.W. Whiteheart, S. Unger, A.D. Frankel, C.O. Pabo, and C. Fenselau), as well as present and former members of the Middle Atlantic Mass Spectrometry Facility (M. Alai, M.F. Bean, D.N. Heller, and P. Demirev) who carried out mass spectral measurements.

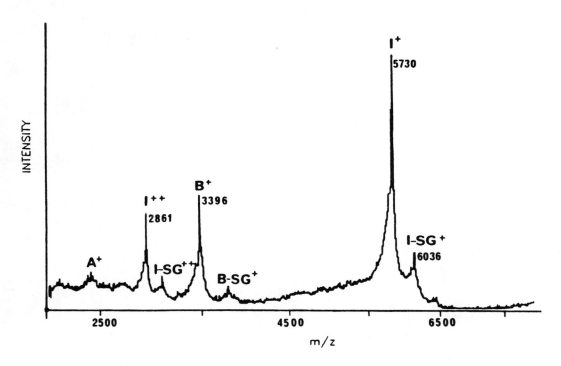

FIGURE 16.   PDMS of sample prepared by adsorption of insulin to a SH-free glutathione-agarose column and elution with reduced glutathione, after 3 column volume washes with 0.1 *M* ammonium acetate, pH 7.5. (From Alai, M. and Fenselau, C., *Biochem. Biophys. Res. Commun.,* 146, 815, 1987. With permission.)

Met	Glu	Pro	Val	Asp	Pro	Arg	Leu	Glu	Pro
Trp	Lys	His	Pro	Gly	Ser	Gln	Pro	Lys	Thr
Ala	Cys	Thr	Asn	Cys	Tyr	Cys	Lys	Lys	Cys
Cys	Phe	His	Cys	Gln	Val	Cys	Phe	Ile	Thr
Lys	Ala	Leu	Gly	Ile	Ser	Tyr	Gly	Arg	Lys
Lys	Arg	Arg	Gln	Arg	Arg	Arg	Pro	Pro	Gln
Gly	Ser	Gln	Thr	His	Gln	Val	Ser	Leu	Ser
Lys	Gln	Pro	Thr	Ser	Gln	Ser	Arg	Gly	Asp
Pro	Thr	Gly	Pro	Lys	Glu				

FIGURE 17.   Amino acid sequences of the tat$_{21-38}$ peptide (upper) and the tat protein (lower). (From Frankel, A. D., Chen, L., Cotter, R. J., and Pabo, C. O., *Proc. Natl. Acad. Sci. U.S.A.,* 85, 6297, 198. With permission.)

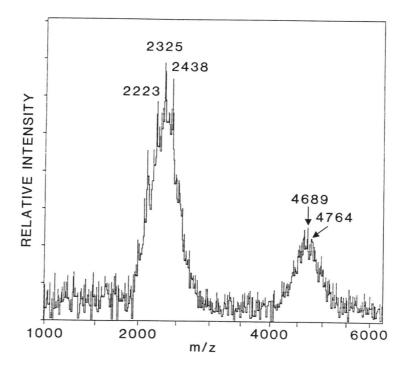

FIGURE 18. PDMS of the tat$_{21-38}$-Cd^{2+} complex. The peaks may be interpreted as follows: 2325 monomer + 2Cd^{2+} - 3H$^+$, 2438 monomer + 3Cd^{2+} - 5H$^+$, 4689 dimer + 4Cd^{2+} + Cl$^-$ - 6H$^+$, and 4764 dimer + 4Cd^{2+} + 3Cl$^-$ - 4H$^+$. (From Frankel, A. D., Chen, L., Cotter, R. J., and Pabo, C. O., *Proc. Natl. Acad. Sci. U.S.A.*, 85, 6297, 1988. With permission.)

REFERENCES

1. **McLafferty, F.W.,** *Interpretation of Mass Spectra*, Benjamin, Reading, MA, 1973.
2. **Beckey, H. D. and Schuelte, D.,** *Z. Instrumenten*, 68, 302, 1960.
3. **Barber, M., Bordoli, R. S., Sedgwick, R. D., and Tyler, A. N.,** *J. Chem. Soc. Chem. Commun.*, 325, 1981.
4. **Posthumus, M. A., Kistemaker, P. G., Meuzelaar, H. L. C., and Ten Noever de Brauw, N. C.,** *Anal. Chem.*, 50, 985.,1978.
5. **Macfarlane, R. D., Skowronski, R. P., and Torgerson, D. F.,** *Biochem. Biophys. Res. Commun.*, 60, 616, 1974.
6. **Cotter, R. J.,** *Anal. Chem.*, 60, 781A, 1988.
7. **Bedarkar, S., Turnell, W. G., Blundell, T. L., and Schwabe, C.,** *Nature*, 270, 5636, 1977.
8. **Jardine, I., Scanlan, G. F., Tsarbopoulous, A., and Liberato, D. J.,** *Anal. Chem.*, 60, 1086, 1988.
9. **Robbins, D. C., Tager, H. S., and Rubenstein, A. H.,** *N. Engl. J. Med.*, 310, 1165, 1984.
10. **Gruppuso, P. A., Gordon, P., Khan, C. R., Cornblath, M., Zeller, W. P., and Schwartz, R.,** *N. Engl. J. Med.*, 311, 626, 1984.
11. **Chan, S. J, Seino, S., Gruppuso, P. A., Schwartz, R., and Steiner, D. F.,** *Proc. Natl. Acad. Sci., U.S.A.*, 84, 2194, 1987.
12. **Schwartz, G. P., Burke, G. T., and Katsoyannis, P. G.,** *Proc. Natl. Acad. Sci., U.S.A.*, 84, 6408, 1987.
13. **Alai, M.,** Ph.D thesis, Johns Hopkins University, Baltimore, 1988.
14. **Minth, C. D., Taylor, W. L., Magazin, M., Tavianini, M. A., Collier, K., Weith, H. L., and Dixon, J. E.,** *J. Biol. Chem.*, 257, 10372, 1982.
15. **Magazin, M., Minth, C. D., Funckes, C. L., Deschenes, R., Tavianini, M. A., and Dixon, J. E.,** *Proc. Natl. Acad. Sci. U.S.A.*, 79, 5152, 1982.
16. **Andrews, P. C. and Dixon, J. E.,** *J. Biol. Chem.*, 256, 8267, 1981.

17. **Andrews, P. C., Pubols, M. H., Hermodson, M. A., Sheares, B. T., and Dixon, J. E.,** *J. Biol. Chem.,* 255, 2251, 1980.
18. **Oyama, H., Bradshaw, R. A., Bates, O. J., and Permutt, A.,** *J. Biol. Chem.,* 255, 2251, 1980.
19. **Andrews. P. C. and Ronner, P.,** *J. Biol. Chem.,* 260, 8128, 1985.
20. **Andrews, P. C.,** unpublished data.
21. **Andrews, P. C., Alai, M., and Cotter, R. J.,** *Anal. Biochem.,* 174, 23, 1988.
22. **Naylor, S., Findeis, A. F., Gibson, B. W., and Williams, D. H.,** *J. Am. Chem. Soc.,* 108, 6359, 1986.
23. **Nielsen, P. F. and Roepstorff, P.,** *Biomed. Environ. Mass Spectrom.,* 18, 131, 1989.
24. **Canova-Davis, E., Chloupek, R. C., Baldonado, I. P., Battersby, J. E., Spellman, M. W., Bosa, L. J., O'Connor, B., Pearlman, R., Quan, C., Chakel, J. A., Stults, J. T., and Hancock, W. S.,** *Am. Biotechnol. Lab.,* 8, May 1988.
25. **Chen, L., Cotter, R. J., and Stults, J. T.,** *Anal. Biochem.,* 183, 190, 1989.
26. **Bogard, W. C., Jr., Kato, I., and Laskowski, M., Jr.,** *J. Biol. Chem.,* 255 (1980) 6569.
27. **Laskowski, M., Jr., Kato, I., Ardelt, W., Cook, J., Denton, A., Empie, M. W., Kohr, W. J., Park, S. J., Schatzley, B. L., Schoenberger, O. L., Tashiro, M., Vichot, G., Whatley, H. E., Wieczorek, A., and Wieczorek, M.,** *Biochemistry,* 26, 202, 1987.
28. **Papamakos, E., Weber, E., Bode, W., Huber, R., Empie, M. W., Kato, I., and Laskowski, M., Jr.,** *J. Mol. Biol.,* 158, 515, 1982.
29. **Wang, R., Cotter, R. J., Lin, T.-Y, and Laskowski, M., Jr.,** *Rapid Commun. Mass Spectrom.,* 2, 71, 1988.
30. **Spiro, R. G.,** *Adv. Protein Chem.,* 27, 349, 1973.
31. **Spiro, R. G. and Bhoyroo, V. D.,** *J. Biol. Chem.,* 249, 5704, 1974.
32. **Nilsson, B., Norden, E., and Svensson, S.,** *J. Biol. Chem.,* 252, 4545, 1979.
33. **Townsend, R. R., Alai, M., Hardy, M. R., and Fenselau, C.,** *Anal. Biochem.,* 171, 180, 1988.
34. **Wold, F.,** private communication.
35. **Vickerman, K.,** *J. Cell Sci.,* 5, 163, 1969.
36. **Cross, G. A. M.,** *Parasitology,* 71, 393, 1975.
37. **Bangs, J. D., Doering, T. L., Englund, P. T., and Hart, G. W.,** *J. Biol. Chem.,* 263 (1988) 17697.
38. **Ferguson, M. A. J., Homans, S. W., Dwek, R. A., and Rademacher, T. W.,** *Science,* 239, 743, 1988.
39. **Bean, M. F., Bangs, J. D., Doering, T. L., Englund, P. T., Hart, G. W., Fenselau, C., and Cotter, R. J.,** *Anal. Chem.,* 61, 2686, 1989.
40. **Miller, M. S., Mas, T. M., and White, H. B., III.,** *J. Biol. Chem.,* 257, 6818, 1982.
41. **Mega, T., Hamazume, Y., Nong Y.-M., and Ikenaka, T.,** *J. Biochem. (Tokyo),* 100, 1109, 1986.
42. **Fenselau, C., Heller, D. N., Miller, M. S., and White, H. B., III.,** *Anal. Biochem.,* 150, 309, 1985.
43. **Vaughn, V. L., Wang, R., Fenselau, C., and White, H. B., III.,** *Biochem. Biophys. Res. Commun.,* 147, 115, 1987.
44. **Whiteheart, S. W., Shenbagamurthi, P., Chen, L., Cotter, R. J., and Hart, G. W.,** *J. Biol. Chem.,* 264, 14334, 1989.
45. **Brange, J., Langkaer, L., Havelund, S., and Sorensen, E.,** presented at the 20th Annual Meeting of the European Association for the Study of Diabetes, London, September 1984.
46. **Brange, J.,** *Galenics of Insulin,* Springer-Verlag, Berlin, 1987.
47. **Unger, S. E., Brange, J., Lauritano, A., Demirev, P., Wang , R., and Cotter R. J.,** *Rapid Commun. Mass Spectrom.,* 2, 109, 1988.
48. **Morin, J. E. and Dixon, J. E.,** *Methods Enzymol.,* 113, 541, 1985.
49. **Varadani, P. R. and Nafz, M. A.,** *Diabetes,* 25, 173, 1976.
50. **Krupp, M. N. and Lane, M. D.,** *J. Biol. Chem.,* 257, 1371, 1982.
51. **Alai, M. and Fenselau, C.,** *Biochem. Biophys. Res. Commun.,* 146 (1987) 815.
52. **Sodroski, J., Rosen, C., Wong-Staal, F., Salahuddin, S. Z., Popovic, M., Arya, S., Gallo, R. C., and Haseltine, W. A.,** *Science,* 227, 171, 1985.
53. **Fisher, A. G., Feinberg, M. B., Josephs, S. F., Harper, M., Marselle, M. L., Reyes, G., Gonda, M. A., Aldovini, A., Debouch, C., Gallo, R. C., and Wong-Staal, F.,** *Nature,* 320, 367, 1986.
54. **Dayton, A. I., Sodroski, J. G., Rosen, C. A., Goh, W. C., and Haseltine, W. A.,** *Cell,* 44, 941, 1986.
55. **Arya, S., Guo, C., Josephs, S. F., and Wong-Staal, F.,** *Science,* 229, 69, 1985.
56. **Sodroski, J., Patarca, R., Rosen, C., Wong-Staal, F., and Haseltine, W. A.,** *Science,* 229, 74, 1985.
57. **Frankel, A. D., Bredt, D. S., and Pabo, C. O.,** *Science,* 240, 70, 1988.
58. **Frankel, A. D., Chen, L., Cotter, R. J., and Pabo, C. O.,** *Proc. Natl. Acad. Sci. U.S.A.,* 85, 6297, 1988.

Chapter 3

THE ANALYSIS OF SYNTHETIC PEPTIDES AND PROTEINS BY ^{252}CF- PLASMA DESORPTION MASS SPECTROMETRY*

Brian T. Chait

TABLE OF CONTENTS

I.   Introduction 42

II.   Verification of the Correctness of the Covalent Structure 42

III.   Verification of the Homogeneity of the Synthetic Product — Identification
       of Unwanted Side-Products 44

IV.   Limitations of the Mass Spectrometric Technique 53

V.   Mass Spectrometry of Fully Protected Synthetic Peptides 53

VI.   Mass Spectrometry of Unusual Synthetic Peptides 55

VII.   Disulfide Bonds in Synthetic Peptides and Proteins 55

VIII. Evaluation and Optimization of the Chemical Procedures
       Used to Synthesize Peptides 61

IX.   Conclusion 61

Acknowledgments 62

References 62

*   The present paper has appeared in large part previously in Chait, B. T., The use of ^{252}Cf plasma desorption mass
    spectrometry for the analysis of synthetic peptides and proteins, in *The Analysis of Peptides and Proteins by Mass
    Spectrometry,* McNeal, C. J., Ed., John Wiley & Sons, Chichester, 1988, 21. Reproduced by permission of John
    Wiley & Sons, Ltd.

I. INTRODUCTION

There is a large and rapidly growing need among members of the biological community for high-purity synthetic peptides and small proteins. The most widely used methods for producing these materials are based on the stepwise solid-phase synthetic procedure devised by Merrifield[1] (for reviews see References 2 to 4). As the size and complexity of the target peptide increase, the opportunity for synthetic errors, unwanted modifications, and cumulative effects arising from incomplete reactions also increases. It is therefore imperative to have available effective means for rapidly verifying the correctness of the covalent structure of these complex materials, establishing their purity, and detecting and identifying undesired peptide byproducts. The numerous methods devised for these purposes have been reviewed.[2-4] The most useful have involved subjecting the products to high resolution high pressure liquid chromatography (HPLC) -separation,[4-5] amino acid analysis,[6] sequence analysis,[7] spectrometric analysis,[8] nuclear magnetic resonance (NMR) analysis,[9] and mass spectrometry.[10-15] Each of these methods has its own particular strengths and weaknesses. Over the past several years, members of the Rockefeller University Mass Spectrometric Research Resource have used [252]Cf-plasma desorption mass spectrometry (PDMS)[16] to analyze in detail more than 1200 synthetic peptides and proteins submitted by members of 16 different laboratories in the U.S. The results of these analyses demonstrated that PDMS provides highly useful information concerning the integrity and purity of these compounds and is a powerful complement to the more established methods. The importance of the use of the PDMS technique can be gauged from our rather staggering finding: almost half of the 1200 purified synthetic peptides and proteins that we examined were found to have either a molecular weight different from that calculated for the desired target material, or to contain significant amounts of unwanted peptide side-products. This chapter describes the utility of [252]Cf-PDMS for the analysis of synthetic peptides and proteins.

II. VERIFICATION OF THE CORRECTNESS OF THE COVALENT STRUCTURE

The most useful and easily obtained single piece of information from the [252]Cf-plasma desorption mass spectrometry is the molecular weight (mol wt) of the compound, as determined from the peaks corresponding to the singly and multiply protonated intact molecule. In a stepwise peptide synthesis, the identity of each added amino acid is known, and therefore a measured mol wt, which is found to agree with the mol wt calculated for the desired synthetic product, provides a relatively dependable verification that the material has been assembled correctly. Conversely, any disagreement observed between the measured and calculated mol wt of the synthetic product indicates immediately that an error or undesired modification has occurred.

Figure 1 shows that the fission fragment time-of-flight (TOF) mass spectrum of a 37-amino acid residue analog of the egg-laying hormone from the mollusk *Aplysia californica* produced in 86% yield with automated stepwise synthesis by Kent and Schiller at the California Institute of Technology. The sample is prepared for mass spectrometry by absorbing approximately 1 nmol of the peptide onto a specially prepared thin film of nitrocellulose as described previously.[17] The measured mol wt of 4441.4 U[*18] agrees well with the mol wt of 4441.2 U calculated from the amino acid sequence shown in Figure 1. This agreement provides a valuable initial verification of the correctness of the synthesized structure. Because the analysis shown in Figure 1 was completed within less than 2 h of receipt of the sample, it can be seen that this mass spectrometric procedure is relatively rapid and straightforward. It should be emphasized that this

---

* The measured molecular weight is taken as the simple average of the values deduced from the singly and doubly protonated molecule peaks.

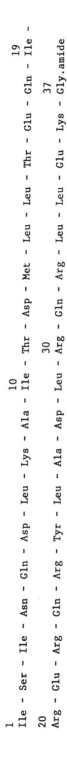

```
 1
Ile - Ser - Ile - Asn - Gln - Asp - Leu - Lys - Ala - Ile -
                                        10
20
Ile - Thr - Asp - Met - Leu - Leu - Thr - Glu - Gln - Ile -
                                                        19
37
Arg - Glu - Arg - Gln - Arg - Tyr - Leu - Ala - Asp - Leu -
                                              30
Arg - Gln - Arg - Leu - Leu - Glu - Lys - Gly.amide
```

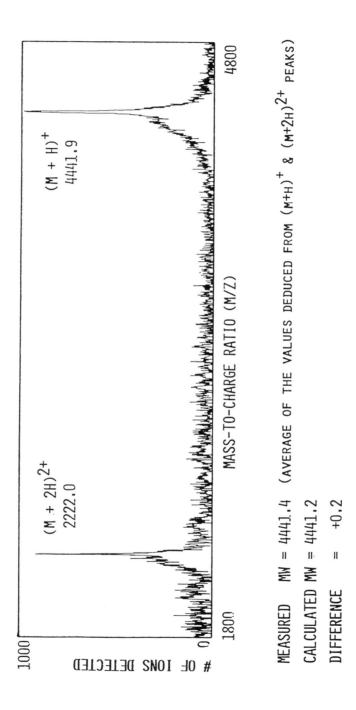

FIGURE 1. Partial ^{252}Cf-PDMS of a synthetic analog of the egg-laying hormone from *Aplysia californica*. M designates the intact molecule. (From Chait, B. T., The use of ^{252}Cf plasma desorption mass spectrometry for the analysis of synthetic peptides and proteins, in *The Analysis of Peptides and Proteins by Mass Spectrometry*, McNeal, C. J., Ed., John Wiley & Sons, New York, 1988, 21. By permission of John Wiley & Sons.)

sample mol wt determination provides a necessary but not a sufficient condition for confirming the correctness of sequence. More stringent confirmation of the proposed sequence can be obtained, for example, by classical Edman sequence analysis[7] or by mass spectrometric sequence analysis.[11-14] Such detailed mass spectrometric sequence information is sometimes available directly from the normal ^{252}Cf-PDMS. Thus, for example, we were able to use PDMS to verify the detailed identity between natural and synthetic alamethicin I, a 20-residue, pore-forming peptide antibiotic.[11] The mass spectrometric verification of sequence was of special value in this case, because alamethicin has a blocked amino terminus. Unfortunately, in many cases, the amount and the nature of the fission fragment bombardment-induced fragmentation produces only weak and incomplete sequence information. In these instances, tandem mass spectrometric sequence determination is of great value.[12-14]

III. VERIFICATION OF THE HOMOGENEITY OF THE SYNTHETIC PRODUCT — IDENTIFICATION OF UNWANTED SIDE-PRODUCTS

^{252}Cf fission fragment ionization mass spectrometry is also a powerful tool for determining the homogeneity of the desired synthetic product. The strength of the method resides in its high resolving power, and it is usually straightforward in the discernment of the mass spectrum side-products that differ from the desired target material by as little as a fraction of a percent. In addition, if such side-products are observed, their masses can be determined accurately to provide an important clue to their identity. Thus, for example, the quasi-molecule ion region of the fission fragment mass spectrum obtained from a synthetic sample of the 35-residue antibacterial peptide cecropin A[19] (Figure 2) showed the presence of a small amount of undesired impurity with a mol wt 28 U higher than the desired material. The relative peak heights indicate that this unwanted side-product is present in approximately 20% abundance. The mass difference of 28 U suggested to the synthetic chemists that the error involved a failure to eliminate fully from the molecule the formyl group, which originally protected the tryptophan residue (shown circled in Figure 2). Once the presence of such an impurity is clearly recognized and its origin established, steps can be taken to eliminate its formation. Figure 3 shows another example of a relatively subtle inhomogeneity in a small methionine-containing synthetic peptide. The impurity peak has a mol wt 15.9 U higher than the desired material, suggesting the occurrence of partial oxidation of the sample, which contains a methionyl sulfur.

It is unfortunate that presently, the most popular technique for assessing the homogeneity of a purified peptide product, i. e., reversed-phase HPLC (RP HPLC), is also the technique that is often used for purifying the desired compound from the crude synthetic peptide product. It is therefore not surprising that undesired materials that co-purify with the compound of interest are also frequently not resolved in the analytical HPLC analysis. Clearly, it is desirable to check for homogeneity using an analytical technique that separates compounds by a different principle from that used in the purification. Mass spectrometry serves this function well. Figure 4A shows the results of an HPLC analysis of a purified 39-residue synthetic cytochrome c fragment where the analysis was made using a C4 RP column. The dominant chromatographic peak looks quite symmetrical and sharp, and one might predict that the synthetic product is homogeneous. Inspection of the mass spectrum obtained from this material (Figure 4B) shows that this conclusion is not correct, however. Although the desired material is present and its experimentally determined mol wt agrees with the predicted mol wt to within 0.1 U, so is a substantial amount of a second compound having mol wt 114.0 U lower. The mass difference between the latter and desired compounds suggests that a deletion(s) involving asparagine occurred during the synthesis. Subsequent HPLC analysis using a high resolution C18 RP column confirmed the presence of the second unwanted compound. The cytochrome c fragment was thus resynthesized carefully, and the mass spectrum from this second synthesis is shown in Figure 4C. The peak corresponding to the deletion peptide is now no longer present in the spectrum, demonstrating that this second synthesis yielded a much purer product.

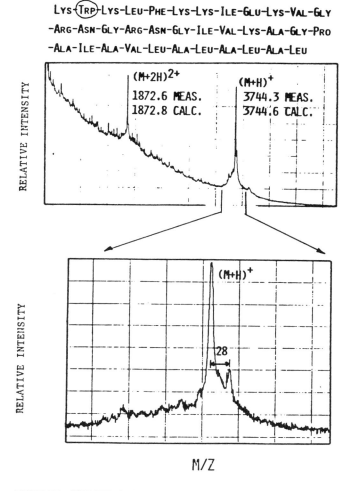

FIGURE 2. ^{252}Cf-PDMS of synthetic cecropin A. The bottom panel shows
a detailed plot of the (M+H)$^+$ ion region. The measured and calculated m/z
values for the (M+H)$^+$ and (M+2H)$^{2+}$ ions are given in the top panel. (From
Chait, B. T., The use of ^{252}Cf plasma desorption mass spectrometry for the
analysis of synthetic peptides and proteins, in *The Analysis of Peptides and
Proteins by Mass Spectrometry*, McNeal, C. J., Ed., John Wiley & Sons, New
York, 1988, 21. By permission of John Wiley & Sons.)

Frequently, the peptide chemist does have an indication that an error may have occurred
during the synthesis, but has little or no information concerning the detailed nature of the error.
The C18 RP HPLC analysis of a purified synthetic sample of the 26-residue bee venom peptide,
melittin, is shown in Figure 5A. Close inspection of the top trace obtained with 214 nm
absorbance shows that the intense peak consists of three unresolved components, indicating the
presence of at least three distinct compounds in the sample. The fission fragment mass spectrum
of this same sample (Figure 5B) confirmed the presence of three main components, designated
M, X, and Y, with abundances of 69, 21, and 10%, respectively. These mass spectrometrically
inferred abundances are consistent with the abundances of the three chromatographic components. Component M has a measured mol wt of 2845.9 U, which corresponds (to within 0.5 U)
to the calculated mol wt of 2846.4 U of melittin. Component X has a measured mol wt 128.0 U
lower, and component Y an mol wt 28.0 U higher than the melittin peak. These mol wt
differences provide valuable clues to the identities and origins of X and Y. The mass difference
of 128 U suggests that X was produced by deletion of a lysine (Lys) or a glutamine (Gln) residue

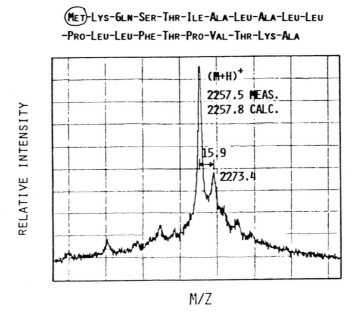

(MET)-Lys-Gln-Ser-Thr-Ile-Ala-Leu-Ala-Leu-Leu
-Pro-Leu-Leu-Phe-Thr-Pro-Val-Thr-Lys-Ala

FIGURE 3.   Detail of the ^{252}Cf-PDMS of a 21-residue synthetic peptide showing the quasi-molecular ion region. (From Chait, B. T., The use of ^{252}Cf plasma desorption mass spectrometry for the analysis of synthetic peptides and proteins, in *The Analysis of Peptides and Proteins by Mass Spectrometry*, McNeal, C. J., Ed., John Wiley & Sons, New York, 1988, 21. By permission of John Wiley & Sons.)

during the stepwise synthesis. The simple mol wt measurement does not, however, provide any definitive information on the actual position(s) of the deletion(s). The mass increment of 28 U suggests that Y was produced by incomplete removal of the formyl group from the tryptophan (Trp) residue at position 19 during the final deprotection step, although again direct confirmation of this possibility cannot be obtained from the simple mol wt measurement. Although this mol wt information is valuable, it is also desirable to have available techniques that focus more tightly onto the precise site in the molecule where the synthetic error or modification has occurred. Tandem mass spectrometry appears to be a good technique for this job, because it provides a powerful means of directly pinpointing sites of error or modification. Indeed, Biemann and Scoble[14] have recently utilized tandem mass spectrometry to positively identify an internal cyclization of an aspartic acid side chain in a short synthetic peptide, and Carr and co-workers[20] have used the technique to ascertain the site of attachment of a formyl group in an undesired side-product of synthetic melittin.

We have developed a complementary approach to tandem mass spectrometry for rapidly extracting information concerning the sites of synthetic errors. The approach, which has been described previously in detail,[17,21,22] involves three sequential steps:

1. Practically nondestructive ^{252}Cf-PDMS of monolayer amounts of the peptide(s) of interest bound to a thin layer of nitrocellulose
2. Enzyme-catalyzed microscale chemical reaction of the surface-bound peptide(s) to produce structurally informative hydrolysis products
3. PDMS of these hydrolysis products

The first step determines the presence and the mol wt of unwanted byproducts, and the subsequent two steps provide information on the location in the peptides where errors have

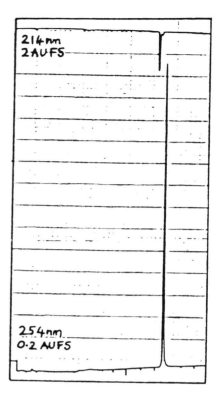

| 1 | 2 | 3 | 4 | 5 | 6 | 7 | 8 | 9 | 10 |
| Glu | Tyr | Leu | Glu | Asn | Pro | Lys | Lys | Tyr | Ile |

| 11 | 12 | 13 | 14 | 15 | 16 | 17 | 18 | 19 | 20 |
| Pro | Gly | Asn | Lys | Met | Ile | Phe | Ala | Gly | Ile |

| 21 | 22 | 23 | 24 | 25 | 26 | 27 | 28 | 29 | 30 |
| Lys | Lys | Lys | Thr | Glu | Arg | Glu | Asp | Leu | Ile |

| 31 | 32 | 33 | 34 | 35 | 36 | 37 | 38 | 39 | |
| Ala | Tyr | Leu | Lys | Lys | Ala | Thr | Asn | Glu | OH |

A

FIGURE 4. (A) RP HPLC of synthetic cytochrome c (66 to 104). Top trace obtained with 214 nm absorbance. Bottom trace obtained with 254 nm absorbance. (B) Partial ^{252}Cf-PDMS of synthetic cytochrome c (66 to 104). First synthesis. (C) Partial ^{252}Cf-PDMS of synthetic cytochrome c (66 to 104). Second synthesis. (From Chait, B. T., The use of ^{252}Cf plasma desorption mass spectrometry for the analysis of synthetic peptides and proteins, in *The Analysis of Peptides and Proteins by Mass Spectrometry*, McNeal, C. J., Ed., John Wiley & Sons, New York, 1988, 21. By permission of John Wiley & Sons.)

occurred. Thus, for example, inspection of the mass spectra of the previously discussed synthetic melittin sample taken prior to (Figure 5B) and after (Figure 5C) incubation with trypsin provides information concerning the regions of the molecule where the errors have occurred. Prior to any reaction, the mass spectrum of 10^{-9} mol of the sample (Figure 5B) exhibited the three peaks discussed previously. After reaction with 10^{-11} mol of trypsin, the mass spectrum (Figure 5C) exhibited a series of additional peaks, which corresponds to the protonated reaction products arising from partial hydrolysis on the carboxy-terminal side of residues 7, 21, 22, 23, and 24. Thus, for example, the peak labeled 1-22 corresponds to the tryptic fragment that includes residues 1 to 22. Because no significant peak is observed 128 U below the 1-22 peak, it can be deduced immediately that the deletion byproduct X does not arise by deletion of either Lys 7 or Lys 21. The error must then have occurred by deletion of Lys 23, Gln 25, or Gln 26. This information was given to the peptide chemists, who found upon close inspection of their records that there was indeed cause for concern during the first four synthetic cycles, which were involved in the incorporation of residues 23 to 26. The compound was therefore resynthesized, and the new preparation was found to give a mass spectrum (Figure 5D) that no longer showed the presence of any significant deletion peptides. The side-product, Y, which was hypothesized to arise from a failure to fully eliminate from the molecule the formyl group that originally protected the Trp residue at position 19, was, however, still present in the sample.

In a separate series of experiments, we have used this technique to investigate the detailed structures of such formyl group-containing byproducts, which are observed frequently when

CYTOCHROME C (66-104)

ION SPECIES	MEASURED MW	CALCULATED MW	Δ
(M+H)$^+$	4586.5	4586.4	+0.1
(X+H)$^+$	4472.5	4586.4	−113.9
(M+2H)$^{2+}$	4586.8	4586.4	+0.4
(X+2H)$^{2+}$	4472.3	4586.4	−114.1

FIGURE 4B.

CYTOCHROME C (66-104)

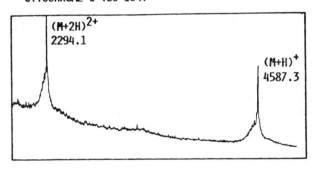

ION SPECIES	MEASURED MW	CALCULATED MW	Δ
(M+H)$^+$	4586.2	4586.4	−0.2
(M+2H)$^{2+}$	4586.3	4586.4	−0.1

FIGURE 4C.

formyl-Trp is used in the synthesis of Trp-containing peptides.[22] Figure 6 shows the RP HPLC analysis of a melittin sample prepared and purified separately than that discussed previously. In this case, a single symmetrical peak was observed, suggesting that the sample was pure. However, inspection of the PDMS obtained from 10^{-9} mol of this melittin sample bound to nitrocellulose (NC) (Figure 7a) again indicated the presence of a formyl group-containing impurity M_1^i with a mol wt 28 U higher than that of melittin. Because formylated Trp was used in the synthesis, the observed impurity was likely the result of either an incomplete deprotection

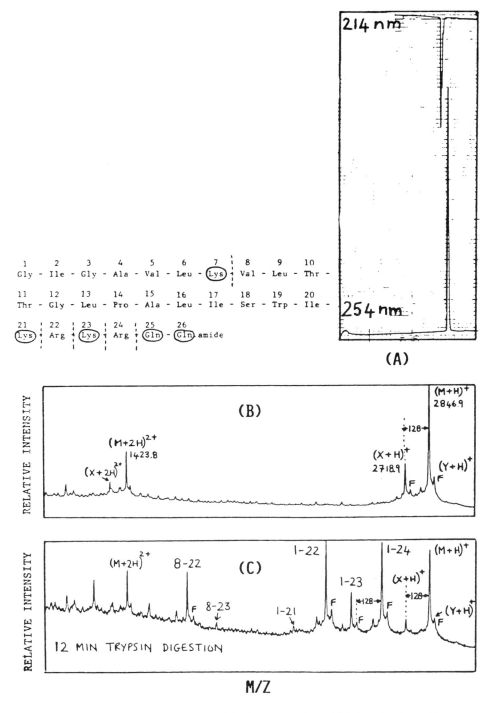

FIGURE 5. (A) C18 RP HPLC of synthetic melittin. Top trace obtained with 214 nm absorbance. Bottom trace obtained with 254 nm absorbance. (B) Partial mass spectrum of melittin. First synthesis. (C) Partial mass spectrum of the same sample of melittin after 12 min trypsin digestion of the nitrocellulose-bound peptide. The small peaks labeled F are each 28 U higher than the large adjacent peak. (D) Partial mass spectrum of melittin. Second synthesis. (From Chait, B. T., The use of ^{252}Cf plasma desorption mass spectrometry for the analysis of synthetic peptides and proteins, in *The Analysis of Peptides and Proteins by Mass Spectrometry*, McNeal, C. J., Ed., John Wiley & Sons, New York, 1988, 21. By permission of John Wiley & Sons.)

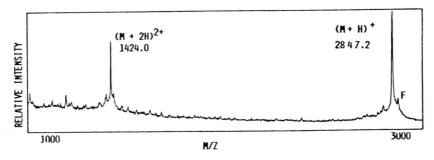

FIGURE 5 (continued).

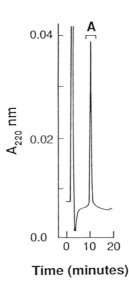

Time (minutes)

FIGURE 6. HPLC trace of the melittin products using a C-18 RP column. (From Chait, B. T., *Anal. Chem.*, 180, 387, 1989. With permission.)

of Trp (For) or a transformylation modification reaction occurring during deprotection.[3,23,24] The deprotection scheme was the low-high two-step hydrogen fluoride (HF) cleavage procedure of Tam et al.[24,25] In order to extract information about the location of the formyl group in the impurity, a series of enzymatic hydrolysis reactions was carried out on the sample mixture containing both the authentic melittin and the impurity. A description of the procedure follows.

The sample used to obtain the spectrum shown in Figure 7a was removed from the mass spectrometer and treated with trypsin. The series of observed tryptic fragments is summarized diagrammatically in Figure 8, together with the results from other enzymatic treatments (discussed later in this chapter). The tryptic fragments observed in Figure 7b occur as paired peaks, 28 U apart. The lower mass component of each pair of peaks (e.g., [1 to 22]$^+$ representing protonated melittin [1 to 22]) originates from the authentic melittin, (M_1), whereas the upper component (e. g., [1 to 22]$_i^+$) originates from the formylated melittin impurity (M_1^i). Because all the fragments are observed as doublets, we can deduce that the formyl group in the impurity, M_1^i, is located between residues 8 and 21, inclusive.

The position of the formyl group in the impurity melittin (M_1^i) was established more accurately by treatment with proteinase K of a freshly deposited sample of the mixture on NC. The mass spectrum of the resulting hydrolysis products is shown in Figure 7c. All of the enzyme-generated fragment ions in the mass spectrum occur as pairs 28 U apart, with the exception of

Gly–Ile–Gly–Ala–Val–Leu–Lys–Val–Leu–Thr–Thr–Gly–Leu–Pro–Ala–Leu–Ile
Ser–Trp–Ile–Lys–Arg–Lys–Arg–Gln–Gln–NH$_2$.

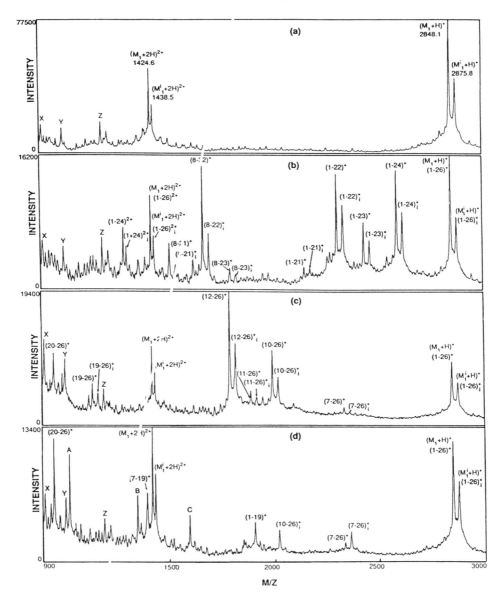

FIGURE 7. ^{252}Cf plasma TOF mass spectra of the purified synthetic melittin product (10^{-9} mol) bound to a thin film of nitrocellulose. (a) Prior to any enzymatic treatment. $(M_1+H)^+$ and $(M_1+2H)^{2+}$ are singly and doubly protonated ions of authentic melittin. $(M_1'+H)^+$ and $(M_1'+2H)^{+2}$ are similar ions of a byproduct with mass approximately 28 U higher than that of melittin. X, Y, and Z result from unidentified impurities. (b) Following a 10-min digestion with trypsin (10^{-11} mol, pH 8.0) at 37°C. The residues contained in each hydrolysis product ion are indicated on top of the corresponding peak. For example, (1 to 22)$^+$ represents the protonated tryptic fragment of melittin, M_1, which contains residues 1 to 22, and (1 to 22)^+_i represents the protonated ion of the corresponding tryptic fragment from the melittin impurity, M_1'. The mass separation between the members of each such pair of ions is 28 U. (c) Following a 10-min reaction with proteinase K (10^{-10} mol, pH 8.5) at 37°C. Ions A, B, and C were not identified. (d) Following a 10-min reaction with α-chymotrypsin (10^{-10} mol, pH 8.5) at 37°C. (From Chait, B. T., *Anal. Biochem.*, 180, 387, 1989. With permission.)

MELLITIN

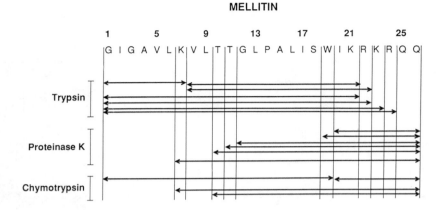

FIGURE 8.   A summary of al the mass spectrometrically observed enzymatically generated fragments of melittin. these results are obtained from the data shown in Figure 7 for the digestion of melittin by trypsin, proteinase K, and α-chymotrypsin, respectively.

the ion at m/z 975.3. We attribute this ion to the fragment (20 to 26)[+], having a calculated m/z 956.1. The absence of a companion ion peak 28 U higher indicates that the formyl group in the impurity melittin is not located between residues 20 and 26, inclusive. On the other hand, the presence of the pair (19 to 26)[+] (observed m/z 1143.2, calculated m/z 1143.3) and (19 to 26)$_i^+$ (observed m/z 1170.8, calculated m/z 1171.3), which are approximately 28 U apart, indicates that the formyl group is present between residues 19 and 26, inclusive, of the impurity melittin. Therefore, the formyl group is located on Trp-19 of the impurity melittin species.

Supporting evidence for the location of the formyl group on Trp-19 was deduced from the mass spectrum (Figure 7d) obtained after α-chymotrypsin treatment (see Figure 8) of the melittin sample bound to NC. Here too, a single peak (20 to 26)[+] with m/z 956.6 (calculated m/z 956.1) was observed with no accompanying (20 to 26)$_i^+$ peak, again indicating the absence of the formyl group in residues 1 to 19. However, this conclusion is incorrect because the presence of the formyl group on the indole nitrogen of the Trp reduces greatly the rate of hydrolysis at Trp by α-chymotrypsin. This result was confirmed in a separate experiment comparing the rates of hydrolysis by α-chymotrypsin of two model substrates, which were identical, except that one contained formylated Trp in place of Trp.[22] We found (data not shown) that the rate of hydrolysis at Trp(For) was more than an order of magnitude slower than that at unprotected Trp. Thus, the nonobservation of (1 to 19)$_i^+$ provides additional evidence that the formyl group is present on Trp-19 in the impurity. Another manifestation of the reduced rate of hydrolysis at Trp(For)-19 is the enhanced intensities observed in Figure 7d of (10 to 26)$_i^+$ and (7 to 26)$_i^+$, compared to those of (10 to 26)[+] and (7 to 26)[+].

A similar strategy was employed to locate the positions of residual formyl groups in an abundant side-product observed in a [*p*-benzoylphenylalanine-8] dynorphin A (1 to 17) preparation in which the removal of the formyl group from Trp-14 was carried out using piperidine. In this case, the impurity was shown to result from migration of the formyl group from Trp-14 to the adjacent Lys-11 or Lys-13 residues.[22]

In addition to the errors describe above, a great many other errors occur during solid phase synthesis.[1-4] These errors range from operator- or machine-related errors to relatively subtle chemical modifications of the peptide during and after synthesis. We have found that the majority of these errors can be detected by PDMS. Thus, we have, for example, been able to detect and identify the elimination of water from amino terminal glutamine as well as from other amino acid residues; S-alkylation of methionyl residues by the tertiary butyl cation; many other examples of incomplete deprotection (see also Reference 26); the production of insertion peptides and deletion peptides; the production of termination peptides; the production of damage products arising during the final deprotection step; and the production of a number of products that we are not yet able to interpret. The importance of the use of the PDMS technique can be gauged from our rather staggering finding that almost half of the 1200 purified synthetic peptides and proteins that we examined were found to have either a mol wt different from that calculated for the desired target material or to contain significant amounts of unwanted peptide side-products.

IV. LIMITATIONS OF THE MASS SPECTROMETRIC TECHNIQUE

Although PDMS constitutes a powerful method for analyzing synthetic peptides and proteins, it should be recognized that the technique has its limitations. One important limitation concerns the ability to determine reliable abundances for the reaction side-product and modification peptides. When these side-products differ from the desired target peptide in their binding properties to the NC surface or in their efficiency for desorption and ionization from such surfaces by fission fragment bombardment, then the relative peak heights observed in the mass spectrum will not accurately reflect the relative abundances of the various components present in the sample. Such may be the case for a termination peptide that is considerably shorter than the desired target peptide, or for a peptide that has not been fully deprotected. On the other hand, if the side-product differs only relatively slightly from the desired peptide, then the mass spectrometric quantitation is expected to be good. An example of this latter case is the single deletion peptide of mellitin discussed previously. (Figure 5B).

Limitations exist at high mass, where the quality of the mass spectral data may be much reduced, and in those cases where a large number of different low-abundance side-products occur. The plasma desorption mass spectrum of a synthetic sample of the 140-residue protein interleukin-3 (IL-3) (Figure 9) provides an extreme illustration of both limitations. The sample IL-3 was produced by Clark-Lewis and co-workers[27] using automated stepwise solid phase syntheses, and represents perhaps the largest biologically active synthetic protein produced to date. It is apparent from the partial TOF mass spectrum shown in Figure 9 that it is not possible to extract the kind of detailed information that we were able to obtain from lower mol wt compounds. The multiply protonated molecule peaks are broad and weak, and have mass uncertainties of 50 to 100 U. Thus, although this mol wt information is superior to that which can be obtained using sodium dodecyl sulfate (SDS) gel electrophoresis — and in this respect is really quite valuable — the peak is so broad and poorly defined that we cannot extract information about the impurities, especially the deletion peptides, which are certainly present in this sample.

V. MASS SPECTROMETRY OF FULLY PROTECTED SYNTHETIC PEPTIDES

Ideally, one would monitor progress at each stage of chain assembly during the synthesis of a peptide. Using mass spectrometry, this goal is a fairly large undertaking at present. However, there are several situations in which limited mass spectrometric monitoring during assembly proves to be highly valuable. One such example concerns the synthesis of small proteins by the segment condensation method.[28] In this technique, protected peptide segments containing fewer than ten amino acid residues are produced by stepwise solid phase synthesis in a fairly homogeneous form and are then purified further by various chromatographic methods. These fully protected segments are then condensed successively to form larger and larger fully protected segments with purification of the resulting segment after each condensation step. Finally, when the whole-protected protein has been assembled, the protecting groups are removed. Because this synthesis is a fairly lengthy procedure involving many purification steps, it is desirable to ensure that the segments are synthesized correctly and that they have not, for

```
ALA SER ILE SER GLY ARG ASP THR HIS ARG LEU THR ARG THR LEU ASN CYS SER SER ILE
1                                                                              20

VAL LYS GLU ILE ILE GLY LYS LEU PRO GLU PRO GLU LEU LYS THR ASP ASP GLU GLY PRO
                                                                               40

SER LEU ARG ASN LYS SER PHE ARG ARG VAL ASN LEU SER LYS PHE VAL GLU SER GLN GLY
                                                                               60

GLU VAL ASP PRO GLU ASP ARG TYR VAL ILE LYS SER ASN LEU GLN LYS LEU ASN CYS CYS
                                                                               80

LEU PRO THR SER ALA ASN ASP SER ALA LEU PRO GLY VAL PHE ILE ARG ASP LEU ASP ASP
                                                                              100

PHE ARG LYS LYS LEU ARG PHE TYR MET VAL HIS LEU ASN ASP LEU GLU THR VAL LEU THR
                                                                              120

SER ARG PRO PRO GLN PRO ALA SER GLY SER VAL SER PRO ASN ARG GLY THR VAL GLU CYS
                                                                              140
```

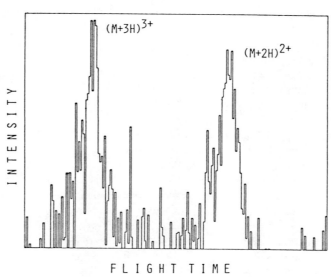

ION SPECIES	MEASURED MW	CALCULATED MW	Δ
$(M+2H)^{2+}$	15,592	15,662	-70
$(M+3H)^{3+}$	15,666	15,662	+4

FIGURE 9.   Partial [252]Cf-PDMS of synthetic IL-3, showing the region containing the doubly and triply protonated molecule peaks. (From Chait, B. T., The use of [252]Cf plasma desorption mass spectrometry for the analysis of synthetic peptides and proteins, in *The Analysis of Peptides and Proteins by Mass Spectrometry*, McNeal, C. J., Ed., John Wiley & Sons, New York, 1988, 21. By permission of John Wiley & Sons.)

example, lost a protecting group. It is also desirable to check that the homogeneity of these difficult-to-purify, fully protected peptide segments is sufficiently good. In mass spectrometry, these materials are also difficult to handle because they are rather insoluble and tend to aggregate and because they are "bristling" with protecting groups that have a tendency to "fall off" upon desorption and ionization in the mass spectrometer. We have found that this tendency to fragment can be minimized by desorbing the protected peptides from nitrocellulose surfaces, and by arranging for the molecule to be ionized by a sodium cation.[29] In order to utilize this technique, it is necessary to find solvents that readily dissolve the protected peptides, but at the same time do not attack the nitrocellulose adsorption surface. We have found trifluoroethanol, trifluoroacetic acid, and very dilute dimethyl sulfoxide (DMSO) to be quite effective for these purposes.

Figure 10 shows examples of plasma desorption mass spectra obtained from three fully protected peptide-intermediates produced during the fragment condensation synthesis of the homeo box protein of *Drosophila*,[30] carried out by Mihara and Kaiser.[31] In each case, the m/z of the dominant spectral peak was found to agree with the calculated m/z value to within the estimated experimental error (see Figure 10). It should be noted (compare Figures 10c and 1) that the higher mol wt protected peptides yield mass spectra that are lower in quality than those obtained from unprotected peptides having an equivalent number of residues. The lower-quality spectra result in somewhat reduced mass determination accuracies for these large, protected peptides. However, even with this reduced quality, the data are very useful for confirming that the condensation reaction proceeded correctly and for checking the degree of homogeneity. These mass spectral data are especially important because no other technique appears to be available for directly analyzing large peptides that are fully protected.

VI. MASS SPECTROMETRY OF UNUSUAL SYNTHETIC PEPTIDES

We have found PDMS to be very useful for the analysis of unusual synthetic peptide-containing molecules that cannot be readily analyzed by techniques such as Edman sequencing or NMR. Examples of such compounds are:

1. Peptides modified at the amino terminus by an acetyl or a myristyl group
2. A large octally branched synthetic peptide (mol wt = 10645 U) containing 8 identical 12 residue terminal peptides (a synthetic malaria vaccine)
3. A mol wt = 8599 U synthetic enzyme mimic consisting of four 15-residue peptides attached to a central porphyrin core
4. A mol wt = 8758 U synthetic mimic of a pore-forming peptide structure containing 4 identical 18-residue peptides attached to a 10-residue cyclic peptide core.

In all these cases, the mass spectrometric determination of the mol wt proved crucial in corroborating the integrity of the synthetic product. For example, Figure 11 shows the mass spectrum of a 13-residue peptide that has been myristylated at the amino terminus. It is readily apparent from the spectrum that, in addition to the desired material designated M_1 (measured mol wt = 1611.7, calculated mol wt = 1611.9), a relatively high-abundance single deletion side-product is present at m/z = 1484.7 in this preparation. In the absence of this mass spectrometric measurement, it would have been difficult to discover and demonstrate convincingly the presence of this side-product.

VII. DISULFIDE BONDS IN SYNTHETIC PEPTIDES AND PROTEINS

When the target synthetic protein contains internal disulfide bonds, the initially produced linear peptide chain must be caused to fold correctly and the thiol groups caused to oxidize and

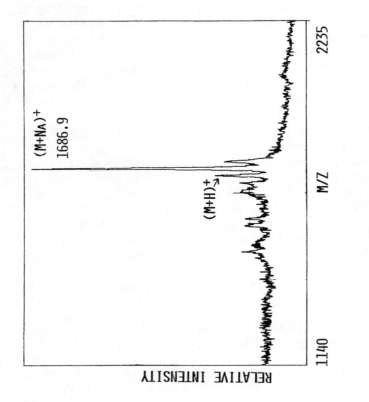

FIGURE 10. Partial ^{252}Cf-PDMS of protonated peptide fragments used in the segment condensation synthesis of the homeo box protein of *Drosophila*. (A) 6-residue protected peptide. (B) 23-residue protected peptide. (C) 40-residue protected peptide. Abbreviations used for the protecting groups are Boc = *tert*-butyloxycarbonyl; Bzl = benzyl; Tos = 4-toluenesulfonyl; MBzl = 4-methylbenzyl; ClZ = 2-chlorobenzyloxycarbonyl; Bon = benzyloxymethyl; and Cl$_2$Bzl = 2, 4-dichlorobenzyl. (From Chait, B. T., The use of ^{252}Cf plasma desorption mass spectrometry for the analysis of synthetic peptides and proteins, in *The Analysis of Peptides and Proteins by Mass Spectrometry*, McNeal, C. J., Ed., John Wiley & Sons, New York, 1988, 21. By permission of John Wiley & Sons.)

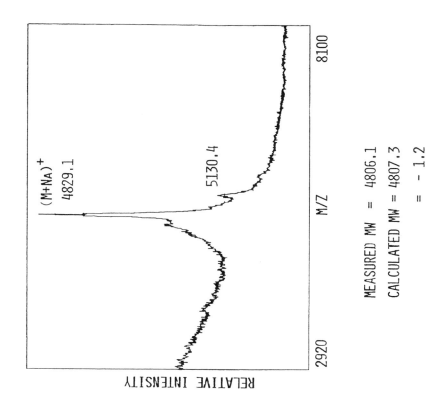

MEASURED MW = 4806.1

CALCULATED MW = 4807.3

= - 1.2

FIGURE 10B.

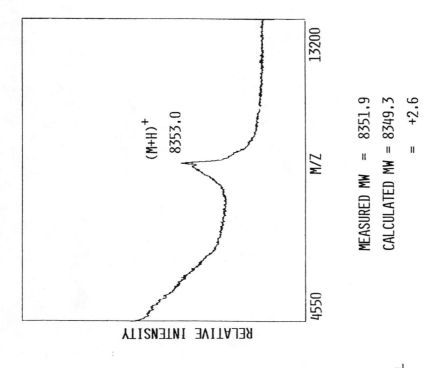

FIGURE 10C.

$CH_3(CH_2)_{12}-CO-NH-Gly-Ser-Ser-Lys-Ser-Lys-Pro-Lys-Asp-Pro-Ser-Gln-Arg$

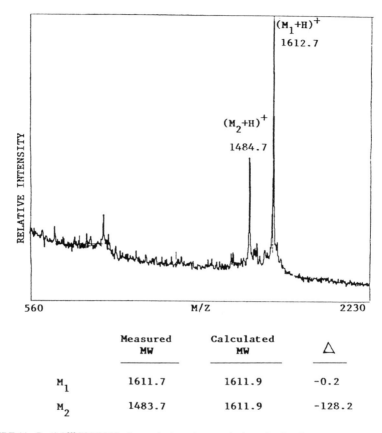

	Measured MW	Calculated MW	△
M_1	1611.7	1611.9	-0.2
M_2	1483.7	1611.9	-128.2

FIGURE 11. Partial [252]Cf-PDMS of a synthetic amino terminal-myristylated peptide. (From Chait, B. T., The use of [252]Cf plasma desorption mass spectrometry for the analysis of synthetic peptides and proteins, in *The Analysis of Peptides and Proteins by Mass Spectrometry*, McNeal, C. J., Ed., John Wiley & Sons, New York, 1988, 21. By permission of John Wiley & Sons.)

form the required disulfide linkages. Even though the folding and oxidation are carried out at low concentrations, we are frequently asked to determine whether the species produced is monomeric or dimeric, i.e., whether intermolecular disulfide linkages have been produced inadvertently. Sometimes this question is prompted by the observation of an atypical migration of the protein during SDS gel electrophoresis. For example a sample of what was thought to be fully oxidized transforming-growth factor-alpha (TGFα) (Figure 12) prepared by Woo et al.[32] migrated upon SDS gel electrophoresis with an apparent mol wt more than three times the expected value of 5546 U. The question of inadvertent intermolecular disulfide formation, which may be difficult to resolve using amino acid or sequence analysis, is readily resolved by simple inspection of the PDMS. The mass spectrum of the TGFα preparation (Figure 12A) exhibits peaks corresponding to the singly, doubly, and triply protonated monomer. The observation of these peaks, together with the absence of peaks corresponding to the singly and triply protonated dimer, confirms that the compound is present exclusively as the monomer. It should be noticed that the doubly charged dimer has the same m/z value as the singly charged monomer, so it is not sufficient to inspect only this latter species in order to resolve the question of dimer formation. The close correspondence between the measured and calculated MWs confirms that this preparation is indeed the fully oxidized form of the growth factor. We obtained

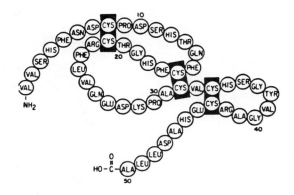

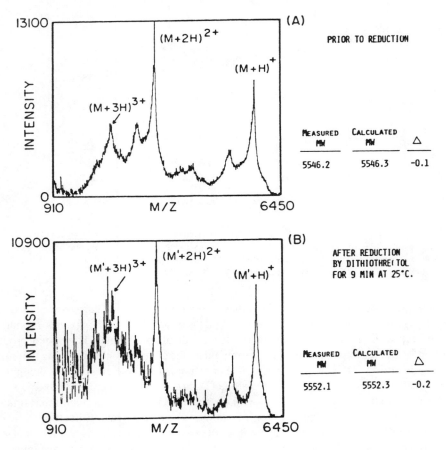

FIGURE 12. Partial [252]Cf-PDMS of TGFα taken (A) prior to and (B) after reduction with dithiothreitol. (From Chait, B. T., The use of [252]Cf plasma desorption mass spectrometry for the analysis of synthetic peptides and proteins, in *The Analysis of Peptides and Proteins by Mass Spectrometry,* McNeal, C. J., Ed., John Wiley & Sons, New York, 1988, 21. By permission of John Wiley & Sons.)

further confirmation that all three disulfide linkages were intact by carrying out an additional experiment on the sample that gave the spectrum shown in Figure 12A. In this experiment, the nitrocellulose-bound sample was removed from the mass spectrometer, reduced by the application of a solution containing dithiothreitol to the nitrocellulose surface, dried, and then reintroduced into the mass spectrometer for reanalysis. Figure 12B shows the resulting spectrum of the reduced TGFα. The observed mass increase of 5.9 U over the oxidized form corresponds closely with the expected increase of 6.0 U if all three disulfides were intact in the folded, oxidized compound. We have not yet used PDMS to assist in the assignment of the disulfide pairings in synthetic peptides, but we have used it in conjunction with Edman sequencing to examine enzymatically generated disulfide-containing peptide fragments to assist in the assignment of the seven previously unknown S-S pairings in neurophysin, a naturally occurring, 95-residue protein.[33-35] Our success with this difficult case leads us to conclude that PDMS used in conjunction with enzyme degradation will be of general utility for the confirmation of the disulfide pairings in synthetic proteins as well.

VIII. EVALUATION AND OPTIMIZATION OF THE CHEMICAL PROCEDURES USED TO SYNTHESIZE PEPTIDES

In collaboration with Merrifield and Singer[36] at the Rockefeller University, we have recently developed a sensitive and precise new technique to evaluate methods for the synthesis of peptides. In this technique, two model oligopeptides containing 10 or 20 alanine residues ($Ala_{10}Val$, $Ala_{20}Val$) were synthesized by automated solid phase methods using a variety of protocols, and the levels of peptide byproducts were measured by PDMS, where the total, unfractionated, synthetic product was deposited on a film of nitrocellulose and analyzed. The use of alanine, which lacks a third functionality, essentially eliminated the production of branched chains or modification peptides. Thus, the observed byproducts were almost exclusively deletion and insertion peptides. The 10 or 20 alanine residues provided a large amplification factor for observing these deletion and insertion peptides. The introduction of D-alanine at every third residue of the model eliminated peptide conformation problems that led to incomplete reactions in an all-L model. Certain of the protocols tested were found to be significantly less efficient in syntheses than others. In particular, a widely used method involving coupling with preformed symmetric anhydrides in dimethylformamide gave rise to significant levels of both deletion and insertion peptides. The best of the protocols examined was a double coupling of alanine by *in situ* activation with dicyclohexylcarbodiimide in dichloromethane. In this case, [D-Ala^{3,6,9,12,15,18}] $Ala_{20}Val$ was synthesized with an average deletion and insertion of only 0.029% per step, which is equivalent to a stepwise yield of 99.93% for the target peptide. We have extended the technique to the study of model oligopeptides containing up to 50 alanine residues, and also to other potentially more problematic amino acids such as lysine. At present, we are using an all L-alanine model to investigate the peptide conformation problems that lead to incomplete reactions in these α-helical peptides. It appears that the use of polyamino acid peptides and mass spectrometry will continue to have considerable utility for the improvement of peptide synthesis.

IX. CONCLUSION

We have found through the examination of a large number of synthetic peptides and proteins originating from many different laboratories that PDMS provides an enormously useful, rapid, easy, and definitive method for assessing the correctness of structure, examining the homogeneity of the final product, identifying and determining the origin of unwanted side-reaction products, and evaluating and optimizing details of the synthetic peptide chemistry. Our experience and findings have convinced us that the mass spectrometer should take its place in the peptide laboratory alongside the more established analytical tools such as the amino acid analyzer and the sequenator.

ACKNOWLEDGMENTS

I wish to thank Mrs. Tanuja Chaudhary and Dr. Swapan Chowdhury for assistance in making the mass spectrometric analyses and Mrs. Gladys McMilleon for typing the manuscript. This work was supported by grants (RR00862 and RO1-GM38274) from the National Institutes of Health.

REFERENCES

1. **Merrifield, R. B.,** Solid phase peptide synthesis. I, The synthesis of a tetrapeptide, *J. Am. Chem. Soc.,* 85, 2149, 1963.
2. **Erickson, W. W. and Merrifield, R. B.,** Solid-phase peptide synthesis, in *The Proteins,* Vol. 2, 3rd ed., Neurath, H. and Hill, R. L., Eds., Academic Press, New York, 1976, 255.
3. **Barany, G. and Merrifield, R. B.,** Solid phase peptide synthesis, in *The Peptides,* Vol. 2, (Part A), Gross, E. and Meienhofer, J., Eds., Academic Press, New York, 1980, 1.
4. **Kent, S. B. H.,** Chemical synthesis of peptides and proteins, *Annu. Rev. Biochem.,* 57, 957, 1988.
5. **Larsen, B., Fox, L. B., Burke, M. F., and Hruby, V. J.,** The separation of peptide hormone diastereoisomers by reverse phase high pressure liquid chromatography, *Int. J. Pept. Protein Res.,* 13, 12, 1979.
6. **Hare, P. E.,** Amino acid composition by column chromatography, in *Protein Sequence Determination,* Vol. 8, Needleman, S. B., Ed., Springer-Verlag, New York, 1975, 204.
7. **Edman, P. and Henschen, A.,** Sequence determination, in *Protein Sequence Determination,* Vol. 8, Needleman, S. B., Ed., Springer-Verlag, New York, 1975, 232.
8. **Merrifield, R. B., Viozioli, L. D., and Boman, H. G.,** Synthesis of the antibacterial peptide cecropin A(1-33), *Biochemistry,* 21, 5020, 1982.
9. **Wuthrich, K.,** *NMR of Proteins and Nucleic Acids,* Wiley-Interscience, New York, 1986, 1.
10. **Bayer, E., Eckstein, H., Hagele, K., Konig, W. A., Bruining, W., Hagenmaier, H., and Parr, W.,** Failure sequences in the solid-phase synthesis of peptide, *J. Am. Chem. Soc.,* 92, 1735, 1970.
11. **Chait, B. T., Gisin, B. F., and Field, F. H.,** Fission fragment ionization mass spectrometry of alamethicin. I, *J. Am. Chem. Soc.,* 104, 5157, 1982.
12. **Biemann, K.,** Mass spectrometric methods for protein sequencing, *Anal. Chem.,* 58, 1288A, 1986.
13. **Biemann, K. and Martin, S. A.,** Mass spectrometric determination of the amino acid sequence of peptides and proteins, *Mass Spectrom. Rev.,* 6, 1, 1987.
14. **Biemann, K. and Scoble, H. A.,** Characterization by tandem mass spectrometry of structural modifications in proteins, *Science,* 237, 992, 1987.
15. **Chait, B. T.,** The use of ^{252}Cf plasma desorption mass spectrometry for the analysis of synthetic peptides and proteins, in *The Analysis of Peptides and Proteins by Mass Spectrometry,* McNeal, C. J., Ed., John Wiley & Sons, New York, 1988, 21.
16. **Macfarlane, R. D. and Torgerson, D. F.,** Californium-252 plasma desorption mass spectrometry. Nuclear particles are used to probe biomolecules, *Science,* 191, 920, 1976.
17. **Chait, B. T. and Field, F. H.,** A rapid, sensitive mass spectrometric method for investigating microscale chemical reactions of surface adsorbed peptides and proteins, *Biochem. Biophys. Res. Commun.,* 134, 420, 1986.
18. **Grace, L. I., Chait, B. T., and Field, F. H.,** A system for collecting high-resolution time-of-flight mass spectrometric data, *Biomed. Environ. Mass Spectrom.,* 14, 295, 1987.
19. **Andreu, D., Merrifield, R. B., Steiner, H., and Boman, H. G.,** Solid-phase synthesis of cecropin A and related peptides, *Proc. Natl. Acad. Sci., U.S.A.,* 80, 6475, 1983.
20. **Carr, S. A.,** private communication.
21. **Chait, B. T., Chaudhary, T., and Field, F. H.,** Mass spectrometric characterization of microscale enzyme catalyzed reactions of surface-bound peptides and proteins, in *Methods in Protein Sequence Analysis,* Walsh, K. A., Ed., Humana Press, Clifton, NJ, 1986, 483.
22. **Chowdhury, S. K. and Chait, B. T.,** A mass spectrometric technique for detecting and identifying by-products in the synthesis of peptides, *Anal. Biochem.,* 180, 387, 1989.

23. **Yamashiro, D. and Li, C. H.,** Protection of tryptophan with formyl group in peptide synthesis, *J. Org. Chem.,* 38, 2594, 1973.

24. **Tam, J. P.,** Acid deprotection reactions in peptide synthesis, in *Macromolecular Sequencing and Synthesis Selected Methods and Applications,* Alan R. Liss, New York, 1988, 153.

25. **Tam, J. P., Heath, F. W., and Merrifield, R. B.,** An S_N^2 deprotection of synthetic peptides with a low concentration of HF in dimethylsulfide: evidence and application in peptide synthesis, *J. Am. Chem. Soc.,* 105, 6442, 1983.

26. **Lindeberg, G., Engstrom, A., Craig, A. G., and Bennich, H.,** Plasma desorption mass analysis in immuno-chemistry, in *The Analysis of Peptides and Proteins by Mass Spectrometry,* McNeal, C. J., Ed., John Wiley & Sons, Chichester, 1988, 1.

27. **Clark-Lewis, I., Aebersold, R., Ziltener, H., Schrader, J. W., Hood, L. E., and Kent, S. B. H.,** Automated chemical synthesis of a protein growth factor for hemopoietic cells, Interleukin-3, *Science,* 231, 134, 1986.

28. **Nakagawa, S. H., Lau, H. S. H., Kezdy, F. J., and Kaiser, E. T.,** The use of polymerbound oximes for the synthesis of large peptides usable in segment condensation: synthesis of a 44 amino acid amphiphilic peptide model of apolipoprotein A-1, *J. Am. Chem. Soc.,* 107, 7087, 1985.

29. **Chait, B. T.,** A study of the fragmentation of singly and multiply charged ions produced by [252]Cf fission fragment bombardment of polypeptides bound to nitrocellulose, *Int. J. Mass Spectrom. Ion Proc.,* 78, 237, 1987.

30. **Gehring, W. J.,** Homeo boxes in the study of development, *Science,* 236, 1987.

31. **Mihara, H. and Kaiser, E. T.,** A chemically synthesized antennapedia homeo domain binds to a specific DNA sequence, *Science,* 242, 925, 1988.

32. **Woo, D. H. L., Clark-Lewis, I., Chait, B. T., and Kent, S. B. H.,** Chemical synthesis in protein engineering: total synthesis, purification and covalent structural characterization of a mitogenic protein, human transforming growth factor-alpha, *Protein Eng.,* 3, 29, 1989.

33. **Burman, S., Breslow, E., Chait, B. T., and Chaudhary, T.,** Partial assignment of disulfide pairs in neurophysins, *Biochem. Biophys. Res. Commun.,* 148, 827, 1987.

34. **Burman, S., Breslow, E., Chait, B. T., and Chaudhary, T.,** Application of high-performance liquid chromatography in neurophysin disulfide assignment, *J. Chromatogr.,* 443, 285, 1988.

35. **Burman, S., Wellner, D., Chait, B., Chaudhary, T., and Breslow, E.,** Complete assignment of neurophysin disulfides indicates pairing in two separate domains, *Proc. Natl. Acad. Sci. U.S.A.,* 86, 429, 1989.

36. **Merrifield, R. B., Singer, J., and Chait, B. T.,** Mass spectrometric evaluation of synthetic peptides for delterions and insertions, *Anal. Biochem.,* 174, 399, 1988.

Chapter 4

ANALYSIS OF PEPTIDES AND PROTEINS BY PLASMA DESORPTION MASS SPECTROMETRY

Peter Roepstorff

TABLE OF CONTENTS

I.   Introduction 66

II.   A Plasma Desorption Mass Spectrum 66
        A.   The Components of the Spectrum 66
        B.   Mass Calibration 68
        C.   Precision of Mass Determination 68
        D.   Accessible Mass Range 69

III.   Sample Preparation 69
        A.   The Nitrocellulose Matrix 70
        B.   Practical Sample Application 73

IV.   Molecular Weight Determination of Intact Peptides and Proteins 74

V.   Peptide Mapping by Plasma Desorption Mass Spectrometry 77
        A.   *In Situ* Reactions 77
        B.   Monitoring Reactions in Solution 77
        C.   Suppression Effects in Mixture Analysis 79

VI.   Application in Protein Sequence Determination 79

VII.   Applications in Protein Biotechnology and Protein Engineering 80

VIII.   Conclusion 82

Acknowledgments 82

References 84

I. INTRODUCTION

Plasma desorption mass spectrometry (PDMS) was invented in 1974 by Macfarlane and co-workers.[1] The principles and fundamentals of the technique are described in Chapter 1. A major achievement with great impact for the subject of this book was the first recording of a mass spectrum of an intact protein, insulin, in 1982 by Håkansson et al.[2] That this observation was not unique was demonstrated by observation of molecular ions of a number of proteins in the following years,[3-5] culminating with a 24 kDa protein, trypsin.[6] At this stage, it was obvious that PDMS was a promising alternative to the traditional methods for molecular weight determination of peptides and proteins, i.e., sodium dodecyl sulfate (SDS)-gel electrophoresis and gel permeation chromatography. Most of the studied proteins were highly purified samples available in relatively large quantities, and attempts to analyze small samples isolated in the course of research projects often failed. It soon became clear that the main reason for failures was that the method hitherto used for preparation of the sample prior to mass spectrometric analysis, the electrospray technique,[7] was inadequate for general protein chemistry work. The sensitivity was not sufficient, low molecular weight contaminants (salts, for example) diminished the quality of the spectra, and only organic solvents could be used for electrospray.

Macfarlane rather early suggested that an alternative sample application technique could be based upon adsorption of the sample to a suitable support and suggested two different materials. One support was a sulfonic acid-containing polymer, Nafion, which could bind the sample by electrostatic interactions,[8] and the other was poly(ethyleneterephthalate), or Mylar, with hydrophobic binding properties.[9] The major breakthrough for the analysis of peptides and proteins came with the introduction of nitrocellulose as a matrix.[10] Nitrocellulose had been used to bind proteins transferred from electrophoretic gels, the so-called Western blotting.[11] As a matrix for PDMS, nitrocellulose has a number of very attractive properties, which will be demonstrated here. Nitrocellulose is now widely used as the standard matrix for peptide and protein analysis by PDMS, and this chapter deals with its properties, sample application techniques, and applications to peptide and protein studies.

II. A PLASMA DESORPTION MASS SPECTRUM

A. THE COMPONENTS OF THE SPECTRUM

An example of a PD-spectrum of a small protein, ribonuclease A, desorbed from a nitrocellulose matrix is shown in Figure 1. The "raw" spectrum in Figure 1A exhibits three major peaks superimposed on a rather intense background continuum. The observed peaks are all molecular ion species representing different charged states, i.e., MH_2^{2+}, MH_3^{3+}, and MH_4^{4+}. The peaks are rather broad, typically 5 to 10% of the apparent mass. A close examination of a peak will reveal that it is composed of two components, a sharp stable component and a broad metastable component.[5] A number of molecular ions, which survive acceleration, decompose during flight in the field-free region. These ions, which arrive at the detector at the same time as the stable molecular ions, but with a larger energy spread due to the metastable decomposition, give rise to the broad base of the peak.

The background continuum is composed of a mixture of contributions from fragment ions, ions, and neutrals formed by metastable decay during acceleration, ions formed by delayed desorption, and ions formed when two desorption events occur within the time window used for recording a single spectrum. Due to the low resolution of the time-of-flight (TOF) mass spectrometer, all of these contributions are unresolved and produce a continuum with increasing abundance as the molecular weight of the sample increases. Thus, the background continuum is an integral part of the plasma desorption spectrum, and cannot be eliminated by instrumental parameters. Instead, an automatic or subjective background subtraction can be performed leading to the spectrum shown in Figure 1B.

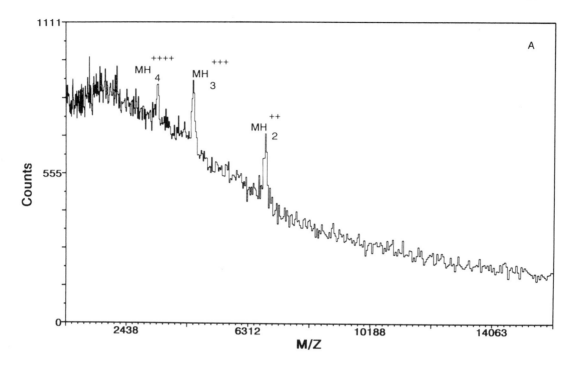

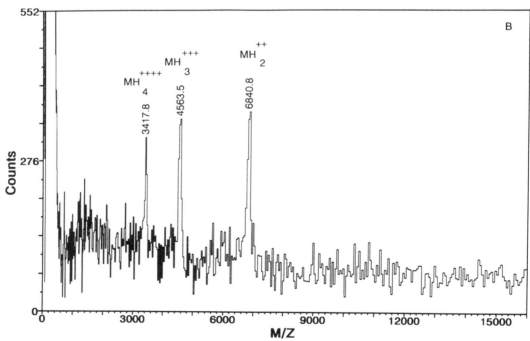

FIGURE 1. Plasma desorption spectrum of bovine ribonuclease. (A) "Raw" spectrum; (B) background-substrated spectrum. Mol wt (measured) 13,678; mol wt (calculated) 13,682.

B. MASS CALIBRATION

The "raw" data are an averaged time spectrum obtained by accumulation of a large number of spectra, each of which is the result of a single fission event. In order to obtain sufficient ion statistics, 10^5 to 10^6 spectra are accumulated for molecules with mol wt up to 5000 to 6000, and 10^6 to 10^7 for larger molecules. Before mass determination can be performed, it is necessary to calibrate the spectrum. In the TOF mass spectrometer, the flight-time (T) and the mass-to-charge ratio (m/z) are correlated by the simple equation $T = a\sqrt{m/z} + b$. The constants a and b can be calculated if the masses and the flight times for two ions in the spectrum are known. The low mass end of each PD-spectrum (Figure 2) always contains a number of known ions among which H^+, Na^+, or NO^+ are usually the most abundant and therefore are used for mass calibration. With the simple short flight-tube TOF mass spectrometer, it is necessary to use empirically determined m/z values for the ions used for calibration. Furthermore, these values depend on the method used for sample preparation, and a set of values must be established for each sample preparation method. This phenomenon is probably related to differences in the initial velocities of the ions used for calibration, among which the initial velocity of the H^+ ion is especially suspected to be sample preparation-dependent. This complication might be overcome by choosing other ions at higher mass for calibration, but such ions may be more difficult to locate and less abundant, resulting in decreased precision in determination of the centroid. Therefore, most laboratories, including ours, still use the low-mass ions mentioned previously. With reflectron-type TOF mass spectrometers, this problem seems to be eliminated, and the theoretical m/z value for the ions can be used independently of the sample preparation method.[12]

C. PRECISION OF MASS DETERMINATION

In a discussion of the precision in mass determination, it is important to separate systematic errors and the precision that can be obtained in determination of the centroid of a given peak.

The systematic errors are related most frequently to the calibration. The use of incorrect empirical values for the mass of the ions used for calibration is often a source of errors, because either these values may not be matched with the sample preparation method or the sample preparation techniques have changed gradually without concomitant adjustment of the values. Determination of the time centroid for the calibration ions is also critical, because even a slight error becomes significant when extrapolated to high mass. Both types of errors usually give rise to a shift of all the determined masses in the same direction, but sometimes the determined masses are too low in the low-mass range, correct in the midrange, and too high in the high mass range, or vice versa. The best way to avoid systematic errors of both types is to frequently analyze a number of standard samples in order to check their centroid determination and to verify the calibration constants.

The precision for determining the centroid of a given peak depends on the peak shape and abundance. Both of these factors are related to the mass, the purity, and the amount of sample as well as to the stability of the ion. Broad peaks dominated by a metastable component give a rather poor precision simply because it is difficult to define reproducibly the peak. As a general trend, the width of the peaks increases with increasing mass, probably because the formed ions are less stable. Moreover, it also seems that protein molecules with many disulfide bridges are more stable, and therefore give narrower peaks than proteins without or with only few disulfide bridges. The influence of impurities seems to be related to two different phenomena. First, many contaminants, especially alkali metal ions, tend to form adduct ions with the sample molecules. The resolution of the TOF mass spectrometer is not sufficient to resolve the adduct ions from MH^+ at high masses, and the determined centroid will thus represent an average of a mixture of different molecular ion species. Second, the presence of impurities seems in many cases to increase the internal energy of the formed ions, thereby causing an increase of the metastable and a decrease of the stable component of the peak.

As a general rule, the mass precision obtained with the commercial plasma desorption mass

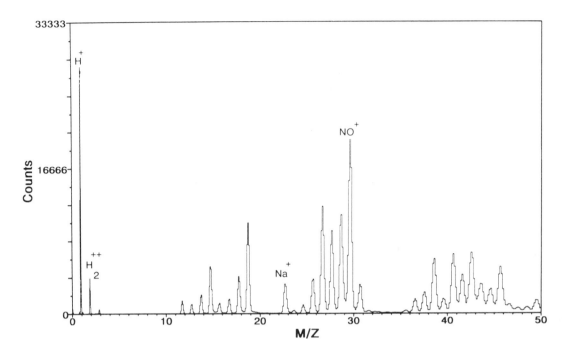

FIGURE 2. Low mass end of the PD-spectrum shown in Figure 1. The peaks corresponding to H^+ and Na^+ or NO^+ are used for calibration.

spectrometer*[13] is approximately 0.1% across the entire mass range. For samples such as insulin (mol wt 5700 to 5800 depending on the origin) and the peptide samples described in this chapter, a deviation of $\leq 0.01\%$ between the calculated and the measured mass is frequently obtained.

D. ACCESSIBLE MASS RANGE

There is no theoretical mass limit for a TOF mass spectrometer in contrast to, for example, magnetic sector or quadrupole instruments. The accessible mass range seems to be related only to the desorption process itself or, in other words, to how large a molecule can be desorbed from the surface upon impact of a single fission fragment. The largest protein molecules observed to date have been porcine pepsin (mol wt 34,600)[14] and ovalbumin (mol wt 45,000).[15] That this pair of proteins does not represent the upper limit at which protein molecules can be brought in the gas phase has been demonstrated by laser desorption TOF mass spectrometry, where molecular ions of proteins between 100 and 200 kDa have been observed.[16]

In practice, most peptides and proteins with mol wt below 10,000 and many proteins between 10,000 and 20,000 are analyzed readily by PDMS, while the success rate beyond 20,000 is limited.

III. SAMPLE PREPARATION

The method for sample preparation is of utmost importance for success in PDMS. The ^{252}Californium (Cf) source is placed behind the sample in most PD mass spectrometers. Therefore, the combined backing and sample layer must be thin enough to allow the fission fragments to penetrate both layers. The backing usually consists of a thin (0.5 to 1 μm) aluminized polyester foil upon which the sample is deposited. The original, and until recently the most popular, method for sample deposition was the electrospray method (see Figure 3,

* Biolon, A. B., Box 14045, S-750 15 Uppsala, Sweden.

left).[7] With this method, molecular ions have been observed for a number of peptides and proteins (reviewed in Reference 17) with trypsin (mol wt 23,463) being the largest. Many samples, however, failed with the electrospray technique. The major reason for this failure seemed to be that the presence of low molecular weight contaminants, especially alkali metal ions, often reduced the spectrum quality and sometimes entirely quenched the spectrum. Another drawback is that aqueous solvents, which are the most appropriate solvents for proteins, are difficult or impossible to electrospray.

A. THE NITROCELLULOSE MATRIX

Complete removal of the alkali metal salts in samples of biological origin proved to be a very difficult task, because numerous sources such as glassware, water, and organic solvents, either liberate or contain small amounts of sodium salts. The introduction of nitrocellulose as a matrix for PDMS solved this problem.[10] The principle is shown in Figure 3 (center). First, a nitrocellulose layer is deposited on the aluminum foil by electrospraying, then the sample is applied onto the nitrocellulose from an aqueous solution and allowed to adsorb, and finally the salts are removed by washing with ultrapure solvents. Removal of the salts by this method is very successful, and unexpectedly, the use of nitrocellulose compared to electrospray also resulted in improved molecular ion yields, increased abundance of multiply charged ions, and sharper peaks (Figures 4 and 5). Qualitatively similar, but not quantitatively equivalent, results were obtained by adding an excess of reduced glutathione to the sample solution prior to sample application by electrospray.[18]

A further improvement of the nitrocellulose technique was obtained by application of the sample onto a spinning nitrocellulose target (Figure 3, right).[19] This method was introduced for the application of low molecular weight peptides onto the nitrocellulose. Such peptides do not bind strongly to nitrocellulose, and are therefore removed during the extensive washing procedure applied originally. The spin-drying method seems to combine the adsorption and washing in a single step, because a simultaneous drying and migration of the sample solution to the periphery of the target is obtained. As a result, a thin sample layer is deposited in the central area (the area exposed to fission fragments), whereas the excess of sample and the more soluble salts are deposited in the periphery.

The spin-drying method has been found very useful, not only for application of small peptides and for other low molecular weight compounds that do not bind to nitrocellulose at all, but also for large peptides and proteins, because a better sensitivity is obtained. Figure 5 shows the molecular ion yield (expressed as the number of molecular ions formed per 100 fission fragments) as a function of the sample amount applied for different sample application techniques. The molecular ion yield is a measure of the combined efficiency for desorption, ionization, and transmission of stable molecular ions to the detector. It is apparent from Figure 5 that the nitrocellulose techniques are superior to the spray techniques by about a factor of five with regard to molecular ion yields and by several orders of magnitude with regard to sensitivity. Comparable maximum molecular ion yields are obtained with the different nitrocellulose techniques, but the best sensitivity is obtained with the spin-drying technique. When larger amounts of sample are applied by the spin-drying technique or if sodium salts are added, then the molecular ion yield decreases to the level obtained with the spray techniques. Simultaneously, the spectra change their qualitative aspect to resemble that of a sprayed sample, i.e., reduced intensity of the multiply-charged ions and wider peaks.[19,20]

These observations suggest that very thin protein layers are best and that protein molecules, which are in contact only with the nitrocellulose matrix, are easier to desorb and form more stable molecular ions after desorption than protein molecules, which interact with other sample molecules. Several experiments support this suggestion. Studies of molecular ion yield as a function of the energy density of the primary ions show that substantially less energy is needed to desorb insulin from nitrocellulose than from an electrosprayed sample.[21] Chait[22] has studied

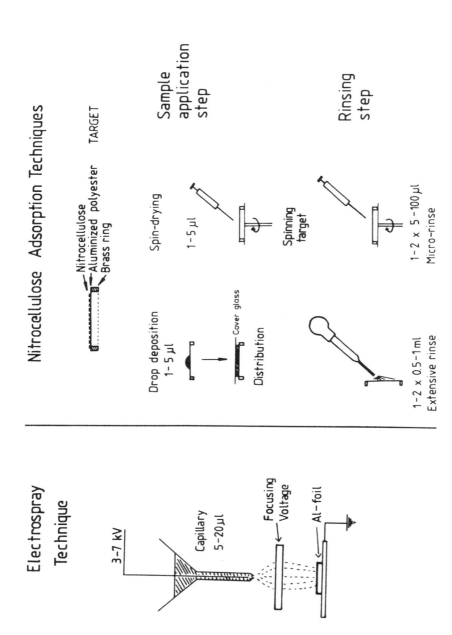

FIGURE 3. Different methods for sample preparation in plasma desorption mass spectrometry.

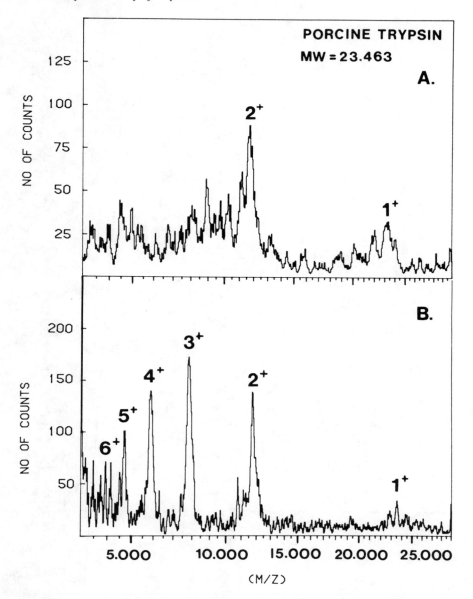

FIGURE 4.   Plasma desorption spectrum of trypsin. (A) Prepared by electrospray; (B) and by adsorption onto nitrocellulose. (From Jonsson, G. P., Hedin, A. B., Håkansson, P. L., Sundqvist, B. U. R., Säve, G. S., Nielsen, P. F., Roepstorff, P., Johansson, K. E., Kamensky, I., and Lindberg, M. S. L., *Anal. Chem.*, 58, 1084, 1986. Copyright 1986 American Chemical Society. With permission.)

the metastable decay of insulin, and has demonstrated that virtually all ions decomposed before arriving at the detector when the sample was electrosprayed, whreras 90% survived after desorption from nitrocellulose. The reduced fragmentation is also observed with the peptide antibiotic, Nisin, which upon electrospraying yields several distinct fragment ions, but only MH$^+$ and MH$_2^{2+}$ when desorbed from nitrocellulose.[23] When the sample amount is increased or if a sodium salt is added to nitrocellulose-bound Nisin, fragmentation is observed again.[20] The narrower peaks[10,24] and the increased abundance of multiply charged ions[10] when the sample is desorbed from nitrocellulose also suggest a lower internal energy of the ions. When the sample amount is reduced, an increased abundance of the multiply charged ions relative to the singly charged

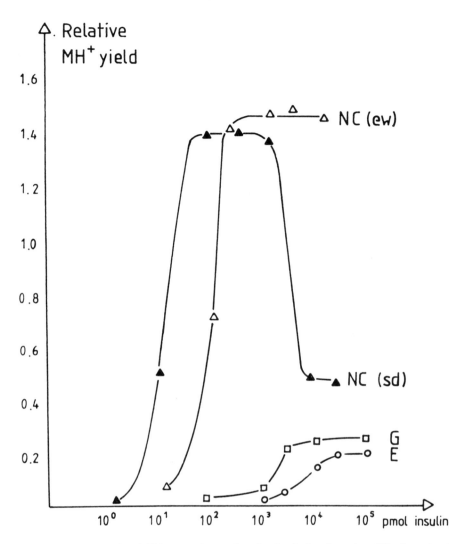

FIGURE 5. Molecular ion yield (expressed as number of molecular ions formed per 100 primary ions) as a function of sample amount and sample application technique. (▲ spin-drying on nitrocellulose (SD), Δ adsorption on nitrocellulose followed by extensive wash (ew), ❑ electrospray with glutathione (G), o electrospray (E). (Adapted from Roepstorff, P., Nielsen, P. F., Klarskov, K., and Højrup, P., in *The Analyses of Peptides and Proteins by Mass Spectrometry*, McNeal, C. J., Ed., John Wiley & Sons, Chichester, 1988, 55. With permission from John Wiley & Sons, Ltd.)

ions is observed,[25-27] also indicating that isolated molecules on the nitrocellulose surface are less excited after desorption.

B. PRACTICAL SAMPLE APPLICATION

As a practical consequence of all the systematic studies in sample preparation methods, nitrocellulose is now the generally used standard matrix for PDMS of peptides and proteins. In the author's laboratory, a sample is applied normally by the spin-drying method from a 0.1% trifluoroacetic acid solution containing 15% ethanol or acetonitrile to decrease the surface tension. Other solvents such as isopropanol, concentrated trifluoroacetic acid, salt-containing buffers, or 8 M urea solutions may also be used, the latter two provided that extensive washing is performed. If a spectrum is of poor quality, the most likely reasons are either a poor sample quality, a sample that desorbs or ionizes poorly, too little or too much sample, or a too-high content of alkali metal ions. The alkali metal content can be judged by comparing the abundance of the peaks for H$^+$ and Na$^+$ (or K$^+$). It is our usual experience that the alkali metal content is too high if the peaks for the latter are more than 50% of the intensity of the H$^+$ peak. As PDMS is practically nondestructive because of the low flux of primary ions, it is often possible to improve a poor result by removing the sample from the mass spectrometer, washing it with water or dilute acid to remove excess sample or alkali metal ions, and reexamining it. The nitrocellulose matrix is an ideal interface between the mass spectrometer and the wet procedures in protein chemistry because it is possible, as will be seen, to apply the sample directly from chromatographic fractions or from the reaction mixture in chemical or enzymatic reactions, or to carry out such reactions directly on the nitrocellulose-bound sample. It is also possible to preferentially enrich the peptides or proteins from extremely dilute solutions[10] and to transfer protein samples from electrophoretic gels to the nitrocellulose-covered target by blotting procedures.[28]

IV. MOLECULAR WEIGHT DETERMINATION OF INTACT PEPTIDES AND PROTEINS

Determination of the molecular weight is always one of the first analyses to characterize a given peptide or protein. In conventional protein chemistry, this determination is usually obtained by gel electrophoresis or gel permeation chromatography. Unfortunately, these methods are rather imprecise (within 5 to 10% at best) and sometimes inaccurate, because they are influenced by parameters that are irrelevant to the molecular weight, such as shape or hydrophobicity of the molecule. In the author's laboratory, several examples of 50 to 100% deviation from the correct molecular weight have been obtained with either very hydrophobic or very hydrophilic proteins.

PDMS and the nitrocellulose matrix are a realistic alternative to the classical methods for molecular weight determination of small proteins. The spectra are very easy to interpret because they normally contain only a series of differently charged molecular ions and no fragment ions. With increasing size of the protein and with decreasing sample amount, reduced abundance of the singly charged molecular ion relative to the doubly charged, triply charged, etc. ions, is observed,[27] and with practical sample quantities (2 to 100 pmol), the singly charged molecular ion is usually absent for proteins with mol wt beyond 10,000. This limit, however, does not create any problems because only one molecular weight can fit the observed series of multiply charged ions, and precision is increased because the mass determination is the average of several different mass values determined for different molecular ion species. Table 1 lists the mass spectrometric data obtained for a number of proteins analyzed in our laboratory.

The following examples illustrate molecular weight determination by PDMS of proteins taken from the daily work in our laboratory.

The first example of a practical application of mass spectrometric molecular weight determination of a protein beyond 10 kDa was in connection with the amino acid sequence determination of an insect cuticle protein.[29] Its molecular weight based on SDS-gel electrophoresis was estimated to 21,600, whereas the determined sequence indicated 15,323. The protein contained several sequence repeats, and lack of recognition of an additional repeat or the presence of an unrecognized prosthetic group could not be excluded. The protein was very hydrophobic. Such proteins are rather difficult to analyze by PDMS, and attempts to analyze the protein using the electrospray method were not successful. After discovering the advantages of the nitrocellulose matrix, the protein was analyzed immediately by this method, and the determined mol wt of 15,329 was consistent with the sequence. Since then, several hydrophobic proteins from locust cuticle and other sources have been analyzed by PDMS in our laboratory.[30-33] It is our experience that hydrophobic proteins need very careful purification prior to PD mass spectrometric analysis, probably because they have a strong tendency to self-aggregate.

TABLE 1
Examples of Data Obtained by Analysis of Various Proteins by PDMS

Protein	Ions observed	Mol wt measured	Mol wt calculated	Δ	Ref[a]
Nisin	MH_2^{2+}, MH^+	3354.3	3354.1	- 0,2	23
Porcine insulin	MH_2^{2+}, MH^+	5777.2	5777.6	+0,4	u
Human IgF1	MH_3^{3+}, MH_2^{2+}, MH^+	7657	7648.6	- 8,4	34
Mouse epidermal growth factor	MH_2^{2+}, MH^+	6040.1	6039.6	-0.5	35
Proinsulin	MH_3^{3+}, MH_2^{2+}, MH^+	9388.0	9388.6	+0,6	42
Acyl-CoA binding protein	MH_2^{2+}, MH^+	9956	9955.3	-0,7	46
β-2-microglobulin	MH_3^{3+}, MH_2^{2+}	11731.3	11729	-2,3	u
Ribonuclease	MH_4^{4+} to MH_2^{2+}	13678	13682	+3	u
Lysozyme	MH_4^{4+} to MH^+	14315	14314	- 1	30
Locust protein 8	MH_3^{3+}, MH_2^{2+}	15230	15224	- 6	31
Locust protein 38	MH_4^{4+} to MH^+	15329	15323	- 6	45
IL 1	MH_3^{3+}, MH_2^{2+}	17358	17359	+1	48
IL 1 precursor	MH_3^{3+}, MH_2^{2+}	17967	17904	-63[b]	48
HGH	MH_4^{4+} to MH_2^{2+}	21122	21125	+3	42
Trypsin	MH_4^{4+} to MH_2^{2+}	23470	23463	- 7	30
Pepsin	MH_4^{4+} to MH^+	34630	34688	+58	14

[a] u means unpublished data from the author's laboratory.
[b] This difference is probably due to an adduct ion instead of protonated ion.

In studies of insulin-like growth factors (IGFs), a number of IGF-variants were isolated from human Cohn fraction IV. Based on their molecular weights (in the 7400 to 7800 mass range), it was possible to classify them in the IGF I and IGF II families, as well as to suggest the most probable modifications relative to IGF I or II.[34] Typical examples of modifications observed were cleavage of a single peptide bond leading to an insulin-like two-chain molecule accompanied by an increase in mol wt of 18 amu.

Studies of β-2-Microglobulin in patients suffering from lung cancer revealed that several modified versions were present. The PD-spectrum of one of these variants showed a decrease in mass relative to authentic β-2-Microglobulin, which indicated a molecule lacking one amino acid residue (Figure 6), but the mass difference did not correspond to loss of either the N- or C-terminal residues. β-2-Microglobulin contains one internal disulfide bridge, and cleavage of a peptide bond between the cysteine residues would result in a two-chain molecule linked by a single disulfide bond. The spectrum (Figure 6B) contained in addition to the singly and doubly charged molecular ions of the intact molecule a number of peaks corresponding to fragmentation of this disulfide bond. These peaks allowed exact positioning of the cleavage point and identification of the excised amino acid residue.[30]

During the amino acid sequence determination of a rat epidermal growth factor, a discrepancy between the determined sequence and amino acid analysis of the isolated protein indicated that a piece was missing. Molecular weight determination of the intact protein showed a mass difference of 522 relative to the determined sequence, corresponding to 3 to 5 amino acid residues. A search in the fractions from high pressure liquid chromatography (HPLC)-separation of the tryptic peptides revealed an early-eluting peptide with a mol wt of 517. Sequence determination of this peptide concluded the structure determination of rat epidermal growth factor.[35]

In general, molecular weight determination by PDMS is a very valuable supplement in protein sequence determination, and this subject is dealt with separately in this account. Another promising field is the use of PDMS to locate processing products from precursor proteins for which the sequence is known. This subject is dealt with in Chapter 12.

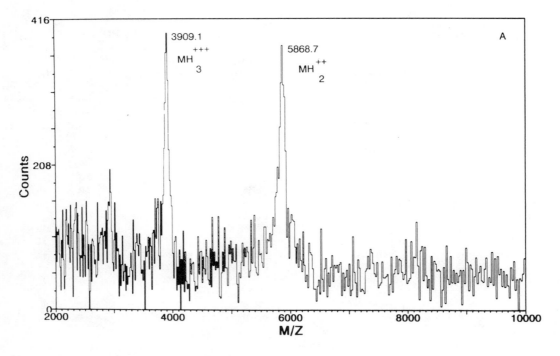

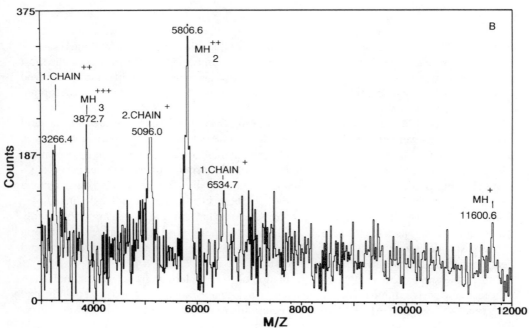

FIGURE 6.   (A) Plasma desorption mass spectrum of β-2-Microglobulin (mol wt measured 11,730. mol wt calculated 11,729); (B) of a post-translationally modified β-2-Microglobulin. (Adapted from Roepstorff, P., Nielsen, P. F., Klarskov, K., and Højrup, P., in *The Analyses of Peptides and Proteins by Mass Spectrometry*, McNeal, C. J., Ed., John Wiley & Sons, Chichester, 1988, 55. With permission from John Wiley & Sons, Ltd.)

V. PEPTIDE MAPPING BY PLASMA DESORPTION MASS SPECTROMETRY

A. *IN SITU* REACTIONS

The PD-spectra, as mentioned previously, generally contain only molecular ions and no structurally meaningful fragment ions. Fortunately, a number of highly specific proteolytic enzymes can be used to generate structural information. A number of specific chemical methods are also available. As suggested by Chait et al.,[36,37] it is possible to take advantage of the non-destructive nature of PDMS to perform such reactions on the nitrocellulose-bound sample after recording a spectrum of the intact molecule. The principle is illustrated in Figure 7, using lysozyme as an example.[30,38] After recording a spectrum of intact lysozyme (Figure 7A), the target is removed from the mass spectrometer and treated successively with a solution of dithiothreitol (DTT) and trypsin. Solvents and reagents are removed between each step by spinning the target, and the sample is reinserted into the mass spectrometer. In the resulting spectrum (Figure 7B), molecular ions for all the expected tryptic peptides, except two dipeptides, are observed. This spectrum is a PD mass spectrometric peptide map of lysozyme, also termed a PD map. The technique can, as described later, be applied with a number of exo- and endopeptidases and also with different chemical reagents (DTT or mercaptoethanol) to reduce disulfide bonds. There are certain limitations in the choice of buffers for the reactions. The use of sodium- or potassium-containing buffers, for example, requires extensive washing after the reaction. This washing might remove some of the small hydrophilic peptides. In contrast, ammonium-containing buffers are "transparent" in PDMS, and can be used without washing.[39] The nitrocellulose targets cannot tolerate prolonged exposure to solvents, and in order to keep the reaction times below approximately 30 min, high concentrations of reagents or enzymes are needed. In general, such *in situ* reactions lead to complete reduction of disulfide bonds, whereas the enzymatic cleavages may only be partial. In spite of these reservations, we have found *in situ* reactions extremely useful because supplementary information is readily obtained without any further sample consumption.

B. MONITORING REACTIONS IN SOLUTION

Preparative or analytical enzymatic and chemical reactions in solution can also be monitored by PDMS with the previously mentioned reservations about buffers and the need for washing procedures. At present, the monitoring cannot be performed on-line in real time, but the time span from sampling a small aliquot, applying it on nitrocellulose by the spin technique, and recording a spectrum is sufficiently short to be useful in monitoring most reactions. In our laboratory, this technique is typically used to follow the time course of preparative enzymatic digestions,[30,40] for C-terminal sequence determination of peptides with carboxypeptidases,[31,41] and for monitoring preparative reduction and alkylation reactions.[29,42] Examples of the first two are given in the section on protein sequence analysis. The reduction and alkylation reactions can be monitored by directly applying small aliquots from the complex reaction mixture onto a nitrocellulose target by the spin technique, followed by extensive washing to remove reagents, buffers, etc. For example, for an aliquot from a reaction mixture containing 8 M urea, sodium acetate, DTT, and iodoacetic acid used for reduction and carboxymethylation of insulin, it could be demonstrated that the B-chain was only partly carboxymethylated.[42] In a reaction to reduce and pyridyl-ethylate insulin, four peaks appeared upon HPLC separation instead of the expected two. PDMS showed that two of those peaks corresponded to the correctly derivatized A- and B-chains, whereas the other two were overalkylated A- and B-chains. Upon *in situ* digestion with *Staphylococcus aureus* protease, the sites of overalkylation could be located to the histidine-containing peptides.[30]

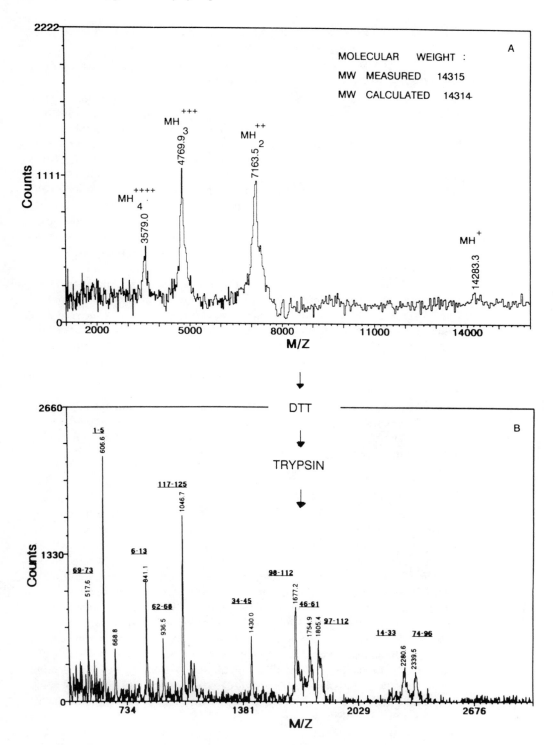

FIGURE 7.   (A) Plasma desorption mass spectrum of lysozyme; (B) of the same sample after successive treatment with DTT and trypsin. The numbers in B indicate the tryptic peptides relative to the sequence of lysozyme. (Adapted from Roepstorff, P., Nielsen, P. F., Klarskov, K., and Højrup, P., in *The Analyses of Peptides and Proteins by Mass Spectrometry*, McNeal, C. J., Ed., John Wiley & Sons, Chichester, 1988, 55. With permission from John Wiley & Sons, Ltd.)

C. SUPPRESSION EFFECTS IN MIXTURE ANALYSIS

It is well known that in mixture analysis by mass spectrometry, suppression of certain peptides may occur. This ion suppression is also true for PDMS. For example, the A-chain of insulin may be suppressed partly or completely in the presence of B-chains,[38] and PD analysis of mixtures of peptides produced by tryptic digestion of interleukin-2 (IL-2) and human growth hormone did not show all the expected peptides.[42] In general, peptide maps obtained by PDMS seem more complete than those maps obtained by fast atom bombardment (FAB) mass spectrometry.[44] In PDMS, the suppression has been demonstrated to be related to the net charge of the peptides.[38] Thus, peptides with a net positive charge are predominant in the positive ion spectra of mixtures, and those with negative net charge in the negative ion spectra. It is interesting that a negatively charged peptide, which is completely suppressed in a positive ion spectrum of a mixture, may, when analyzed alone, give a high positive ion yield. It therefore seems that there is a competition for protons favoring the peptides with highest proton affinity. The hydrophobicity, which is the dominant discriminatory factor in FAB mass spectrometry of peptide mixtures, has no influence in PDMS.[38] PDMS and FAB mass spectrometry therefore may often be complementary in mass spectrometric peptide mapping.

VI. APPLICATION IN PROTEIN SEQUENCE DETERMINATION

In our laboratory, analysis by PDMS has been used extensively for the last few years as a supplement to automatic Edman degradation in protein sequence determination. Based on the experience gained during sequence determination of a number of proteins,[31,32,45-47] the general strategy shown in Figure 8 has been developed.[40] Some of the more important features of this strategy are described in the following section.

After recording the molecular weight, the *in situ* digest is useful in checking, without further consumption of sample, whether the first choice of digesting enzyme will result in peptides of a size suitable for automatic Edman degradation. The time course of the enzymatic digest establishes the optimal digestion time. For example, with a locust cuticle protein, the time-course showed that a small peptide was lost in the tryptic digest when applying digestion times exceeding 15 min.[31] As a byproduct of the time course digest, it is often possible to align the peptides because of different hydrolysis rates of the susceptible bonds.[30,31]

Molecular weight determination of the peptides isolated by HPLC serves as a purity check, gives a reliable estimate of how many steps of Edman degradation are needed, serves to confirm the assigned sequence, and often serves to solve ambiguities or fill in any gaps in the determined sequence.

It is often difficult to determine the sequence of the last few C-terminal residues of a peptide by Edman degradation. We therefore routinely perform *in situ* digestion with carboxypeptidases after the molecular weight determination described in the previous step. Carboxypeptidase sequentially removes the C-terminal amino acid residues leading to mixture of shortened peptides. The resulting PD spectrum (Figure 9) contains MH+ ions corresponding to the different molecular species present, and the sequence can be deduced from the mass difference between these ions. With *in situ* digestion, it is normally possible to identify the sequence of one to four C-terminal residues without extra sample consumption. If further sequence information from the C-terminus is needed, it can be obtained by monitoring with PDMS the time course of carboxypeptidase digestions in solution. With this technique, it has been possible to determine the sequence of up to 10 residues from the C-terminus.[41]

At this point of the strategy, the complete sequence is often established based on the combined data from Edman degradation and mass spectrometry. The protein is now digested with a different enzyme in order to verify the sequence and, if necessary, to find any missing overlaps between the peptides from the first digest. It is normally sufficient to match the molecular weight determined for the second set of peptides with the postulated sequence. If unambiguous results are not directly obtained, supplementary information on the peptides in the second digest can

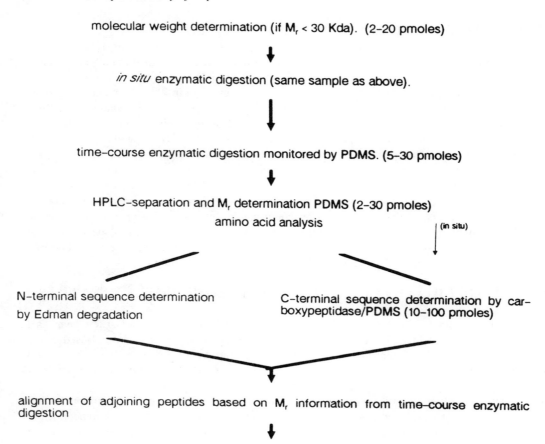

molecular weight determination (if M_r < 30 Kda). (2–20 pmoles)

↓

in situ enzymatic digestion (same sample as above).

↓

time–course enzymatic digestion monitored by PDMS. (5–30 pmoles)

↓

HPLC–separation and M_r determination PDMS (2–30 pmoles)
amino acid analysis
(in situ) ↓

N-terminal sequence determination by Edman degradation

C-terminal sequence determination by carboxypeptidase/PDMS (10–100 pmoles)

alignment of adjoining peptides based on M_r information from time–course enzymatic digestion

↓

verification of results by a second, different enzymatic digestion. Location of S–S bonds, if any (5–30 pmoles).

Total sample consumption for PDMS analysis is 100–300 pmoles.

FIGURE 8.   General strategy for protein sequence determination supported by PDMS. The amount of sample used in each step is indicated in parentheses. (From Roepstorff, P., Klarskov, K., and Højrup, P., in *Methods in Protein Sequence Analysis*, Wittman-Liebold, B., Ed., Springer-Verlag, Heidelberg, 1989, 191. With permission.)

be obtained by *in situ* digestion with endo- or exopeptidases, by a few steps of Edman degradation, or by amino acid analysis. Such a strategy has been used extensively (see reference 31). If the protein contains disulfide bridges, their position can be established by a similar strategy based on analysis by PDMS of an enzymatic digest of the protein before and after *in situ* reduction with dithiothreitol. If the peptide mixture is too complex, it may be advantageous to include an HPLC separation step prior to the mass spectrometric analysis.

The advantages of using the scheduled strategy are that the overall sample consumption for sequence determination is reduced significantly, the sequenator use is approximately halved and the costs are reduced correspondingly, and not least that PDMS provides a method for independent confirmation of the results.

VII. APPLICATIONS IN PROTEIN BIOTECHNOLOGY AND PROTEIN ENGINEERING

The growth in biotechnological protein production has created a demand for techniques that can provide a quick, reliable, and accurate characterization of proteins. It is important to identify

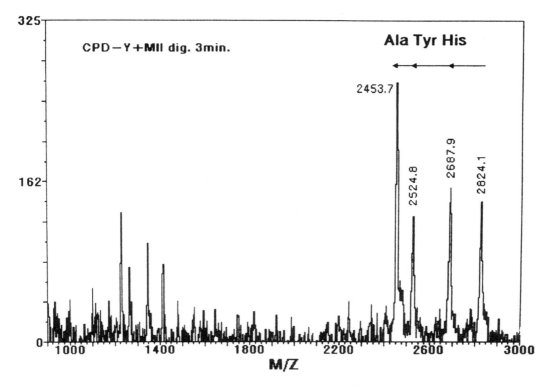

FIGURE 9. Plasma desorption spectrum obtained after *in situ* digestion for 3 min. of the C-terminal peptide from protein 8 of locust cuticle with a mixture of carboxypeptidase Y and MII. The sequence of the three C-terminal residues can be determined from this spectrum.

the protein, to verify that it is not mutated, and to ensure that desired post-translational modifications have taken place and that undesired post-translational modifications are absent. For proteins in the accessible mass range, PDMS has been found very suitable for this purpose.

The first example of a biotechnological application of PDMS was the molecular weight determination of the intermediate and final products in the semi-synthetic production of human insulin based on porcine insulin.[6] Since then, molecular weight determination by PDMS has been used to characterize numerous proteins produced by DNA-recombinant techniques. For biosynthetic human proinsulin, the mol wt was measured as 9388.0 (calculated value 9388.6). The sample contained traces of a dimeric molecule, as indicated by a small peak at m/z 6240 located between the singly and doubly charged ion of the monomer. This ion, which corresponded to the triply charged molecular ion of the dimer, was the only evidence of the dimer in the mass spectrum because the singly charged ion was absent, and the doubly and quadruply charged ions would superimpose upon the singly and doubly charged molecular ions of the monomer, respectively.[42] Successful removal of an N-terminal methionine in human growth hormone (HGH)[42] and of a four amino acid residue-long leader sequence in interleukin-1 (IL-1)[48] was evident from the determined mol wt 22,122 and 17,358 (calculated 22,125 and 17,358), respectively.

An alternative to molecular weight determination of the intact protein is based on PD mapping, as described for recombinant HGH and IL-2.[42] Here, the proteins were cleaved with either trypsin or cyanogen bromide followed by mass spectrometric peptide mapping. A crude HPLC separation of the digestion mixture was required prior to mass spectrometric analysis to overcome suppression. Both methods have their advantages. The direct molecular weight determination is by far the fastest alternative, and the dimer of proinsulin, for example, would not have been disclosed by peptide mapping. When poor molecular ion yields are obtained for the intact protein, peptide mapping can be applied, but it requires careful identification of all the peptides, and must therefore, due to suppression, be nearly always preceded by an HPLC separation. The position of the disulfide bonds cannot be verified by molecular weight determination. This verification requires a cleavage of the protein, followed by partial separation of the peptides and recording of mass spectra before and after *in situ* reduction of the disulfide bonds.

The activity and/or stability of peptides and proteins may be altered by exchanging, deleting, or adding amino acid residues. The rapidly increasing capability to perform such changes on the genetic level, either by DNA synthesis or site-directed mutagenesis, has created a new field, protein engineering. PDMS has been successfully used to characterize numerous products from industrial protein engineering. A general strategy for analysis of such products by PDMS is based on molecular weight determination of the intact molecule, followed by *in situ* reduction and enzymatic cleavage with recording of the intermediate PD maps.[42] The strategy has been applied to a number of variants of human insulin including several mini-proinsulins, i.e., proinsulins in which the connecting peptide has been shortened drastically or simply replaced by a peptide bond between the C-terminus of the B-chain and the N-terminus of the A-chain. In most cases, the molecular weight alone was sufficient to demonstrate that the desired modification was introduced (cf. Table 2[48]), but sometimes more precise information was needed. In such cases, PD mapping was used to locate more precisely the site of modification, and also to increase the precision of the mass determination in the regions of interest (Figure 10). For Glu-to-Gln or Asp-to-Asn mutations, which result in only a 1 amu mass difference, the mass accuracy was still insufficient for an unambiguous identification. Here, it was possible to take advantage of the specificity of *S. aureus* protease, which cleaves only at acid residues.[30,49] An alternative method for distinguishing between the acidic and the amidated residues is to methyl-esterify the carboxyl groups. This method is also useful for analyzing partial deamidation, and for counting the number of carboxyl groups in a peptide.[50]

VIII. CONCLUSION

PDMS has in the last few years been demonstrated to be a very useful technique in protein chemistry. The mass spectrometer is simple in concept, requires little maintenance, is easy to operate, and does not require highly specialized operators. It therefore finds a natural place in the protein chemistry laboratory next to the HPLC equipment and the sequenator. The nitrocellulose matrix is compatible with most solvents used in protein chemistry, and the sensitivity in the low picomole range is adequate for most purposes. Because of the speed and ease with which valuable information can be obtained, PDMS is now in our laboratory the prime technique used to characterize any HPLC fraction, peptide, or protein, and in our sequence studies, it is an indispensable supplement to the sequenator.

ACKNOWLEDGMENTS

Sincere gratitude is expressed to students and postdoctoral collaborators, whose daily use of PDMS has contributed ideas and applications to our general concept of the applicability of PDMS in protein chemistry. Their names are cited in the references. A special thanks is extended to Professor B. Sundqvist, Uppsala University, who introduced me to the world of sputtering of large biomolecules. Support for our work has been provided by the Danish Technical Science and Natural Science Research Councils.

TABLE 2
Modified Insulins Analyzed by PDMS

Modified human insulin	Mol wt measured	Mol wt calculated	Δ
(B25-Tyr)-des(B26-30)-insulin-B25-amide	5232.2	5231.9	+0.3
des(B30)-insulin	5706.3	5706.5	-0.2
des(B23-30)-insulin	4866.0	4865.6	-0.4
(B21-Gln)-insulin	5806.5	5806.6	+0.1
(AO-Arg)-des(B30)-insulin	5862.5	5862.7	+0.2
(A4-Gln)-des(B30)-SCI[a]	5688.0	5687.5	-0.5
(B21-Gln)-des(B30)-SCI[a]	5688.1	5687.5	-0.6
des(B30)-SCI[a]	5688.8	5688.5	-0.3
(B31-Lys, B32-Arg)-SCI[a]	6073.7	6073.9	+0.2

[a]   Mini-proinsulin.

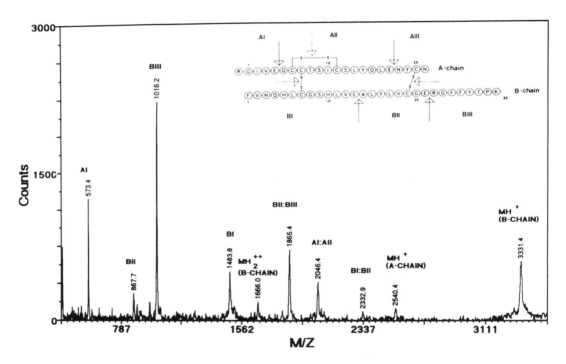

FIGURE 10.   Plasma desorption spectrum of A_0-Arg-des-B_{30} insulin reduced and digested with *S. aureus* protease *in situ*. (Reproduced from Nielsen, P. F., Roepstorff, P., Clausen, I. G., Jensen, A. B., Jonassen, I., Svendsen, A., Balschmidt, P., and Hansen, F. B., *Protein Eng.*, 2, 449, 1987. With permission from IRL Press.)

REFERENCES

1. Torgerson, D. F., Showronski, R. P., and Macfarlane, R. D., New approach to the mass spectrometry of non-volatile compounds, *Biochem. Biophys. Res. Commun.*, 60, 616, 1974.

2. Håkansson, P., Kamensky, I., Sundqvist, B., Fohlman, J., Peterson, P., McNeal, C. J., and Macfarlane, R. D., 127-I Plasma desorption mass spectrometry of insulin, *J. Am. Chem. Soc.*, 104, 2948, 1982.

3. Håkansson, P., Kamensky, I., Kjellberg, J., Sundqvist, B., Fohlman, J., and Peterson, P., The observation of molecular ions from a protein neurotoxin (MW 7821) using 127-I plasma desorption mass spectroscopy, *Biochem. Biophys. Res. Commun.*, 110, 519, 1983.

4. Kamensky, I., Håkansson, P., Kjellberg, J., Sundqvist, B., Fohlman, J., and Peterson, P., The observation of quasi-molecular ions from a tiger snake venom component (MW 13309) using ^{252}Cf plasma desorption mass spectroscopy, *FEBS Lett.*, 155, 113, 1983.

5. Sundqvist, B., Håkansson, P., Kamensky, I., Kjellberg, J., Salehpour, M., Widdiyaesekera, S., Fohlman, J., Peterson, P., and Roepstorff, P., Californium-252 plasma desorption time of flight mass spectroscopy of proteins, *Biomed. Mass Spectrom.*, 11, 242, 1984.

6. Sundqvist, B., Roepstorff, P., Hedin, A., Håkansson, P., Kamensky, I., Lindberg, M., Salehpour, M. and Säve, G., Molecular weight determinations of proteins by Californium plasma desorption mass spectrometry, *Science*, 226, 696, 1984.

7. McNeal, C. J., Macfarlane, R. D., and Thurston, E. L., Thin film deposition by the electrospray method for Californium-252 plasma desorption studies of involatile molecules, *Anal. Chem.*, 51, 2036, 1979.

8. Jordan, E. A., Macfarlane, R. D., Martin, C. R., and McNeal, C. J., Ion-containing polymers: A new matrix for ^{252}Cf-plasma desorption mass spectrometry, *Int. J. Mass Spectrom. Ion Phys.*, 53, 345, 1983.

9. Macfarlane, R. D., McNeal, C. J., and Martin, C. R., Mass spectrometric study of ion adsorption on poly(ethylene) terephtalate and polypropylene surfaces, *Anal. Chem.*, 58, 1091, 1986.

10. Jonsson, G. P., Hedin, A. B., Håkansson, P. L., Sundqvist, B. U. R., Säve, G. S., Nielsen, P. F., Roepstorff, P., Johansson, K. E., Kamensky, I., and Lindberg, M. S. L., Plasma desorption mass spectrometry of peptides and proteins adsorbed on nitrocellulose, *Anal. Chem.*, 58, 1084, 1986.

11. Towbin, H., Staehelin, T., and Gordon, J., Electrophoretic transfer of proteins from polyacrylamide gels to nitrocelluose sheets: Procedures and some applications, *Proc. Natl. Acad. Sci. U.S.A.*, 78, 4350, 1979.

12. Sundqvist, B.U.R., personal communication, 1988.

13. Kamensky, I. and Craig, A. G., ^{252}Cf-plasma desorption mass spectrometry: recent advances and applications, *Anal. Instr.*, 16, 71, 1987.

14. Craig, A. G., Engstrøm, A., Bennich, H., and Kamensky, I., Enhancement of molecule ion yields in plasma desorption mass spectrometry, presented at the 35th Am. Soc. Mass Spectrom. Conference on Mass Spectrometry and Allied Topics, Denver, May, 1987.

15. Jonsson, G., Hedin, A., Håkansson, P., Sundquist, B. U. R., Bennich, H., and Roepstorff, P., Compensation of non-normal ejection of large molecular ions in plasma desorption mass spectrometry, *Rapid Commun. Mass Spectrom.*, 3, 190, 1989.

16. Hillenkamp, F., Laser desorption mass spectrometry: mechanisms, techniques and applications, presented at the 11th International Mass Spectrometry Conference, Bordeaux, August 1988.

17. Roepstorff, P. and Sundqvist, B., Plasma desorption mass spectrometry of high molecular weight molecules, in *Mass Spectrometry in Biomedical Research*, Gaskell, S. J., Ed., John Wiley & Sons., Chichester, 1986, 269.

18. Alai, M., Demirev, P., Fenselau, C., and Cotter, R. J., Glutathione as a matrix for plasma desorption mass spectrometry of large peptides, *Anal. Chem.*, 58, 1903, 1986.

19. Nielsen, P. F., Klarskov, K., Højrup, P., and Roepstorff, P., Optimization of sample preparation for plasma desorption mass spectrometry of peptides and proteins, *Biomed. Environ. Mass Spectrom.*, 17, 355, 1988.

20. Nielsen, P. F. and Roepstorff, P., Sample-preparation dependent fragmentation in plasma desorption mass spectrometry of nisin, *Biomed. Environ. Mass Spectrom.*, 17, 137, 1988.

21. Roepstorff, P., Nielsen, P. F., Sundqvist, B. U. R., Håkansson, P. and Jonsson, G., The influence of sample preparation on molecular ion formation in plasma desorption mass spectrometry of peptides and proteins, *Int. J. Mass Spectrom. Ion Proc.*, 78, 229, 1987.

22. Chait, B. T., A study of the fragmentation of singly and multiply charged ions produced by ^{252}Cf fission fragment bombardment of polypeptides bound to nitrocellulose, *Int. J. Mass Spectrom. Ion Proc.*, 78, 237, 1987.

23. Roepstorff, P., Nielsen, P. F., Kamensky, I., Craig, A. G. and Self, R., ^{252}Cf-plasma desorption mass spectrometry of a polycyclic peptide antibiotic nisin, *Biomed. Environ. Mass Spectrom.*, 15, 305, 1988.

24. Silly, L., Fenselau, C., and Cotter, R. J., Aggregation, glycosylation and glutathione: peptide behaviour on and off the foil, in *The Analysis of Peptides and Proteins by Mass Spectrometry*, McNeal, C. J., Ed., John Wiley & Sons, Chichester, 1988, 203.

25. Nielsen, P. F. and Roepstorff, P., Sensitivity as function of sample preparation method in plasma desorption mass spectrometry, in *Proceedings from the Fourth Workshop on Molecular Ion Formation from Organic Solids, IFOS IV*, Benninghoven, A., Ed., J. Wiley & Sons, Chichester, 1989, 87.

26. **Jonsson, G., Hedin, A., Håkansson, P., and Sundqvist, B. U. R.,** Competition between protein-protein interactions and protein-substrate interactions studied by plasma desorption mass spectrometry, *Rapid Commun. Mass Spectrom.*, 2, 154, 1988.

27. **Craig, A. G. and Bennich, H.,** Influence of sample concentration and adsorption time on the yield of biomolecule ions in plasma desorption mass spectrometry, *Anal. Chem.*, 61, 375, 1989.

28. **Klarskov, K. and Roepstorff, P.,** PDMS-analysis of proteins after separation on polyacrylamide gels, in *Proceedings of the Fifth Workshop on Molecular Ion Formation from Organic Solids, IFOS V*, Hedin, A., Sundquist, B. U. R., and Benninghoven, A., Eds., John Wiley & Sons, Chichester, 1990, 69.

29. **Roepstorff, P., Højrup, P., Sundqvist, B. U. R., Jonsson, G., Håkansson, P., Andersen, S. O., and Johansson, K. E.,** Application of plasma desorption mass spectrometry to molecular weight determination of a structure protein from insect cuticle, *Biomed. Environ. Mass Spectrom.*, 13, 689, 1986.

30. **Roepstorff, P., Nielsen, P. F., Klarskov, K., and Højrup, P.,** Optimization and use of nitrocellulose as a matrix for peptide and protein analysis by mass spectrometry, in *The Analyses of Peptides and Proteins by Mass Spectrometry*, McNeal, C. J., Ed., John Wiley & Sons, Chichester, 1988, 55.

31. **Klarskov, K., Højrup, P., Andersen, S. O., and Roepstorff, P.,** Plasma desorption mass spectrometry as an aid in protein sequence determination. Application of the method on a cuticular protein from the migratory locust, *Biochem. J.*, 262, 923, 1989.

32. **Talbo, G., Højrup, P., Andersen, S. O., Nielsen, H. R., and Roepstorff, P.,** Determination of the covalent structure of an N- and C- terminally blocked glycoprotein from endocuticle of *Locusta migratoria*, submitted.

33. **Schindler, P., Van Dorselaar, A., Klarskov, K., and Roepstorff, P.,** manuscript in preparation.

34. **Van den Brande, J. L., Hoogenbrugge, C. M., Beyreuther, K., Roepstorff, P., Jansen, X. J., and Buhl-Offers, S. C.,** Isolation and partical characterization of somatomedin-like peptides from cohn fraction IV of human plasma, *Acta Endocrinol.*, 122, 683, 1990.

35. **Nexø, E., Jørgensen, P., Thim, L., and Roepstorff, P.,** High and low molecular weight epidermal growth factor in rat urine, purification and primary structure, *Biochim. Biophys. Acta*, 1037, 38, 1990.

36. **Chait, B. T. and Field, F. H.,** A rapid sensitive mass spectrometric method for investigating microscale chemical reactions on surface adsorbed peptides and proteins, *Biochim. Biophys. Res. Commun.*, 134, 420, 1986.

37. **Chait, B.T., Chaudhary, T., and Field, F. H.,** Mass spectrometric characterization of microscale enzyme catalyzed reactions on surface bound peptides and proteins, in *Methods in Protein Sequence Analysis 1986*, Walsh, K. A., Ed., Humana Press, Clifton, N. J., 1987, 483.

38. **Nielsen, P. F. and Roepstorff, P.,** Suppression effects in peptide mapping by plasma desorption mass spectrometry, *Biomed. Environ. Mass Spectrom.*, 18, 138, 1989.

39. **Mann, M., Rahbek-Nielsen, H., and Roepstorff, P.,** Practical aspects of calibration and effect on non-protein compounds on spectrum quality in protein analysis by PDMS, in *Proceedings of the Fifth Workshop on Molecular Ion Formation from Organic Solids, IFOS V*, Hedin, A., Sundquist, B. U. R., and Benninghoven, A., Eds., John Wiley & Sons, Chichester, 1990, 47.

40. **Roepstorff, P., Klarskov, K., and Højrup, P.,** Strategy for the use of plasma desorption mass spectrometry in protein sequence analysis, in *Methods in Protein Sequence Analysis, 1988*, Wittman-Liebold, B., Ed., Springer-Verlag, Berlin, 1989, 191.

41. **Klarskov, K., Breddam, K., and Roepstorff, P.,** C-terminal sequence determination of peptides degraded with carboxypeptidases of different specificities and analyzed by [252]Cf plasma desorption mass spectrometry, *Anal. Biochem.*, 180, 28, 1989.

42. **Roepstorff, P., Nielsen, P. F., Klarskov K., and Højrup, P.,** Application of plasma desorption mass spectrometry in protein chemistry, *Biomed. Environ. Mass Spectrom.*, 16, 9, 1988.

43. **Tsarbopoulos, A., Becker, G. W., Occolowitz, J. L., and Jardine, I.,** Peptide and protein mapping by [252]Cf-plasma desorption mass spectrometry, *Anal. Biochem.*, 171, 113, 1988.

44. **Keough, T., Lacey, M. P., Lucas, D. S., Grant, R. A., Smith, L. A., and Jeffries, T. C.,** The limited proteolysis of polypeptides, in *The Analysis of Peptides and Proteins by Mass Spectrometry*, McNeal, C. J., Ed., John Wiley & Sons, Chichester 1988, 115.

45. **Højrup, P., Andersen, S. O., and Roepstorff, P.,** Primary structure of a structural protein from the cuticle of the migratory locust, *Locusta migratoria, Biochem. J.*, 236, 713, 1986.

46. **Mikkelsen, J., Hørup, P., Nielsen, P. F., Roepstorff, P., and Knudsen, J.,** Amino acid sequence of acyl-CoA-binding protein from cow liver, *Biochem. J.*, 245, 857, 1987.

47. **Højrup, P., Gerola, P., Mikkelsen, J., Hansen, H. F., El-Shahed, A., Knudsen, J., Roepstorff, P., and Olsson, J. M.,** The primary structure of the bacteriochlorophyll C-binding protein from chlorosomes of *Chlorobium limicola f. thiosulfatophilum*, submitted.

48. **Nielsen, P.F. and Roepstorff, P.,** presented at the 11th International Mass Spectrometry Conference, Bordeaux, August 1988.

49. **Nielsen, P. F., Roepstorff, P., Clausen, I. G., Jensen, A. B., Jonassen, I., Svendsen, A., Balschmidt, P., and Hansen, F. B.,** Plasma desorption mass spectrometry, an analytical tool in protein enginering: characterization of modified insulins, *Protein Eng.*, 2, 449, 1987.

50. **Højrup, P., Knudsen, J., Olsson, J. M., Ødum, L., and Roepstorff, P.,** Use of peptide mapping by [252]Cf-plasma desorption mass spectrometry in protein structure elucidation, in *Advances in Mass Spectrometry,* Vol. 11, Longevialle, P., Ed., Heyden & Sons, 1989, 1388.

Chapter 5

LASER-INDUCED MULTIPHOTON IONIZATION OF PEPTIDES IN SUPERSONIC BEAM/MASS SPECTROMETRY

David M. Lubman

TABLE OF CONTENTS

I.   Introduction 88

II.   High-Pressure Fluid Injection 89
      A.   High-Pressure Fluid Experimental Setup 90
      B.   High-Pressure Ammonia Injection 92
      C.   High-Pressure Liquid Injection 93

III.   Pulsed Laser Desorption..95
      A.   Experimental Description of LD/MPI Method 95
      B.   Laser Desorption/MPI: Results 99

Acknowledgments 103

References 105

I. INTRODUCTION

Multiphoton ionization (MPI) has been shown to be an ionization method with unique capabilities for mass spectrometry (MS).[1-19] MPI occurs when ionization follows the absorption of several photons by a molecule upon irradiation with an intense visible or ultraviolet (UV) light source. When the laser frequency is tuned to a real intermediate electronic state, resonance enhanced MPI (REMPI) occurs and the cross-section for ionization increases dramatically. When the laser wavelength is not tuned to a real electronic state, the probability for MPI becomes small. Thus, although ions are produced as the final product for detection in MS, the ionization cross-section reflects the excitation spectrum of the intermediate state. The truly unique property of MPI is that one can use specific optical excitation to ionize selectively an analyte species in a background matrix, prior to mass analysis.

The MPI method that has found the most extensive application in analysis and which is discussed here is resonant two-photon ionization (R2PI) (see Figure 1). In this process, one photon excites a molecule to an excited electronic state (i.e., $S_0 \rightarrow S_1$) and a second photon ionizes the molecule. The sum of the two photon energies must exceed the ionization potential (IP) of the molecule in R2PI, although the two photons may have either the same (one-color) or different (two-color) frequencies. Because most organic species have IPs between 7 and 13 eV, R2PI is produced using near-UV pulsed laser sources. In contrast, direct photoionization, which has been used as an ionization source for MS, requires vacuum ultraviolet (VUV) radiation (<100 nm) to induce ionization. In R2PI, the presence of an intermediate state allows for spectroscopic selectivity not available in direct photoionization. In addition, the use of high-powered near UV lasers provides for high efficiency in the two photon process, whereas VUV lamps are noncoherent light sources, which produce low-ion yields.

R2PI can serve as a versatile ionization source for MS, in which either soft ionization or extensive fragmentation is produced for identification and/or structural analysis. In many cases, R2PI can provide very efficient soft ionization of molecules yielding the molecular ion with little or no fragmentation. Soft ionization occurs at modest laser energies (<10^6 W/cm^2) for a wide range of organic species[3,8-12,20-22] and, as described in this chapter, can produce strong molecular ion signals for relatively fragile and thermally labile biological molecules. Some systems, such as $Fe(CO)_5$, photodissociate by absorption of the first photon;[2] however, these systems appear to be the exception rather than the rule. Although soft or hard ionization can also be produced by electron ionization (EI), soft ionization occurs at low-beam energies resulting in a loss of ionization efficiency of several orders of magnitude. R2PI can produce soft ionization with efficiencies often approaching several percent of the molecules in the laser beam, while the beam is on.[6-8]

Extensive fragmentation can be produced via MPI by increasing the laser power or by varying the wavelength chosen for ionization.[4-6,8,9] In the process described in this work, R2PI occurs initially to produce the molecular ion M^+. In turn, the M^+ absorbs additional photons that have sufficient energy to cause ionic fragmentation.[3] For example, in previous experiments using 266 nm to ionize benzene, the laser power density was varied to produce fragments including $C_6H_n^+$, $C_5H_n^+$, $C_4H_n^+$, $C_3H_n^+$, $C_2H_n^+$, and CH_n^+, whose relative intensity also changed as a function of laser power.[4,5] The fragmentation produced by MPI is similar to that observed in EI because both involve fragmentation of the radical cation, M^+;[14,19] the fragmentation is very different than that produced in chemical ionization (CI), collision induced dissociation (CID), or particle-induced desorption methods where MH^+ is formed. It should be noted that fragmentation in R2PI is produced by light ~300 nm in wavelength as opposed to EI at 70 eV, which corresponds to radiation of ~0.15 nm wavelength. The result is that, although the fragmentation produced may be similar for R2PI/MPI and EI, differences arise from the different selection rules for the two processes.

The R2PI method is used to best effect in combination with supersonic jet introduction.[20,21]

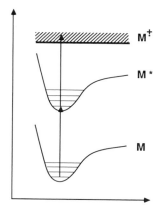

FIGURE 1.   Resonant two-photon ionization process.

In this technique, the analyte molecule is seeded into a monatomic carrier gas at typically 1 atm and expanded through a pinhole orifice into vacuum at $< 10^{-5}$ Torr. In the expansion, the internal energy of the analyte molecules is relaxed through two-body collisions with the carrier gas. The internal energy of the molecules is converted ultimately into the translational energy of the carrier, resulting in ultracold molecules. At room temperature, large polyatomic molecules exhibit broad unresolvable absorption spectra. This phenomenon is due to the thermal population of a large manifold of rovibronic states in the ground electronic state, which results in spectral congestion in the electronic transition. However, in a jet expansion the cold molecules that result exhibit sharp features in R2PI spectroscopy. These sharp spectral absorptions can be used for selective ionization and identification. In addition, the narrowing of the translational energy distribution that occurs in the jet provides enhanced resolution in a time-of-flight (TOF) mass spectrometer.[22,23] The R2PI method is particularly compatible even with high density jets because the near UV wavelengths are transparent to the expansion gas and only interact with the analyte species.

II. HIGH-PRESSURE FLUID INJECTION

The R2PI and supersonic jet methods have been applied quite extensively to molecules that are heated easily into the gas phase. However, only recently has methodology been developed to extend these techniques to nonvolatile and thermally labile biomolecules. One strategy involves dissolving the nonvolatile molecules into a high-pressure fluid, which then transports the sample into vacuum in a jet expansion. In previous work, supercritical fluids of CO_2 and N_2O at several hundred atmospheres have been used as a means of solubilizing nonvolatile species for introduction into a mass spectrometer.[24] These high-density supercritical fluids have the advantage that they can dissolve molecules with low volatility at modest temperatures; however, they convert into a beam of molecules upon introduction into a vacuum. These nonpolar fluids dissolve nonpolar molecules very effectively. However, the problem here is that these fluids exhibit very low bulk solubility for highly polar biological molecules, such as peptides.[25]

An alternate method of injecting polar molecules into a mass spectrometer involves using direct high-pressure liquid injection (DLI).[26] The use of liquid injection avoids decomposition of labile molecules at elevated temperature or the necessity of derivatization for gas-phase introduction into vacuum. The drawback of DLI is that the large amount of liquid injected into a MS produces pressure conditions that are conductive to soft ionization methods (CI or IE) but not to electron impact (EI) ionization. The result is that fairly simple mass spectra are obtained, which are dominated by the protonated molecular ion. However, these soft ionization mass spectra are often inadequate for structural analysis and identification. In addition, at low temperature extensive formation of solvent cluster ions is observed below m/z 100. This background will overwhelm many analytes or important fragments in this mass region. The use of elevated temperature will dissociate these clusters, but may also result in thermal decomposition of the analyte.

In this section, we discuss a method utilizing a combination of high-pressure pulsed liquid injection followed by R2PI as a method of introducing and ionizing small biological molecules to MS with great versatility. Because R2PI depends upon absorption by a resonant electronic state, one can selectively ionize the analyte on the basis of wavelength of the laser source while discriminating against the liquid solvent. The solvent remains essentially transparent to the laser radiation, so that a separation of analyte and liquid carrier can be achieved by the use of laser selectivity without the need for actual physical separation. The result is that R2PI/MPI can be used to produce selectively either soft ionization or extensive fragmentation of the analyte without interference by the solvent background. The fragmentation in turn can be used for identification and structural analysis of the analyte.

The unique feature of the high-pressure liquid method described here is that it utilizes a pulsed injection source. Because a pulsed laser source is used to produce R2PI, a pulsed injection source is used to optimize the amount of sample interrogated by the laser pulse over that time possible with the use of continuous injection methods. The use of a pulsed source also reduces the duty cycle required to pump the liquid jet so that a larger orifice can be used with a resulting increase in the on-axis density and thus the sensitivity of the technique. Most significantly, the use of pulsed injection allows one to take advantage of the kinetics of the liquid pulse formation to minimize clustering. Under the conditions in the experiments described here, high-pressure CH_3OH, NH_3, and H_2O are injected into vacuum at relatively low temperature without significant clustering.

A. HIGH-PRESSURE FLUID EXPERIMENTAL SETUP

The experimental setup for high-pressure fluid injection consists of a differentially pumped vacuum system with a TOF mass spectrometer (TOFMS) mounted vertically on top of the chamber (see Figure 2). The laser beam enters the chamber through a quartz window, and R2PI is produced in the acceleration region of the TOFMS. The high-pressure fluid is introduced with the use of a pulsed solenoid valve in the first chamber, which is pumped by a 4-in. diffusion pump and cryobaffle. The main load of pumping the liquid jet is performed with four liquid nitrogen (LN_2) traps, which extend from the top of the chamber into vacuum. Because CH_3OH and H_2O are condensable at liquid nitrogen temperatures, the pressure in the first chamber does not rise above 1×10^{-4} Torr, even with a backpressure of 300 atm. The second chamber is pumped by a 6-in diffusion pump and cryobaffle, and is separated from the first chamber by a liquid-N_2-cooled partition. Attached to this partition is a skimmer, which was designed with a sharp edge to slice the beam and a shape to deflect away molecules that are not transmitted through the orifice. The skimmer uses a large orifice (0.95 cm) to allow efficient transmission of the beam into the TOF acceleration region. The pressure in the second chamber never rises above 2×10^{-5} Torr.

The key to this work is the development of a pulsed injection source for high-pressure fluid injection. This valve consists of a custom-designed solenoid pulser with a metal-to-metal seal.[27] The valve delivers pulses of ~200 μs FWHM at 10 Hz, so that the duty cycle is reduced by a factor of ~500. The pulsed injection mode allows the use of a 200-μm orifice at the methanol back-pressure of 200 atm with a resulting flow rate of ~1.5 ml/min. In a continuous flow, a 10-μm orifice would be required in order to handle the pumping load. Because the on-axis density increases as D^2, where D = diameter of the orifice, a significant increase in sensitivity should be possible with the use of pulsed injection. In addition, the expansion is compacted into pulses that can be effectively probed by the 3-mm-diameter laser beam when the laser and molecular beam

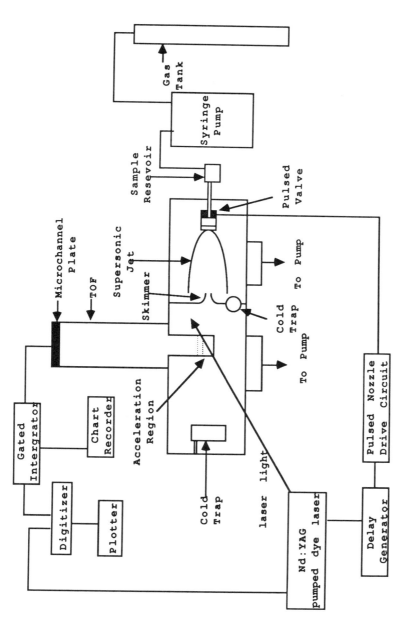

FIGURE 2. Experimental setup for high-pressure fluid injection.

are synchronized correctly. At a jet velocity of 4×10^4 cm/s, it is estimated that ~4% of the pulse on-axis is probed by the laser as compared with $7.5 \times 10^{-3}\%$ in a continuous flow.

The ionization source is a Quanta-Ray frequency-doubled Nd:YAG pumped dye laser, which produces tunable radiation in the near ultraviolet. In addition, the fourth harmonic of the Nd:YAG laser (226 nm) was also used for single-wavelength ionization. The typical laser power used was ~2×10^6 W/cm^2. A positive-negative lens combination (positive lens, 30-cm focal length; negative lens, 10-cm focal length) was used to collimate the laser into a 2- to 3-mm beam. The delay between the pulsed valve and the laser source was optimized with the use of a delay generator at ~600 μs. The mass spectra were recorded with a LeCroy 9400 transient waveform recorder, and all spectra were averaged over 100 laser pulses.

The analytes used in these experiments were obtained from Sigma Chemical Co. (St. Louis, MO) and used without further purification. The methanol was obtained from Burdick and Jackson (Muskegon, MI) and was specified with a UV cutoff of 203 nm. The water solvent was distilled water and the NH$_3$ was obtained from Scott Specialty Gases (Plumsteadville, PA). The high-pressure fluid was provided by a Varian 8500 syringe pump. In order to obtain mass spectra, the solid sample was dissolved in methanol and placed into a high-pressure stainless steel reservoir between the pulsed valve and the syringe pump. Pure methanol delivered at 200 atm pressure transported the sample to the nozzle for introduction. In all experiments, the nozzle and tubing were heated to between 70° and 80°C for methanol injection and ~140°C for water and NH$_3$ injection.

B. HIGH-PRESSURE AMMONIA INJECTION

The use of high-pressure liquid NH$_3$ as a solvent has been shown to provide a promising sample introduction method for polar biologicals into a mass spectrometer under relatively mild conditions. Figure 3 contains an R2PI mass spectrum of adenosine introduced into a TOFMS using pulsed liquid NH$_3$ injection at 120 atm and 150°C. R2PI at 266 nm produces the molecular ion, $M^{+\cdot}$, with relatively high abundance compared to other easily formed fragments. In this experiment, liquid NH$_3$ was used to initially dissolve the sample, and the liquid was converted subsequently to supercritical conditions upon injection into vacuum. Soft ionization is often difficult to achieve for the nucleosides even with relatively soft desorption/ionization methods such as CI, fast atom bombardment (FAB), and field desorption (FD) MS. The salient point here is that liquid NH$_3$ extraction offers a means to inject relatively labile polar molecules into a MS without extensive decomposition prior to ionization.

Figures 4A and B show R2PI mass spectra of the dipeptide isomers Leu-Tyr and Tyr-Leu injected into the TOFMS using high-pressure liquid NH$_3$ (150 atm). The pulsed nozzle was held at 150°C in order to produce a supercritical fluid expansion with subsequent R2PI at 266 nm. In these experiments, the power was raised to 5×10^6 W/cm^2 to increase the observed signal. In Figure 4, the molecular ion is observed as well as several fragments distinctly characteristic of each isomer. These fragments are characteristic of MPI and are induced by the increased laser power required to enhance the signal because the solubility of these compounds in liquid NH$_3$ may be the limiting factor here. The fragments are not those characteristic of thermal decomposition, which might be otherwise produced due to the high temperature in the valve.[28] The short residence time in the valve allows injection under supercritical fluid conditions without significant decomposition. The result is that R2PI/MPI of small peptides can be studied for identification and structural analysis by liquid NH$_3$ injection without significant interference of the solvent.

A number of unique features are observed in the spectra of Figures 3 and 4. In the mass spectrum of adenosine, a molecular ion $M^{+\cdot}$ at m/z 267 and base ion, BH$^+$, at m/z 135 both appear. In addition, a peak at m/z 18 is observed at NH$_4^+$. It has been shown that NH$_3$ does not ionize directly upon irradiation at 266 nm at the power densities used in these experiments. The ammonia peak is due to charge exchange, and its height varies as a function of the background.

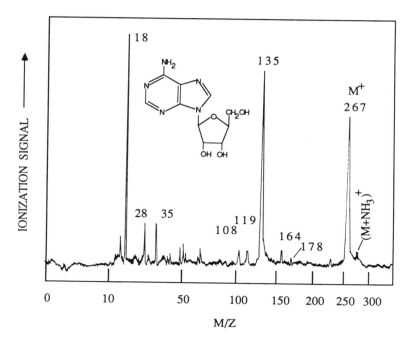

FIGURE 3. The R2PI mass spectrum of adenosine obtained by supercritical NH_3 expansion at 120 atm and 150°C. (From Pang, H. M. and Lubman, D. M. , *Rev. Sci. Instr.*, 59, 2460, 1988. With permission.)

Ammonia is present in the TOF in a concentration overwhelming the analyte, yet it appears as a relatively small peak in the mass spectrum. This relationship is due to the low residence time and relatively low pressure in the TOFMS so that the probability for charge exchange is minimal.

A second important feature is the relatively low amount of clusters formed due to solute-solvent clustering. Enhanced cluster formation might be expected due to the strong tendency for hydrogen bonding to occur between NH_3 and the highly polar solute compounds. The low degree of clustering is observed when the laser pulse is timed to intercept the rising edge of the molecular beam pulse. As the laser pulse probes further into the molecular beam pulse, the degree of clustering becomes much more extensive.[27] This relationship appears to be due to a kinetic effect, where at the beginning of the molecular beam pulse, insufficient time exists for three-body collisions to occur in the high density region of the expansion to produce clustering. Thus, by timing our laser probe pulse accordingly, the clustering effect can be minimized.

C. HIGH-PRESSURE LIQUID INJECTION

Liquid solvents such as methanol and water are often excellent solvents for polar molecules. However, the temperature at which these liquids convert to supercritical conditions [T_c (methanol) = 240.5, T_c, (water) = 374.4] is too high to be of practical use for jet injection of thermally labile compounds. DLI or thermospray methods can be used to inject liquid into a mass spectrometer; however, temperatures in excess of 200°C may be needed to prevent extensive cluster formation upon expansion into vacuum.[26] Nevertheless, using high-pressure pulsed liquid injection in combination with R2PI, labile molecules dissolved in methanol solvent can be introduced into a MS with minimal clustering observed.

Figure 5 and Table 1 show results for injection of various small polar molecules into a TOFMS using high-pressure methanol injection (200 atm) at 80°C and R2PI detection at 280 nm.[29] In each case the molecular ion $M^{+\cdot}$ is observed, and ion fragments can be produced by increasing the laser power. The methanol solvent remains transparent to the laser radiation at 280

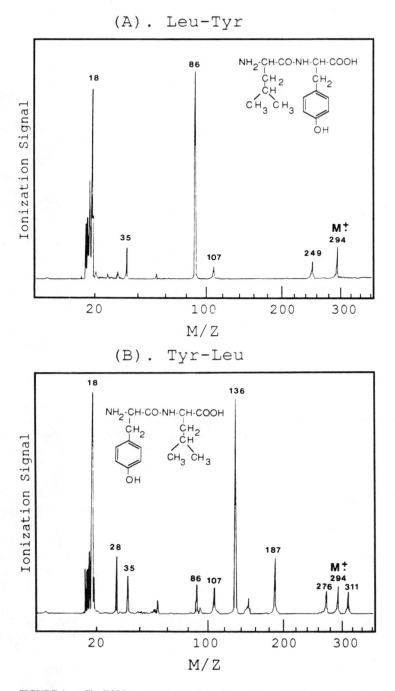

FIGURE 4.   The R2PI mass spectrum of (A) Leu-Tyr and (B) Tyr-Leu obtained by supercritical NH₃ expansion at 150 atm and 150°C.

nm so that either soft or hard ionization can be produced without interference from the solvent. It would be difficult otherwise to use EI or other fragmentation methods under the relatively high-liquid densities introduced into the mass spectrometer. Also, minimal solute-solvent cluster ions are observed, even at the relatively low temperature used in these experiments. This lack of clustering is observed provided the laser is timed to probe only the front edge of the molecular beam pulse. As in the high-pressure NH_3 injection, significant clustering is observed as the laser probes further into the pulse where clusters have had sufficient time to form. Thus, relatively low-temperature liquid methanol (80°C) or water (140°C) injection can be used to introduce labile compounds into a jet expansion for R2PI detection by taking advantage of the unique capabilities of high-pressure pulsed detection.

One of the key properties of R2PI is the high efficiency for soft ionization. We therefore undertook quantitation measurements in order to determine the limits of sensitivity of this method. A range of solution concentrations in methanol was prepared over three orders of magnitude for tryptamine and 4-hydroxy-3methoxy mandelic acid (HMA) according to the successive dilution technique. Each solution was placed in the syringe pump cylinder, injected into the system, and detected by R2PI at 266 nm at a laser power of 2×10^6 W/cm². A resulting log-log plot of the ionization signal as a function of concentration for HMA is shown in Figure 6. A linear response is obtained over almost three orders of magnitude. The detection limit was found to be ~2 ppm for tryptamine, ~50 ppb for tyramine, and ~2 ppm for HMA, all three with an S/N of 2. The solution flow-rate through the nozzle was found to be 10.0 ml/3.0 min or ~5.6 μl/pulse at a 10-Hz repetition rate. In the case of tryptamine, the detection limit of 2 ppm is equivalent to 2 ng/μl. Because at least 10 laser pulses are needed to obtain an S/N of 2, the minimum detection limit is ~110 ng. In the case of tyramine, the detection limit is ~2.8 ng. The detection limit could be increased significantly by performing R2PI as close to the nozzle orifice as possible. The on-axis density in the jet decreased as $1/x^2$ where x = distance from the orifice to the ionization region, which is 22 cm in these experiments due to the constraints of our experimental apparatus.

III. PULSED LASER DESORPTION

A second method used for volatilizing labile biomolecules into supersonic jet expansions for detection by MPI is pulsed laser desorption (LD).[13-19] The LD technique involves using a pulsed laser to induce a rapid heating effect, in which molecules desorb from a surface before they have time to decompose. In the desorption process, ions and neutrals are both formed in a ratio that depends to a first approximation on the surface temperature induced by the laser. Most investigators to date have focused their attention on the small percentages of ions produced during the desorption event. However, in our work a two-step process involving pulsed laser desorption of neutral molecules followed by R2PI of the neutral plume has been used.[13-19] This two-step method has several distinct advantages including:

1. Enhanced sensitivity because the bulk of the species desorbed are neutral
2. The possibility of selectivity in the ionization process based upon R2PI
3. Controlled fragmentation, where either soft or hard ionization is possible based upon MPI
4. Lack of cationization, which is ubiquitous in other desorption methods

In this section, we describe the application of LD/MPI to the detection of small peptides. In these experiments, pulsed laser desorption produces a neutral plume, which is then entrained into a supersonic jet expansion. The jet transports the material into a TOFMS where laser-induced MPI ionizes the molecules. The capability of R2PI for producing soft ionization in labile peptides as a function of power and wavelength is explored. It is demonstrated that either molecular ions or minor fragmentation only can be produced at modest laser energies under the proper conditions. In addition, the ability to obtain wavelength selectivity based on the ultracooling effect of the supersonic jet expansion is explored.

A. EXPERIMENTAL DESCRIPTION OF LD/MPI METHOD

The supersonic beam TOFMS setup is similar to that described for pulsed liquid injection, but the specific features are different because the LD/MPI experiment has different experimental

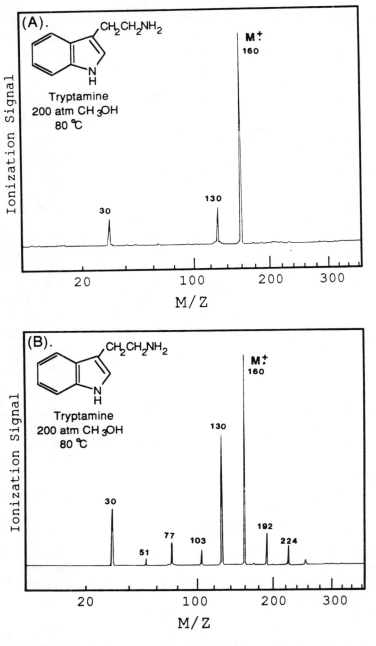

FIGURE 5.   R2PI mass spectra of tryptamine with the use of methanol injection at 200 atm backpressure at 80°C with the use of (A) 280 nm radiation at 8×10^5 W/cm^2, (B) 266 nm radiation at 2×10^6 W/cm^2. (From Pang, H. M., Sin, C. H., and Lubman, D. M., *Appl. Spec.*, 42, 1200, 1988. With permission.)

requirements. The system consists of a TOFMS mounted vertically in a stainless steel six-port source cross-pumped by a 6-in. diffusion pump (see Figure 7). The pulsed supersonic molecular beam expands into the acceleration region of the TOF, and a laser beam perpendicular to both the supersonic jet and flight-tube ionizes the sample. The acceleration region and TOF flight-tube are enclosed in a liquid-N$_2$-cooled cryoshield, and are differentially pumped by a 4-in. diffusion pump. In this configuration, a background pressure of ~10^{-7} Torr is obtained in the

TABLE 1
R2PI of Biological Indole-Based Compounds Injected by Liquid Methanol

Compounds	Mol. wt.	Solvent	Pressure (atm)	Nozzle temp.	m/z	Relative intensity[a]	
						266 nm	280 nm
Tryptamine	160	CH₃OH	200	80°C	160	100	100
					130	54	15
					103	7	—
					77	8	—
					51	3	—
					30	20	7
Serotonin	176	CH₃OH	200	80°C	176	100	100
					146	37	18
					63	6	—
					39	8	—
					30	16	—
Harmaline	214	CH₃OH	200	90°C	214		
Melatonin	232	CH₃OH	100	80°C	232	100	100
					174	25	9
					160	31	7
Indole-3-acetic acid	176	CH₃OH	200	80°C	175	100	100
					130	26	22
Tryptophan	204	H₂O	300	140°C	204		

[a]   Laser power: 266 nm, 2×10^6 W/cm²; 280 nm, 2×10^6 W/cm².

From Pang, H. M., Sin, C. H., and Lubman, D. M., *Appl. Spec.,* 42, 1201, 1988. With permission.

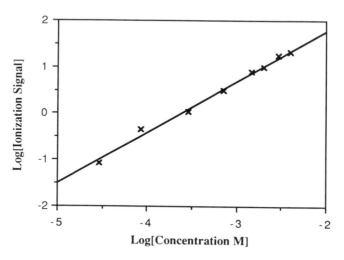

FIGURE 6.   The log ionization signal vs. log (concentration M) plot of HMA with the use of 266 nm radiation with 2×10^6 W/cm² laser power using 200 atm methanol injection at 80°C. (From Pang, H. M., Sin, C. H., and Lubman, D. M., *Appl. Spec.,* 42, 1200, 1988. With permission.)

region, and the laser R2PI-induced background due to pump oil and other contaminants becomes negligible due to the presence of the cryotrap. The supersonic beam source is a stainless steel pulsed valve (0.5-mm aperture diameter), based on the magnetic repulsion principle (R. M. Jordan, Co., Grass Valley, CA), which can provide gas pulses of ~55 μs FWHM at "choked flow" at a 10-Hz pulse rate, thus reducing substantially the duty cycle needed for pumping. In

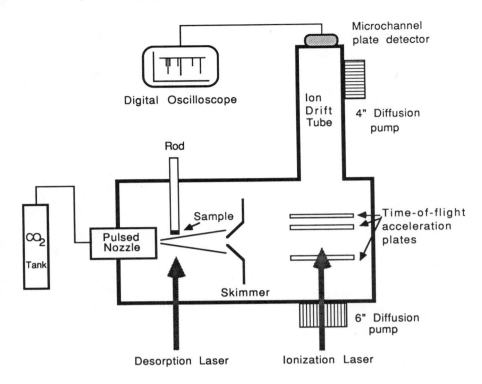

FIGURE 7.   Laser desorption/laser multiphoton ionization setup. (From Li, L. and Lubman, D. M., *Rev. Sci. Instr., 59, 558, 1988. With permission.)*

these experiments, in which CO_2 is used as the carrier gas, the average background pressure remains at $\sim 1 \times 10^{-6}$ Torr, because CO_2 is cryopumped efficiently by the LN_2 trap.

Sample introduction was performed by desorption of the compound of interest from the surface of an $1/8$ in. diameter rod made of machinable Macor® ceramic (Corning Glass Works, Corning, NY), which was found to provide minimum decomposition for even the most labile compounds studied. An important aspect of these experiments is laser shot-to-shot desorption stability, which is achieved by dissolving the material of interest in a high-viscosity fluid such as glycerol or silicone diffusion pump fluid. Generally, ~ 100 μg of sample is used in these experiments, and the glycerol matrix causes the sample to form a very even, thin layer from which pulse-to-pulse stability appears quite good, i.e., somewhat less than $\pm 5\%$. The rate of desorption can then be controlled by adjusting the laser power on the surface so that total desorption can be produced in seconds or over an extended length of time, i.e., more than 0.5 h. In addition, the presence of a glycerol matrix serves as a heat sink, which prevents thermal decomposition in the desorption process.[15]

A critical parameter for obtaining optimal rovibronic cooling is the position of the ceramic rod with respect to the jet expansion. Two dimensions must be considered here (1) the distance of the rod from the molecular beam axis and (2) the distance of the rod from the nozzle orifice. In the first case, the best cooling was observed when the rod was as close to the supersonic beam as possible without causing destruction of the jet. In our work, we achieved this goal by including a volatile sample such as aniline in the jet and moving the rod toward the beam until the molecular ion peak intensity started to decrease. As the rod was moved further, a dramatic decrease generally occurred as the rod began to interfere with the jet. In the second case, the distance between the center of the rod and the nozzle in our experiments was ~ 5.5 mm. This distance provided excellent cooling; if this distance was decreased, we observed a significant decrease in signal with no noticeable increase in cooling. This response appeared to occur because at shorter distances, where the carrier density was very high, it was difficult to penetrate efficiently into the beam without causing shock waves. At a longer expansion distance, where the carrier density was lower, sample penetration appears to be much more efficient.[15] The use of CO_2 vs. Ar or He provides an increased number of collisions at longer expansion distance for enhanced collisional cooling.[30]

The desorption was performed with a pulsed CO_2 laser at 10.6 μm (Quanta-Ray EXC-1, Spectraphysics, Mountain View, CA). The infrared (IR) beam was focused softly with a 10-cm focal-length biconvex germanium lens to a ~2- to 3-mm spot for desorption, although we adjusted the focus for each sample molecule to optimize our results. The amount of sample desorbed per pulse depends on the power density on the surface and on the properties of the sample — in particular, the melting point. An estimate of the desorption power density on the surface can be made by measuring the beam energy using a power meter (Scientech Model 365, Boulder, CO) and by referring to the known beam temporal profile. Typically, a power density of ~5×10^6 W/cm² was used for desorption. By careful control of the desorption laser power, the sample can be made to last for an extended period of time for spectroscopic scans. In several spectroscopic scans in our work, a 100-μg sample has lasted for ~0.5 h at a repetition rate of 10Hz. This desorption rate is equivalent to ~3×10^{13} molecules or ~0.1 monolayers desorbed per CO_2 laser pulse.[15]

The laser ionization (R2PI) was performed with the use of the doubled dye output from a Quanta-Ray DCR-2A Nd:YAG pumped dye laser system. The 6-mm output beam was collimated with a combination 30-cm focal-length positive lens and a 10-cm focal-length negative lens to produce a laser beam 2 to 3 mm in diameter. We adjusted the power density to obtain the desired signal level, which depended upon the amount of material entrained in the jet and the efficiency of ionization of each particular compound. The nozzle-to-excitation distance was generally ~16 cm.

The actual sequence of events was controlled by several delay generators, by which the pulsed CO_2 laser fires first, to produce desorption, and is subsequently followed by the pulsing of the valve. The two events were synchronized, so that the desorbed plume was entrapped into the jet expansion of the CO_2 and carried into the acceleration region of the TOFMS. The flight time of the jet from the pulsed valve to this region was ~300 μs, and the laser was therefore set to pulse as the gas pulse arrived at this point. Laser R2PI was produced, and a LeCroy 9400 digital oscilloscope was used to record the mass spectrum. The wavelength spectrum was obtained by the use of a SRS 250 gated integrator to monitor only the molecular ion as the dye laser was scanned.

B. LASER DESORPTION/MPI: RESULTS

In our work, pulsed laser desorption has been used to desorb neutral small peptides into the gas phase. The resulting plume is entrained into a supersonic jet that transports the molecules into a TOFMS, where R2PI is produced by a UV laser source. R2PI is readily produced in peptides containing an aromatic group, which can absorb the laser radiation used at either 280 or 266 nm through the p-p* transition, i.e., tyrosine-, tryptophan-, and phenylalanine-containing compounds. In this section, results specifically pertaining to tyrosine-containing peptides are described.

Table 2 shows the mass spectra of various tyrosine-containing dipeptides produced using the LD/MPI technique.[31] The laser-induced mass spectra of the isomers of Ala-Tyr and Tyr-Ala are shown in Figures 8A and B, and wavelength ionization spectra obtained by monitoring the molecular ion as a function of wavelength of the dipeptides are shown in Figures 9A and B.[31] A key feature of the R2PI method is that soft ionization with production of $M^{+\cdot}$ is observed in almost every case at modest laser intensities (P = 5×10^5 W/cm²). Each of the mass spectra in Table 2 was obtained by optimizing the jet expansion cooling following desorption and entrainment. The frequency of the doubled dye laser output is then tuned to the optimal

TABLE 2
R2PI/MPI Mass Spectra of Tyrosine-Containing Dipeptides

Compound	Mol wt	Wavelength (nm)	Major fragments [m/z (rel. int. %)]	
			Soft[a]	Hard[b]
Tyr-Gly	238	281.40	238 (100)	238 (100), 221 (7), 147 (5), 136 (62), 131 (45), 107 (13), 85 (9), 74 (6), 57 (17), 30 (54)
Gly-Tyr	238	280.70	238 (100)	238 (87), 221 (6), 180 (5), 165 (35), 131 (7), 107 (41), 74 (12), 73 (65), 47 (7), 30 (100)
Tyr-Ala	252	281.60	252 (100)	252 (100), 145 (38), 136 (67), 107 (13), 99 (8), 44 (52)
Ala-Tyr	252	280.40	252 (100)	252 (97), 164 (16), 107 (17), 89 (32), 44 (100)
Tyr-Val	280	281.50	280 (100)	280 (66), 173 (30), 156 (11), 147 (7), 136 (100), 127 (10), 116 (5), 107 (20), 85 (18), 72 (70)
Val-Tyr	280	280.60	280 (100) 72 (100)	280 (27), 107 (5), 72 (100)
Tyr-Leu	294	281.25	294 (100)	294 (100), 277 (5), 187 (41), 136 (52), 107 (5), 86 (100), 43 (5)
Leu-Tyr	294	280.80	294 (100) 86 (35)	294 (64), 132 (7), 107 (5), 86 (100), 43 (5)
Ile-Tyr	294	280.80	294 (100)	294 (47), 164 (5), 132 (6), 107 (9), 86 (100)
Pro-Tyr	278	280.60	278 (100)	278 (45), 70 (100)

[a] $P \approx 5 \times 10^5$ W/cm^2.
[b] $P \approx 4 \times 10^6$ W/cm^2.

From Li, L. and Lubman, D. M., *Appl. Spec.*, 43, 543, 1989. With permission.

absorption in the spectral contour as in Figure 9, and wavelength-specific ionization is obtained for each dipeptide. By optimizing the jet entrainment, and cooling and exciting at the specific optimal wavelength for absorption, the signal is maximized and relatively low laser power can be used for ionization. Soft ionization, in which only the molecular ion is observed, can be obtained in almost every case under these conditions at a laser energy of 250 μJ ($P = 5 \times 10^5$ W/cm^2). In earlier work, where only incomplete cooling was obtained and nonspecific wavelength ionization was used, a laser energy of about 5 mJ, or nearly 20 times that of the present studies, was required for efficient ionization. Under these conditions, it was difficult to prevent fragmentation. By optimizing the jet-cooling and exciting on-resonance in these present studies, the partition function for absorption is enhanced so that a relatively low ionization laser power density can be used, resulting in minimal fragmentation.

As the laser power is increased, extensive fragmentation can be produced in these tyrosine-containing dipeptides for structural studies, as shown in Table 2.[31] This fragmentation is induced at the given wavelength, which corresponds to the peak of the dipeptide absorption at a laser

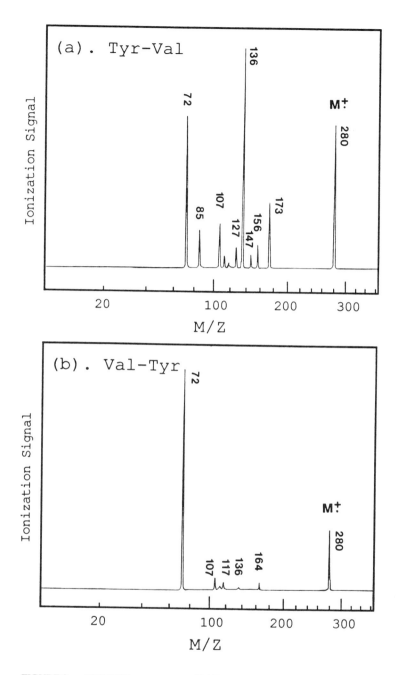

FIGURE 8.   R2PI/MPI mass spectra of (a) Tyr-Val at 280.50 nm and (b) Val-Tyr at 280.60 nm ($P = 4 \times 10^6$ W/cm^2) using laser desorption into a supersonic jet.

energy of 2 mJ ($P = 4 \times 10^6$ W/cm^2). This laser power is relatively low compared to that used in other studies for inducing fragmentation, and is a result of obtaining efficient jet-cooling in the expansion followed by excitation at the optimal absorption. It should be noted that M$^{+\cdot}$ is still present in each case, although it may or may not be the dominant ion peak, depending on the laser power. Also, isomeric dipeptides exhibit different fragmentation by which they may be uniquely identified. For example, mass spectra of Tyr-Val and Val-Tyr are compared in Figure 8.[31] In both

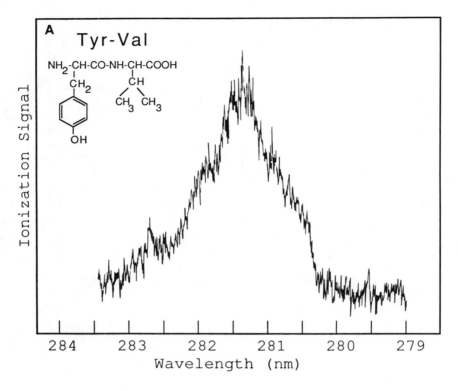

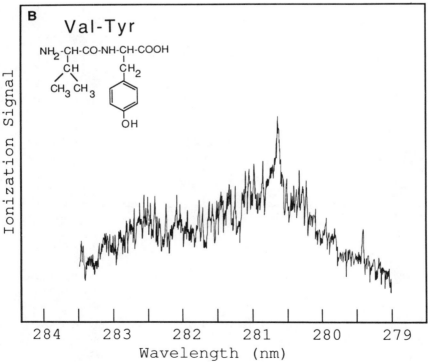

FIGURE 9.   Supersonic jet-cooled R2PI spectra of (A) Tyr-Val and (B) Val-Tyr obtained with the use of pulsed laser desorption and subsequent entrainment in a jet expansion of CO_2 carrier. (From Li, L. and Lubman, D. M., *Appl. Spec.*,43, 543, 1989. With permission.)

cases, an abundant molecular ion peak is observed in the mass spectrum. The base peak in each case is the A_1 fragment (in the Roepstorff-Fohlman notation[32]), which has generally been observed to be a dominant fragment product in many of the dipeptides. If the tyrosine is on the N-terminal end such as in Tyr-Val, then the A_1 fragment results in an ion at m/z 136; conversely, if the tyrosine is on the C-terminal end as in Val-Tyr, the fragment ion is at m/z 72. In the Val-Tyr, the other cleavages result in only minor fragments at the given laser energy. In Try-Val, simple β-cleavage in the tyrosine moiety results in a m/z 107, which corresponds to a hydroxybenzyl ion and m/z 173, which corresponds to the dipeptide minus this 107 species. A strong m/z 72 ion arises from Y_1 cleavage followed by loss of the COOH group with an accompanying H migration.

Such differences in sequence ions are observed in all the isomeric dipeptides upon laser-induced fragmentation by R2PI/MPI. Note also that Leu-Tyr and Ile-Tyr have different fragmentation patterns with the appearance of a m/z 43 peak for Leu-containing peptides but not for Ile-containing peptides. This difference is significant because the presence of Ile or Leu is difficult to establish with present MS sources.[31]

The LD/MPI method has also been extended to larger peptides. This extension is illustrated in Figure 10 for met-enkephalin.[33] Figure 10A contains the MPI-induced spectrum of met-enkephalin at low ionization energy ($P = 1.0 \times 10^6$ W/cm^2, $\lambda = 266$ nm). The molecular ion $M^{+\cdot}$ is observed accompanied by a small number of fragment ions. As the power is raised to high ionization energy ($P = 1.5 \times 10^7$ W/cm^2, $\lambda = 266$ nm) (see Figure 10B), extensive fragmentation is observed with production of many of the classical A, B, C, and X, Y, Z fragment ions. Most notably, cleavage occurs at the –CO–NH– peptide bond to produce a series of acylium ions. In addition, a series of aldimine ions is also produced. Aldimine ions are often difficult to produce with electron impact ionization, and are significant here because they provide the information complementary to that of the acylium ions to reconstruct the sequence of the peptide. Also, abundant ions are observed due to formation of acylimmonium ions (A fragments) due to loss of CO from the corresponding acylium ions. In addition, a series of fragments due to cleavage at every N-C bond is observed, where charge retention can occur on the N or C fragment.

It should be noted that the fragmentation observed in LD/MPI is very different from that observed by FABMS or LDMS. In LD/MPI, the radical cation is produced (as in EI), as opposed to these other methods, in which MH$^+$ is produced. The fragmentation observed from the species produced by these methods is thus different; however, the information obtained will be complementary for structural analysis. One obvious advantage of LD/MPI vs. FAB is that a glycerol matrix is not required for LD/MPI, and if it is used and desorbed by the laser, it will remain transparent. Thus, the problem of interfering background matrix peaks at low mass in FAB mass spectra can be avoided. In addition, LD/MPI requires minimal sample preparation and avoids many of the sample preparation problems of other MS methods.

ACKNOWLEDGMENTS

DML would like to thank Roger Tembreull, Liang Li, Ho Ming Pang, and Chung Hang Sin for their various contributions to this work. We acknowledge financial support of this work under NSF Grants CHE 8419383 and CHE 8720401. We also acknowledge support by NSF Grant DMR 8418095 for acquisition of the Chemistry and Materials Science Laser Spectroscopy Laboratory. DML is a Sloan Foundation Research Fellow.

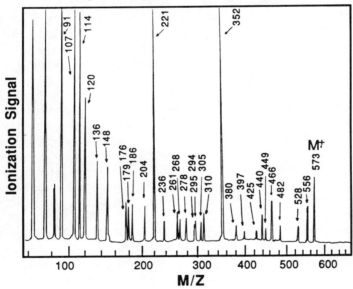

FIGURE 10. LD-MPI ionization-fragmentation pattern at $\lambda = 266$ nm of (A) met-enkephalin at high laser ionization energy (P = 1.5×10^7 W/cm^2) and (B) at low laser energy (P = 1.0×10^6 W/cm^2). (From Li, L. and Lubman, D. M., *Anal. Chem.*, 60, 1409, 1988. Copyright American Chemical Society. With permission.)

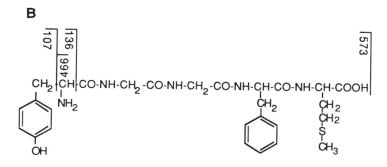

Tyr-Gly-Gly-Phe-Met

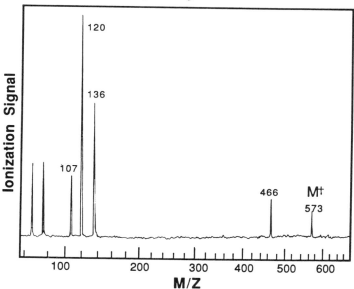

FIGURE 10 (continued).

REFERENCES

1. **Lubman, D. M.,** Optically selective molecular mass spectrometry, *Anal. Chem.,* 59, 31A, 1986.
2. **Duncan, M. A., Dietz, T. G., and Smalley, R. E.,** Efficient multiphoton ionization of metal carbonyls cooled in a pulsed supersonic beam, *Chem. Phys.,* 44, 415, 1979.
3. **Boesl, U., Neusser, H. J., and Schlag, E. W.,** Visible and UV multiphoton ionization and fragmentation of polyatomic molecules, *J. Chem. Phys.,* 72, 4327, 1980.
4. **Zandee, L. and Bernstein, R. B.,** Laser ionization mass spectrometry: extensive fragmentation via resonance-enhanced multiphoton ionization of a molecular benzene beam, *J. Chem. Phys.,* 70, 2574, 1979.
5. **Bernstein, R. B.,** Systematics of multiphoton ionization-fragmentation of polyatomic molecules, *J. Phys. Chem.,* 86, 1178, 1982, and references cited therein.
6. **Lubman, D. M., Naaman, R., and Zare, R. N.,** Multiphoton ionization of azulene and naphthalene, *J. Chem. Phys.,* 72, 3034, 1980.

7. **Dietz, T. G., Duncan, M. A., Liverman, M. G., and Smalley, R. E.,** Efficient multiphoton ionization of jet-cooled aniline, *Chem. Phys. Lett.,* 70, 346, 1980.
8. **Boesl, U., Neusser, H. J., and Schlag, E. W.,** Multiphoton ionization in the mass spectrometry of polyatomic molecules: cross sections, *Chem. Phys.,* 55, 193, 1981.
9. **Lubman, D. M. and Kronick, M. N.,** Mass spectrometry of aromatic molecules with resonance-enhanced multiphoton ionization, *Anal. Chem.,* 54, 660, 1982.
10. **Sack, T. M., McCrery, D. A., and Gross, M. L.,** Gas chromatography multiphoton ionization Fourier transform mass spectrometry, *Anal. Chem.,* 57, 1290, 1985.
11. **Tembreull, R. and Lubman, D. M.,** Use of resonant two-photon ionization with supersonic beam mass spectrometry in the discrimination of cresol isomers, *Anal. Chem.,* 56, 1962, 1984.
12. **Dobson, R. L. M., D'Silva, A. P., Weeks, S. J., and Fassel, V. A.,** Multidimensional, laser-based instrument for the characterization of environmental samples for polycyclic aromatic compounds, *Anal. Chem.,* 58, 2129, 1986.
13. **Tembreull, R. and Lubman, D. M.,** Resonant two-photon ionization of small peptides using pulsed laser desorption in supersonic beam mass spectrometry, *Anal. Chem.,* 59, 1003, 1987.
14. **Tembreull, R. and Lubman, D. M.,** Pulsed laser desorption of biological molecules in supersonic beam mass spectrometry with two-photon ionization detection, *Anal. Chem.,* 59, 1082, 1987.
15. **Li, L. and Lubman, D. M.,** Analytical jet spectroscopy of tyrosine and its analogs using a pulsed laser desorption volatilization method, *Appl. Spec.,* 42, 418, 1988.
16. **Engelke, F., Hahn, J. H., Henke, W., and Zare, R. N.,** Determination of phenylthiohydantoin-amino acids by two-step laser desorption/multiphoton ionization, *Anal. Chem.,* 59, 909, 1987.
17. **Grotemeyer, J., Boesl, U., Walter, K., and Schlag, E. W.,** Biomolecules in the gas phase. II. Multiphoton-ionization mass spectrometry of angiotensin I, *Org. Mass Spectrom.,* 21, 595, 1986.
18. **Grotemeyer, J., Boesl, U., Walter, K., and Schlag, E. W.,** Biomolecules in the gas phase. I. Multiphoton-ionization mass spectrometry of native chlorophylls, *J. Am. Chem. Soc.,* 108, 4233, 1986.
19. **Grotemeyer, J. and Schlag, E. W.,** Multiphoton-ionization mass spectrometry (MUPI-MS), *Angew. Chem.,* 27, 447, 1988.
20. **Smalley, R. E., Wharton, L., and Levy, D. H.,** Molecular optical spectroscopy with supersonic beams and jets, *Acc. Chem. Res.,* 10, 139, 1977.
21. **Levy, D. H.,** The spectroscopy of supercooled gases, *Sci. Am.,* 250, 96, 1984.
22. **Lubman, D. M. and Jordan, R. M.,** Design for improved resolution in a time-of-flight mass spectrometer using a supersonic beam and laser ionization source, *Rev. Sci. Instr.,* 26, 1150, 1985.
23. **Opsal, R. B., Owens, K. G., and Reilly, J. P.,** Resolution in the linear time-of-flight mass spectrometer, *Anal. Chem.,* 57, 1884, 1985.
24. **Sin, C. H., Pang, H. M., and Lubman, D. M.,** Supercritical carbon dioxide injection in supersonic beam mass spectrometry, *Anal. Chem.,* 58, 487, 1986.
25. **Pang, H. M., Sin, C. H., and Lubman, D. M.,** Supercritical fluid jet expansions of polar aromatic carboxylic acids using simple derivatization with detection by resonant two-photon ionization, *Spectrochim. Acta,* 43B, 671, 1988.
26. **Covey, T. R., Lee, E. D., Bruins, A. P., and Henion, J. D.,** Liquid chromatography/mass spectrometry, *Anal. Chem.,* 58, 1451A, 1986.
27. **Pang, H. M. and Lubman, D. M.,** Design of a pulsed valve for high-pressure NH_3 injection into a supersonic beam/mass spectrometry, *Rev. Sci. Instr.,* 59, 2460, 1988.
28. **Li, L. and Lubman, D. M.,** Resonant two-photon ionization for identification of thermal decomposition products in the laser desorption of small peptides, *Rapid Commun. Mass Spectrom.,* 3, 12, 1989.
29. **Pang, H. M., Sin, C. H., and Lubman, D. M.,** Pulsed high-pressure liquid injection of biological molecules into supersonic beam mass spectrometry with resonant two-photon ionization detection, *Appl. Spec.,* 42, 1200, 1988.
30. **Lubman, D. M., Rettner, C. T., and Zare, R. N.,** How isolated are molecules in a molecular beam, *J. Phys. Chem.,* 86, 1129, 1982.
31. **Li, L. and Lubman, D. M.,** Resonance enhanced multiphoton ionization jet spectroscopy and mass spectrometry of tyrosine-containing dipeptides using a pulsed laser desorption/volatilization method, *Appl. Spec.,* 43, 543, 1989.
32. **Roepstorff, P. and Fohlman, J.,** Proposal for a common nomenclature for sequence ions in mass spectra of peptides, *Biomed. Mass Spectrom.,* 11, 601, 1984.
33. **Li, L. and Lubman, D. M.,** Resonant two-photon ionization of enkephalins and related peptides volatilized by using pulsed laser desorption in supersonic beam mass spectrometry, *Anal. Chem.,* 60, 1409, 1988.

Section II: Instrumental Developments

Part A: Magnetic Sector Instruments
Chapter 6 109
Chapter 7 121

Part B: Quadrupole Fourier Transform
Chapter 8 139

Part C: Time-of-Flight Instruments
Chapter 9 159

Chapter 6

THE MOLECULAR WEIGHT DETERMINATION OF LARGE PEPTIDES BY MAGNETIC SECTOR MASS SPECTROMETRY

B. N. Green and R. S. Bordoli

TABLE OF CONTENTS

I.   Introduction 110

II.   Techniques and Instrumentation 110

III. Scanning Methods 111
      A.   Survey Scans 111
      B.   Narrow Magnetic Scans 111
      C.   Narrow Voltage Scans 112

IV.   Spectrum Characteristics 113

V.   Molecular Weight Determination 114

VI.   Summary 116

References 119

I. INTRODUCTION

Prior to the introduction of fast atom bombardment (FAB) in early 1981,[1] the analysis of peptides by magnetic sector mass spectrometry (MS) was restricted to low molecular weights and then often only after derivatization. Although field desorption (FD) had been applied with some success, it was not in general use mainly because of difficulties in handling the delicate emitters. Plasma desorption mass spectrometry (PDMS) was already showing promise, and has since overtaken sector MS in the size of underivatized peptides that can be analyzed.

Almost from its inception, FABMS gained general acceptance as a technique for analyzing peptides as well as many other previously intractable compounds. With the parallel development of FAB and higher mass range, more sensitive instrumentation, the size of peptides that gave intact molecular-type ions increased rapidly. The analyses of glucagon and the oxidized β-chain of bovine insulin (3.5 kDa)[2] followed by bovine insulin itself (5.7 kDa)[3,4] and human proinsulin (9.4 kDa)[5] were all achieved by early 1983.

By 1985, however, progress to molecular weights much above 10,000 had slowed down, mainly because the ion yields became increasingly small.[6] Compact cesium ion guns had been developed as early as 1982/83[7,8] as an alternative to the xenon atom guns in common usage by that time, and a higher energy version[9] promised higher sensitivity at the higher masses. By mid-1987, the application of a 35-kV cesium ion gun had produced molecular-type ions at usable levels from peptides as high as 24 kDa mol wt.[10]

Furthermore, although the use of the higher-energy cesium ions as the primary ion source gave the most dramatic improvement on peptides over 10 kDa mol wt, a worthwhile sensitivity gain at lower mass also increased the analytical utility in the 3 to 10 kDa range. Because ions, instead of atoms, are now used as the bombarding particles, the term liquid secondary ion mass spectrometry (LSIMS), instead of FAB, is used subsequently in this chapter to describe the technique.[11]

The enhanced sensitivity given by high energy LSIMS, together with improved data acquisition and processing facilities, now allow reliable molecular weight assessments to be made in a few minutes from sub-nanomole quantities of many peptides with mol wts up to 24 kDa. It is the purpose of this chapter to illustrate this perhaps relatively small facet of magnetic sector MS performance applied to peptides in the mol wt range of 6 to 24 kDa, detailing the accuracy of the molecular weight assessment and the quantity of material required.

II. TECHNIQUES AND INSTRUMENTATION

In order to carry out an LSIMS analysis, the peptide is usually dissolved in an appropriate solvent, and approximately 1 μl of the resulting solution is applied to the sample holder on which about 2 μl of a liquid matrix has been previously placed. The choice of a suitable liquid matrix is crucial to a successful analysis by LSIMS; for a recent review that discusses the role of the matrix, see Hemling.[12] In spite of its relatively high volatility, the most generally applicable matrix for high molecular weight peptides is still 1-thioglycerol (TG) containing 1% trifluoroacetic acid (TFA). Although *m*-nitrobenzyl alcohol (MNBA) containing 1% TFA produces more intense spectra with a lower background with many large peptides, there have been several examples of failure to produce a spectrum at all.

After loading the sample, the holder is introduced into the MS source through a vacuum lock, where most of the sample solvent is evaporated; an operation that takes 1 to 2 min. Cesium ions from a gun at 30 to 35 kV potential are directed at the sample and release, among other particles, ions that are characteristic of the molecular weight of the peptide. These ions, either positive or negative, are accelerated to between 5 and 10 keV, where they are separated according to their mass and detected, usually after being accelerated through a further 15 to 20 keV energy to increase the detection efficiency. The mass spectrum is obtained over a period of about 2 min after beginning bombardment of the sample by the cesium ion beam.

All the data in this chapter were obtained on a VG Analytical ZAB-2SE (mass range 15,000 at 8 keV ion energy), the essential ion optics of which have been given previously.[6] In order to obtain the highest sensitivity, the spectrometer is generally operated at relatively low resolution, approximately 1000. Hence, the average or chemical molecular weight of the peptide is measured.[13] The relative merits of operating at high (unit mass separation) and low resolution have been discussed in some detail.[6] For peptides above 6 kDa mol wt, the extra effort required to produce and interpret unit mass-resolved spectra, together with the reduced sensitivity and mass range, is rarely reflected in the value of the additional information obtained.

Of course, exceptions exist. In one instance, unit mass-resolved data gave conclusive evidence that hirudin (mol wt 6905) was monomeric and not dimeric as sodium dodecylsulfate (SDS) gel electrophoresis had suggested.[14] In this case, the occurrence of unit mass separated isotope peaks in the m/z 6905 region showed unequivocally that the molecule had a mol wt of 6905. Had it been a dimer with a mol wt circa 13,810, the response at m/z 6905 would have been due to the doubly charged dimer and the isotope peaks separated only by 0.5 Da.

III. SCANNING METHODS

A. SURVEY SCANS

Often, the first approach in analyzing a high molecular weight peptide is to obtain a survey spectrum by magnetic exponential scans covering a large fraction of the available mass range. With this method, an accurate knowledge of the molecular weight is not a prerequisite, and there is a reasonable chance of discovering several components in a mixture, provided they are at sufficient concentration. Typically, such spectra may be obtained over a range of 5- or 10-to-1 in mass at an instrument resolution of about 1000 as measured on cesium iodide clusters. Data are acquired with the data system operating as a multichannel analyzer (MCA), and several scans are summed to give the final spectrum. Given sufficient sample, survey spectra will generally give the average molecular weight to within 1 Da on compounds up to 10 kDa and to within 5 Da up to 15 kDa. Measurements on the multiply charged ion peaks, in addition to those on the singly charged ions, can be used to improve the reliability of the mass determination.

Figures 1A to C are survey spectra, shown as "raw" MCA data, from 500, 25, and 10 pmol, respectively, of the *N*-acetyl derivative of eglin C (mol wt 8133.1). These data were obtained by scanning from 10 to 1.5 kDa at 20 s/decade with a resolution of 1000.

Figure 1D shows the cesium iodide spectrum that was used to calibrate the mass scale. There were 33,000 data channels used to record each of these spectra over a total acquisition time for each analysis of 1.6 min (5 scans). The protonated molecular ion peaks are expanded in the corresponding insets to show that, even at the 10 pmol level (Figure 2C), the signal-to-noise ratio is about 10:1.

B. NARROW MAGNETIC SCANS

More accurate measurements of the molecular weight than are achieved with the survey spectra may be obtained by scanning over a narrower mass range. This technique is particularly useful in cases in which the amount of sample is limited or the peptide does not produce a very intense spectrum. Provided the molecular weight is known to within a few hundred daltons, the spectrum may be scanned typically over a range that encompasses five cesium iodide reference ions or about 1200 Da. In this way, the number of ions collected in the molecular ion peak is about ten times greater than in the typical survey spectrum, giving rise to the improved reproducibility. Reproducibilities of the molecular weight determination are usually better than ±0.3 Da standard deviation on peptides up to 8 kDa mol wt.

Figures 2A to C show examples of narrow scan data obtained on 10, 5, and 1.5 pmol, respectively, of the *N*-acetyl derivative of eglin C, illustrating a lower detection limit than with the survey spectra. Figure 2D shows the cesium iodide reference spectrum. These spectra were

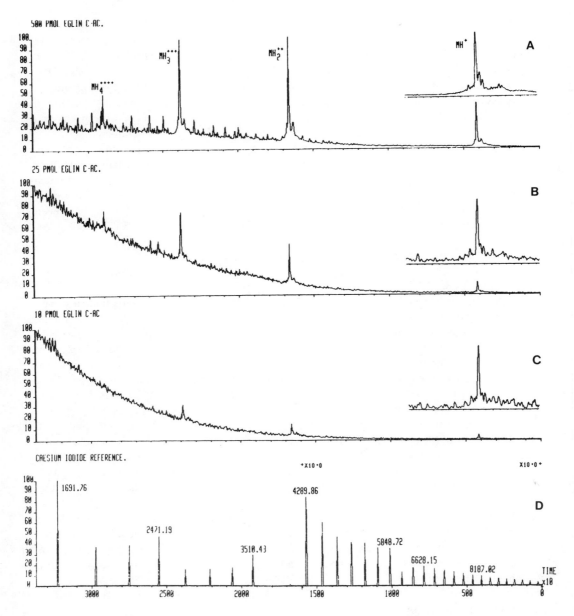

FIGURE 1.   Survey spectra from (A) 500 pmol, (B) 25 pmol, and (C) 10 pmol of *N*-acetyl eglin C with (D) the cesium iodide reference spectrum.

obtained by scanning from 8.8 to 7.5 kDa, at 200 s/decade and 800 resolution. The cycle time was 16 s, and five scans were summed over a total acquisition time of 1.3 min for each spectrum. The data were recorded in 5900 channels.

Exponentially scanned MCA data are processed by semiautomatic procedures that smooth and convert the raw data into peaks. Calibration and mass measurement of the processed data are then obtained with computer routines similar to those used at low mass, employing spectra obtained from a separate introduction of cesium iodide.

C. NARROW VOLTAGE SCANS

A slightly more sensitive method is to scan over a mass range just greater than the spacing

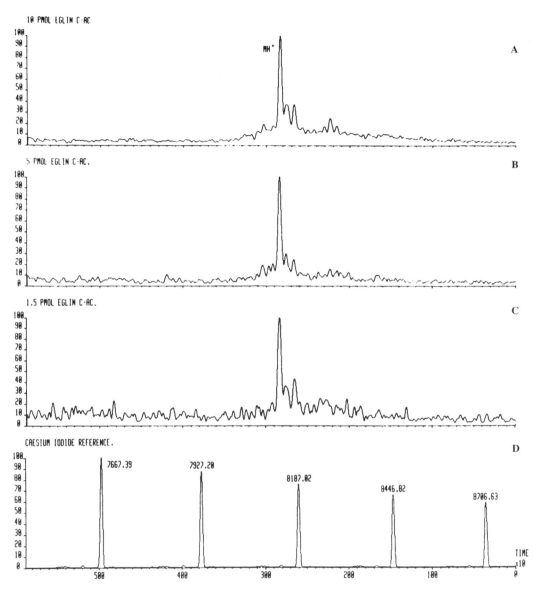

FIGURE 2.   Narrow magnetic scans over the molecular ion region from (A) 10, (B) 5, and (C) 1.5 pmol of *N*-acetyl eglin C with (D) the cesium iodide reference spectrum.

between two cesium iodide cluster peaks (259.8 Da) by linear voltage scanning. The disadvantage of this method is that the molecular weight has to be known fairly accurately and, of course, components in a mixture may be missed. Nevertheless, it offers the most accurate molecular weight determination on the least amount of sample. Mass measurement in this case is usually carried out interactively on the visual display unit, again using cesium iodide clusters from a separate introduction of cesium iodide to calibrate the mass scale.

IV. SPECTRUM CHARACTERISTICS

In many ways, the spectra shown in Figures 1A to C are typical of an 8 kDa mol wt peptide in that they show prominent peaks corresponding to the singly charged ion and multiply charged ions with up to four charges. In addition, they show a background, mainly due to the matrix, which increases toward low mass, particularly as the quantity of sample is reduced. The spectrum from 500 pmol of sample is atypical in that it shows substantial interpretable fragmentation almost up to the singly charged ion. It is perhaps significant in regard to this that eglin C contains no cysteine bridges. Other peptides of this size that show fragmentation are ubiquitin (mol wt 8.5 kDa), which again has no cysteine bridges and several hirudins (mol wt 6.9 kDa), where the part of the molecule that contains no bridges gives interpretable fragments up to around 3 kDa.[14]

When looked at in more detail, (Figure 2A), the molecular ion region is seen to be composed of a prominent narrow peak, usually the protonated molecular ion (MH+), superimposed on a relatively broad peak, which extends further above the MH+ ion peak than below it. The shape of the MH+ ion peak is essentially as expected from the isotopic distribution in the molecule convoluted with the mass spectrometer resolution. On the high mass side of the MH+ ion, where peaks can be distinguished from the broad background peak, they can be interpreted from their masses as adducts of the matrix with the peptide, often with the addition of sodium and sometimes potassium to the peptide.

On the low mass side, distinguishable peaks can be interpreted as the loss of side chains from the peptide molecule. After allowing for the number of charges, the multiply charged ion peaks, including the associated broad background peaks, closely mimic in shape the singly charged ion peak.

As the molecular weight of the peptide is increased, the MH+ peak becomes less prominent above the broad peak. This trend can be seen by comparing the spectra of eglin C (mol wt 8.1 kDa) and lysozyme (mol wt 14.3 kDa) shown in Figures 2A and 3A, respectively. With many peptides in the mol wt range of 15 to 20 kDa, the MH+ peak becomes indistinguishable from the broad background peak, rendering the molecular weight determination relatively imprecise (± 50 Da). Peptides above 20 kDa mol wt rarely give an MH+ ion peak that is distinguishable from the broad background peak. The highest molecular weight peptide that has given an interpretable spectrum by LSIMS is trypsinogen (mol wt 24 kDa) shown in Figure 4.

V. MOLECULAR WEIGHT DETERMINATION

Survey spectra of the type illustrated in Figure 1 generally give the molecular weight to within 1 Da on peptides up to 8 kDa. For example, in the cases shown in Figure 1, the molecular weights calculated from the mean of the measurements on the singly, doubly, and triply charged ions were all within 0.5 Da of the expected value. Similar accuracies (within 1 Da) were obtained on three recombinant hirudins (mol wt 6.9 kDa).[14]

More precise values of the molecular weight are obtained from narrow scans. In the case of eglin C, the results of repeat measurements at the three different sample levels shown in Figure 2 are summarized in Table 1. These data show that, provided 5 pmol or more of eglin C is available, the molecular weight can be established to within a fraction of a dalton. Similar reproducibility and accuracy were obtained on the three recombinant hirudins,[14] although larger amounts of sample were needed because the sensitivity of hirudin was about a factor of ten lower than that of eglin C. In general, provided sufficient sample is available, narrow scans will give the molecular weight with a reproducibility of 0.3 Da standard deviation on peptides of this size.

On higher molecular weight peptides, the molecular weight determination becomes less precise, and more sample is needed. For example, repeat survey scans on 1 nmol of lysozyme (Figure 5) gave a mean difference between the measured and expected molecular weight of −4.4 Da and a standard deviation of ± 1.2 Da. At the 0.2 nmol level, the mean error was 3.8 Da with a standard deviation of ± 8.7 Da.

On narrow scans, such as those illustrated in Figure 3, the results of three sets of repeat measurements at two sample levels are summarized in Table 2. These data indicate a standard deviation of better than 2 Da with a systematic error of about 2 Da.

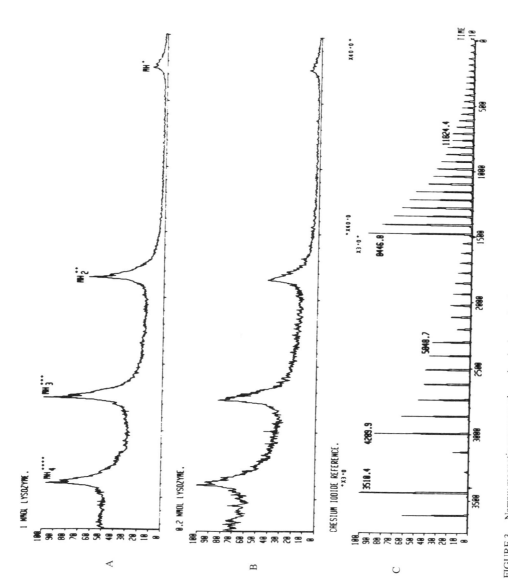

FIGURE 3. Narrow magnetic scans over the molecular ion region from (A) 1 nmol and (B) 0.2 nmol of lysozyme with (C) the cesium iodide reference spectrum.

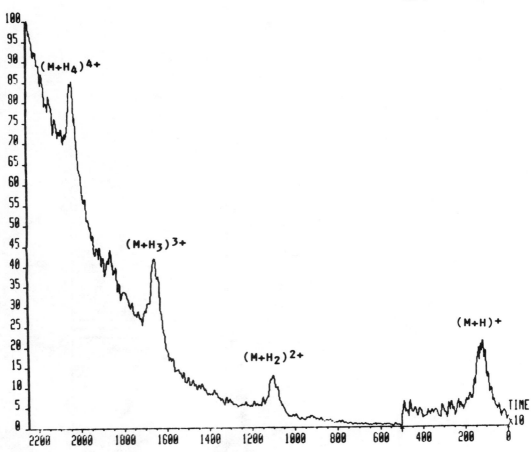

⁺X10·0   X10·0⁺

FIGURE 4. Survey spectrum from 0.2 nmol of trypsinogen (mol wt 23,978) taken over the mass range 5 to 26 kDa in 1.3 min.

TABLE 1
Results of Repeat Mol Wt Determinations on *N*-Acetyl Eglin C
(Expected Mol Wt 8133.1)

Quantity (pmol)	Molecular weight (Da)	Error (Da)	No. of measurements
10	8132.9 ± 0.2	–0.2	3
5	8133.2 ± 0.2	0.1	5
1.5	8133.7 ± 0.5	0.6	2

The accuracies achieved in analyzing 13 small proteins in the molecular weight range of 10 to 18 kDa are reproduced in Table 3, which is part of the table given in Reference 10, extended to include some more recent measurements.

VI. SUMMARY

High energy LSIMS on appropriate magnetic sector instrumentation enables peptides up to 24 kDa mol wt to be analyzed. In the molecular weight range up to 8 kDa, the accuracy of mass

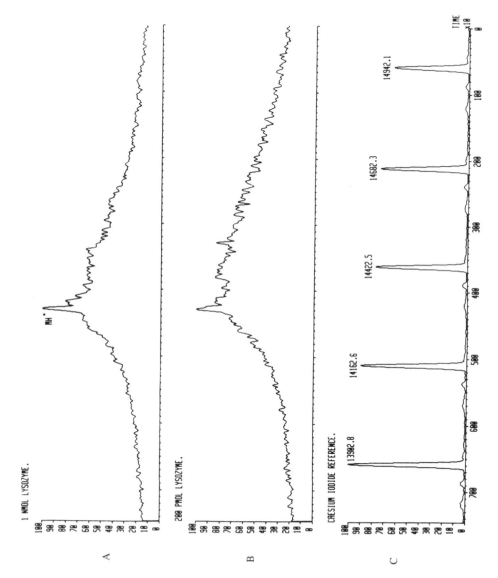

FIGURE 5. Survey spectra from (A) 1 and (B) 0.2 nmol of lysozyme with (C) the cesium iodide reference spectrum. Acquisition times were 1.8 min.

TABLE 2
Results of Repeat Mol Wt Determinations on Lysozyme (Expected Mol Wt 14305.1)

Quantity (nmol)	Molecular weight (Da)	Error (Da)	No. of measurements
1.0	14303.0 ± 1.5	−2.1	5
1.0	14304.0 ± 1.0	−1.1	6
0.2	14302.2 ± 1.0	−2.9	5

TABLE 3
Mol Wt Determination of 13 Large Peptides

	Molecular weight		Error		
	Measured	Expected	Da	%	
Peptide 1	10833.6 ± 2.1	10832.5	1.1,	0.010	(4)
Thioredoxin E.c	11672.2 ± 1.3	11673.4	−1.2	−0.010	(4)
Cytochrome C					
Tuna heart	12028.7 ± 1.6	12028.6	0.1	0.001	(5)
Rat heart	12131.5 ± 3.3	12132.8	−1.3	−0.011	(2)
Pigeon breast	12191.9 ± 5.4	12190.9	1.0	0.008	(2)
Horse heart	12359.4 ± 0.7	12360.1	−0.7	−0.006	(4)
Peptide 2	13112.8	13110.9	1.9	0.015	(1)
Lactalbumin (bovine)	14177.3 ± 0.9	14175.0	2.3	0.016	(2)
Lysozyme (hen egg)	14302.8 ± 1.6	14305.1	−2.3	−0.017	(22)
Hemoglobin (normal α chain)	15126.2 ± 7.1	15126.4	−0.2	−0.001	(4)
Superoxide dismutase	15585.8 ± 4.6	15591.4	−5.6	−0.036	(4)
Lactoglobulin B	18278.8 ± 2.3	18277.2	1.6	0.009	(5)
Lactoglobulin A	18373.9 ± 8.6	18363.3	10.6	0.058	(6)

Note: The figures in parentheses refer to the number of analyses made.

From Barber, M. and Green, B. N., *RCM*, 1(5), 80, 1987. With permission.

measurement is sufficient to define the molecular weight to well within 0.5 Da, generally on sample quantities between 10 and 100 pmol.

With higher molecular weight peptides, the mass measurement accuracy becomes less precise and more sample is required. Nevertheless, the example of lysozyme shows that the molecular weight determination is within 3 Da at 14305.1 Da on a few hundred picomoles of material. Such analyses, including processing the data, can be completed in less than 15 min.

Peptides in the range up to 16 kDa have generally given sharp protonated molecular ion peaks, which allow precise molecular weight determination. The majority above 16 kDa have not produced well-defined protonated molecule ion peaks, and in these cases the molecular weight determination is substantially less precise (±50 Da). Nevertheless, the molecular weight determination is substantially better than can be obtained by the non-mass spectrometric methods commonly used.

REFERENCES

1. **Barber, M., Bordoli, R. S., Sedgwick, R. D., and Tyler, A. N.,** Fast atom bombardment of solids (FAB): a new ion source for mass spectrometry, *J. Chem Soc. Chem. Commun.,* 325, 1981.

2. **Barber, M., Bordoli, R. S., Sedgwick, R. D., Tyler, A. N., Garner, G. V., Gordon, D. B., Tetler, L. W., and Hider, R. C.,** Fast atom bombardment mass spectrometry of the large oligopeptides melittin, glucagon and the B chain of bovine insulin, *Biomed. Mass Spectrom.,* 9, 265, 1982.

3. **Dell, A. and Morris, H. R.,** Fast atom bombardment, high field magnet mass spectrometry of 6000 dalton polypeptides, *Biochem. Biophys. Res. Commun.,* 106, 1456, 1982.

4. **Barber, M., Bordoli, R. S., Elliott, G. J., Sedgwick, R. D., Tyler, A. N., and Green, B. N.,** Fast atom bombardment mass spectrometry of bovine insulin and other large peptides, *J. Chem Soc. Chem. Commun.,* 936, 1982.

5. **Barber, M., Bordoli, R. S., Elliott, G. J., Horoch, N. J., and Green, B. N.,** Fast atom bombardment mass spectrometry of human proinsulin, *Biochem. Biophys. Res. Commun.,* 110, 753, 1983.

6. **Green, B. N. and Bordoli, R. S.,** High mass capabilities of sector mass spectrometers, in *Mass Spectrometry in Biomedical Research,* Gaskell, S. J., Ed., John Wiley & Sons, Chichester, 1986, 235.

7. **Aberth, W., Straub, K. M., and Burlingame, A. L.,** Secondary ion mass spectrometry with cesium ion primary beam and liquid target matrix for analysis of bio-organic compounds, *Anal. Chem.,* 54, 2029, 1982.

8. **McEwan, C. N.,** Source mounted cesium ion gun for obtaining liquid matrix secondary ion mass spectra of organics, *Anal. Chem.,* 55, 967, 1983.

9. **Elliott, G., Cottrell, J. S., and Evans, S.,** A comparison of 8keV FAB and 30keV cesium ion bombardment for the desorption of large molecules, presented at the 34th ASMS Conference on Mass Spectrometry and Allied Topics, Cincinnati, June 8-13, 1986.

10. **Barber, M. and Green, B. N.,** The analysis of small proteins in the molecular weight range 10-24kDa by magnetic sector mass spectrometry, *RCM,* 1, 80, 1987.

11. **Burlingame, A. L., Baillie, T. A., and Derrick, P. J.,** Mass spectrometry, *Anal. Chem.,* 58, 165R, 1986.

12. **Hemling, M. E.,** Fast atom bombardment mass spectrometry and its application to the analysis of some peptides and proteins, *Pharm. Res.,* 4, 5, 1987.

13. **Yergey, J., Heller, D., Hansen, G., Cotter, R. J., and Fenselau, C.,** Isotopic distributions in mass spectra of large molecules, *Anal. Chem.,* 55, 353, 1983.

14. **Van Dorsselaer, A., Lepage, P., Bitsch, F., Whitechurch, O., Riehl-Bellon, N., Fraisse, D., Green, B., and Roitsch, C.,** Mass spectrometry analyses of recombinant hirudins (7kDa), *Biochemistry,* 28, 2949, 1989.

Chapter 7

FOUR-SECTOR TANDEM MASS SPECTROMETRY OF PEPTIDES

Alison E. Ashcroft and Peter J. Derrick

TABLE OF CONTENTS

I.   Introduction 122

II.   Methods of Ionization 123

III.   Collision Gas and Collision Energy 124

IV.   Mechanisms of Excitation and Fragmentation 126

V.   Nomenclature of Fragmentation Pathways of Protonated Linear Peptides 127

VI.   Published Tandem Mass Spectra of Peptides 132

VII.   Conclusion 135

Acknowledgments 136

References 136

I. INTRODUCTION

Only a very short time ago, the 4-sector tandem mass spectrometer seemed to be a type of instrument that had been found to be of limited usefulness. The prognosis seemed to be that the odd 4-sector mass spectrometer might be built for experiments of a chemical physics or ion chemistry nature, but that such instruments had little future for analytical application in biology. Opinions have changed significantly in the past few years. The 4-sector tandem mass spectrometer is now seen as an important instrument, one which may become as commonplace as the high-resolution double-focusing mass spectrometer in the 1960s. Indeed, the 4-sector mass spectrometer may prove to be the true successor of the 2-sector double-focusing mass spectrometer. This turnabout in opinion is due in no small measure to the demonstrated capability of the combination of kiloelectronvolt particle bombardment discussed in this chapter and 4-sector tandem mass spectrometry for the sequencing of peptides.[1] Under favorable circumstances, the 4-sector tandem mass spectrometer now gives sensitivity comparable to that of the wet-chemistry stepwise Edman degradation method for sequencing peptides. There are, of course, qualifications and conditions that must be considered, but the fact remains that the amounts of material required to sequence a peptide with about 10 or 20 residues by 4-sector mass spectrometry are small by any standards. With parallel detection of about 5% of the mass range, a peptide of mass about 1200 Da can in favorable circumstances be sequenced given 1 pmole of sample.[2] This figure for minimum sample size will probably fall significantly during the next few years as a result of technical developments. The third-generation (discussed later in this chapter) of 4-sector mass spectrometers should benefit from detecting in parallel a higher percentage ($\geq$50%) of the mass range, from more efficient fragmentation of mass-selected parent ions, and from more efficient collection of the fragment ions produced. Two orders of magnitude in sensitivity are likely to be gained in these ways, and if potential improvements in the efficiency of parent ion formation are brought into the picture, the attainment of femtomole sensitivities becomes a real possibility. Whatever the future of 4-sector mass spectrometers, they are unquestionably setting the standards that any new mass spectrometric technique, such as electrospray ionization,[3] must surpass.

The first generation of 4-sector tandem mass spectrometers (referred to as "4-sectors" from here on) were built in-house. These instruments were the tandem mass spectrometers built by physicists in the 1950s; later, and more significant in terms of applications to peptides, there was the 4-sector built by McLafferty et al.[4-6] at Cornell University (New York) in the late 1970s. The second generation of 4-sectors are instruments commercially available at this writing, which various companies have constructed by "bolting together" two of their 2-sector double-focusing mass spectrometers. This is, of course, an oversimplification, but the point is that the two sectors in the front-half (parent ion selection) and the two sectors in the back part (fragmentation analysis) were designed originally to meet criteria set by 2-sector double-focusing mass spectrometry. The creativity and innovation in the design of the second-generation 4-sectors is confined largely to the junction of the two halves (collision cells and lenses) and to the final detector (array detectors). Third-generation 4-sectors will in all probability be commercially available in the near future, and these instruments will be designed as "total" 4-sector instruments. The greatest differences from existing instruments should arise with the ion optical characteristics of the third and fourth sectors and with the accelerating potentials employed. High degrees of parallel detection should make possible the exploitation of pulsed ionization techniques, such as laser desorption and post-ionization of desorbed neutrals.

As of Autumn 1989, 4-sector tandem mass spectrometers are located in about a dozen laboratories, including the following:

- Department of Chemistry, Cornell University, Ithaca, NY (McLafferty, F. W. et al. Home built)

- Laboratory of Molecular Biophysics, National Institute of Environmental Health Sciences, Research Triangle Park, NC, (Tomer, K. B. et al. VG-(ZAB)[2])
- Department of Chemistry, Massachusetts Institute of Technology, Cambridge, MA (Biemann, K. et al. Joel-(HX 110)[2])
- SmithKline and French Laboratories, King of Prussia, PA (Carr, S. A. et al. VG-(ZAB)[2])
- Protein Chemistry Department, Genentech Inc., San Francisco, CA (Stultz, J. et al. Joel-(HX 110)[2])
- School of Chemical Sciences, University of Illinois, Urbana-Champaign, IL (Rinehart, K. L., Jr. et al. VG-(70-70)[2])
- Department of Pharmaceutical Chemistry, University of California, San Francisco, CA (Burlingame, A. L. et al. Kratos-(Concept)[2])
- Laboratory of Mass Spectrometry, University of Lille Flandres Artois, Lille, France (Fournet, B. et al. Kratos-(Concept)[2])
- Department of Chemistry and Warwick Institute of Mass Spectrometry, University of Warwick, Coventry, U.K. (Derrick, P. J., Jennings, K. R. et al. Kratos-(Concept)[2])
- Department of Chemistry, University of Maryland, Baltimore, MD (Fenselau, C. et al. Joel-(HX 110)[2])
- Ciba-Geigy, Basel, Switzerland (Richter, W. G. et al. VG-(70-70)[2])
- Department of Chemistry, University of Manchester Institute of Science and Technology, Manchester, U.K. (Barber, M. et al. Kratos-(Concept)[2])
- Genetics Institute, Massachusetts (Martin, S. A., Scoble, H. A. et al. Joel-(HX 110)[2])
- Department of Chemistry, Technical University of Berlin, W. Germany (Schwarz, H. et al. VG-ZAB/Maurer design)

II. METHODS OF IONIZATION

In almost all studies to date of peptides by 4-sector tandem mass spectrometry, kiloelectronvolt particle bombardment has been employed to form the parent ions. The particles have sometimes been atoms (xenon is commonly employed) and sometimes atomic ions (usually Cs^+). The peptide in almost all cases has been dissolved and presented to the particle beam in a liquid matrix (glycerol/thioglycerol, for example). Some of the earliest studies of peptides by tandem mass spectrometry made use of field desorption (FD), but these studies were on 2-sector instruments.[7-10]

Kiloelectronvolt ions are generally considered to give significantly higher sensitivities than kiloelectronvolt atoms in the cases of higher mass peptides (M> about 2000). As will be seen, 4-sector tandem mass spectrometric studies of peptides have to date been restricted almost entirely to peptides with relative molecular masses M well below 3000. Whether ion bombardment is significantly superior to atom bombardment for the purposes of tandem mass spectrometry as practiced today is therefore a moot point, because the advantages of ion bombardment tend to lie in a mass range not readily accessible to tandem mass spectrometry. The prolonged sample lifetime provided by ion bombardment can, however, be an advantage in some circumstances. FD has been used for tandem mass spectrometry of industrial polymers, again with 2-sector instruments.[10] FD will very probably be used in the future for 4-sector studies of industrial polymers, and given that FD sources are available on 4-sectors, there are likely to be FD studies of peptides. Tandem mass spectrometry and FD of peptides are well matched (up to M ~ 2000), because with FD most of the total ion current is concentrated in abundant molecule-ion peaks. The papers covered in this review have concerned predominantly kiloelectronvolt atom bombardment (known as fast atom bombardment [FAB] or liquid secondary ion mass spectrometry [liquid SIMS]).

The number of peptide ions in which tandem mass spectra have been measured by 4-sector mass spectrometry and published is still fewer than 200. Most of these ions have masses M below

2000. The bar graph (Figure 1) represents the distribution of peptide-ion masses obtained on 4-sectors up until about the end of 1988. These data have been extracted from papers published in the years up to and including 1988 and part of 1989, and are reasonably complete. The height of a bar represents the number of peptide ions whose 4-sector tandem mass spectra have been published, in a particular 100 mass unit range. Thus, the highest bar (figure) represents the 14 peptide ions with mass-to-charge ratios in the range of 1200 to 1300. It is clear that the distribution peaks at around 1200, and that almost all peptide-ions run to date by 4-sector tandem mass spectrometry have masses below 2500 Da. Of the peptide ions included in the survey, 98% have masses below 2500 Da, 94% have masses below 2000 Da, and 82% have masses below 1500 Da. The mean of the distribution (figure) falls at m/z 1142.0. There are no multiply charged ions included in the survey, although these ions are amenable to tandem mass spectrometry[10] and should be much studied in the future.[3,12,13] No negative ions have been included in the survey, although again these types of ions are amenable to tandem mass spectrometry (see discussion to follow).

III. COLLISION GAS AND COLLISION ENERGY

In all of the studies reviewed, collisional activation has been the method used to excite mass-selected parent ions. Relatively few papers have contained much discussion of collision regimes, although the numerous peptides analyzed and sequences deciphered indicate underlying appreciation of practical factors involved. Papers by Martin et al.[14] and Carr et al.[15] are notable exceptions. Our conclusion is that, once successful collision cell conditions have been established for a particular mass range on an individual instrument, they tend to be adhered to in the laboratory until modifications are dictated by the demands of different techniques or sample types.

It is agreed that the use of a collision gas is essential if a significant number of reasonably intense product ions is to be observed from the chosen precursor ion. Tandem mass spectra acquired in the absence of a collision gas have been found to contain product ions of generally much lower intensity, with some peaks missing and others attenuated in comparison to collisionally activated decomposition (CAD) spectra.[14,15] For peptides up to 2000 Da, the overall intensity of the product ion spectrum increases with increasing gas pressure until 70% attenuation of the precursor ion intensity is reached; further attenuation causes the intensities of the low-mass product ions to increase at the expense of the more structurally informative high-mass fragments.[14,16] The latter effect is more marked for peptides about 2000 Da in mass.[14] The collision gas reported most frequently to date by all groups working in this area is helium, which has been found to exhibit a higher fragmentation efficiency than argon. Argon seems to be the alternative choice. Fragmentation efficiency is defined here as the total number or sum of the intensities of all fragment ions collected as a percentage of the initial number or intensity of parent or precursor ions, i. e., number or intensity prior to collision. The fragmentation efficiency so defined reflects both the efficiency of CAD (in broader terms, the collision dynamics) and the efficiencies of focusing and transmitting fragment ions (in broader terms, the ion optics of the collision cell region and second mass analyzer). The fragmentation efficiency with helium tends to be in the region of 1 to 4 % for peptide ions below m/z 2000. These efficiencies are somewhat lower values than those observed for certain inorganic molecules, glycopeptides, and carbohydrates. Argon might be expected to produce more effective collisions than helium due to the fact that it is heavier, and indeed it has been found to produce detectable product ions at lower masses. The lower fragmentation efficiency of argon is generally proposed to originate from scattering, which presumably increases with increasing mass of the collision gas.[17] The product ion spectra resulting from decompositions induced by argon or helium differ mainly in regard to the intensities of the fragments that require multiple bond cleavages for their formation. For example, d-ions (see discussion to follow), the partial sidechain losses resulting from the a-series

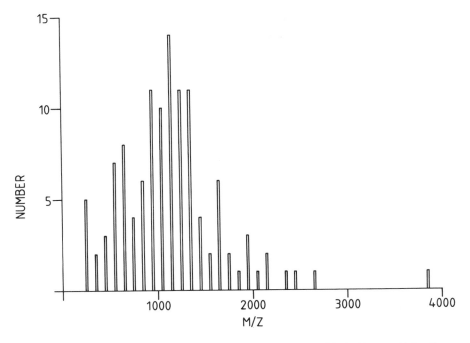

FIGURE 1. Numbers of published tandem mass spectra against the mass of the incident peptide ion. Papers up to the first few months of 1989 have been included.

peptide-backbone cleavages, are enhanced in argon-induced decompositions.[15] Other than scattering effects, the factor favoring helium has been said to be a greater efficiency in terms of the proportion of the center-of-mass collision energy converted to internal energy.[18,19] Greater efficiency would offset the lower center-of-mass collision energy, so that the energy depositions might not differ greatly in the two cases at incident ion energies around 10 keV. At lower incident ion energies, as would be found in electrically floating cells, heavier target gases would eventually become necessary. Electrically floating cells also reduce the disadvantages of greater scattering with heavier target gas.

Laboratory collision energies, meaning incident ion energies E_i, have been discussed in some detail,[14,15,20-23] particularly with regard to the dependence of fragmentation on laboratory collision energy E_i. Studies using 4-sector instruments have been restricted to date to collision energies E_i of 10 keV and below, and collision energies have been changed by floating the collision cell at positive potentials (with positive incident ions).

The product ion spectra from the protonated molecules [M+H]$^+$ of the model peptides chosen have shown little variation in fragment ion types for collision energies ranging from 5 keV (collision cell floated at 5 kV) at 10 keV (grounded cell). An enhancement in the transmission of low-energy product ions without any attenuation of the high-energy product ions, has been apparent at the lower-energy end of this range for parent ions below m/z 2000. For laboratory collision energies between 0.2 and 2 keV, the relative abundances of the sidechain cleavage ion types, namely d-, v-, and w-ions, have been reduced severely. For peptide ions above m/z 2000, the use of lower collision energies has been found to be counterproductive because the percentage of product ions being formed in the collision cell seems to be much lower. An increase in product ion mass resolution has been shown to be another advantage of electrically floating the collision cell and hence lowering the collision energy E_i. This phenomenon is a consequence of decreasing the fractional energy spreads of the product ions, following reacceleration from the collision cell into the mass analyzer. Electrically floating the collision cell has other significant ion optical advantages in regard to focal place detection.[24,25] Achieving a focal plane of good quality, encompassing a high proportion of the mass range, becomes simpler and easier if the range of ion energies to be focused is reduced by acceleration of the product ions subsequent to their formation.

Results of experiments involving electrically floating the collision cell of a 4-sector instrument reflect closely ion optical effects arising from the cell acceleration into the second mass analyzer. Effects due to changes in collision energy E_i are not necessarily clearly revealed. Nevertheless, it does seem clear that reducing collision energies E_i from 10 to 5 keV, and perhaps a little lower, does not greatly, if at all, reduce the efficiency of CAD (M <2000). There are results[20] which suggest that raising the collision energy E_i from 10 to 15 keV significantly increases the efficiency of CAD of peptide ions around m/z 2000. This being the case, it follows that efforts should be made to work at high collision energies (>10 keV), while at the same time retaining acceleration subsequent to product ion formation for the ion optical benefits so obtained.

IV. MECHANISMS OF EXCITATION AND FRAGMENTATION

Excitation of parent ions has, in all studies reviewed, been by collisional activation at keV ion energies. The mechanisms of excitation of collisional activation of large ions are currently attracting much attention, and opinions differ as to the relative importance of various possibilities.[19,20,26-28] The energetics of fragmentation of large ions represent an equally active area of inquiry which, in the context of 4-sector tandem mass spectrometry, should not be separated from the question of excitation. What is not known well enough is how much internal energy is deposited in a large peptide ion by collisional activation and how such energy contributes to fragmentation. Some of the questions, some of the basic facts, and such results as do exist are briefly reviewed here.

A large peptide ion might contain 300 atoms, or about 1000 internal degrees of freedom. Guessing the internal temperature of such a large ion to be 300 K, the total vibration energy of a peptide ion of approximately 2000 Da could be about 10 eV prior to collision, even if the peptide ion were relatively cool. 10 eV is sufficient energy to break almost any bond in the peptide ion; however, this hypothetical ion would probably be stable with respect to unimolecular decomposition. This stability arises from the same cause as the high total amount of energy, namely the large number of internal degrees of freedom. Internal energy can be distributed over the very large number of vibrations, and the amount of energy in any one bond at any one time is unlikely to much exceed hundredths of an eV. If, therefore, internal energy deposited in an ion collisional activation were spread over the large number of vibrations prior to fragmentation, the amount of deposited energy (Q) needed to induce fragmentation within a reasonable time (μs) would be considerable. An eV or two of deposited energy would seem to be a more reasonable guess as to the energy necessary to induce extensive and rapid fragmentation of a very large ion (M = 2000, for example).

The unknown factor in this scenario is the extent to which internal energy Q deposited in collisional activation is randomized over all degrees of freedom. If the internal energy Q were highly localized, it would be possible for much smaller amounts of energy to induce sufficiently rapid fragmentation. There is no compelling experimental evidence bearing on this question of the extent of energy randomization. Assuming that the initial deposition of internal energy were localized (see later in chapter), a certain time would be required for randomization of that energy. Given 1000 degrees of freedom, it would be unreasonable to assume that, following localized energy deposition, randomization could typically occur in much less than a nanosecond. This number is a rough estimate, perhaps in error by being too short. The time available for a large ion to fragment following collisional activation in a 4-sector instrument is of the order of microseconds. The fragmentation time is set by the need for fragmentation to occur within the collision cell in order for the product ions to be properly focused. It is seemingly possible, therefore, for internal energy to be deposited in one region of a large ion and to be subsequently randomized, prior to fragmentation describable by statistical rate theories. On the other hand, it is possible that fragmentation occurs within very short times (<< nanoseconds) following initial excitation, that energy is not randomized, and that the description of the fragmentation by statistical rate theories is less easy or inappropriate. It is quite conceivable that, following collisional activation of a large ion, energy randomization and unimolecular fragmentation would occur at comparable rates and would in this sense be competitive processes.

Experimental evidence concerning the deposition of internal energy into large ions by collisional activation derives from measurements of the translational energy losses ΔE suffered by the ions as a result of collision.[29] The energy loss ΔE represents the internal energy taken up by the incident ion, plus the translational and any other energy gained by the target gas.[26-28] This relationship assumes that the ion does survive collision, and does not dissociate during the collisional interaction.[17] The energy losses ΔE can be of the order of tens of eV, and possibly hundreds of eV, and detailed consideration of their dependences on ion target gas masses suggests that, with large ions, energy deposition may occur via direct momentum transfer, i.e., direct vibrational excitation.[16-19,26-29] Electronic excitation may be less relevant to CAD of large ions. An impulsive collision treatment (ICT) describes a number of the characteristics of collisional activation of large ions.[17] One significant prediction of ICT is that measured energy losses ΔE are a reliable guide as to the magnitudes of associated internal energy uptakes Q. As a very rough rule of thumb: with a large organic ion and helium as target gas, Q is predicted to be about half of ΔE, or perhaps a little less. Another prediction is that internal energy is deposited in a localized manner, i.e., the large ion is excited in that small region interacting directly with a target gas.

V. NOMENCLATURE OF FRAGMENTATION PATHWAYS OF PROTONATED LINEAR PEPTIDES

A communal code for the description and explanation of peptide fragments resulting form high-energy collisionally activated dissociations using 4-sector instrumentation appears to have evolved as data and statistics have accumulated during the course of the numerous peptide studies conducted to date.

The basic nomenclature for mass spectral peptide sequence ions proposed by Roepstorff and Fohlman[30] appears to have been a focal starting point. One group has adopted the use of lower case rather than capital letters to denote the type of fragmentation and also the notation +1, +2, −1, −2 instead of superscript hyphens to indicate the addition or subtraction of one or two hydrogen atoms to the fragment generated by simple cleavage of a bond.[31] Some have followed suit,[21] while others have maintained the use of capital letters.[32,33] Suffice it to say that these minor variations in nomenclature do not affect the interpretations or cross-referencing of the fragmentation patterns that have been deciphered. We have adopted the lower case and the numerals (+1, +2, etc.), not subscripted. The prominent numerals seem to us to better emphasize that chemically, the presence or absence of a hydrogen atom is highly significant. There also seems to be a tendency in the literature for the superscripted hyphens to be omitted, either deliberately or through carelessness.

Table 1, showing names of different types of fragment ions, has been compiled from a mixture of articles,[14,30,33] and the information contained therein should have a general applicability for interpretation of tandem mass spectra from 4-sectors. Tandem mass spectra of peptides obtained by high-energy collisional activation have been observed to be highly reproducible not only on the same instrument, but between instruments from different manufacturers operated under different conditions.[33] This is not to say that the spectra are not strongly dependent on instrumental conditions, but rather that the reproducibility is a sign that similar experimental conditions can be produced on the various second-generation 4-sectors. The structures shown for fragment ions in Table 1 are chemically plausible, but do not enjoy the support of strong

TABLE 1
Common Types of Fragment Ions Arising From Protonated Linear Peptides

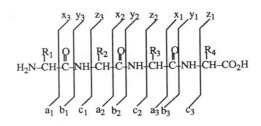

Name	Type of cleavage	Terminal contained	Assumed structure
a_n	Backbone	N	$H(NHCHRCO)_{n-1} \overset{+}{N}H=CHR$
b_n	Backbone	N	$H(NHCHRCO)_{n-1} NH-CHR-\overset{+}{C}\equiv O$
c_{n+2}	Backbone	N	$[H(NHCHRCO)_n NH_2]H^+$
x_n	Backbone	C	$\overset{+}{O}\equiv C(NHCHRCO)_n OH$
y_{n+2}	Backbone	C	$[H(NHCHRCO)_n OH]H^+$
z_{n+1}	Backbone	C	$[CHR-CO(NHCHRCO)_{n-1} OH]H^+$
d_n	Sidechain	N	$[H(NHCHRCO)_{n-1} NH-CH=CHR']H^+$
v_n	Sidechain	C	$[HN=CH-CO(NHCHRCO)_{n-1} OH]H^+$
w_n	Sidechain	C	$[R'CH=CH-CO(NHCHRCO)_{n-1} OH]H^+$
Internal acyl	Backbone	Neither	$H(NHCHRCO)_{n-1} NH-CHR-C\equiv O^+$
Internal immonium	Backbone	Neither	$H(NHCHRCO)_{n-1}-\overset{+}{N}H=CHR$
Amino acid immonium	Backbone	N($a1$), C or neither	$H_2\overset{+}{N}=CHR$

Note: R and R′ represent a sidechain and the β-substitutent of a sidechain, respectively.

evidence. The pathways and structures should be regarded as useful, indeed essential, working aids to the interpretation of spectra for the purposes of sequence determination. There are six main types of cleavage of the amide backbone (a, b, c, x, y, z) of a linear peptide ion, each of which may or may not be accompanied by hydrogen rearrangement, and each of these cleavages can give rise to a series of peaks whose adjacent members differ in mass by that of an amino acid residue. In addition, there are three sidechain fragmentations (d, v, w) and three nonspecific sequence-type fragmentations.

The mechanisms of formation of a_n, b_n, and y_{n+2} ions, which appear to be the most commonly observed ions resulting from backbone cleavages, are currently accepted to involve initial protonation of the appropriate amide nitrogen atom[34] (Scheme 1). Again, it is pointed out that these mechanisms are plausible rationalizations, whose status might best be described as that of untested hypotheses.

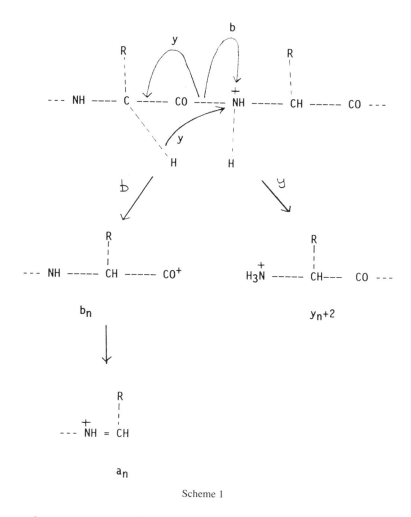

Scheme 1

d_n-Ions are fragment ions retaining the positive charge at the N-terminus, and correspond nominally to the loss of an alkene from the sidechain of the other terminus (i.e., not the N-terminus) of an a_n-ion.[34] There is evidence, however, that at least in certain ions, the precursors of d_n fragments are a_{n+1} ions, rather than a_n-ions.[33,34] In this case, radical loss occurs from the a_n +1 radical cation (this type of ion is not included in Table 1), perhaps as in Scheme 2. d_n-Ions have been observed where the amino acid composing the other terminus has a sidechain of at least two carbon atoms in length without being cyclic. Alanine, glycine, and proline are thus ruled out. d_n-Ions have been found to have low intensities when the amino acid in question is aromatic. This observation is consistent with homolytic cleavage of a phenyl bond being unfavorable. d_n-Ions provide a means of differentiating between the isomeric amino acids leucine and isoleucine, because the product ions retain different parts of the sidechains. If leucine were the amino acid at the terminus, an *i*-propyl radical would be eliminated. In the same situation, isoleucine eliminates either a methyl or an ethyl radical.[34-36]

The C-terminal v_n-ions are formed nominally by loss of the sidechain from the amino acid composing the other terminus (i.e., not the C-terminus) of either y_{n+2} or x_{n+1} ions, according to the proposed mechanisms (Scheme 3).[34] The formation of a v_n fragment has been found to require the presence at the other terminus of the precursor of an aromatic amino acid (H, W, Y, or F), or an amino acid with a β-substituent (S, V, I, or T), or asparagine.[34] The x_{n+1} ions are again a type of ion not included in Table 1.

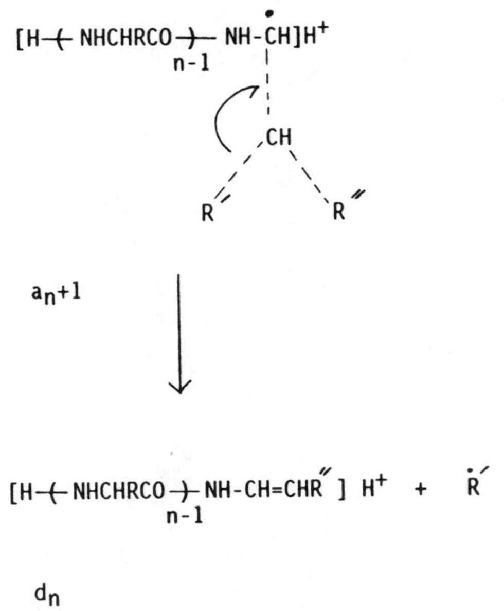

$$[H\!\!-\!\!(\,NHCHRCO\,)\!\!-\!\!NH\overset{\bullet}{-}\overset{|}{C}H]H^+$$
$$_{n-1}$$

$$CH$$
$$R \qquad R''$$

a_{n+1}

$$[H\!\!-\!\!(\,NHCHRCO\,)\!\!-\!\!NH\!\!-\!\!CH\!\!=\!\!CHR''\,]\ H^+\ +\ \overset{\bullet}{R'}$$
$$_{n-1}$$

d_n

Scheme 2

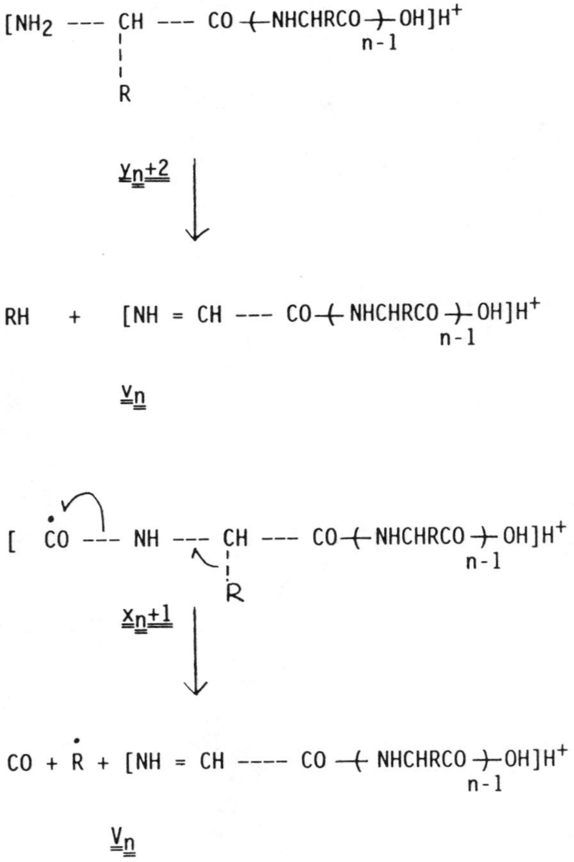

$$[NH_2 \text{---} \overset{|}{\underset{R}{CH}} \text{---} CO\!\!-\!\!(\,NHCHRCO\,)\!\!-\!\!OH]H^+$$
$$_{n-1}$$

$\underline{\underline{y}}_{n+2}$

$$RH\ +\ [NH\!=\!CH\text{---}CO\!\!-\!\!(\,NHCHRCO\,)\!\!-\!\!OH]H^+$$
$$_{n-1}$$

$\underline{\underline{y}}_n$

$$[\ \overset{\bullet}{CO}\text{---}NH\text{---}\overset{|}{\underset{R}{CH}}\text{---}CO\!\!-\!\!(\,NHCHRCO\,)\!\!-\!\!OH]H^+$$
$$_{n-1}$$

$\underline{\underline{x}}_{n+1}$

$$CO\ +\ \overset{\bullet}{R}\ +\ [NH\!=\!CH\text{----}CO\!\!-\!\!(\,NHCHRCO\,)\!\!-\!\!OH]H^+$$
$$_{n-1}$$

$\underline{\underline{v}}_n$

Scheme 3

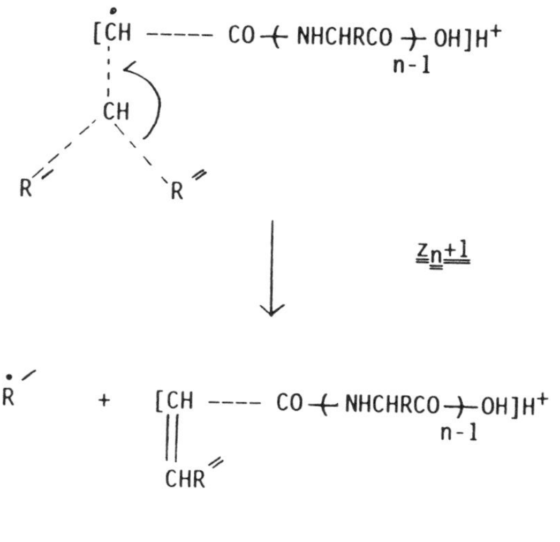

$$[\overset{\bullet}{CH} ----- CO \overset{}{+} NHCHRCO \overset{}{+} OH]H^+$$
$$n-1$$

$$\underline{\underline{z_n+1}}$$

$$\overset{\bullet}{R} \quad + \quad [\overset{}{CH} ---- CO \overset{}{+} NHCHRCO \overset{}{+} OH]H^+$$
$$\quad\quad\quad\quad \| \quad\quad\quad\quad\quad\quad\quad\quad\quad n-1$$
$$\quad\quad\quad\quad CHR$$

$$\underline{\underline{w_n}}$$

Scheme 4

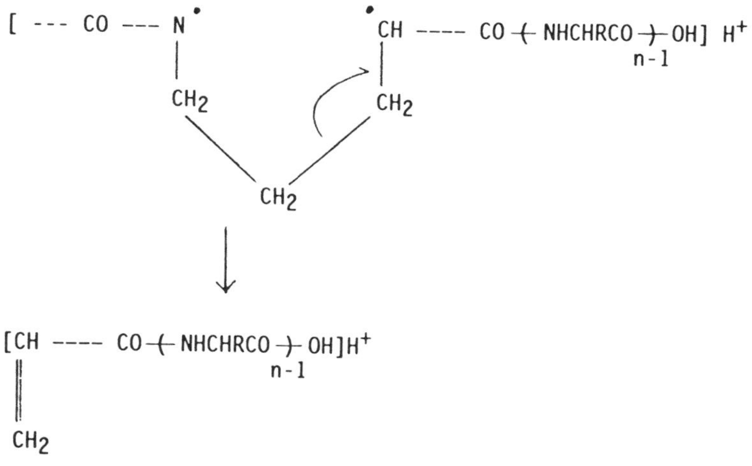

$$[--- CO --- \overset{\bullet}{N} \quad\quad \overset{\bullet}{C}H ---- CO \overset{}{+} NHCHRCO \overset{}{+} OH] \; H^+$$
$$\quad\quad\quad\quad | \quad\quad\quad\quad\quad\quad | \quad\quad\quad\quad\quad\quad\quad\quad n-1$$
$$\quad\quad\quad CH_2 \quad\quad\quad\quad CH_2$$
$$\quad\quad\quad\quad\quad\quad CH_2$$

$$[\overset{}{CH} ---- CO \overset{}{+} NHCHRCO \overset{}{+} OH]H^+$$
$$\| \quad\quad\quad\quad\quad\quad\quad\quad\quad n-1$$
$$CH_2$$

$$\underline{\underline{w_n}}$$

Scheme 5

There is evidence to suggest that the w_n fragment ions are derived from z_{n+1} precursors.[33,37,38] Cleavage between the β- and γ-carbons in certain amino acid sidechains is said to be implicated, as shown in Scheme 4. Similar to d_n-ions, w_n-ions are not favored when aromatic amino acids, glycine, or alanine are present in the other terminus (i.e., not the C-terminus) of the z_{n+1} precursor. w_n-Ions have, however, been observed in cases in which proline has been in the key position. The explanation put forward[37] is that when homolytic cleavage of the proline N-C bond occurs, dissociation to a z_{n+1} ion cannot take place because of the remainder of the ring, but further cleavage gives the w_n-ion as shown in Scheme 5. If the β-carbon of a sidechain carries two groups

TABLE 2
Peptides Whose Tandem Mass Spectra are Characterized by N-Terminal
Fragment Ions (Positive Ions)

Sequence	M	Name	Fragment ions	Ref.
RPPGFSPER	1059.6	Bradykinin	a, some y	32, 64
RPKPQQFFGLM	1346.7	Substance P	a, some d	39, 41, 42, 60
PKPQQFFGLM	1190.6	Des-arg substance P	a, b	41
AGCKNFFWKTFTSC	1636.7	Somatostatin (ox.)	little a, b	41
AGCKNFFWKTFTSC	1638.7	Somatostatin (red.)	b, some a	41
UADPNKFYGLM	1264.6	Physalaemin	a, b	34, 40
LGG	245.1		a, b	48, 49
ELAGAPPEPA	1063.7	β-lipotropin	a, d, some b, y	14
DRVYIHPFHLLVYS	1757.9	Renin substrate	a, some d	14, 33, 43
LEATINELL	1014.6	Thioredoxin	b, some a, y	44,
APDVLIPGQQTTJW	1694.9	Dopamine	a, b, c,	33, 45, 46
RYVF	663.3		a, some d	45
YIIKGVFWDPAC (sub)	1628.6	a^z mating factor	a, b	62
RPVKVYPNGAEDE	2464.2	ACTH peptide	a, d	47
SAEAFPLEF				
RVYIHPF	930.5	Angiotensin III	a, b	49
PNWGT	572.7	HTF-II peptide	b	63

Note: y is used as an abbreviation for y_{n+2}. (sub) is a long chain hydrocarbon.

of different masses, as, for example, with threonine or isoleucine, two different w_n-ions can be formed, depending upon which group is lost. It has been suggested that the two fragments be labeled w_{na} and w_{nb}, with the former being the lower in mass.[37] Given the mechanisms put forward for their formation (Schemes 2, 4, and 5), d_n- and w_n-ions are both formed following homolytic cleavage of the backbone, and both should provide information concerning the identity of the amino acid cleaved. w_n-Ions, like d_n-ions, provide a means of distinguishing between the isomeric leucine and isoleucine.[35,37]

Due to their dependence on the individual structure of the relevant amino acids, d_n-, v_n-, and w_n-ions tend not to give rise to extended series of peaks in tandem mass spectra, thereby contrasting with the backbone cleavages. The sidechain fragmentations d_n, v_n, and w_n are found generally only in tandem mass spectra from high-energy (several keV and higher) CAD and their recognition is important in the interpretation of such spectra.

VI. PUBLISHED TANDEM MASS SPECTRA OF PEPTIDES

Most of the 4-sector tandem mass spectra of peptides published as of mid-1989 are included in the lists in Tables 2, 3, and 4, where an attempt has been made to separate the spectra according to whether the protonated molecules fragment to give mainly N-terminal ions (a, b, c, d), C-terminal ions (v, w, x, y, z), or a mixture of both. The distinctions are not clear-cut, and there are many marginal cases. The name given is, in some cases, that of the peptide measured, and in others that of the protein from which the peptide was obtained. The sequences given do not take account of any substitution or derivatization (e.g., acetylation).

There have been one or two negative ion tandem mass spectra,[45,49] which are not included in the tables. Tandem mass spectra have been reported for cyclic peptides,[68] which by definition have no termini and are not included in the tables. A tandem mass spectrum has been reported for a peptide ASVGFKAGVL forming a radical-cation, presumably because Lys[6] is a trimethyllysine residue[69] This spectrum contains strong peaks due to a-, y-, x-, z-, w-, and v-ions, but there are no peaks of significant intensity below m/z 600. The conclusion drawn[69] was that the trimethyllysine was the charge site and that any fragments not containing that site would not be charged (i.e., product ions would contain that site).

TABLE 3
Peptides Whose Tandem Mass Spectra are Characterized by C-Terminal
Fragment Ions (Positive ions)

Sequence	M	Name	Fragment ions	Ref.
ADSGEGDFLAEGGGVR	1535.7	Fibrinopeptide A	y, some w, x	37, 41, 34
FSATWCGPCK	1214.4	Thioredoxin	y, some w, b	44
M*APV*DEIADEYAGR	1761.9	Thioredoxin	y, x	50
VAYSNLSTCLOGTYLQ YLKNFHMFPGINFGPEIP	3863.5	Val-ala-β-calcitonin	z, y some b	43
AGEETTTFTALTEK	1540.6	Prostatropin	z, y, some x, b	51
GIQLR	585.4	SBP	z, y, x, w, v some a, b	47, 52
CYVTELPDGFPR	1454.6	Dopamine	v, w, x, y, z some a, b	46
VTELPDGFPR	1129.6	Dopamine	v, w, x, y, z some b	45
AEGETTTFTALTEK	1497.7		w, y, z, some b	33
ANALLADGVELR	1240.6		v, w, x, y, z some b	33,45
SEGAGTFRM	1009.5	Retinaldehyde-binding protein	w, x, y, z, b	54
SGGASHSELIHNLR	1476.7		y, z	21

Note: y is used as an abbreviation for y_{n+2}.

* Signifies I or L (not distinguishable).

The incident ions for all of the tabulated spectra were protonated molecules, so questions arise about the sites of protonation[70] and the influence of site of protonation on fragmentation.[10] Johnson et al.[34] have recently addressed these questions and concluded that charge-remote fragmentation[71] is important in the collisionally activated dissociation of protonated peptides. The general observation from published spectra (Tables 2 to 4) is that the presence of basic amino acids near the N-terminus tends to favor the formation of N-terminal fragment ions, whereas basic amino acids near the C-terminus tend to favor the formation of C-terminal fragment ions. The implication is that the proton is located at the basic site, and when the peptide dissociates into two parts, the charge remains with that moiety containing the protonated base on account of energetic considerations. The fragmentation is thus not triggered by the proton, given that its location is fixed elsewhere. Thus, the conclusion is that the fragmentation is charge-remote and is essentially "chemistry of neutral species". This same picture has been put forward for the collisionally activated dissociation of polymer ions.[72,73] With protonated peptides, it has been proposed[34] that the charge-remote fragmentation tends to lead to a- and d-ions if the charge is fixed at the N-terminus, and to v-, w-, and y-ions if the charge is fixed at the C-terminus. The so-called conventional mechanism, in which the amide nitrogen atom in the backbone is protonated, possibly becomes important if there is no strongly basic sidechain site in the peptide. The possibility must also be considered that the proton is mobile to a greater or lesser extent, bearing in mind that energetic differences between different sites of protonation are likely to be much less than the energies required to fragment the molecules.

The proposed charge-remote fragmentations would seem to be higher-energy processes, which are not induced by low-energy collisions.[34] The possibility must be considered that the homolytic cleavages proposed as initiating at least certain of these fragmentations are nonergodic processes. That is to say, the kiloelectronvolt-collision deposits energy in a localized region of the molecule and bond cleavage occurs before that energy is dissipated throughout the rest of the molecule. The probability of initial bond cleavage then becomes dependent on the site of impact and energy deposition in the collision, the dissociation energies of bonds in that region,

TABLE 4
Peptides Whose Tandem Mass Spectra are Characterized by Mixed Fragmentation
(Both C- and N- Terminal Fragment Ions)

Sequence	M	Name	Fragment ions	Ref.
SEGAGTFR	865.4	Retinaldehyde-protein	x, y, z, b	53, 54
ENILGNNVGTY	1174.7	SBP	b, y	33, 47, 52
ADSGEGDFLAEGGGV	1379.6	Des-arg fibrino-peptide A	b, some y	41, 34
NFQTE**DSDK	1308.6	Thioredoxin	b, y	56
AMVSEFLK	965.4	Lipocortin I	y, b, some w	38
GIPTLLLFK	1006.6	Thioredoxin	y, b, w	1, 37
GIPTLMIFK	1018.6	Thioredoxin	y, b, w	1
GIPTLMLFR	1046.2	Thioredoxin	y, b, some w	1, 50
SDEEVEHVEE	1322.4	Troponin T	b, y, a	38
RASQGIRNDLG	1185.5	Amino acyl tRNA synthetase	a, b, some x, y	38
ELAFAPPEPA	950.4	β-lipotropin	b, y, some d, v, w	14, 35
VDAFLGTWK	1077.6		b, x, y, z	35
VDAFLGTWK	1105.6		b, x, y, z	35
ELAGAPPEPA	991.6	β-lipotropin	b, y, a, d	14
EHWSYGLRPG	1181.6	LHRH	v, y, z, b	14, 34
EHWSYALRPG	1195.6	D-Ala^6-LHRH	a, b, v, x, y, z	
EHWSYALRPG	1209.7	D-Ala^6-N-MeLeu^7-LHRH	a, b, x, w, y, z	14, 34
K*DEHIG	810.4		b, y	56
S*PTM***K	1014.5		b, y, some z, x	56
KRPPGFSPER	1187.6	Lys-bradykinin	a, y	31, 34, 22
DDPMLLSSGRVQLVVG	1684.9	PCMT	b, a, x	21
FPTIPLSR	978.5		a, b, w, x, y,z	55
DLEEGIQTLMGR	1360.7	Human growth hormone (hGH)	b.,w, x, y	55
LHQLAFDTYQEFEE AYIPK	2341.1		a, b, c, w, y x, z	55
ELYSALANKCCHV GCTKRSLARFC	2656.3		a, b, c, w, y	55
GGGG	246.1		y, b	49
YGGFM	573.2	Met-enkephalin	a, b, y	49, 22, 64
YGGFL	555.3	Leu-enkephalin	a, b, y	49
VVAGVANALAHK	1149		w, b, z, x, y	57
VVAGVANALAHK	1269	Xenobiotically modified	w, b, z, x, y	57
ADAFVGTWK	1036.5		b, y	58
GFVISHmsW	809		y, b	59
GFVISDW	823		y, b	59
SVVTLDGGK	875		y, w, b, x, z	60
ADAFVGTWK	1036		z, y, b	60
SLNTARNALTRAE	1416.8		d, w	61
SLRISPRE	957.5		a, w	61
ELTFT	573.3	HTF-II peptide	b, y	63
MLF	409		a, b, y	64, 67
YAGFL			a, b, y	65
<EVDPINQA	866.4	Physalaemin-like	b, y	66
<EVDPNIQA	866.4	Physalaemin-like	b, y	66

Note: y is used as an abbreviation for y_{n+2}; Hms signifies homoserine.

* Signifies I or L.

and the characteristics of intramolecular energy transfer. Once radical species have been created by homolytic cleavages, subsequent decompositions and rearrangements are likely to proceed with lower-energy barriers. What is certain is that the mechanistic understanding of the unimolecular fragmentation of protonated peptides and hence the ability to interpret spectra have both been advanced greatly by the advent of the 4-sector instrumentation, due in large measure to the careful detailed studies of Biemann and colleagues.[35] Computer programs are becoming available to aid interpretation of peptide tandem mass spectra,[74] and the extraction from tandem mass spectra of the maximum amounts of information contained therein concerning amino acid sequences promises to become a well-understood procedure.

VII. CONCLUSION

Tandem mass spectrometry of peptides using 4-sector instrumentation has altered qualitatively the role of mass spectrometry in the determination of amino acid sequences. Key characteristics of the 4-sector instrumentation are the abilities to select the ^{12}C-only component of the $[M+H]^+$ isotope cluster and to achieve unit mass resolution in the tandem mass spectra with mass assignments of the order of 0.1 a.m.u. These advantages make the difference between being able to interpret a spectrum reliably, and being unable to interpret with confidence, as is the case with other mass spectrometric techniques giving only low mass-resolution in mass-selection and production analysis. Sensitivity is the other key characteristic, with the array detectors having already improved sensitivity levels for synthetic peptides to the picomole domain.[24,25,75]

The general advantage of 4-sector tandem mass spectrometry, which is shared with all types of mass spectrometry, is that no implicit assumptions need be made concerning the range and type of molecular structure. This absence of restrictions is in sharp contrast to wet-chemistry procedures such as the Edman degradation. Thus, amino acid modifications pose no fundamental problem for tandem mass spectrometry, and, of course, the type of molecule accessible is not restricted to peptides. Confirmation of the structures of nisins (including the positions of the sulfur-bridged rings), which are a group of closely related highly-modified peptides, has been achieved by tandem mass spectrometry.[76] Components within actinomycin complexes with complicated ring structures incorporating sarcosine groups have been characterized by tandem mass spectrometry.[77] Tandem mass spectrometry has been employed to define the structures of peptidoglycan monomers[78] and to characterize structurally recombinant human interleukin-1α (IL-1α) in the context of a high-resolution X-ray crystallography study.[79] On-line interfaces with chromatographs are being developed, and interfacing with capillary zone electrophoresis has been reported as giving low femtomole sensitivities with high (hundreds of thousands of plates) electrophoretic separation efficiencies.[67] The expectation should be that 4-sector tandem mass spectrometry will be applied much more widely in the near future and that applications will develop quickly in carbohydrate fields, for biological surfactants, and for synthetic organic and inorganic polymeric materials.

Delineation of post-translational modification is a class of achievements of 4-sector tandem spectrometry in the characterization of protein structure.[80] The extensive studies of the thioredoxins by Biemann and collaborators[81] represent a clear example of how 4-sector tandem spectrometry can be used to investigate the amino acid sequence of high molecular weight (12 kDa) proteins. The effectiveness of 4-sector tandem spectrometry in the determination of sequence of modified peptides is well illustrated by the study of yeast-mating hormone a-factor, in which a farnesyl group attached to the sulphur of a cysteine residue and a terminal methyl ester were identified.[62] Xenobiotic modifications are amenable to study by 4-sector tandem spectrometry,[82] and their characterization can be effected unambiguously. Burlingame and colleagues[57] have demonstrated the power of the technique in an investigation of human hemoglobin treated *in vitro* with styrene-7-8-oxide. 4-Sector tandem spectrometry allows the specific residue modified to be identified reliably as β His-143. All indications are that the next few years will bring a great many more biologically significant applications of 4-sector tandem spectrometry as instrumentation is improved and more instruments become more widely available.

ACKNOWLEDGMENTS

We are pleased to acknowledge transmittance of preprints and reprints by Klaus Biemann, Al Burlingame, Barry Bycroft, Steven Carr, E. Kubota, Kenneth Rinehart, Jr., John Stults, and Ken Tomer. We are particularly grateful to and appreciative of Shani Ingram for her careful preparation of the manuscript, and we are grateful to Alex Colburn for the help with the drawings.

REFERENCES

1. **Biemann, K.,** *Biomed. Environ. Mass Spectrom.,* 16, 99, 1988.
2. **Burlingame, A. L.,** American Society for Mass Spectrometry, Miami, May 1989.
3. **Meng, C. K., Mann, M., and Fenn, J. B.,** *Z. Phys. D.,* 10, 361, 1988.
4. **McLafferty, F. W.,** in *Analytical Pyrolysis,* Jones, C. E. R. and Cramers, C. A., Eds., Elsevier, Amsterdam, 1977, 39.
5. **McLafferty, F. W., Todd, P. J., McGilvery, D. C., and Baldwin, M. A.,** *J. Am. Chem Soc.,* 102, 3360, 1980.
6. **Amster, I. J., Baldwin, M. A., Cheng, M. T., Proctor, C. J., and McLafferty, F. W.,** *J. Am. Chem Soc.,* 105, 1654, 1983.
7. **Weber, R. and Levsen, K.,** *Biomed Mass Spectrom.,* 7, 314, 1980.
8. **Desiderio, D. M. and Sabbatini, J. Z.,** *Biomed Mass Spectrom.,* 8, 565, 1981.
9. **Matsuo, T., Matsuda, H., Katakuse, I., Shimonishi, Y., Maruyama, Y., Higuchi, T., and Kubota, E.,** *Anal. Chem.,* 53, 416, 1981.
10. **Neumann, G. M. and Derrick, P. J.,** *Aust. J. Chem.,* 37, 2261, 1984.
11. **Craig, A. G. and Derrick, P. J.,** *Aust. J. Chem.,* 39, 1421, 1986.
12. **Mann, M., Meng, C. K., and Fenn, J. B.,** *Anal. Chem.,* 61, 1702, 1989.
13. **Covey, T. R., Bonner, R. F., Shushan, B. I., and Henion, J.,** *Rapid Commun. Mass Spectrom.,* 2, 249, 1988.
14. **Martin, S. A., Johnson, R. S., Costello, C. E., and Biemann, K.,** in the *Analysis of Peptides and Proteins by Mass Spectrometry,* McNeal, C. J., Ed., John Wiley & Sons, New York, 1988, 135.
15. **Carr, S. A., Green, B. N., Hemling, M. E., Roberts, G. D., Anderegg, R. J., and Vickers, R.,** American Society for Mass Spectrometry, Cincinnati, 1987.
16. **Sheil, M. M. and Derrick, P. J.,** *Org. Mass Spectrom.,* 23, 429, 1988.
17. **Uggerud, E. and Derrick, P. J.,** *J. Phys. Chem.,* to be published.
18. **Uggerud, E. and Derrick, P. J.,** *Z. Naturforsch.,* 44a, 245, 1989.
19. **Sheil, M. M. and Derrick, P. J.,** *J. Am. Chem. Soc.,* to be published.
20. **Sheil, M. M., Uggerud, E., and Derrick, P. J.,** *Adv. Mass Spectrom.,* 11, 1012, 1989.
21. **Henzel, W. J., Aswad, D. W., and Stults, J. T.,** in *Techniques in Protein Chemistry,* Hugli, T. E., Ed., Academic Press, New York, 1989.
22. **Elliott, G., Gallagher, R., Evans, S., Gooch, G., Wright, A. D., Burlingame, A. L., Walls, F. C., and Gillece-Castro, B.,** American Society for Mass Spectrometry, San Francisco, 1988.
23. **Martin, S. A.,** American Society for Mass Spectrometry, Miami, May 1989.
24. **Cottrell, J. and Evans, S.,** *Anal. Chem.,* 59, 1990, 1987.
25a. **Biemann, K.,** American Society for Mass Spectrometry, Miami, May 1989.
25b. **Itagaki, Y., Ishihara, M., Otsuka, K., Kammei, Y., Kubota, E., Musselman, B. D., and Mehlman, B. D.,** American Society for Mass Spectrometry, Miami, May 1989.
26. **Neumann, G. M. and Derrick, P. J.,** *Z. Naturforsch.,* 39a, 584, 1984.
27. **Bricker, D. L. and Russell, D. H.,** *J. Am. Chem. Soc.,* 108, 6174, 1986.
28. **Guevremont, R. and Boyd, R. K.,** *Rapid Commun. Mass Spectrom.,* 2, 1, 1988.
29. **Neumann, G. M. and Derrick, P. J.,** *Org. Mass Spectrom.,* 19, 165, 1984.
30. **Roepstorff, P. and Fohlman, J.,** *Biomed. Mass Spectrom.,* 11, 601, 1984.

31. **Biemann, K. and Martin, S. A.,** *Mass Spectrom. Rev.,* 6, 1, 1987.
32. **Taylor, L. C. E. and Poulter, L.,** American Society for Mass Spectrometry, San Francisco, 1988.
33. **Carr, S. A.,** *Adv. Drug. Deliv. Rev.,* 1989.
34. **Johnson, R. S., Martin, S. A., and Biemann, K.,** *Int. J. Mass Spectrom. Ion Proc.,* 86, 137, 1988.
35. **Biemann, K.,** in *Some Recent Applications of Mass Spectrometry to Biochemistry,* 62nd Biomedical Society Meeting, Oliver, R. W. A. and Thompson, I. S., Eds., University of Sheffield, 1988.
36. **Stults, J. T. and Watson, J. T.,** *Biomed. Environ. Mass Spectrom.,* 14, 583, 1987.
37. **Johnson, R. S., Martin, S. A., Biemann, K., Stults, J. T., and Watson, J. T.,** *Anal. Chem.,* 59, 2621, 1987.
38. **Biemann, K. and Scoble, H. A.,** *Science,* 237, 992, 1987.
39. Jeol Product Information, MS25, Jeol Ltd., Tokyo.
40. **Sato, K., Asada, T., Ishihara, M., Kunihiro, F., Kammei, Y., Kubota, E., Costello, C. E., Martin, S. A., Scoble, H. A., and Biemann, K.,** *Anal. Chem.,* 59, 1652, 1987.
41. **Martin, S. A. and Biemann, K.,** *Int. J. Mass Spectrom. Ion Proc.,* 78, 213, 1987.
42. **Scoble, H. A., Martin, S. A., and Biemann, K.,** *Biomed. J.,* 245, 621, 1987.
43. **Carr, S. A., Green, B. N., Hemling, M. E., Roberts, G. D., Anderegg, R. J., and Vickers, R.,** American Society for Mass Spectrometry, Denver, 1987.
44. **Johnson, R. S., Matthews, W. R., Biemann, K, and Hopper, S.,** *J. Biol. Chem.,* 263, 9859, 1988.
45. **Carr, S. A., Roberts, G. D., and Hemling, M. E.,** in *Mass Spectrometry of Biological Materials,* McEwen, C. N. and Larsen, B., Eds., Marcel Dekker, New York, 1989.
46. **DeWolf, W. E., Jr., Carr, S. A., Varrichio, A., Goodhart, P. J., Mentzer, M. A., Roberts, G. D., Southan, C., Dolle, R. E., and Kruse, L. T.,** *Biochemistry,* 27, 9093, 1988.
47. **Carr, S. A., Anderegg, R. J., and Hemling, M. E.,** in *The Analysis of Peptides and Proteins by Mass Spectrometry,* McNeal, C. J., Ed., John Wiley & Sons, Chichester, 1988, 95.
48. **Tomer, K. B., Guenat, C. R., and Deterding, L. J.,** *Anal. Chem.,* 60, 2232, 1988.
49. Jeol Product Information, MS25, Jeol Ltd., Tokyo.
50. **Johnson, R. S. and Biemann, K.,** *Biochemistry,* 26, 1209, 1987.
51. **Crabb, J. W., Armes, L. G., Carr, S. A., Johnson, C. M., Roberts, G. D., Bordoli, R. S., and McKeehan, W. L.,** *Biochemistry,* 25, 4988, 1986.
52. **Anderegg, R. J., Carr, S. A., Huang, I. Y., Hupakka, R. A., Chang, C., and Liao, S.,** *Biochemistry,* 27, 4214, 1988.
53. **Carr, S. A., Hemling, M. E., and Roberts, G. D.,** in *Macromolecular Sequencing and Synthesis Selected Methods and Applications,* Alan R. Liss, New York, 1988.
54. **Crabb, J. W., Johnson, C. M., Carr, S. A., Armes, L. G., and Saari, J. C.,** *J. Biol. Chem.,* 263, 18678, 1988.
55. **Stults, J. T. and Chakel, J. A.,** American Society for Mass Spectrometry, San Francisco, 1988.
56. **Matthews, W. R., Johnson, R. S., Cornwall, K. L., Johnson, T. C., Buchanan, B. B., and Biemann, K.,** *J. Biol. Chem.,* 262, 7537, 1987.
57. **Kaur, S., Hollander, D., Haas, R., and Burlingame, A. L.,** *J. Biol. Chem.,* in press.
58. **Gibson, B. W., Yu, A., Aberth, W., Burlingame, A. L., and Bass, N. M.,** *J. Biol. Chem.,* 263, 4182, 1988.
59. **Sanders, D. A., Gillence-Castro, B. L., Burlingame, A. L., and Koshland, D. E., Jr.,** *J. Biol. Chem.,* 264, 21770, 1989.
60. **Ginson, B. W., Yu, Z., Gillence-Castro, B. L., Aberth, W., Walls, F. C., and Burlingame, A. L.,** in *Techniques in Protein Chemistry,* Hugli, T., Ed., Academic Press, New York, 1989, 135.
61. **Falick, A. M., Mel, S. F., Stroud, R. M., and Burlingame, A. L.,** in *Techniques in Protein Chemistry,* Hugli, T., Ed., Academic Press, New York, 1989, 152.
62. **Anderegg, R. J., Betz, R., Carr, S. A., Crabb, J. W., and Duntze, W.,** *J. Biol. Chem.,* 263, 18236, 1988.
63. **Gade, G. and Rinehart, K. L., Jr.,** *Biol. Chem., Hoppe-Seyler,* 368, 67, 1987.
64. **Deterding, L. J., Moseley, M. A., Tomer, K. B., and Jorgenson, J. W.,** in press.
65. **Deterding, L. J., Tomer, K. B., and Spatola, A. F.,** in press.
66. **Harvan, D. J., Haas, J. R., Wilson, W. E., Hamm, C., Boyd, R. K., Yajima, H., and Klapper, D. G.,** *Biomed. Environ. Mass Spectrom.,* 14, 281, 1987.
67. **Moseley, M. A., Deterding, L. J., and Tomer, K. B.,** *Rapid Commun. Mass Spectrom.,* in press.
68. **Rinehart, K. L., Thompson, A. G., Rong, L., Milberg, R. M., and Curtis, J. M.,** American Society for Mass Spectrometry, San Francisco, 1988.
69. **Houtz, R. L., Stults, J. T., Mulligan, R. M., and Tolbert, N. E.,** *Proc. Natl. Acad. Sci. U.S.A.,* in press.
70. **Bojesen, G.,** *J. Am. Chem. Soc.,* 109, 5557, 1987.
71. **Jensen, N. J., Tomer, K. B., and Gross, M. L.,** *J. Am. Chem. Soc.,* 107, 1863, 1985.
72. **Craig, A. G. and Derrick, P. J.,** *J. Chem. Soc. Chem. Commun.,* 981, 1985.
73. **Agma, M.,** Ph.D. thesis, University of New South Wales, Sydney, 1989.
74. **Scoble, H. A., Biller, J. E., and Biemann, K.,** *Fres. A. Anal. Chem.,* 327, 239, 1987.
75. **Hill, J. A., Martin, S. A., Biller, J. E., and Biemann, K.,** *Biomed. Environ. Mass Spectrom.,* 17, 147, 1988.

76. **Barber, M., Elliott, G. J., Bordoli, R. S., Green, B. N., and Bycroft, B. W.,** *Experientia,* 44, 266, 1988.
77. **Barber, M., Bell, D., Morris, M., Tetler, L., Woods, M., Bycroft, B. W., Monaghan, J. J., Morden, W. E., and Green, B. N.,** *Talanta,* 35, 605, 1988.
78. **Martin, S. A., Rosenthal, R. S., and Biemann, K.,** *J. Biol. Chem.,* 262, 7514, 1987.
79. **Hassell, A. M., Johanson, K. O., Goodhart, P., Young, P. R., Holskin, B. P., Carr, S. A., Roberts, G. D., Simon, P. L., Chen, M. J., and Lewis, M.,** *J. Biol. Chem.,* 264, 4948, 1989.
80. **Burlingame, A. L.,** in *Techniques in Protein Chemistry,* Hugli, T., Ed., Academic Press, New York, 1989, 121.
81. **Johnson, T. C., Yee, B. C., Carlson, D. E., Buchanan, B. B., Johnson, R. S., Matthews, W. R., and Biemann, K.,** *J. Bacteriol.,* in press.
82. **Deterding, L. J., Srinivas, P., Mahmood, N. A., Burka, L. T., and Tomer, K. B.,** *Anal. Biochem.,* in press.

Chapter 8

PEPTIDE SEQUENCE ANALYSIS BY TRIPLE QUADRUPOLE AND QUADRUPOLE FOURIER TRANSFORM MASS SPECTROMETRY

D. F. Hunt, T. Krishnamurthy, J. Shabanowitz, P. R. Griffin, J. R. Yates, III, P. A. Martino, A. L. McCormack, and C. R. Hauer.

TABLE OF CONTENTS

I.   Introduction 140

II. Methods 140

III. Results and Discussion 140
     A.   Oligopeptide Sequence Analysis on the TSQ-70 Triple Quadrupole
          Mass Spectrometer 140
     B.   Oligopeptide Sequence Analysis on the Quadrupole Fourier
          Transform Mass Spectrometer 142
     C.   Characterization of the Protein Society Symposium Test Peptide-3 145
          1.   Summary of Chemical Experiments and Molecular Weight
               Determinations 148
          2.   Summary of Results Obtained from Chemical and Enzymatic
               Digests 149
          3.   Linkage Assignment of the Two Peptide Chains 151
     D.   Characterization of Toxic Cyclic Peptides from Blue-Green Algae 152

References 157

I. INTRODUCTION

Since 1981,[1] a major part of our research effort has focused on continued development of tandem mass spectrometry and the chemical methods that allow us to utilize this instrumental technique for characterizing the primary structures of both peptides and proteins.

Here, we (1) provide an update on the performance level of our present instrumentation, (2) discuss the approach used to sequence symposium test peptide-3 (STP-3), a novel synthetic peptide distributed by the Protein Society to test existing instrumental methods of sequence analysis, and (3) describe efforts to characterize the amino acid sequence of a toxic cyclic peptide from blue-green algae.

II. METHODS

Mass spectra were recorded on either a triple quadrupole instrument assembled at the University of Virginia with components provided by Finnigan-MAT, the TSQ-70 triple quadrupole (Finnigan-MAT, San Jose, CA),[2] or a quadrupole Fourier transform instrument constructed at the University of Virginia.[3] Operation of these instruments has been described previously.[2,3] Sample ionization/volatilization on the TSQ-70 was performed by particle bombardment with an Antek (Palo Alto, CA) cesium ion gun operated at 6 keV. Ion detection was accomplished with the conversion dynode operated at 15 keV. Samples to be ionized by particle bombardment were prepared by adding 0.5 μl of 0.1% trifluoroacetic acid (TFA) solution containing peptide at the 1 to 50 pmol in 0.5 ml of thioglycerol matrix on a gold-plated stainless steel probe.

III. RESULTS AND DISCUSSION

A. OLIGOPEPTIDE SEQUENCE ANALYSIS ON THE TSQ-70 TRIPLE QUADRUPOLE MASS SPECTROMETER

Limitations of the tandem mass spectrometry approach to peptide sequence analysis as practiced previously on the triple quadrupole mass spectrometer built at the University of Virginia[4] included (1) an inability to produce fragment ions under low-energy conditions that facilitate differentiation of the two amino acids, Leu and Ile; (2) a sample requirement on the order of 100 to 800 pmol for collision activated dissociation experiments; and (3) a mass range limitation of 1800 Da imposed by design of both the quadrupole mass filters and instrument electronics. Use of the Finnigan Model TSQ-70 triple quadrupole instrument has now made it possible to reduce the sample quantity to the 5 to 100 pmol level and to expand the useful mass range for sequence analysis to at least 2600 Da.

Shown in Figure 1 is the collision activated dissociation mass spectrum recorded on 5 pmol of renin-substrate tetradecapeptide. To obtain these data, the first mass filter of the instrument, quadrupole 1, was set to pass all ions within a 6 mass unit around the (M+H)$^+$ ion at m/z 1759. On transmission to the collision chamber, a bent-quadrupole mass filter operated in the radio frequency (rf) only mode, the (M+H)$^+$ ions suffer as many as 10 low energy (15 to 40 eV) collisions with an Ar/Xe (2/1) mixture present at a total pressure of 3 mTorr. During this process, kinetic energy is converted to vibrational energy and the peptide (M+H)$^+$ ions undergo fragmentation, more or less randomly, at the various amide bonds along the backbone of the linear peptide chain. A collection of neutrals and ions is produced by this process. Neutrals are pumped away by the vacuum system, and the charged fragments are transmitted to a second mass filter, quadrupole 3, which then separates them according to mass. To maximize sensitivity, the second mass filter is operated at less than unit resolution. Detection of the ions is accomplished with a newly implemented 15 keV conversion dynode, a device that contributes at least a factor of 5 to the observed improvement in sensitivity.

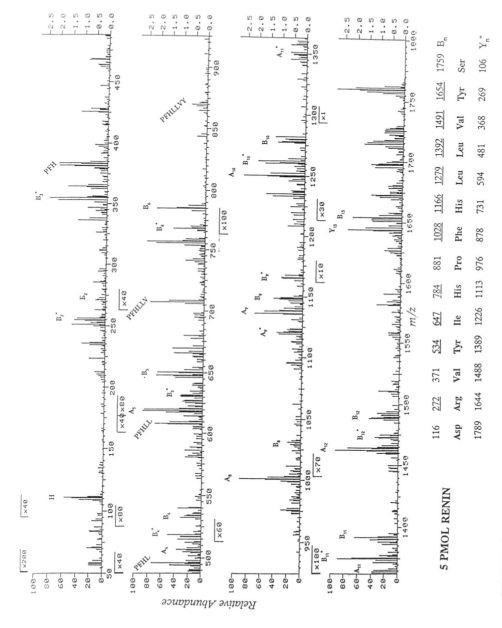

5 PMOL RENIN

116	272	371	534	647	784	881	1028	1166	1279	1392	1491	1654	1759	B_n
Asp	Arg	Val	Tyr	Ile	His	Pro	Phe	His	Leu	Leu	Val	Tyr	Ser	
1789	1644	1488	1389	1226	1113	976	878	731	594	481	368	269	106	Y_n''

FIGURE 1. Collision activated dissociation mass spectrum recorded on (M+H)⁺ (mz 1759) derived from a 5-pmol sample of renin-substrate tetradecapeptide.

To generate the spectrum in Figure 1, the mass range was scanned five times and the resulting spectra were then summed. Total instrument time to acquire the data was 20 s. Predicted fragment ions[5] of type b for the tetradecapeptide are shown above the structure in Figure 1, and those ions observed in the spectrum are underlined. These ion types all contain the amino terminal reside plus one or more additional residues. Subtraction of m/z values for any two fragments of type b that differ by a single amino acid, NHCH(R)CO, generates a value that identifies the extra residue present in the larger fragment.

Predicted masses m/z values, for ions of type y, all of which contain the C-terminal residue plus one, two, three, four, etc., additional residues,[5] are shown below the structure. Subtraction of m/z values for any two fragments that differ by a single amino acid, NHCH(R)CO, generates a value that specifies the mass and thus the identity of the extra residue in the larger fragment. Note that the sequence of all but the first two amino acids in the tetradecapeptide can be deduced easily from fragment ions of type b. Appearence of the y_{13} ion at m/z 1644 defines the order of the first two residues as Asp-Arg rather than Arg-Asp.

Ions of type a, those that differ from type b by loss of carbon monoxide,[5] and a set of ions (labeled with *), formed by loss of ammonia from either type a or type b ions is also prominent in the spectrum. Fragments, resulting from multistep cleavage of the peptide chain at Pro-7, are labeled with the single letter code for the amino acid residues contained within the fragment.

Shown in Figure 2 is the collision activated dissociation mass spectrum recorded on $(M+H)^+$ ions derived from the 20-residue, C-terminal tryptic peptide from ribonuclease. Total instrument time to obtain these data from a 20 pmol sample was only 40 s. Predicted fragment ions of type b and y for the C-terminal peptide are shown, respectively, above and below the structure in Figure 2. Those ions observed in the spectrum are underlined. Ions of type b again provide the information needed to deduce the sequence of all but the first two residues. Presence of an abundant ion of type y at m/z 2088 dictates that the order of the first two amino acids be His-Ile rather than Ile-His.

Use of the TSQ-70 instrument now makes it possible to sequence oligopeptides containing more than 20 residues on a routine basis. The methyl ester of ACTH CLIP, a 22-residue peptide, $(M+H^+ = 2549)$ can be sequenced completely at the 20 pmol level. At present, data are insufficient to determine whether the useful range of the instrument can be extended beyond m/z 2600.

B. OLIGOPEPTIDE SEQUENCE ANALYSIS ON THE QUADRUPOLE FOURIER TRANSFORM MASS SPECTROMETER

The quadrupole Fourier transform mass spectrometer (Q-FTMS), shown schematically in Figure 3, consists of a standard ion source, two sets of quadrupole rods, an ion cyclotron resonance (ICR) cell housed within a 7 Tesla, superconducting magnet, and an argon fluoride excimer laser for photodissociation of trapped ions.

Sample analysis on the Q-FTMS is performed by adding 0.5 μl of 0.1% trifluoroacetic acid solution containing peptides at the 5 to 30 pmol level to 0.5 μl of a matrix of glycerol: monothioglycerol (1:3) on a gold-coated, stainless steel probe tip that is inserted into the ion source of the instrument. Mass spectra are acquired by using a series of pulses controlled by the data system. The first, or beam pulse, lowers the voltage on the extractor lens of the Cs^+ ion gun and allows 10 keV Cs^+ ions to impact on the sample matrix for 4 ms. The same pulse places rf potential on the quadrupole rods and lowers the dc potential on trapping plate 1 from +3 to 0 V. This process allows efficient transport of the sputtered sample ions through the fringing fields of the magnet and into the ICR cell. The ions become trapped when the voltage on trapping plate 1 is raised back to a value of +3 V. After a 10 ms delay, the data system can also trigger a 10-ns pulse of light from the ArF excimer laser to photodissociate trapped ions. In the next step, 10 ms later, a rf pulse of 35 V, peak-to-peak, is applied to the transmitter plates (top) for 2.5 ms. This pulse contains a range of frequencies that accelerates all ions of all masses trapped in the cell into large

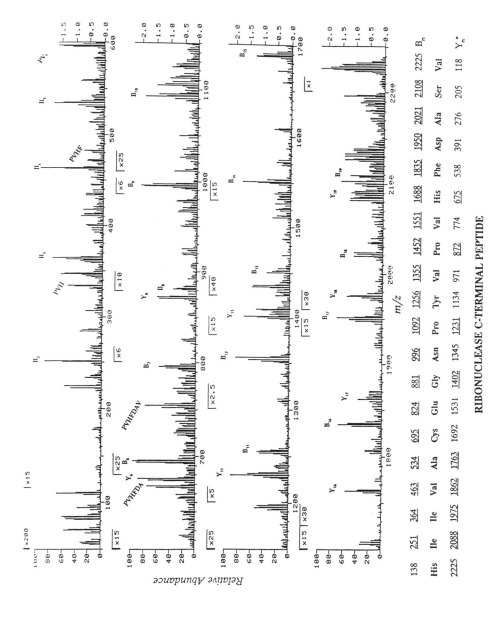

138	251	364	463	534	695	824	881	996	1092	1256	1355	1452	1551	1688	1835	1950	2021	2108	2225	B_n
His	Ile	Ile	Val	Ala	Cys	Glu	Gly	Asn	Pro	Tyr	Val	Pro	Val	His	Phe	Asp	Ala	Ser	Val	
2225	2088	1975	1862	1763	1692	1531	1402	1345	1231	1134	971	872	774	675	538	391	276	205	118	Y_n''

RIBONUCLEASE C-TERMINAL PEPTIDE

FIGURE 2. Collision activated dissociation mass spectrum recorded on (M+H)⁺ derived from a 20-pmol sample of ribonuclease C-terminal tryptic peptide.

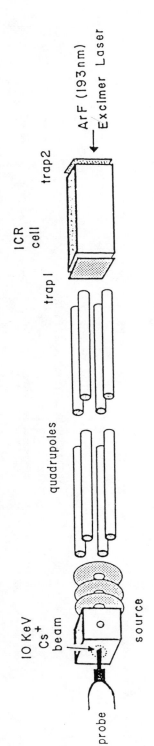

FIGURE 3.   Quadrupole Fourier transform mass spectrometer. (From Hefta, S. A., et al., in *Techniques in Protein Chemistry*, Hugli, T. E., Ed., Academic Press, New York, 1989, 560. With permission.)

orbits and causes them to move coherently at their characteristic cyclotron frequencies, values that are inversely proportional to the mass of the ion. Ion image currents induced by these orbiting ions on the ICR receiver (side) plates are monitored for 8 to 64 ms, amplified, and then digitized. Data from 30 to 200 such experiments are acquired, summed together, and the resulting time-averaged signal is then converted to the desired mass spectrum by using Fourier transform analysis. The total experiment time to produce the required data is approximately 30 s, and the total sample ionization or consumption time is less than 1 s. In general, mass spectra on samples in the mass range between 2000 and 6000 Da can now be obtained routinely at the 5 to 20 pmol level. Detection limits for many peptides below mass 2000 are in the high femtomole range.

A particularly useful feature of the Q-FTMS is that it is often possible to obtain spectra of oligopeptide samples, with and without fragmentation, by simply altering the time that the ions spend in the cell prior to excitation with the external rf pulse. Spectra recorded immediately after the ions enter the cell are usually devoid of fragmentation and are dominated by signals due to $(M+H)^+$ ions. Spectra recorded after the ions have been stored for several hundred milliseconds often undergo extensive metastable fragmentation, particularly if the internal energy of the ions is increased by using a very thin film of thioglycerol as the matrix for ionization by particle bombardment.

Shown in Figure 4 is a metastable fragmentation spectrum recorded on a 10-pmol sample of an oligopeptide methyl ester derived from a recently characterized, 76 residue, human monocyte chemoattractant, a putative mediator of cellular immunity.[6] Predicted fragment ions of types b and y are shown above and below the structure at the top of the figure. Those ions observed in the spectrum are underlined. In this particular case, the complete amino acid sequence was obtained by simply energizing the $(M+H)^+$ ions and then letting them fragment in the cell over a 100-ms time period before the spectrum was recorded.

Laser photodissociation of $(M+H)^+$ ions can also be employed to generate the fragments required for sequence analysis.[4,7] Shown in Figure 5 is the photodissociation mass spectrum recorded of the methyl ester of a hexadecapeptide isolated from an endo-Lys C digest of a 61 kDa calmodulin-dependent cyclic nucleotide phosphodiesterase isozyme.[8] Sample at the 10 pmol level was bombarded with 10 keV Cs^+ ions for 4 ms, and the resulting $(M+H)^+$ ions were then stored in the ICR cell and exposed to a single 10 ns pulse of radiation at 193 nm from the ArF excimer laser. Light of this wavelength is absorbed efficiently by amide bonds in the peptide backbone. Absorption of a photon places 6.42 eV of additional energy in the $(M+H)^+$ ion, which then dissociates, primarily at the amide linkages, to produce the fragments required for structural characterization. Data from 50 such experiments on the same 10 pmol sample were acquired and summed in less than 20 s to produce the spectrum shown in Figure 5.

Edman degradation failed to release an amino acid from this peptide. This result, along with the known specificity of endo-Lys-C, suggested that the peptide was blocked at the N-terminus and that the C-terminus contained a Lys residue. Note that the spectrum is dominated by b-type ions and/or ions of type b minus 18 Da. The latter ions are formed by loss of water from fragments containing threonine. Predicted fragment ions of types b and y for this peptide are shown above and below the structure in Figure 5. Note that several ions of type y also appear in the spectrum. Information contained in b and y type ions is sufficient to specify the complete amino acid sequence in this hexadecapeptide.

C. CHARACTERIZATION OF THE PROTEIN SOCIETY SYMPOSIUM TEST PEPTIDE-3

More than 150 different research labs accepted samples of the STP-3. Only three groups succeeded in obtaining the correct amino acid sequence. Two of these groups employed a combination fo standard Edman degradation for sequence analysis and mass spectrometry for molecular weight determination.[5,6] Only the authors employed tandem mass spectrometry to generate the complete sequence.[7] The following is a description of this effort.

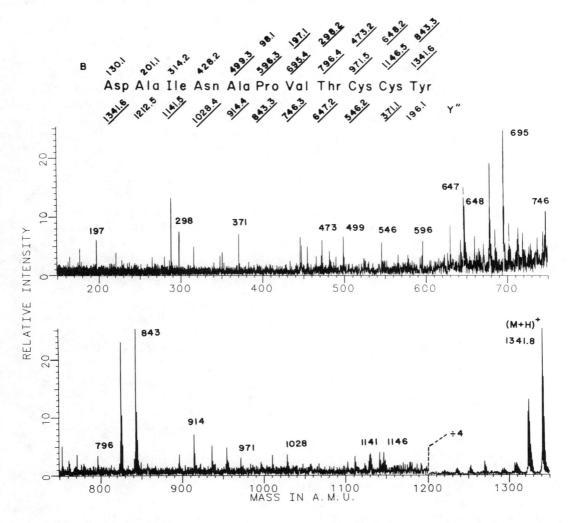

FIGURE 4. Metastable decomposition mass spectrum recorded on the methyl ester (10 pmol) of a peptide fragment from human monocyte chemoattractant. (From Robinson, E. A., et al., *Proc. Natl. Acad. Sci. U.S.A.*, 86, 1850, 1989. With permission.)

STP-3 was shown to consist of a single Lys-amide residue connected through both ε- and α-amino groups to 2 peptide chains, 9 and 17 residues in length, respectively. As shown here, the longer of the two chains contains an intramolecular disulfide linkage.

$$\text{NH}_2\text{–A G S H F C N G M L C N A G L S G–NH–CH–CONH}_2$$

$$(\text{CH}_2)_4$$

$$\text{NH}_2\text{-A G H F Y H L G M–NH}$$

STP-3, ≈3 nmol, was received as a lyophilized solid in the bottom of a plastic Eppendorf tube. Salts and other contaminants, if present, were removed from the sample by reversed-phase high pressure liquid chromatography (RP HPLC) on an Applied Biosystems Model 130A separations system.

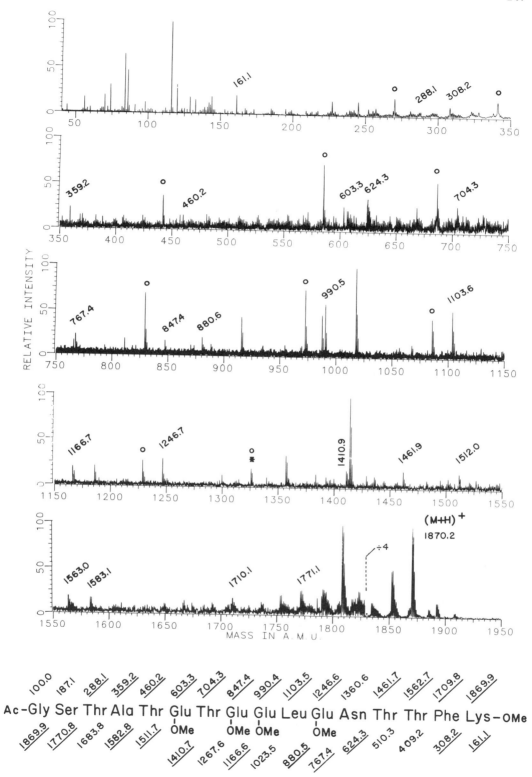

FIGURE 5. Laser photodissociation mass spectrum recorded on (M+H)⁺ from a 10-pmol sample of the N-terminal peptide methyl ester from a calmodulin-dependent cyclic nucleotide phosphodiesterase isozyme.

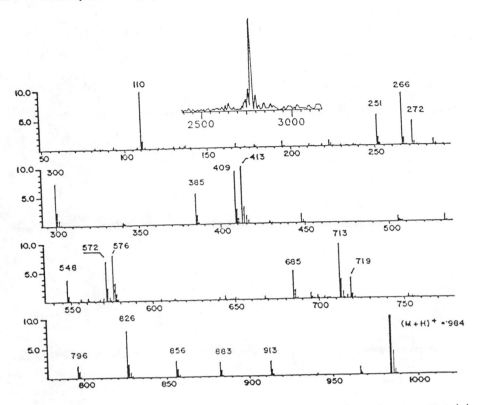

FIGURE 6.   (A) (M+H)⁺ ion at m/z 2778.3 in the low-resolution Fourier transform mass spectrum recorded on 40 to 60 pmol of STP-3; (B) collision activated dissociation mass spectrum recorded on the homoserine lactone (M+H)⁺ ion derived from the CNBr fragment of molecular mass 984. (From Hefta, S. A., et al., in *Techniques in Protein Chemistry*, Hugli, T. E., Ed., Academic Press, New York, 1989, 560. With permission.)

1. Summary of Chemical Experiments and Molecular Weight Determinations

Figure 6A shows the low-resolution mass spectrum (16 K transform) recorded on 40 to 60 pmol of HPLC-purified STP-3. The unresolved (M+H)⁺ multiplet appears at an m/z of 2778.3. Accordingly, STP-33 was assigned an average of 2777.3.

Acetylation of STP-3 was performed on the solids probe of the mass spectrometer by addition of 0.5 μl of 3/1 (v:v) MeOH/acetic anhydride and a trace of pyridine to the sample. Mass spectra recorded 30 s later showed that the (M+H)⁺ ion had shifted by 126 Da to higher mass (m/z 2906). Acetylation in the absence of pyridine occurs only on amine groups. In the presence of pyridine, Tyr residues are also acetylated. Introduction of a single acetyl group increased the mass of the molecule by 42 mass units. The shift of 126 mass units observed for the (M+H)⁺ ion of STP-3 suggested that the molecule contained a total of three amino and/or phenolic groups.

Performic acid oxidation of STP-3 was performed on the solids probe of the mass spectrometer by addition of 0.5 μl of standard reagent (15 μl of 30% hydrogen peroxide plus 85 μl of 100% formic acid) to the sample matrix. Mass spectra recorded several minutes later showed the presence of new (M+H)⁺ ions at m/z 2908, 2924, and 2940. The observed mass shifts of 130, 146, and 162 U are consistent with the incorporation of two hydrogens (1 mass unit each) plus eight, nine, and ten oxygen atoms (16 mass units each), respectively. Oxidation of two disulfide-linked, Cys residues to cysteic acid groups and two Met residues to Met sulfoxide and/or sulfone groups explains these data.

STP-3 (60 pmol) was converted to the corresponding methyl ester on treatment with 2N HCl/MeOH for 1 h at room temperature. A mass spectrum recorded on the product of this transformation showed a new (M+H)⁺ ion at m/z 2808. Conversion of sidechain or C-terminal

COOH groups to the corresponding methyl esters (–COOCH$_3$) increases the mass of the molecule by 14 Da/COOH. Two explanations were considered for the observed mass shift of 30 Da: (1) conversion of two acid-sensitive Asn or Gln residues to the corresponding methyl esters (15 mass units/CONH$_2$) or (2) oxidation of 1 Met residue to the sulfoxide and conversion of 1 COOH group to the corresponding methyl ester. Only the first explanation is consistent with the structure proposed for STP-3.

2. Summary of Results Obtained from Chemical and Enzymatic Digests

Treatment of STP-3 at the 100 pmol level with trypsin, Asp-N, Glu-C, and Lys-C failed to cleave the molecule. Mass spectra recorded on an aliquot from each reaction mixture showed that the (M+H)$^+$ ion remained at m/z 2778.3.

Treatment of STP-3 (500 pmol) with cyanogen bromide afforded two fragments. (M+H)$^+$ ions for the two species were observed at m/z 984 and 1735, respectively. Because treatment of the two fragments with 2N HCl in MeOH shifted each (M+H)$^+$ ion to higher mass by 32 Da, it was concluded that both molecules contained homoserine lactones, and that neither had a free carboxyl group. It was also noted that the molecular weights of the two fragments (containing Met rather than homoserine lactone residues) minus water summed to give the original molecular weight of STP-3. This relationship could be the case only if one of the two Met residues were part of a ring system.

Shown in Figure 6B is the collision activated dissociation mass spectrum recorded on the (M+H)$^+$ ion at m/z 984. The amino acid sequence deduced from the spectrum is shown here. Fragment ions of types b and y predicted for the sequence are shown above and below the structure. Those ions observed in the spectrum are underlined. Note that the fragmentation patterns observed on the triple quadrupole mass spectrometer do not allow differentiation of Leu and Ile, two residues of identical mass. Residue 7 in the cyanogen bromide fragment is assigned as Leu because amino acid analysis performed on STP-3 showed that Ile was not present in the molecule.

72	129	<u>266</u>	<u>413</u>	<u>576</u>	<u>713</u>	<u>826</u>	<u>883</u>	<u>984</u>	y
Ala	Gly	His	Phe	Tyr	His	Leu	Gly	hSer	
<u>984</u>	<u>913</u>	<u>856</u>	<u>719</u>	<u>572</u>	<u>409</u>	<u>272</u>	159	102	b

Digestion of STP-3 (500 pmol) with chymotrypsin produced two small peptides, the mass spectra of which showed (M+H)$^+$ ions at m/z 728 and 993. Treatment of the smaller peptide with methanolic HCl failed to shift the (M+H)$^+$ to higher mass. On-probe acetylation added two acetyl groups to the peptide, and thus shifted the (M+H)$^+$ to higher mass by 84 Da. These results indicated that the smaller chymotryptic fragment contained two amino groups and a blocked C-terminus. These structural features could be accomodated by either a conventional linear peptide containing a Lys residue and a blocked C-terminus, or an unconventional peptide conaining two chains linked to the α- and ε-amino groups of a Lys residue blocked at its C-terminus. Only the latter possibility agreed with the fragmentation pattern obtained in the collision activated dissociation mass spectrum recorded on the (M+H)$^+$ ion at m/z 728 (Figure 7). The amino acid sequence deduced from this spectrum is shown below. Predicted fragment ions containing His and Ser as the N-termini are shown above the below structure, respectively. Those ions observed in the spectrum are underlined.

	138	<u>251</u>	<u>308</u>	<u>439</u>	<u>584</u> CONH2	<u>641</u>	<u>728</u>
	NH$_2$–His	Leu	Gly	Met	| Lys	Gly	Ser–NH$_2$
	<u>728</u>	<u>591</u>	<u>478</u>	<u>421</u>	<u>290</u>	145	88

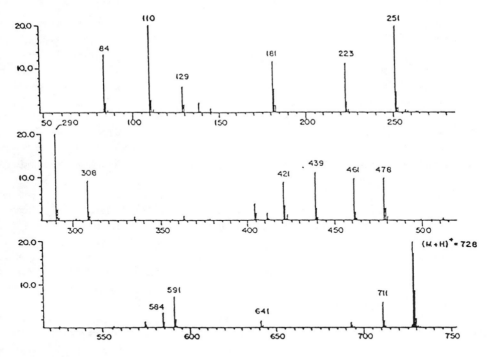

FIGURE 7. Collision activated dissociation mass spectrum recorded on the (M+H)⁺ion, m/z 728, derived from a chymotryptic fragment of STP-3. (From Hefta, S. A., et al., in *Techniques in Protein Chemistry*, Hugli, T. E., Ed., Academic Press, New York, 1989, 560. With permission.)

Esterification and acetylation of the larger chymotryptic fragment shifted the (M+H)⁺ ion at m/z 993 to higher mass by 14 and 42 mass units, respectively. These are the expected results for a peptide containing a single amino group and a single carboxyl group. A collision activated dissociation mass spectrum recorded on the (M+H)⁺ ion for the methyl ester (m/z 1007) provided the necessary fragmentation to deduce the sequence of this peptide. Confirmation of the deduced sequence was obtained by treating the peptide with dithiothreitol and iodoacetic acid. Both Cys residues were reduced and carboxymethylated under the above conditions, and the (M+H)⁺ at m/z 993 shifted to m/z 1111 as expected.

Cys Asn Gly Met Leu Cys Asn Ala Gly Leu

Information to align these sequences within the structure of STP-3 was obtained by analysis of a laser photodissociation mass spectrum recorded on 50 pmol of the intact parent molecule. Photodissociation of the (M+H)⁺ ion at m/z 2778.3 gave the spectrum shown in Figure 8. Observed fragment ions are consistent only with attachment of the two chymotryptic peptides and CNBr fragment in the order shown in the following.

		1658	1544	1473	1416	1303	1216	1159
Cys Asn Gly Me Leu Cys		Asn	Ala	Gly	Leu	Ser	Gly	Lys–CONH₂

Ala	Gly	His	Phe	Tyr	His	Leu	Gly	Met
72	129	266	413	576	713	826	883	1014

Subtractive Edman degradation performed at the 500 pmol sample level was employed to deduce the order of the missing amino acids at the N-terminus of the larger peptide chain in STP-3. In this experiment, mass measurements were made on the (M+H)⁺ ion observed for the parent molecule, before and after removal of an amino acid by manual Edman degradation. Results

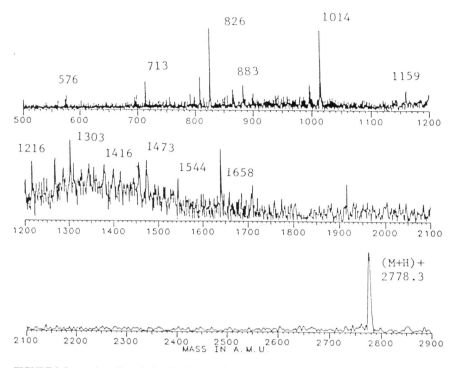

FIGURE 8. Laser photodissociation Fourier transform mass spectrum recorded at the 50 pmol sample level on the (M+H)⁺ ion, m/z 2778.3, of STP-3. (From Hefta, S. A., et al., in *Techniques in Protein Chemistry*, Hugli, T. E., Ed., Academic Press, New York, 1989, 560. With permission.)

TABLE 1

Data Obtained on Intact STP-3 by Manuel Edman Degradation Plus Mass Spectrometry

Manual Edman Degradation and Mass Spectrometry

Cycle no.	Observed (M+H)⁺	Mass lost	Amino acids identified	
0	2778	—		
1	2636	142	A	A
2	2522	114	G	G
3	2298	224	H	S
4	2014	284	F	H
5	1704	310	Y	F

From Hefta, S. A., et al., in *Techniques in Protein Chemistry*, Hugli, T. E., Ed., Academic Press, New York, 1989, 560. With permission.)

obtained after each of five such cycles of Edman degradation are summarized in Table 1. Because STP-3 contains two amino termini, two amino acids were released during each cycle of Edman degradation. As expected, one of the two amino acids removed in each cycle matched the residue predicted from the sequence of the CNBr fragment discussed previously. The other amino acid released in each cycle was assigned to the amino terminus of chain 2. Addition of amino acids AGSHF- to the N-terminal side of the first Cys residue brought the mass of the deduced structure up to that measured for the (M+H)⁺ ion of STP-3.

3. Linkage Assignment of the Two Peptide Chains

To determine which peptide chain was linked to the α- and ε-amino groups of Lys, STP-3 (200 pmol) was reduced, carboxymethylated, and cleaved with CNBr. The fragment containing

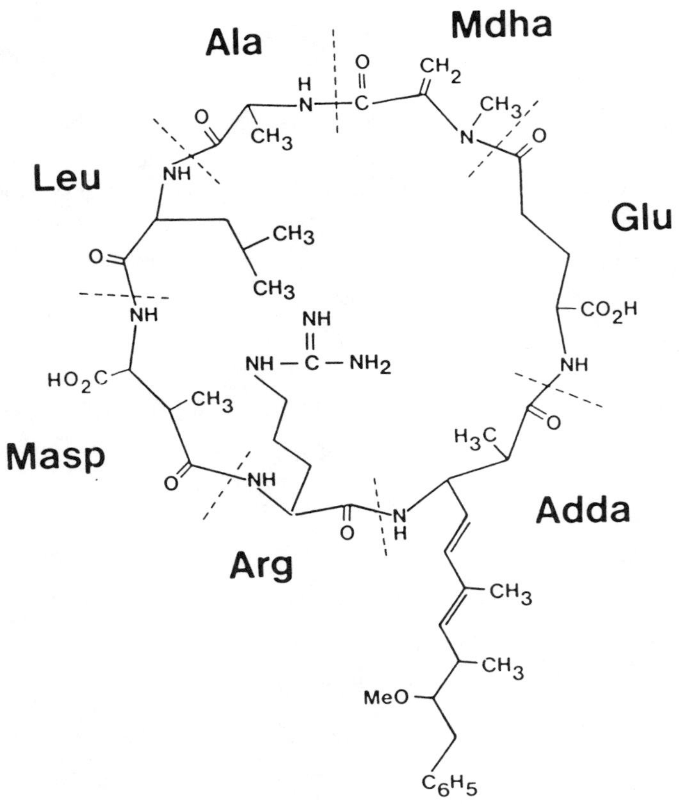

FIGURE 9.   Structure of toxic cyclic peptide from blooms of blue-green algae, *M. aeruginosa.* (From Krishnamurthy, T. A., et al., *Proc. Natl. Acad. Sci. U.S.A.,* 86, 770, 1989. With permission.)

Lys at the C-terminus, $(M+H)^+ = 919$, was then acetylated under experimental conditions that failed to introduce an acetyl group at the ε-amino group of model peptides (3/1 v:v $MeOH/Ac_2O$ containing 2% TFA).

Leu Cys Asn Ala Gly Leu Ser Gly Lys–$CONH_2$ $(M+H)^+ = 919$

Acetylation of the N-terminus of the model peptides occurred to the extent of 40% under these conditions. Acetylation of the cyanogen bromide fragment from STP-3 with the above reagent incorporated a single acetyl group and shifted the observed $(M+H)^+$ ion from m/z 919 to 961. This result suggests strongly that the cyanogen bromide fragment does not contain two α-amino groups. Accordingly, it was concluded that the Lys residue must have a free ε-amino group in the CNBr fragment and, therefore, that the longer of the two peptide chains is linked to the α-amino group of Lys.

D. CHARACTERIZATION OF TOXIC CYCLIC PEPTIDES FROM BLUE-GREEN ALGAE

Production of a family of toxic cyclic peptides in fresh-water supplies by the cyanobacterium *Microcystis aeruginosa* has been linked to the death of livestock in both Australia and South Africa.[12,13] Here we describe the use of tandem mass spectrometry to deduce the amino acid sequences in several members of this family.[14] The approach is illustrated for the toxin, cyclo-D-Ala-L-Leu-D-Masp-L-Arg-Adda-D-isoGlu-Mdha, shown in Figure 9.

Analysis of toxin by liquid secondary ion mass spectrometry (SIMS) afforded a spectrum containing a strong signal for an $(M+H)^+$ ion at m/z 995. Amino acid analysis on the purified peptide indicated the presence of Ala, Arg, Glu, Leu, and β-methylaspartic acid (β-Masp) in equimolar quantities. Only Leu and Arg were found to have the L-configuration by HPLC analysis of diasteriomeric derivatives.[15] The summed masses of the above residues total 598 Da. Previous work on toxic peptides from blue-green algae suggested that the sample might be cyclic and contain two unusual amino acid residues, Adda (3-amino-9-methoxy-2,6,8-trimethyl-10-phenyldeca-4,6-dienoic acid) and Mdha (N-methyldehydroalanine),[16] not detected in the normal amino acid analysis scheme. Addition of masses for Adda (313 Da) and Mdha (83 Da) to the number deduced from amino acid composition data afforded a molecular mass of 994 that is in aggreement with that determined by mass spectrometry. Thus, it seemed likely that the isolated toxin was a member of the previously mentioned family of cyclic peptides.

Collision activated dissociation mass spectra recorded on the toxin $(M+H)^+$ ion contained a large number of fragment ions that could not be assigned unambiguously to a single sequence of amino acids. To convert the cyclic peptide to a linear structure, toxin was treated directly with neat TFA, and the progress of the reaction was monitored by mass spectrometry. Two products formed in about equal abundance. The $(M+H)^+$ ion for product I occurred at m/z 1013, a number consistent with that mass expected for a linear structure resulting from the addition of water to the cyclic peptide of molecular mass 994. The mass spectrum of product II showed an $(M+H)^+$ ion at m/z 930, the expected m/z value for a peptide formed by loss of the Mdha residue (83 Da) from the linear peptide. Unfortunately, collision activated dissociation spectra recorded on the above two products also failed to produce a distribution of fragment ions that would define unambiguously the amino acid sequence of the linear peptides.

To circumvent this problem, products I and II (obtained from acid hydrolysis of the cyclic peptide) were converted to the corresponding methyl esters. Product II was also derivatized by adding a prolyl moiety to the unblocked N-terminus.[4] Collision activated dissociation mass spectra of $(M+H)^+$ ions corresponding to the methyl ester and prolyl derivative of the linear peptide of molecular mass 929 are shown in Figures 10A and 10B, respectively. Note that Figure 10B contains a complete series of type b ions. Introduction of the strongly basic prolyl group at the N-terminus of the peptide promotes formation of fragment ions containing the amino terminus, and allows the complete sequence of the peptide to be determined. The abundant ion at m/z 169 dictates that the order of the first two amino acids in underivatized product II be Ala-Leu rather than Leu-Ala.

Conversion of product I to the corresponding methyl ester shifted the $(M+H)^+$ from m/z 1013 to 1087, an increase of 74 Da. Incorporation of 3 methyl groups and 1 molecule of methanol into the linear peptide accounts for the observed mass shift. The collision activated dissociation mass spectrum of m/z 1087 is shown in Figure 10C. Ions of type y provide the necessary information to deduce the complete sequence of this modified linear peptide. We conclude that product I is formed by protonation of the enamine moiety in the Mdha residue as shown in Figure 11. This reaction is then followed by acid-catalyzed hydrolysis of the resulting Schiff base to afford a linear peptide blocked at the N-terminus and C-terminus with α-ketoacyl and N-methyl amide groups, respectively. Treatment of this molecule with 2N methanolic HCl would be expected to esterify the two carboxylic acid side chains in Masp and Glu and also to convert the keto group to the corresponding dimethyl ketal. The result of these transformations would be to increase the mass of the molecule by the observed 74 Da.

Once this sequence information had been obtained, additional experiments were performed to determine which of the two carboxyl groups in Masp and Glu were involved in the amide linkages along the backbone of the cyclic peptide. Base-catalyzed incorporation of deuterium selectively onto carbons adjacent to all free α-carboxyl groups in the peptide was employed to obtain this information.[17] Labeling of Masp and Glu is expected even when they are in non-C-terminal positions, provided that these residues are connected by an *iso*-linkage and thus have a free α-carboxyl group.

When this experiment was performed on the mixture of two products obtained in the acid-

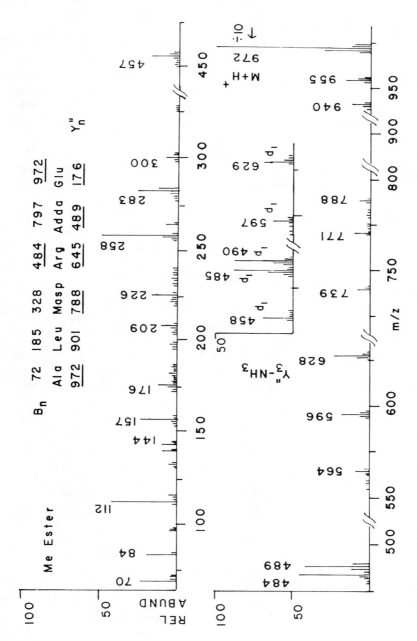

FIGURE 10.   Collision activated dissociation spectra recorded on (M+H)⁺ ions from (A) the methyl ester of product II, (B) the prolyl derivative of product II, and (C) the methyl ester derivation of product I. (From Krishnamurthy, T. A., et al.*Proc. Natl. Acad. Sci. U.S.A.*, 86, 770, 1989. With permission.)

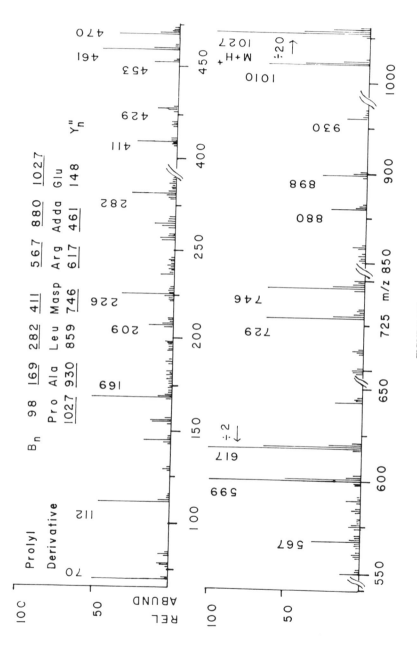

FIGURE 10B.

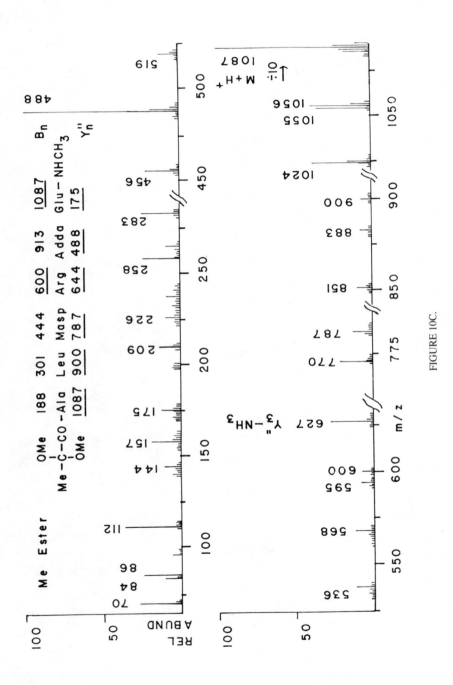

FIGURE 10C.

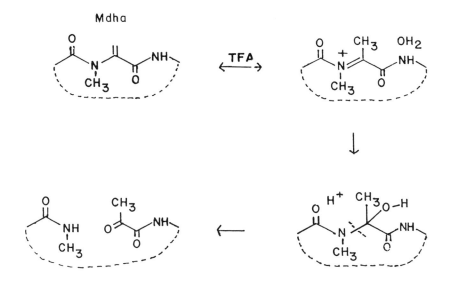

FIGURE 11. Scheme for the acid-catalyzed cleavage of cycloheptapeptide toxins from blue-blue green algae. TFA = trifluoroacetic acid. (From Krishnamurthy, T. A., et al., *Proc. Natl. Acad. Sci. U.S.A.,* 86, 770, 1989. With permission.)

catalyzed hydrolysis of the toxin and the labeled peptides were then converted to the corresponding methyl esters, the m/z values for the (M+H)$^+$ ions of products I and II shifted from m/z 1087 and 972 to 1091 to 1092 and 974, respectively. Products I and II incorporate up to five and two deuterium atoms, respectively. The insert in Figure 10B shows a portion of the collision activated dissociation mass spectrum recorded on the (M+H)$^+$ at m/z 974 for product II. Note that m/z values for the major fragment ions containing either Masp or Glu in Figure 10B all shift by 1 Da to higher mass in the spectrum of the labeled peptide. This result requires that Masp be connected to Arg by an *iso*-linkage. Incorporation of deuterium into Glu results because this residue is C-terminal in the 6-residue fragment. That this residue is also part of an *iso*-linkage in the intact cyclic peptide is suggested strongly by the observation that the methyl ester of the linear peptide in Figure 10C incorporates up to five deuterium atoms, three on the carbon alpha to the ketoacyl moiety and one each alpha to the free carboxyl groups of Masp and Glu. Further support for the presence of *iso*-Glu was obtained from a liquid secondary ion mass spectrum of the deuterium-labeled methyl ester of product I recorded on the tandem quadrupole Fourier transform instrument. Fragment ions corresponding to y$_1$-y$_3$ all showed deuterium incorporation to an extent greater than 25%. We concluded that the toxic heptapeptide isolated from blooms of the blue-green algae *M. aeroginosa* has the structure cyclo-D-Ala-L-Leu-D-β-methylisoAsp-L-Arg-Adda-D-isoGlu-*N*-methyldehydroAla.

REFERENCES

1. **Hunt, D. F., Bone, W. M., Shabanowitz, J., Rhodes. J., and Ballard, J. M.,** Sequence analysis of oligopeptides by secondary ion/collision activated dissociation mass spectrometry, *Anal. Chem.,* 54, 1704, 1981.
2. **Hunt, D. F., Yates, J. R., III, Shabanowitz, J., Winston, S., and Hauer, C. R.,** Protein sequencing by tandem mass spectrometry, *Proc. Natl. Acad. Sci. U.S.A.,* 83, 6233, 1986.

3. **Hunt, D. F., Shabanowitz, J., Yates, J. R., III, Zhu, N. Z., Russell, D. H., and Castro, M. E.,** Tandem quadrupole Fourier-transform mass spectrometry of oligopeptides and small proteins, *Proc. Natl. Acad. Sci. U.S.A.,* 84, 620, 1987.

4. **Hunt, D. F., Shabanowitz, J., Yates, J. R., III, Griffin, P. R., and Zhu, N. Z.,** Protein sequence analysis by tandem mass spectrometry: new methods and instrumentation, in *Analysis of Peptides and Proteins,* McNeal, C., Ed., John Wiley & Sons, New York, 1988, 151.

5. **Biemann, K.,** *Biomed. Environ. Mass Spectrom.,* 23, 1988, 99.

6. **Robinson, E. A., Yoshimura, T., Leonard, E. J., Tanaka, S., Griffin, P. R., Shabanowitz, J., Hunt, D. F., and Appella, E.,** Complete amino acid sequence of a human monocyte chemoattractant, a putative mediator of cellular immunity, *Proc. Natl. Acad. Sci. U.S.A.,* 86, 1850, 1989.

7. **Brinegar, A. C., Cooper, G., Stevens, A., Hauer, C. R., Shabanowitz, J., Hunt, D. F., and Fox, J. E.,** Characterization of a benzyladenine binding-site peptide isolated from a wheat cytokinin-binding protein: sequence analysis and identification of a single affinity-labeled histidine residue by mass spectrometry, *Proc. Natl. Acad. Sci. U.S.A.,* 85, 5927, 1988.

8. **Charbonneau, H., Kumar, S., Novack, J. P., Blumenthal, D. K., Stover, D., Griffin, P. R., Shabanowitz, J., Hunt, D. F., Beavo, J. A., and Walsh, K. A.,** Identification of the calmodulin-binding domain and determination of the complete amino acid sequence of the 61 kDa calmodulin-dependent cyclic nucleotide phosphodiesterase from bovine brain, *Biochemistry,* 1989, submitted.

9. **Hefta, S. A., Besman, M. J., Lee, T. D., Shively, J. E., and Paxton, R. J.,** Structure determination of STP-3, in *Techniques in Protein Chemistry,* Hugli, T. E., Ed., Academic Press, New York, 1989, 560.

10. **Yuen, S. W., Otteson, K. M., Colburn, J. C., Moore, W. T., Schlabach, T. D., Dupont, D. R., and Mattaliano, R. J.,** Characterization and synthesis of STP-3, in *Techniques in Protein Chemistry,* Hugli, T. E., Ed., Academic Press, New York, 1989, 589.

11. **Hunt, D. F., Griffin, P. R., Yates, J. R., III, Shabanowitz, J., Fox, J. W., and Beggerly, L. K.,** Characterization of symposium test peptide-3, in *Techniques in Protein Chemistry,* Hugli, T. E., Ed., Academic Press, New York, 1989, 580.

12. **Carmichael, W. W. and Mohmood, N. A.,** *Seafood Toxins,* Ragelis, E., Ed., American Chemical Society Symp. Ser. No. 262, Washington, D.C., 1984, 377.

13. **Botes, D. P., Druger, H., and VIljoen, C. C.,** Isolation and characterization of four toxins from the blue-green alga, *Microcystis aeruginosa, Toxicon,* 20, 945, 1982.

14. **Krishnamurthy, T., Szafraniec, L., Hunt, D. F., Shabanowitz, J., Yates, J. R., III, Hauer, C. R., Carmichael, W. W., Skulberg. O., Codd, G. A., and Missler, S.,** Structural characterization of toxic cyclic peptides from blue-green algae by tandem mass spectrometry, *Proc. Natl. Acad. Sci. U.S.A.,* 86, 770, 1989.

15. **Marfey, P.,** *Carlsberg Res. Commun.,* 49, 591, 1984.

16. **Botes, D. P., Wessels, P., Kruger, H., Runnegar, M. T. C., Santikarn, S., Smith, R. J., Barna, J. C. J., and Williams, D. H.,** Structural studies on cyanoginosins –LR, –YR, –YA, and –TM, peptides toxins from *Microcystis aeruginosa, J. Chem. Soc. Perkin Trans. 1,* 2747, 1985.

17. **Holcomb, G. N., James, S. A., and Ward, D. N.,** *Biochemistry,* 7, 1291, 1968.

Chapter 9

CORRELATION MEASUREMENTS IN A REFLECTING TIME-OF-FLIGHT MASS SPECTROMETER

K. G. Standing, W. Ens, X. Tang, and J. B. Westmore

TABLE OF CONTENTS

I. Introduction 160

II. The Reflecting TOF Mass Spectrometer as a Tandem Instrument 160
    A. The Parent Ion Spectrometer MS-1 161
    B. The Product Ion Spectrometer MS-2 162

III. Correlated Mass Measurements 165
    A. Efficiency of the Correlated Measurement 165
    B. Product Ion Resolution 166

IV. Conclusions 168

Acknowledgments 169

References 169

I. INTRODUCTION

There is an urgent need for more sensitive methods of determining the amino acid sequence of peptides and proteins, as these biomolecules may often be available only in sub-picomole quantities.[1] Mass spectrometry is capable of giving sequence information that is difficult to obtain by any other method, particularly for modified peptides.[2,3] However, the sensitivities of most types of mass spectrometers suffer from the need to scan the mass spectrum; only one mass can be examined at a time. Various methods of alleviating this problem are being developed, particularly the use of detector arrays to allow simultaneous measurement of a range of masses.[4-6]

Alternatively, one may choose a type of mass spectrometer that does not require scanning. In particular, time-of-flight (TOF) instruments can examine the full mass range simultaneously if a suitable data system is available.[7] A TOF spectrometer also has excellent transmission, so high sensitivity is obtained. Secondary ion mass spectra have been reported from TOF instruments for 2×10^{-13} mol leucine enkephalin,[8] 5×10^{-14} mol bradykinin,[9] and 10^{-15} mol gramicidin D (synthetic peptides).[10] Thus, these spectrometers appear capable of providing sensitivities in the required range.

Traditionally, TOF methods have been regarded as low resolution techniques. Indeed, linear TOF spectrometers[11,12] are limited in resolution because of the spread in velocities of the ions ejected from the target. They also suffer from large backgrounds arising from the products of unimolecular decay in the flight tube. Figure 1A shows the spectrum of synthetic substance P in the molecular ion region, obtained on our own instrument.[12] Although some isotopic structure is visible, the dominant features of the spectrum are the broad peaks resulting from decay of the parent ions.

The resolution is improved and the background is reduced by the introduction of an ion mirror to make a reflecting TOF spectrometer, as originally proposed by Mamyrin.[13] Figure 2 shows a schematic diagram of our version of the instrument,[14] which has produced the spectrum of the $(M+H)^+$ ions from substance P shown in Figure 1B. The isotopic pattern is now clearly visible and the background is greatly reduced. In addition to these improvements, the mirror also enables the study of the decay patterns of the various parent ions, thus yielding information about the molecular structure.[14-23] In the following sections we discuss this technique.

II. THE REFLECTING TOF MASS SPECTROMETER
AS A TANDEM INSTRUMENT

Measurement of the product ion mass spectrum arising from decomposition of a given parent ion provides useful structural information about the parent.[2,3] Such measurements are made by various methods, often referred to as tandem mass spectrometry, or MS/MS.[24,25] True tandem instruments use two separate mass-resolving systems with a decomposition or collision region between them; the first selects a given parent ion, and the second measures the spectrum of its decomposition products. These instruments can provide very high resolution for both parent and product, but their efficiency is reduced by the necessity of scanning in both spectrometers. Only one parent/product pair is examined at a time, so the instrument must often operate at lower resolution to provide adequate ion statistics.

A reflecting TOF mass spectrometer in the configuration of Figure 2 possesses some of the characteristics of a tandem instrument. A parent ion that decays in the region between the target and the mirror normally gives rise to one neutral and one charged product, both of which have approximately the same velocity as the parent. When the mirror voltage is turned on, the neutral fragment still continues on into detector 1, its TOF identifying the parent ion. Thus, detector 1 and the flight path between it and the target constitute a linear TOF spectrometer (MS-1) that measures the mass of the parent. The charged product is reflected by the mirror into detector 2,

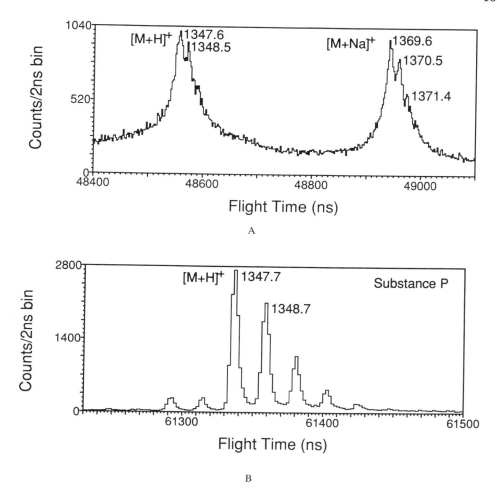

FIGURE 1. TOF spectra of the peptide substance P: (A) with the linear spectrometer (showing the (M+H)⁺ and (M+Na)⁺ ions), and (B) with the reflecting spectrometer (showing only the (M+H)⁺ isotopic distribution); in (B) the scale is expanded, so (M+Na)⁺ ions are outside the range of the figure.

and its mass is determined by the total flight time, if the mass of the parent is known. Detector 2, together with the mirror and the overall flight path, constitute a reflecting TOF spectrometer (MS-2) that measures the mass of the charged product ion.

In this description, the two "spectrometers", MS-1 and MS-2, have elements in common, and the decay region is part of both. Thus, the analogy with normal tandem instruments is not perfect, but we believe that the description may be useful for comparison with other varieties. Properties of MS-1 and MS-2 are described in the following sections.

A. THE PARENT ION SPECTROMETER MS-1

Figure 3 shows secondary ion mass spectra observed in detector 1 from bombardment of the tripeptide glycyl-glycyl-phenylalanine (GGF) by ~10 keV Cs⁺ ions. The direct spectrum (A) is taken with the mirror voltage off, so the peaks include undissociated parent ions as well as charged and neutral products. The neutral spectrum (B) is taken with the mirror voltage on, so the peaks include only the neutral fragments. These peaks are wider because of the variable velocity gained by the neutral fragment in the decay, but their centroids coincide with the peak positions in the direct spectrum, as expected. Thus, the flight time of a neutral fragment identifies its parent ion.

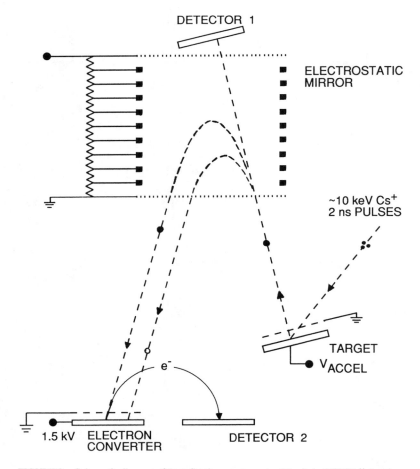

FIGURE 2.   Schematic diagram of the reflecting spectrometer (Manitoba TOF II),[14] showing typical flight paths of a parent ion • and a product ion o. The angle between the secondary ion path and the spectrometer axis has been exaggerated for the sake of clarity; the actual angle is ≈ 1.4°.

The effective resolution in MS-1 is determined by the velocity spread of the neutral products, i.e., by the energy released in the decay. Measured resolving powers $M/\Delta M_{FWHM}$ in MS-1 for various parent ions from several peptides vary from 50 to 200,[19] so clearly MS-1 is not a high-resolution device. However, the resolution is usually sufficient to distinguish between different cationized species (see Figure 3), or between different components of simple mixtures.[21]

The instrument has other useful features. The masses of all parent ions whose neutral products are detected are measured at the same time; no scanning is necessary. MS-1 also serves to identify those parent ions that decompose between the target and the mirror, because these ions are the only ones that can give a neutral fragment in detector 1.

B. THE PRODUCT ION SPECTROMETER MS-2

Figure 4A shows a portion of the overall reflected spectrum of GGF measured in detector 2. The spectrum includes parent ions of every mass together with the charged products resulting from the parent ion decomposition. The spectrometer of Figure 2 utilizes a single-stage mirror; in that case, a simple relation exists between the mass m′ of the product ion from decay in free flight, the mass of its parent m, and their respective flight times t* and t[14]:

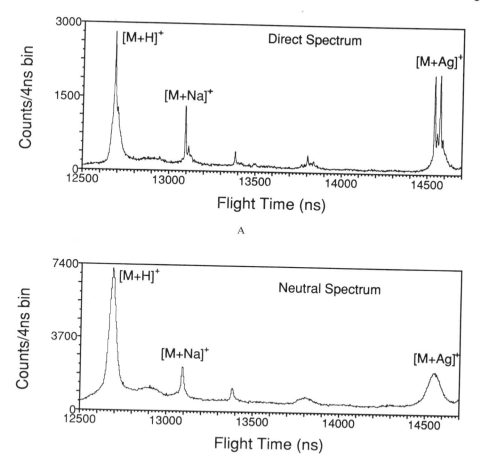

FIGURE 3. Secondary ion mass spectra from GGF (mass ≈ 279 u) in detector 1 (A) with the mirror voltage off, and (B) with the mirror voltage on. The spectrum (B) was recorded for a time about three times as long as the spectrum (A). In (A) the peak labeled (M+H)⁺ includes parent (M+H)⁺ ions as well as charged and neutral decay products, but in (B) the peak labeled (M+H)⁺ includes only the neutral products from decay of the (M+H)⁺ parent; a similar relation holds for the other peaks shown.

$$2t^*/t = (m'/m) + 1 \tag{1}$$

The flight time of a product ion depends on the mass of its parent as well as its own mass. Although this relationship complicates the spectrum, it is useful for separating the product ions arising from decay in free flight from ions of the same mass produced directly at the target. For example, the peak at ≈ 23,500 ns in Figure 4A corresponds to a product ion of mass ≈ 166 u from decay of the (M+H)⁺ ion, whereas an ion of the same mass produced at the target appears at ≈22,800 ns.

As mentioned previously, the mirror produces a considerable improvement in resolution for the parent ions. A smaller improvement is seen in the product ion peaks, particularly at low mass, because the products do not spend enough time within the mirror to provide full compensation for their velocity spread.[14]

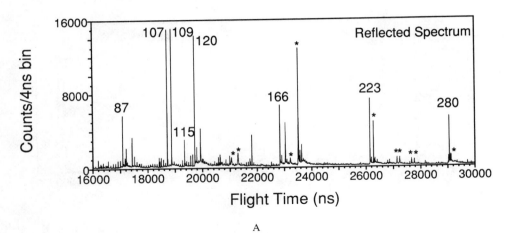

A

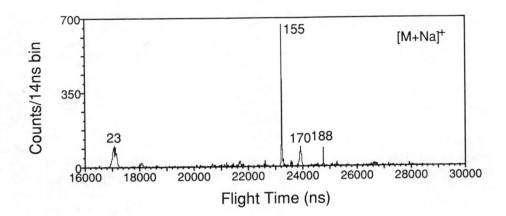

B

C

FIGURE 4.   TOF spectra in detector 2: (A) total spectrum; peaks shown with asterisks correspond to product ions; (B) spectrum correlated with (M+H)⁺ neutrals; (C) spectrum correlated with (M+Na)⁺ neutrals.

III. CORRELATED MASS MEASUREMENTS

The spectrometer is operated as a tandem instrument by measuring the product ions in MS-2 that are correlated with a given neutral fragment in MS-1. In order to avoid false correlations caused by random events, a low primary beam current is used, so that on average only about one ejected secondary ion decay is observed for each primary ion pulse. During the observation period ($\sim$100 μs) following the arrival of a given primary ion pulse at the target, the flight times of neutrals and ions are both digitized and stored in a memory buffer in the time digitizer (TDC). At the end of the observation period, these numbers are transferred to a computer, which examines the spectrum of the neutral fragments to see whether a fragment has arrived within a set of preselected time windows.[26] For example, in the GGF neutral spectrum of Figure 3B, the computer first searches the region between 12,640 and 12,750 ns for a neutral event corresponding to an (M+H)$^+$ decay; if such an event is found, the flight times of any charged particles detected in that observation period are recorded in a dedicated section of computer memory. The computer then searches between 13,050 and 13,120 ns for a neutral event corresponding to an (M+Na)$^+$ decay; if one is found, the computer stores the product ion flight times in a second section of memory. The process is repeated for the (M+Ag)$^+$ parent and for any other parent ions of interest. The calculations are executed while the time digitizer is gathering data during the next observation period, so no time is lost.[7,26]

After many observation periods, product ion spectra arising from the decays of the selected parent ions have been accumulated in computer memory. The product ion spectra corresponding to (M+H)$^+$ and (M+Na)$^+$ decays of GGF are shown in Figures 4B and 4C, respectively. Because the spectra are much simpler than the overall reflected spectrum of Figure 4A, unambiguous assignment of decay pathways can be made. It is also clear that this method is fairly efficient at separating useful events from background; the peak corresponding to the 155 u decay ion is prominent in the correlated spectrum of Figure 4C, but is barely visible in Figure 4A.

A. EFFICIENCY OF THE CORRELATED MEASUREMENT

The correlated data can also be used to estimate the probability of detecting a decay event. For a given decay, the number of events when both neutral and charged fragments are detected (i.e., the number of "coincidences" [N_{coin}], is measured by the number of counts in that peak in the correlated spectra (Figures 4B and 4C). The total number of product ions detected from this decay (N_+) is measured by the number of counts in the corresponding peak in the overall reflected spectrum (Figure 4A). The ratio N_{coin}/N_+ is therefore a direct measure of the probability of detecting a neutral fragment when the corresponding product ion has been detected, i.e., the efficiency ε_n for detection of the neutral fragment. Figure 5A shows the ratio N_{coin}/N_+ for (M+H)$^+$ decays in GGF and leucine-enkephalin. As might be expected, the ratio decreases as the energy E_n of the neutral fragment becomes smaller — from $\varepsilon_n \sim 50\%$ at $E_n = 6$ keV to $\varepsilon_n \sim 20\%$ at $E_n = 1$ keV.

The total number of neutrals detected from decay of a given parent ion is given by the number of counts in that peak in the neutral spectrum (Figure 3B). By a similar argument, the ratio of N_{coin} to the total number of neutrals detected gives the average efficiency ε_+ for detecting charged products from the decay of a given parent when the corresponding neutral fragment has been observed, although in this case we cannot determine the efficiency for individual product ions. Figure 5B shows the average efficiency ε_+ for detecting charged products as a function of secondary ion acceleration voltage for (M+H)$^+$ and (M+Na)$^+$ decay in GGF. The large difference in ε_+ between the (M+H)$^+$ and (M+Na)$^+$ products at 5 kV arises because of the different decay patterns; a substantial fraction of the (M+Na)$^+$ ions decays by emission of Na$^+$, which has a small detection efficiency because of its low energy. No information is available on the variation of efficiency with product ion mass, but the effect is expected to be much smaller than with the neutrals, because the ions are accelerated across 1 kV before striking the detector.

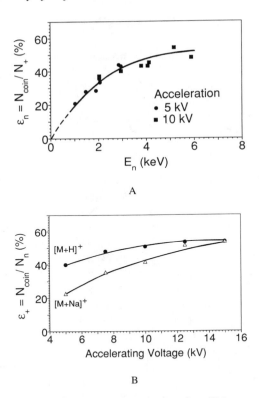

FIGURE 5.   (A) Neutral product detection efficiency;
(B) charged product average detection efficiency.

It should be noted that these efficiencies are overall figures. They take into account all losses in the spectrometer and in the detection system, such as attenuation in the mirror grids, particles that miss the detectors, and particles that strike the detectors but are not recorded. It is clear from Figure 5 that strong correlations exist between ion and neutral efficiencies as a function of m′/m. These effects are included in the measured ε_n and ε_+. If we assume that other correlations are small, we may estimate the efficiency of detecting *both* charged and neutral fragments from any decay as the product ($\varepsilon_n\varepsilon_+$), or ~25% for a typical decay where m′ = m/2. This efficiency is obtained simultaneously for all parent ions.

Since our earlier reports,[19,20] we have enlarged the front aperture of the ion mirror and increased the accelerating voltage. These improvements have given an increase in the overall efficiency ($\varepsilon_n\varepsilon_+$) by an order of magnitude. They have also enabled observation of additional decay channels, as shown in Figure 4C. Comparison of the data in Figures 3 and 4 with earlier measurements[19] shows the beneficial effect of the increased efficiency.

The efficiencies may be compared with those observed in other types of tandem instrument. In other spectrometers, stable ions may be lost at the entrance slit of MS-1 or en route through the spectrometer, and metastable parents may decay before reaching the decomposition region. If more than one parent ion is to be examined, the efficiency of MS-1 is reduced correspondingly. Similar effects occur in MS-2, where scanning produces large losses unless an array detector is used.

B. PRODUCT ION RESOLUTION

Unit mass resolution is obtained for parent ions of masses up to several thousand u by the use of the reflector, as illustrated in Figure 1. However, as remarked previously, product ions do not spend enough time in the mirror to reap full benefit from its effect, so the resolution for product

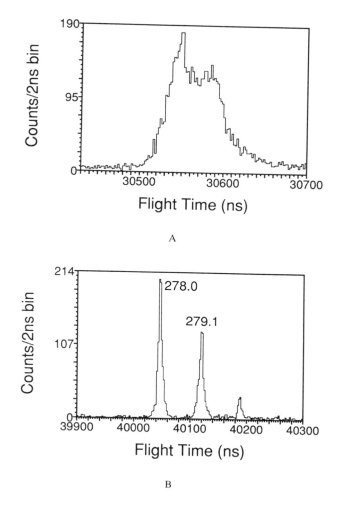

FIGURE 6.   Product ion doublet from the (M+H)$^+$ parent ion of leucine-enkephalin (mass ≈ 555 u) with the mirror optimized (A) for the parent ion and (B) for the product.

ions deteriorates significantly as m'/m decreases. Figure 6A illustrates this effect for the mass 278/279 u product ion doublet from leucine enkephalin;[20] the two components are not resolved in the normal product ion spectrum from (M+H)$^+$ decay. For the same reason, the Na$^+$ product ion peak in Figure 4C is very wide.

Fortunately, there is a simple solution to this problem. It is possible to optimize the reflector for examination of a particular product ion of mass m' by reducing the electric field in the mirror to a value that gives the same flight time t for the product as obtained previously for the parent ion. This procedure requires the reduction of the mirror voltage to a fraction m'/m of its previous value. The mass m'' of another product ion can then be determined from its flight time t** by an equation similar to Equation (1):[14]

$$2t^{**}/t = m''/m' + 1 \tag{2}$$

Under this condition, products of mass >m' pass through the reflector, so only part of the spectrum can be examined at one time.

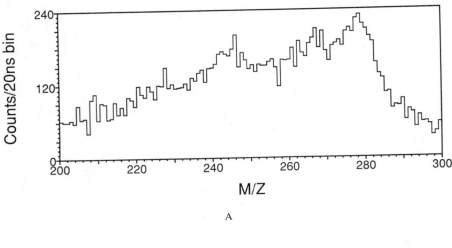

A

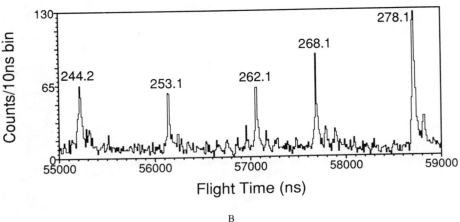

B

FIGURE 7.   Portions of the product ion spectrum from (M+H)⁺ decay of α-neoendorphin with the mirror optimized (A) for the parent ion and (B) for the product of mass 278 u.

The results of this technique are indicated clearly in Figure 6B, where reduction of the mirror voltage to optimize the resolution for mass 279 u yields a well-resolved doublet.[23] The improvement is even more dramatic when examining products from parent ions of large mass. Figure 7A shows a portion of the normal product ion spectrum from ~240 to 280 u from the decay of the (M+H)⁺ ion of α-neoendorphin (mass ≈ 1230 u);[23] no separate product masses can be resolved. When the mirror voltage is optimized for mass 278 u, the clearly resolved spectrum of Figure 7B is obtained. Of course, this procedure requires examination of the product ion spectrum in several segments, but it would appear that the extra measuring time can usually be justified by the additional information produced.

IV. CONCLUSIONS

A reflecting TOF mass spectrometer can be used to provide useful sequence information about peptides by examination of the product ion spectra from metastable decay in the first leg of the flight path. Parent ions are selected at low resolution ($M/\Delta M_{FWHM} \sim 100$), but the overall efficiency is high (~25%). Unit mass resolution can be obtained for the product ions by examining the spectrum in several segments.

ACKNOWLEDGMENTS

This work was supported by grants from the U.S. National Institutes of Health (Institute of General Medical Sciences GM 30605) and from the Natural Sciences and Engineering Research Council of Canada.

REFERENCES

1. **Morris, H. R.,** Strategies for analysis of high mass biopolymers, in *Mass Spectrometry in the Analysis of Large Molecules,* McNeal, C. J., Ed., John Wiley & Sons, Chichester, England, 1986, 121.
2. **Biemann, K. and Martin, S.,** Mass spectrometric determination of the amino acid sequence of peptides and proteins, *Mass Spectrom. Rev.,* 6, 1, 1986.
3. **Biemann, K.,** Contributions of mass spectrometry to peptide and protein structure, *Biomed. Environ. Mass Spectrom.,* 16, 99, 1988.
4. **Boerboom, A. J. H.,** Simultaneous ion detection, in *Mass Spectrometry of Large Molecules,* Facchetti, S., Ed., Elsevier, Amsterdam, 1985, 53.
5. **Cottrell, J. S. and Evans, S.,** Characteristics of a multichannel electrooptical detection system and its application to the analysis of large molecules by fast atom bombardment mass spectrometry, *Anal. Chem.,* 59, 1990, 1987.
6. **Hill, J. A., Martin, S. A., Biller, J. E., and Biemann, K.,** Use of a microchannel array detector in a four-sector tandem mass spectrometer, *Biomed. Environ. Mass Spectrom.,* 17, 147, 1988.
7. **Macfarlane, R. D.,** Californium-252 plasma desorption mass spectrometry, *Anal. Chem.,* 55, 1247A, 1983; **Ens, W., Beavis, R., Bolbach, G., Main, D., Schueler, B., and Standing, K. G.,** A data system for time-of-flight measurements, *Nucl. Instrum. Methods,* A245, 146, 1986.
8. **Chait, B. T. and Field, F. H.,** A highly sensitive pulsed ion bombardment mass spectrometer, in *Am. Soc. Mass Spectrom., Proc. 32nd Annual Conference,* San Antonio, TX, 1984, 237.
9. **Benninghoven, A., Niehuis, E., Friese, T., Greifendorf, D., and Steffens, P.,** Secondary ion mass spectrometry of biomolecules in the pico- and femto-mol range, *Org. Mass Spectrom.,* 19, 346, 1984.
10. **Lange, W., Greifendorf, D., van Leyen, D., Niehuis, E., and Benninghoven, A.,** Analytical applications of high-performance TOF-SIMS, *Springer Proc. Phys.,* 9, 67, 1986.
11. **Macfarlane, R. D. and Torgerson, D. F.,** ^{252}Cf-plasma desorption time-of-flight mass spectrometry, *Int. J. Mass Spectrom. Ion Phys.,* 21, 81, 1976.
12. **Chait, B. T. and Standing, K. G.,** A time-of-flight mass spectrometer for measurement of secondary ion mass spectra, *Int. J. Mass Spectrom. Ion Phys.,* 40, 1981, 185.
13. **Mamyrin, B. A., Karataev, V. I., Schmikk, D. V., and Zagulin, V. A.,** The mass reflectron, a new nonmagnetic time-of-flight mass spectrometer with high resolution, *Sov. Phys. JETP,* 37, 45, 1973.
14. **Standing, K. G., Beavis, R., Bolbach, G., Ens, W., Lafortune, F., Main, D., Schueler, B., Tang, X., and Westmore, J. B.,** Secondary ion time-of-flight mass spectrometers and data systems, *Anal. Instrum.,* 16, 173, 1987; **Tang, X., Beavis, R., Ens, W., Lafortune, F., Schueler, B., and Standing, K. G.,** A secondary ion time-of-flight mass spectrometer with an ion mirror, *Int. J. Mass Spectrom. Ion Proc.,* 85, 43, 1988.
15. **Della-Negra, S. and LeBeyec, Y.,** A ^{252}Cf time-of-flight mass spectrometer with improved mass resolution, *Int. J. Mass Spectrom. Ion Proc.,* 61, 21, 1984; New method for metastable ion studies with a time-of-flight mass spectrometer. Future applications to metastable structure determinations, *Anal. Chem.,* 57, 2035, 1985; Metastable ion studies with a ^{252}Cf time-of-flight mass spectrometer, *Springer Proc. Phys.,* 9, 42, 1986.
16. **Della-Negra, S., LeBeyec, Y., and Tabet, J. C.,** Etude des decompositions metastables des pentapeptides met et leu enkephaline par la méthode des coincidences avec un nouveau spectrometre à temps de vol ^{252}Cf, Orsay Report IPNO-DRE-85-22, Institut de Physique Nucléaire, Orsay, France, 1985.
17. **Standing, K. G., Ens, W., Beavis, R., Bolbach, G., Main, D., Schueler, B., and Westmore, J. B.,** Ion-neutral correlations following metastable decay, *Springer Proc. Phys.,* 9, 37, 1986.
18. **Standing, K. G., Beavis, R., Ens, W., Lafortune, F., Schueler, B., Tang., X., and Westmore, J. B.,** Unimolecular decay measurements with an ion mirror, in *Mass Spectrometry of Involatile Materials - IFOS IV,* Benninghoven, A., Ed., John Wiley & Sons, Chichester, England, 1989.
19. **Standing, K. G., Beavis, R., Ens, W., Tang, X., and Westmore, J. B.,** Time-of-flight measurements of product ions from small peptides, in *The Analysis of Peptides and Proteins by Mass Spectrometry,* McNeal, C. J., Ed., John Wiley & Sons, Chichester, England, 1988, 267.

20. **Tang, X., Ens, W., Standing, K. G., and Westmore, J. B.,** Product ion mass spectra from cationized molecules of small oligopeptides in a reflecting time-of-flight mass spectrometer, *Anal. Chem.*, 60, 1791, 1988.

21. **Standing, K. G., Ens, W., Mao, Y., Mayer, F., Tang, X., and Westmore, J. B.,** Measurements of ion dissociation in a reflecting time-of-flight mass spectrometer, in *Advances in Mass Spectrometry,* Vol. 11, Longevialle, P., Ed., Heyden & Son, London, 1989, 736.

22. **Standing, K. G., Ens, W., Mao, Y., Lafortune, F., Mayer, F., Poppe, N., Schueler, B., Tang, X., and Westmore, J. B.,** Measurements of ion dissociation in a reflecting time-of-flight mass spectrometer, *J. Phys. (Paris),* 50 C2, 163, 1989.

23. **Standing, K. G., Ens, W., Mayer, F., Tang, X., and Westmore, J. B.,** Measurement of product ion spectra in a reflecting time-of-flight mass spectrometer, in *Ion Formation from Organic Solids - IFOS V,* Hedin, A., Sundqvist, B. U. R., and Benninghoven, A., Eds., John Wiley & Sons, Chichester, England, 1990, 93.

24. **McLafferty, F. S., Ed.,** *Tandem Mass Spectrometry,* John Wiley & Sons, New York, 1983.

25. **Busch, K. L., Glish, G. L., and McLuckey, S. A.,** *Mass Spectrometry/Mass Spectrometry,* VCH Publishers, New York, 1988.

26. **Ens, W., Mao, Y., Tang, X., and Standing, K. G.,** A data system for daughter ion measurements with a reflecting time-of-flight mass spectrometer, *Am. Soc. Mass Spectrom., Proc. 37th Annual Conference,* Miami Beach, FL, 1989, 1059.

Section III: Analysis of Peptides

Part A: Sample Preparation
    Chapter 10 173

Part B: LC-MS Analysis
    Chapter 11 201

Part C: Analysis of Protein Products
    Chapter 12 223
    Chapter 13 257

Part D: Protein Cross-Linkages
    Chapter 14 275

Part E: Peptide Interactions with Metal Ions
    Chapter 15 289

Part F: Permethylated Peptides
    Chapter 16 315

Part G: Neuropeptides
    Chapter 17 327
    Chapter 18 347

Part H: Quantification of Neuropeptides
    Chapter 19 367

Chapter 10

SAMPLE PREPARATION AND MATRIX SELECTION FOR ANALYSIS OF PEPTIDES BY FAB AND LIQUID SIMS

Kenneth L. Busch

TABLE OF CONTENTS

I.   Introduction 174

II.   Matrix Selection 174
      A.   Solubility Considerations 175
      B.   Acid/Base Characteristics 177
      C.   Oxidizing and Reducing Matrices 180
      D.   Sample Surfactancy 181
      E.   Solvents for Flow-FAB Analysis of Peptides 188

III. Sample Derivatization 190
      A.   N-Terminal Peptide Molecule Derivatizations 190
      B.   C-Terminal Peptide Molecule Derivatizations 191
      C.   Internal Derivatizations 191
      D.   The Importance of Disulfide/Sulfhydryl Transformations 192
      E.   Monitoring of Degradation Reactions 195

IV.   Conclusions 195

References 197

I. INTRODUCTION

Fast atom bombardment (FAB) and liquid secondary ion mass spectrometry (LSIMS) have been at the forefront of the applications of mass spectrometry to biotechnological problems. As evidenced by the breadth and diversity of applications detailed in this volume, interest in the area of peptide analysis is sufficient to support not only FAB and LSIMS, but a host of other desorption ionization techniques. FAB is defined as the sputtering of sample molecules and ions from a liquid solution by an impinging beam of fast neutral particles. LSIMS is the same experiment carried out with a primary beam of ions rather than neutral atoms. To a peptide molecule sputtered from the surface of a liquid solution, the presence or absence of a charge on the primary particle is irrelevant. In historical terms, the application of SIMS to organic problems preceded the introduction of FAB; the explicit contribution of the latter was the first use of the liquid matrix from which a stable and persistent secondary ion emission could be maintained. The difference between FAB and LSIMS, for the purposes of this chapter, will be semantic. "FAB" guns produce a substantial number of ions, and "SIMS" guns produce large fluxes of neutral particles. Further, on the current generation of commercial instrumentation, first generation FAB guns operating at 10 kV have been replaced by 25 to 30 keV SIMS guns to enhance sensitivity; practical instrumental differences have thus also subsided.

This overview of matrix selection and sample preparation in peptide analysis therefore applies equally well to LSIMS and FAB. Detailed strategies for the interpretation of the mass spectra for sequence information, or the use of mass spectrometry/mass spectrometry (MS/MS) for peptide analysis is not covered. Numerous reviews for peptide spectral interpretation and MS/MS experiments have appeared in the literature.[1-5] Although such experiments will never be simple or easy, the elucidation of repetitive information through manual interpretation is being replaced with automated data system interpretation. Relief from the duty of extracting information from the mass spectrum provides us with more opportunities to examine the various aspects of sample preparation and the fundamental mechanisms of sputtering and ionization that underlie the phenomenal success of FAB and LSIMS. As distinctions in hardware and instrument variations have been reduced, attention has refocused on the key parameters that determine the success or failure of the FAB or LSIMS analysis of a peptide — the characteristics of the sample molecules themselves (acidity/basicity, surfactancy, and hydrophobicity/hydrophilicity), and the properties of the liquid matrix chosen for preparation of the sample solution. The emphasis of this chapter is on the solution chemistry of peptide samples, an area in which many mass spectrometrists may not be sufficiently versed. Further, the solution chemistry of the samples in question is not particularly easy to study — the chemistry of particular interest is that of a few microliters of sample solution bombarded continuously by an energetic particle beam, and the sample (solute) is often in limited supply and is of undetermined purity. Few wide-ranging studies define the solution chemistry for peptides with respect to successful FAB or LSIMS experiments. The large number of applications can be reviewed, however, to condense the procedures that are successful in practice, and which may serve to sketch out general boundaries of sample preparation and solution chemistry. This short review is an overview of those procedures; it deals with, in turn, properties of the solvent of importance for FAB and LSIMS, and properties of the peptide molecules themselves and of their derivatives.

II. MATRIX SELECTION

The various matrices commonly used in FAB and LSIMS have been reviewed by Gower[6] and by De Pauw.[7] The physical parameters of the matrices, to the extent that they are known, appear in the compilation of Cook et al.[8] These reviews, however, neglect the specifically chosen solvents or mixture of solvents used by many laboratories for samples commonly encountered in that laboratory. Some of these specific matrices ultimately appear in the literature. With respect to the analysis of peptides by FAB and liquid SIMS, it is likely that:

1. Anything that is as polar as glycerol can serve as a liquid matrix for the analysis of peptides
2. Sensitivity enhancements with any "new" matrix are likely to be within a factor of ten of the response from glycerol
3. Glycerol, thioglycerol, dithiothreitol/dithioerythritol mixtures, and perhaps one or two aromatic solvent matrices, with appropriate cosolvents and modifiers, will continue as the most widely used solvents in the general analysis of peptides by FAB and LSIMS

The high dielectric constants of hydroxylated liquid matrices favor the dissociation of ion pairs that are formed within them (generally by acid/base chemical reactions), lowering the Coulombic energy that must otherwise be surmounted prior to sputtering. The acid/base chemistry of these matrices can be changed over a broad range with the use of acidic or basic additives. For neat liquid solvents, thiols provide a higher solvent acidity than glycerol or other compounds that are more neutral, whereas the amino alcohols provide a more basic solvent when required, as in negative ion analysis. Surface properties of the hydroxylic liquid matrices can be changed by a systematic variation of the ratio of hydroxylic to hydrocarbon portions of the molecule; control of solubility parameters for various peptides is determined empirically. These general rules are considered in more specific detail for peptides in the following sections.

A. SOLUBILITY CONSIDERATIONS

In FAB and LSIMS, conventional wisdom holds that the sample molecules must be soluble in the selected matrix to ensure a stable and persistent secondary ion emission from the sputtered surface. In choosing a solvent/solute pair, two issues are considered: (1) is the solute soluble? and (2) how soluble is it? It is clear that peptides *are* soluble in matrices such as glycerol and thioglycerol; the success of the experiments argues that this attribute is generally true. However, there are no compilations of peptide solubility in common FAB matrices. Consider a sample size of 100 ng of a purified peptide with a mass of 2000 amu; this 50 pmol of sample will be dissolved in 5 μl of solvent (about 5 mg) for a solution concentration of 0.002% w/w, or 10^{-5} M. Peptides will usually dissolve to this extent in slightly polar solvents (water, glycerol, dimethylsulfoxide, and acidified or basified solutions of the same). Problems may arise when larger sample sizes are used, because the solubility of the peptide sample will be bounded by some upper limit, especially in the more organic solvents. Therefore, a concentration of 10^{-5} M seems to be an appropriate upper limit for most FAB and SIMS analyses.

Concentrations greater than 10^{-5} M are not usually encountered due to the limited amount of sample available for analysis. The spectral behavior expected when the solvent is overloaded with peptide sample is demonstrated in Figure 1. The signal for the protonated molecule of angiotensin reaches a plateau level above a certain concentration.[9] Although initial concentrations may not reach such high levels, as solvent evaporates within the mass spectrometer, the concentration of sample in the remaining solvent increases. The fact that ion currents from depleted samples can be reestablished by the addition of solvent shows that the sample is still present. Figure 2 illustrates this concentration effect, using the sample and solvent amounts described in the previous paragraph with reasonable assumptions about initial sample and solvent amounts, rates of evaporation of the solvent, and sample consumption rates.

Solubility of a peptide sample may also be significantly affected by the presence of other components in the solvent. "Salting out" (the precipitation of a solute upon an increase in the ionic strength of the solvent) is one undesirable result. A greater concern is the relative surface activity of the peptide(s) of interest in a mixture, as differences in the surfactancy of samples in a mixture will lead to changes in concentration of the sample at the surface and in the bulk of the solution. These effects are discussed fully in Section II.D.

Other aspects of sample preparation and solvent selection are concerned with additives or impurities in the samples as prepared and in the solvent as selected. Each solvent carries its own characteristic impurities, which may be carried along into the final FAB or LSIMS solvent if the

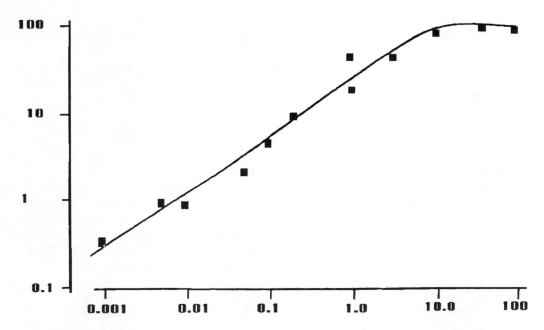

FIGURE 1.   The absolute signal level for the protonated molecule of angiotensin (vertical axis) is plotted against the amount of sample added to a constant volume of glycerol in a FAB experiment. (Adapted from Reference 11.)

TIME/CONCENTRATION PROFILES

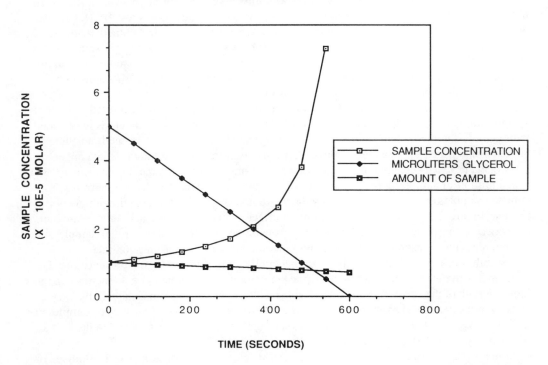

FIGURE 2.   Time/concentration profile for sample concentration assuming typical rates of solvent evaporation and sample consumption.

solvents selected are miscible as the sample is prepared and concentrated. Miscibility is generally a desirable characteristic to prevent sample partitioning, and so the problem is insidious. Glycerol, for example, is hygroscopic, and the water is extremely difficult to remove. A surface layer of water will persist even in the vacuum of the mass spectrometer. Many samples of biological origin have relatively large endogenous amounts of salt, mostly in the form of sodium and potassium chlorides. Samples that have been exposed to buffers may contain an even more complex mixture of anions and cations. Complexation reactions, both in solution and in the mass spectrum, can be expected, and the transferral of ion current to a mass other than that expected (for instance, to $[M+K]^+$ rather than $[M+H]^+$) may provide anomalous results in quantitative and high sensitivity work. Large amounts of adventitious salts may set up gradients within the prepared sample solutions; these gradients may also result in biased sampling and erroneous quantitative values. Finally, diffusion constants of peptide molecules in the liquid solution will change with sample and salt concentration. A common-sense approach suggests that the sample concentration in FAB and LSIMS be kept as low as possible, yet consistent with the detection limits of the mass spectrometer. Furthermore, the salt concentration of the solutions should be established at a low and reproducible value.

New matrices for FAB and LSIMS are described regularly in the literature, with special solvent mixtures providing increases in sensitivity by factors of two to ten over the typical analysis in glycerol.[10] Most of the new matrices are developed for specific analyses of particular compounds or compound classes, and their potential contributions to peptide analysis by FAB or LSIMS are not established. Most work continues to use either glycerol, thioglycerol, or a dithiothreitol/dithioerythritol mixture for positive ion analysis, and di- or triethanolamine for negative ion analysis, provided that the amount of sample is sufficient to permit the preparation of two separate sample solutions. Use of such simple matrices also minimizes the chances for uncharacterized sample transformations, or sample/solvent interactions, that would make the FAB or LSIMS spectra more complex. Those interactions that do occur, for example, the formation of a Schiff base, have been described in the literature.[11] Selection of an acidic or a basic matrix is considered in the next section.

B. ACID/BASE CHARACTERISTICS

For positive ion analysis, an acidic matrix is preferred to ensure that the sample molecule will be maintained in a preformed $(M+H)^+$ state. For negative ion analysis, for analogous reasons, a basic matrix is preferred to maximize the intensity of the $(M-H)^-$ ion, the deprotonated form of the neutral molecule. Given those fundamental tenets, more sophisticated models describe the processes of FAB and LSIMS sputtering and ionization as either condensed or gas phase protonation and deprotonation reactions.[12-15] In some instances, the predictions from a quantitative evaluation of such a model may be useful, and discussion of the relative importance of condensed vs. gas phase mechanisms continues.[16] Such models tend to include equilibrium-based arguments, which poorly describe a decidedly nonequilibrium sputtering phenomenon. For every sample/solvent system accurately described by the model, one will be found that holds exception.

For peptides, it was recognized that positive ion FAB mass spectra are produced readily from peptides that contain arginine, lysine, and histidine residues. Conversely, aspartic acid and glutamic acid seem to promote the formation of negative ions of small peptides. An early rule of thumb stated that the mass spectrum should be recorded for ions of polarity that correspond to the net charge of the peptide established by electrophoresis.[17] The correlation fades as the peptide becomes larger with an increasing number of basic and acidic residues, and as the solvent/molecule interaction becomes more removed from the specific solvent interactions that form the basis of the electrophoretic tests. Compounding predictions of ion abundances from acid/base considerations are the relative surfactancies of the peptides involved, as is discussed in a later section.

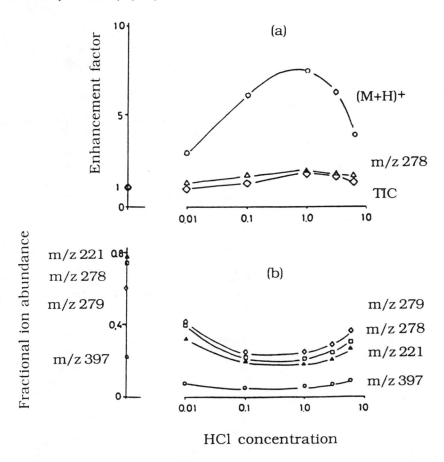

FIGURE 3. Variations in the intensity of the protonated molecule of leucine enkephalin, a fragment ion (m/z 278), and the total ion current (TIC) are plotted vs. the addition of 0.1 μl of HCl of various concentrations to a constant glycerol volume (A) variations in the fractional ion abundances in the same experiment are noted; (B) the order of fractional ion abundances in glycerol is given on the vertical axis to show a shift in the relative abundances. (Adapted from Reference 18.)

The acid or base strengths of liquid matrices used in FAB and LSIMS are based on equilibrium measurements of the proton affinities of these compounds. The proton affinities of the commonly used FAB and LSIMS matrices can be found in the recent compilation of Cook et al.[8] For numerous matrices, the values of the proton affinities (PA) are not known. Triethanolamine (PA of 975 kJ/mol) and diethanolamine (PA of 954 kJ/mol) are obviously basic matrices. Ethylene glycol (PA of 828 kJ/mol) has the same acid strength as glycerol (828 kJ/mol), although a range of values is associated with the latter measurement. Thioglycerol is more acidic than glycerol, but a value for the proton affinity of thioglycerol is not found in the literature.

The acidity or basicity of these matrices is augmented by the addition of acids to ensure that protonation of the sample peptide will occur in the positive ion mode of analysis and the addition of bases for negative ion analysis. A number of acids have been used as additives to glycerol to increase the quality of the positive ion mass spectrum of peptides. Several papers report the variation in (M+H)+ and fragment ion abundances as the "pH of the matrix" is varied. It is worthwhile to remember that a glycerol matrix cannot be described in terms of pH; "pH" applies only to aqueous solutions. Rather, the mass spectral behavior as a function of the number of equivalents of acid or base added to (and retained in) the matrix must be followed. Figure 3 illustrates effects noted in the positive ion LSIMS mass spectra for leucine enkephalin as differing amounts of 0.01 N HCl are added to a glycerol solution of the sample peptide.[18] Of particular interest is the increase in the intensity of the $(M+H)^+$ ion as the acid is added, followed by measurement of a plateau intensity. Figure 3B shows that the intensity of the fragment ions relative to that of the protonated molecule decreases, but the ratio also reaches a constant value.

Acids added to a glycerol solution are added usually as aqueous solutions, creating a glycerol/water mixture that may have physical properties significantly different than a matrix of pure glycerol, and a mixture in which the surfactant properties of the dissolved peptides may also be altered. Although the acid strengths of the common additives are well known in aqueous solutions, the relative acid strengths in glycerol solutions (with a much lower dielectric constant) are not known. In addition, the vapor pressure of the acid itself must be considered. A commonly used acid is HCl, which will react with any base in solution to cause protonation. In most instances, the preformed ion is not volatile, and will remain in solution. However, excess HCl will evaporate eventually into the vacuum of the mass spectrometer. A large proportion of the excess added HCl is lost in the rough vacuum of the direct insertion probe. Any addition of HCl beyond the equivalence point of the potential bases in solution is superfluous (see Figure 3).

p-Toluenesulfonic acid is a relatively strong acid with a lower vapor pressure than HCl, and therefore a long residence time in a liquid matrix. The toluenesulfonate counterion may provide some increase in the surface concentration of the protonated species, as reported by De Pauw.[19] Camphorsulfonic acid has also been used as an acid and as a surface-active counterion in this regard. The use of *p*-toluenesulfonic acid stems from the early reports of this acid to enhance the quality of field desorption (FD) and solid SIMS mass spectra.[20]

Other acids in common use as additives to glycerol solutions include trifluoro- and trichloroacetic acids. In some work, the peptide sample is dissolved in 0.1% acid, and an aliquot is added to the glycerol matrix and mixed thoroughly on the surface of the direct insertion probe. A 2 N solution of acetic acid can also be used to dissolve the sample before transferral into the liquid matrix. In other work, the sample is dissolved in glycerol/30% acetic acid in a 3:1 v/v ratio. Acetic acid is volatile, of course, and will evaporate as the free acid into the vacuum of the mass spectrometer. In determining the amount of acid to be added, conditions must be avoided that would lead to acid hydrolysis of the sample.[21] A typical acid hydrolysis of a protein requires 5% HCl and prolonged heating at 90°C, so the conditions used to ensure protonation are usually not sufficient to induce hydrolysis. Initial approaches that dissolved the peptide samples in 50% formic acid have since been modified to decrease possibilities for acid-catalyzed hydrolyses.

Assuming that the added acid is not pumped away, and assuming that the acids used are much stronger than glycerol (G), the strength of any added acid is leveled to that of protonated glycerol. In this regard, the choice of added acid centers on its volatility and surface properties, and the possibility that it will create background ions in the FAB or LSIMS mass spectra. With the addition of HCl, the reaction producing protonated molecules of a peptide is $M + GH^+ \rightarrow (M+H)^+ + G$, and not $M + HCl \rightarrow (M+H)^+ + Cl^-$.

Far fewer reports appear in the literature on the use of additives to liquid solvents to increase the quality of the negative ion FAB or LSIMS mass spectra.[22] Table 1 illustrates effects noted in the negative ion FAB analysis of methionine enkephalin from a glycerol solution with various additives.[23] The data are reported as the ratio of intensities of the $(M-H)^-$ ion of the methionine enkephalin at m/z 572 relative to the ion signal at m/z 275, corresponding to the deprotonated trimer of glycerol. Ammonium hydroxide is added as a 20% aqueous solution. Glutathione (GSH) is added to displace the sample molecules from the bulk of the solution to the surface.

In general, the effects of acid/base additives on absolute signal intensities or sensitivity (described rigorously as C/μg) are not given. However, Tolun et al.[23] report on limits of detection for leucine and methionine enkephalins in both positive and negative ion FAB experiments using a number of liquid solvents and additives. In addition to the changes in the signal intensity for the ion of interest, the level of the background signal at all masses becomes a limiting factor

TABLE 1
Variations in Ion Intensity Ratio for the (M–H)⁻ Ion of Methionine Enkephalin vs. the Glycerol Trimer (3G-H)⁻ as the Matrix is Varied in FAB Mass Spectra[23]

Matrix	Ratio of intensities
Glycerol	37
Glycerol + 20% NH$_4$OH	50
Glycerol + glutathione/H$_2$O	46
Glycerol + glutathione/NH$_4$OH	74

Note: Glutathione solutions were 1 mg/ml in the listed solvents, and 0.5 µl of solution were added to a constant volume of glycerol on the tip of the direct insertion FAB probe.

in signal-to-noise determinations. Tolun et al. note that solvent and solvent mixtures that increase the level of the signal sometimes also increase the level of the noise, resulting in no gain in sensitivity.

C. OXIDIZING AND REDUCING MATRICES

Besides the acid/base character of the various FAB and LSIMS liquid matrices in common use, the oxidizing/reducing nature of the matrices is also of interest, especially in the analysis of peptides containing disulfide bonds that can be reduced with dissolution and/or bombardment. Reducing matrices are those that cause or permit reduction of the sample molecule while molecules of the liquid matrix are themselves oxidized. Oxidizing matrices are those that oxidize (or preserve in an oxidized state) the sample molecules, and are themselves relatively easy to reduce. Commonly used reducing liquid matrices in FAB and LSIMS include thioglycolic acid, thioglycerol, and glycerol, as well as dimethylsulfoxide, and the dithiothreitol/dithioerythritol mixture. Indeed, Cleland[24] described the original use of dithiothreitol as a compound designed to maintain sulfhydryl groups of peptides in their reduced form. Common oxidizing matrices include nitrobenzylalcohol and nitrophenyloctylether. The redox properties of the solvents in FAB and LSIMS depend also on the primary beam parameters that serve to increase the extent of oxidation and reduction processes, and thus no definitive ranking of the relative redox propensities of these solvents can be undertaken. Further, oxidation or reduction processes require a redox pair, and the redox properties of the sample molecules themselves determine whether the matrix can act as an oxidizer or a reducer.

Several reports investigate the production of reduced forms of a peptide sample molecule in various matrices, but also containing a matrix modifier. For instance, although thioglycerol was found to more readily reduce oxytocin under standard LSIMS condition than glycerol, the addition of several drops of dilute HCl to the liquid matrix containing the sample was found to prevent the reduction of the sample molecule by thioglycerol.[25] Once the oxytocin was reduced, however, the addition of the acid would not reoxidize the sample, in contrast to earlier observations with other peptides.[26] A full discussion of disulfide/sulfhydryl transformations is found in Section III.D.

Several workers have also noted reduction processes in FAB and LSIMS that result from irradiation of the sample with light, often in conjunction with primary beam bombardment. For instance, Wirth et al.[27] have reported on the reduction of azo-group-containing peptides in FAB. Irradiation of a glycerol solution of such peptides with the primary beam from a xenon discharge source produces reduction far in excess of the reduction observed with irradiation of the same sample with the primary beam from a cesium thermionic source. This difference was attributed to irradiation of the sample with the short wavelength ultraviolet (UV) radiation and X-rays emitted from the xenon atom discharge source, which by definition must be in a line-of-sight with the sample probe. The redox properties of a solution also change as cesium ions are added to it from the primary beam bombardment, whereas an equivalent change will not be observed with the implantation of inert noble gas atoms. The presence of cesium ions in the sample solution may serve to reduce the extent of reduction of some sample molecules. Changes in the redox properties of sputtered surfaces induced by chemical effects of the primary beam itself are well known in classical SIMS experiments. These effects include increases in ion yields for some species upon bombardment with Cs$^+$ or O$^-$.[28] A discussion of such atomic effects can be found in the monograph by Benninghoven et al.[29] The role of implanted cesium ions, and of photochemistry in general in facilitating reduction reactions, has been discussed for other organic compounds studied by FAB and LSIMS.[30,31]

Organic compounds with especially low reduction potentials, such as the quinones, undergo beam-induced reductions in FAB and LSIMS, and even in SIMS studies in which neat samples are analyzed.[32] In peptides, the disulfide bond is most susceptible to reduction, and shifts in measured molecular masses in the FAB and LSIMS spectra resulting from such reductions were noted early. In some instances, control of the matrix and the conditions used to retard or promote reduction can be used to manipulate the redox chemistry, and to derive detailed information about the number and location of the disulfide linkages in the peptide. Interpretation of the FAB or LSIMS mass spectra of disulfide-containing peptides is complicated by the fact that the ratio of the reduced to the oxidized forms of the peptides can often change dramatically with time, in addition to changing as a function of the matrix from which the sample is sputtered. This tendency is most troublesome in matching predicted isotopic envelope patterns to those actually observed, leading to uncertainties in the derivation of the true molecular weight of the peptide of higher mass unless specific steps are taken to control the reduction chemistry.

An early example of the complications introduced with disulfide reduction chemistry is illustrated in Figure 4, adapted from Buko and Fraser.[26] Vasopressin (containing a single disulfide bond) was noted to undergo reduction upon bombardment of its glycerol solution, and further, the extent of the reduction was dependent on the time of the irradiation. The two spectra shown in the figure are those recorded after 31 and after 93 s of continuous bombardment by the primary particle beam. Of course, different primary beam fluxes will result in different time dependences for the reduced/nonreduced ion ratio. The use of a more-reducing matrix will result generally in a higher degree of reduction, whereas an oxidizing matrix will preserve the disulfide bond. Reports in the literature, however, must be very carefully evaluated. Oxytocin, reported by several workers to reduce with FAB sputtering from a glycerol matrix, as described previously, does[33] or does not[26] appear to reduce when dissolved in a matrix of dithiothreitol/dithioerythritol, a more-reducing matrix than glycerol. Such inconsistencies are due to changes in the irradiation time, sample concentration, and primary ion flux rather than to the intrinsic properties of the redox couple.

Reduction of disulfide bonds in peptides is an important reaction, but addition of hydrogen (formally a reduction) also occurs for peptides that do not contain disulfide bonds. Fujita et al.[34] discuss the formation of (M+2H)$^+$ and (M+3H)$^+$ for several peptides not containing disulfide bonds. Using a matrix of glycerol (acidified with aqueous trichloroacetic acid), the extent of reductions for compounds such as eledoisin, bradykinin, angiotensin I, and α-endorphin were about half the extent observed for oxytocin and somatostatin, but the occurrence of such reduction processes was still confirmed. The site of reduction in such instances has not been established. One also notes that *losses* of hydrogen also contribute to the signal in the isotopic envelope in the molecular ion region, and that the actual measured signal is due to a combination of (M-H)$^+$, M$^+$, (M+H)$^+$, and (M+2H)$^+$, requiring very accurate intensity measurements for proper deconvolution of the ion species present.

D. SAMPLE SURFACTANCY

Many of the initial successes of FAB and LSIMS for generation of the mass spectra of peptides were for sample molecules that exhibited a moderate degree of surfactancy in the matrices used. A few simple examples of the effect of surfactancy on the FAB and LSIMS mass

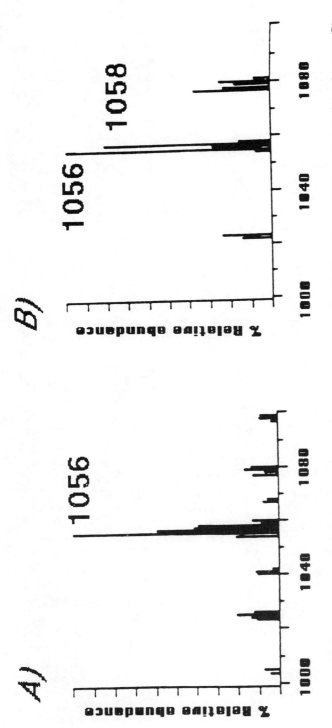

FIGURE 4.   Changes in the molecular ion region of vasopressin occur with prolonged bombardment by a FAB beam. Spectrum (A) is recorded after 31 s and spectrum (B) after 93 s in a glycerol matrix. (Adapted from Reference 26.)

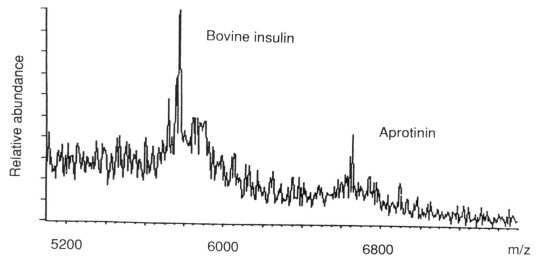

FIGURE 5.   Positive ion FAB mass spectrum of an equimolar mixture of bovine insulin and aprotinin in glycerol. (Adapted from Reference 35.)

spectra of peptides are helpful in clarifying the magnitude of the observed effects. Figure 5 gives an example of the surfactant effect for some relatively large peptides.[35] Bovine insulin is a hydrophobic protein that will be enriched at the surface of a glycerol-thioglycerol (1:1 v:v) liquid matrix relative to the more hydrophilic aprotinin. The figure documents the measured ion intensities for equal amounts (15 nmol) of each peptide in the solution. The greater response for the insulin is apparent. These surfactant effects are intrinsic to the peptides themselves, and are not interpreted with respect to sample derivatizations that modify the situation (discussed in Section III).

Larger peptides (more than 15 or 20 residues) tend to dissolve without problem in any of the commonly used FAB and LSIMS liquid matrices. The hydrophobic and hydrophilic character of these peptides can vary, but large excursions away from a mean based on average residue compositions will be rare. Differential sample responses for larger peptides are expected to be smaller in magnitude than for medium-sized peptides, or for peptides created in digestions or degradations of larger proteins. Very small peptides (fewer than five or so residues) will exhibit some differences in surfactancy, because the contribution of any individual amino acid relative to the charge on the termini will be reduced. However, for medium-sized peptides (peptides with 8 to 20 residues), large deviations in hydrophobic or hydrophilic character may be expected based on differences in their compositions, leading to drastic spectral effects. It is for this class of compounds that large changes in responses and suppression effects have been noted in the literature.

The Bull and Breese[36] indices for the individual amino acids are used to provide an overall hydrophobic or hydrophilic estimate for the peptide. These values, given in Table 2 for the individual amino acids, are determined by investigation of the surface concentrations of the amino acids at a water/air interface. Straightforward extension to, for instance, a glycerol/vacuum interface is assumed implicitly; however, no values directly germane to any FAB matrix have been measured. Furthermore, secondary and higher order structures for large peptides will certainly play a significant role in determining the overall hydrophobicity or hydrophilicity. A peptide in a water-like solvent will assume a conformation such that the more hydrophobic residues are sequestered in the interior of a three-dimensional structure. A number of papers in the literature describe a more sophisticated approach to determination of the overall hydrophobicity/hydrophilicity of a peptide,[37-39] and some reports of spectral effects (in MS/MS at least) generated as a result of peptide secondary structure have appeared.[40]

TABLE 2
Bull and Breese Index Values
for Amino Acid Residues

Ala	+610
Arg	+690
Asn	+890
Asp	+610
Cys	+360
Gln	+970
Glu	+510
Gly	+810
His	+690
Ile	-1450
Leu	-1650
Lys	+460
Met	-660
Phe	-1520
Pro	-170
Ser	+420
Thr	+290
Trp	-1200
Tyr	-1430
Val	-750

Note: Negative values indicate a hydrophobic residue, and positive values
indicate hydrophilicity.

The values given in Table 2 are used as follows. The individual values for the amino acids in a peptide are summed, and then the total value is divided by the number of amino acids in the peptide, providing a net positive or negative value. A negative value corresponds to a hydrophobic peptide that would presumably exhibit an excess surface concentration in a liquid matrix such as glycerol. Conversely, a positive value indicates a more hydrophilic peptide that would not, given the same experimental and sample preparation conditions, exhibit a surface concentration excess. All other things being equal, the FAB or LSIMS mass spectrum of a mixture of peptides would show an increased spectral response for the more hydrophobic peptide. In some instances, a complete suppression of the signal from the second, more hydrophilic peptide is observed. Changes in the mass spectra with time are also observed, as the surface-enriched peptide is sputtered preferentially in the first few minutes of the analysis, with signals increasing with time for peptides not enriched originally in surface concentration.

Several early investigators commented on differential sensitivity of FAB and LSIMS for different peptides. The paper by Naylor et al.[41] in 1986 catalyzed a great deal of study in the effects of peptide hydrophobicity and hydrophilicity and the correlation of values of the Bull and Breese index value with the signal intensity for the peptide in the mass spectrum. The conclusions of that study were that the analysis of peptides by FAB and LSIMS is facilitated by the preparation of derivatives that increase the hydrophobicity of the analyte molecules (see Sections IIIB and IIIC), that time-dependent response curves were the result of differential sputtering of peptides at the surface of a glycerol solvent, and finally, that liquid chromatographic separation of the hydrophobic from the hydrophilic peptides should ameliorate suppression effects, and could ease the task of quantitation. The value of high pressure liquid chromatography (HPLC) separation of peptide mixtures and digests of peptides is well recognized, and the use of such separations in conjunction with digestion methods has been particularly useful. Not foreseen was the increasingly widespread use of flow-FAB[42] to avoid such suppression effects by removing most of the solvent.

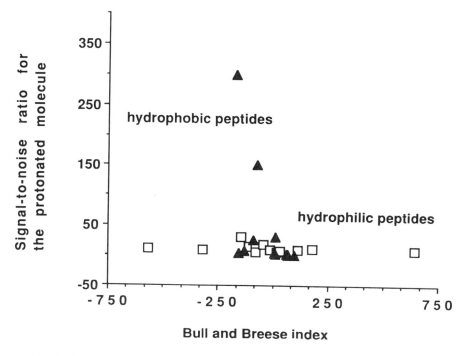

FIGURE 6. Signal-to-noise ratios for the protonated molecules of a number of peptides are plotted versus the calculated Bull and Breese index values. Triangles indicate data measured by Naylor et al.[35] and open squares represent data measured by Pucci et al.[43] In each set of data, equal concentrations of peptides were used.

The correlation of hydrophobicity/hydrophilicity with signal intensity must be discussed with reference to solute behavior in a single solvent, because the responses of a series of peptides in other solvents will certainly be different. Furthermore, the use of any solvent additives may mask the intrinsic characteristics of the sample molecules themselves by changing the parameters of the solvent and, in some cases, the chemical nature of the sample molecules. The model advanced by Naylor[41] describes the surface enrichment of peptide molecules based on their hydrophobic or hydrophilic nature. By definition, then, changes in the relative intensities of both $(M+H)^+$ and $(M-H)^-$ ion signals with Bull and Breese index values are required to assess rigorously the degree of correlation. Suppression of $(M+H)^+$ signals from one peptide in the presence of another have been shown by Allmaier,[9] but the negative ion mass spectra in this instance show no signs of surface discrimination in favor of one peptide over another. This control experiment has not been performed for most of the work reported in discussions of the Bull and Breese correlations.

Figure 6 plots data derived from the work of Naylor[41] (triangles) and from Pucci et al.[43] (open squares). These data reflect signal-to-noise values for a number of peptides as a function of the Bull and Breese index values. As the figure shows, more hydrophobic peptides can provide much higher signal-to-noise values. The general correlation of a negative index value with observation of the signal is helpful, but not predictive. Akashi et al.[44] point out that hydrophilic peptides are observed in some cases in the positive ion FAB mass spectrum, even when equal amounts of hydrophobic peptides are also present in the 1:1 (v:v) glycerol:thioglycerol liquid solvent. Figure 7 is adapted from Pucci,[43] and shows the signal-to-noise values obtained for peptide fragments in equal amounts taken from a tryptic digest of a human hemoglobin. The index values for each of the indicated peptides is shown on the figure. Even peptides with positive values are observed without problem from this mixture. As our understanding grows,

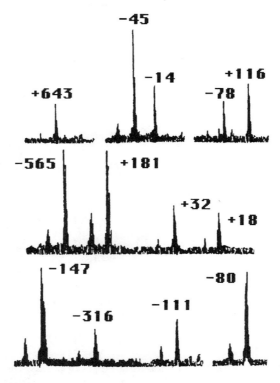

FIGURE 7. Positive ion FAB mass spectrum of the tryptic digest of normal human b-globin chain. The calculated Bull and Breese index values are shown adjacent to each signal that represents a protonated peptide molecule. (Adapted from Reference 43.)

it may be possible eventually to factor into the correlation the mass of the peptide (both with respect to decreased transmission and detection efficiencies at higher masses), the ability of larger mass peptides to enclose hydrophobic residues in the interior of a three-dimensional structure, and the propensity for molecular ions to fragment.

Allmaier[9] has provided an important assessment of the influence of surface concentration on the FAB mass spectrum of hydrophobic peptides. Figure 8 plots the signals for the protonated molecule of substance P (Bull and Breese index value of –163), expressed as the inverse ratio with respect to the signal from the protonated tetramer of glycerol, and surface concentration as a function of bulk solution concentration. Surface tension measurements were made explicitly with a ring tensiometer, and the excess surface concentration for the peptide was derived through the Gibbs equation for dilute solutions. If the surface concentration is sufficiently high to displace all of the glycerol molecules from the surface, then the ratio of signal-to-matrix intensity is very high. As the bulk concentration decreases, the surface excess decreases, and the signal from the glycerol tetramer relative to that from the peptide increases accordingly. Behavior of this sort is reasonable with solutions in which the primary particle beam does not excessively mix the surface, and in which diffusional processes occur rapidly to reestablish stable surface/bulk concentration ratios. For disruptive primary particle fluxes and energies, glycerol cluster ions would be observed no matter what the concentration of the dissolved peptide.

Naylor[45] has extended earlier work to examine the fragmentation of peptides in FAB as a function of derivatization of the peptide, acid additives in the matrix, and the hydrophobicity of the fragment sequence ions. The latter issue is dealt with here. No correlation appears to exist

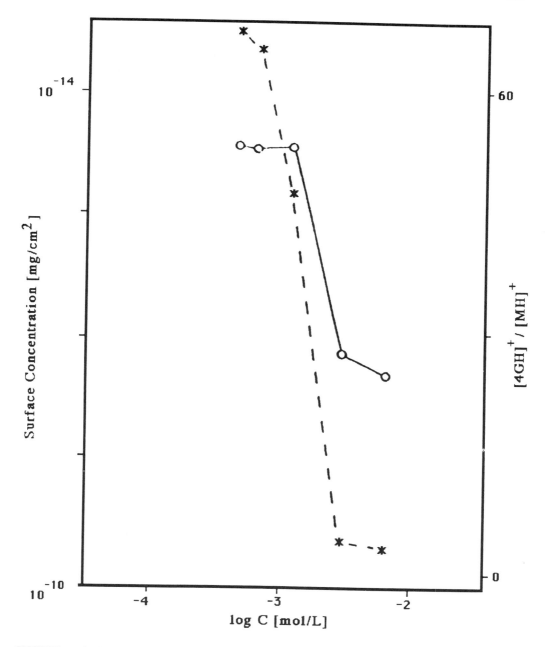

FIGURE 8. Surface concentration values (open circles) and signal for the protonated molecule of substance P relative to the tetramer of glycerol (stars) are plotted vs. bulk peptide concentration. (Adapted from Reference 9.)

between the hydrophobicity of any one particular fragment ion and its relative abundance in the FAB mass spectrum. These results suggest that the fragmentation occurs in the "gas phase", as opposed to the bulk solution, from which a sampling bias might be expected. That the dissociations of sputtered molecules occur in the selvedge, the region immediately above the surface, has been at the core of mechanistic theories in FAB and LSIMS for at least the past decade.[13]

Finally, the literature contains a number of compilations that provide values that can be used instead of the Bull and Breese index values.[37-39] Alternative lists are based on other physical properties, including partition coefficients in water:octanol, and column chromatographic retention indices. Cross correlations of most of the available indices have also been reported.[39] The assignment of a particular value of hydrophilicity or hydrophobicity to an amino acid residue is not without contention.[38] For larger peptides, the use of a *hydropathic* index value has been suggested.[38] This value includes an assessment of not only the values for each individual amino acid residue, but also uses a moving average technique to determine the hydrophilic or hydrophobic regions of a large peptide or protein. The concerted use of such values requires an assessment of peptide secondary structure and includes the probability that any given residue will be solvent-accessible.

Along with the issue of the surface activity of peptides in the particular liquid matrix selected for a FAB or an LSIMS analysis, the case of adsorption onto the sample platform must also be addressed. A number of metal surfaces have been used to support the sputtered sample solution. Copper is cleaned easily with nitric acid, and was used widely in early work, but has been replaced for the most part with sample platforms machined from stainless steel. Irreversible adsorption or reaction of peptides with a metal surface is a possibility as the amount of liquid matrix is reduced and as the concentration of the sample increases. A recent paper by van Breemen[46] reports increased sensitivity in peptide analysis by FAB with the use of nitrocellulose as a substrate. A homogeneous layer of nitrocellulose was placed on the surface of a standard stainless steel FAB probe tip by deposition of 1.0 µl of an acetone solution of nitrocellulose (concentration of 3.5 µg/µl). Peptide solutions in dimethylsulfoxide were then applied to the probe, followed by 1.0 µl of thioglycerol. Positive ion FAB mass spectra for a series of peptides ranging in molecular weight from 1600 to 8000 amu were studied, and a sensitivity increase of at least 2-fold for each peptide was noted. van Breemen suggests that the nitrocellulose deactivates the surface of the metal platform, decreasing the extent of peptide adsorption to that surface. In such a situation, more peptide remains in solution for sputtering by the primary particle beam. If such an explanation is correct, the implication is that approximately half of the sample molecules from the solutions are absorbed to the metal substrate. In some instances, 50 µg of a sample yielded no signal in the positive ion FAB spectrum without nitrocellulose, but provided a signal when the metal surface was deactivated. In other cases, before-and-after enhancement experiments were demonstrated at the 1 ng sample level. It seems unlikely that the capacity of the metal substrate varies by 10^4, depending on the sample used. Changes in the adsorption/desorption behavior might be responsible for some changes, but an equally likely explanation is based on changes in the surface concentration in the thioglycerol solution with the use of nitrocellulose as an additive to the liquid matrix. If nitrocellulose is soluble in thioglycerol, deactivation of the metal surface would be temporary.

Suppression effects have also been noted in peptide mapping of mixtures by plasma desorption mass spectrometry (PDMS).[47] The suppression effects noted in plasma desorption ionization differ in origin from those observed in FAB mass spectra since there is no solvent in which different sample molecule surface activities can be exhibited. The relative signal levels for different peptide molecules can instead be attributed to the relative net charges of the peptides in the solution from which they are deposited onto the surface for subsequent analysis.

E. SOLVENTS FOR FLOW-FAB ANALYSIS OF PEPTIDES

In flow-FAB, the solvent containing the analyte flows through a capillary inserted in a direct insertion probe and terminating at a platform in the source of the mass spectrometer, depositing a liquid on the surface of a bullet-shaped FAB probe (Figure 9). The solvent composition chosen must be such that an appropriate flow through the capillary can be maintained to transport sample into the source without precipitation, that enough solvent remains on the tip for adequate measurement of the FAB mass spectrum, and that the evaporation of the more volatile constituents of the solvent does not lead to excessive pressure inside the source of the mass spectrometer. Several solvent systems have been used with flow-FAB introduction of samples

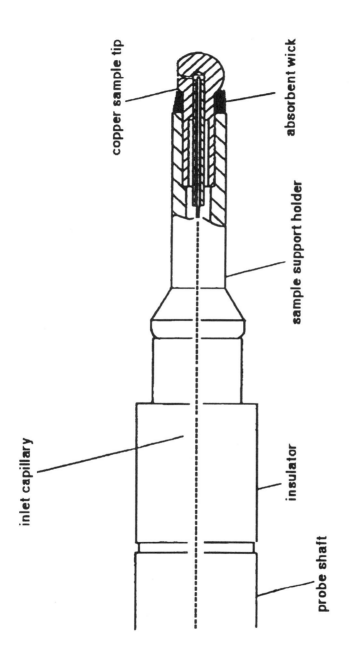

FIGURE 9.   Schematic of a continuous flow fast atom bombardment direct insertion probe.

TABLE 3
Representative Solvent Systems Reported for Use in Continuous Flow-FAB Studies of Peptides and Peptide Mixtures

Component 1	Component 2	Component 3
95% Water	5% Glycerol	0.1% CF_3COOH
95% Water	5% DTE/DTT[a]	0.1% CF_3COOH
95% MeOH:H_2O (75:25 v:v)	5% Glycerol	0.1% CF_3COOH
60% Acetonitrile/ water gradient	5% Glycerol	0.1% CF_3COOH
90% Methanol	10% Thioglycerol	
80% Water	20% Glycerol	

[a] DTE is dithioerythritol; DTT is dithiothreitol.

as outlined in Table 3, an overview of solvent systems reported at recent meetings. If the flow probe is used in a direct interface to a liquid chromatograph, other solvent systems are chosen, and the solvent reflects the gradient chosen for most effective column separation. In almost all systems, glycerol is present at a low percentage (5 to 20% by volume) and 0.1% trifluoroacetic acid (TFA) is present as an additive, acting to increase the probability that sample molecules will be sputtered as $(M+H)^+$ ions.

Sensitivity is increased in flow-FAB (or flow-SIMS) relative to the analysis of discrete samples because the contribution to noise from the matrix is reduced. The instantaneous concentration of sample in the solution is increased significantly, and the ratio of signal from the analyte to signal from the solvent rises. The solvent flows continuously to the probe tip, but the role of sample diffusion for glycerol is replaced by transport of excess sample and radiation products away from the point of particle beam impact to another part of the probe. Many of the second-generation probes contain a wick material (often cellulose) in which the excess glycerol solvent is absorbed until the probe is removed from the source and the wick is replaced.

Because the role played by the solvent has changed significantly in flow-FAB from that involved in discrete sample FAB analysis, the issue of the relative surface concentrations of sample, including peptides, is reduced in importance. Peptide molecules of different surface activity terminating on the tip at the same time will still attempt to establish the appropriate differences in surface concentration, but the combination of flow and reduced sample volume precludes the establishment of significant differences. As a result, the mass spectral response per given amount of sample is much more uniform.

III. SAMPLE DERIVATIZATION

A. N-TERMINAL PEPTIDE MOLECULE DERIVATIZATIONS

Derivatization reactions have been developed to increase the sensitivity of peptide analysis. Most of these reactions have concentrated on changes in the character of the N-terminus of the peptide. Naylor et al.[41] described the use of an esterification reaction (1.0 *M* HCl in 2-propanol at 37°C for 1 h) to level the hydrophobicities of peptides and ensure a more uniform response in FAB. Care must be taken, however, because higher concentrations of HCl were observed to cause the conversion of some amide side chains to the corresponding isopropyl esters. With such a derivatization reaction, Naylor et al. report, for instance, that all of the expected digest peptides from a sample of glucagon or horse cytochrome *c* were observed in the FAB spectrum recorded during the first few minutes of bombardment and spectral measurement. Reactions for the preparation of N-terminally acylated peptides were also described using acetic anhydride as the derivatization reagent.

Kidwell et al.[48] have described a method for the derivatization of small peptides, developed for LSIMS, that involves several steps. If the N-terminus of a peptide is labeled with a charged group, the peptide then cleaved with an acid, and the fragments finally esterified and acylated, the positive ion FAB or LSIMS spectrum should contain primarily the ions with the charged group attached to them. The mass spectrum should therefore contain ions that originated only from the N terminus, and the sequence of the peptide can be discerned through multiple reaction steps. Quaternization with methyl iodide was explored as a method for introducing a charge label on the N-terminus of a peptide, but the conditions for reaction are severe and the yields low and variable. Acylation with chloracetyl chloride was adopted as an alternative procedure. After acylation, the peptide is cleaved with acid (6 N HCl at 110°C for 5 to 10 minutes), esterified with 3 M HCl in methanol, and then reacted with triethylamine (60°C in methanol overnight) to form the ammonium salt.

The N-terminus of a peptide molecule can be reacted with a trisubstituted pyrylium salt to form a pyridinium salt. Reaction of a pyrylium salt with the lysine residues of chymotrypsin was described in 1971 by O'Leary and Samberg[49] and was described first for the study of peptides by FAB and LSIMS by Busch et al.[50,51] Good quality FAB and LSIMS mass spectra are obtained for the derivative. Multiply derivatized forms of the sample molecule provide multiple checks of the molecular weight of the peptide, and also serve to distinguish lysine from glutamine because of the reaction of the sidechain amino group with the reagent. Low energy collision-induced dissociation of the peptide pyridinium derivatives yields product ion MS/MS spectra that contain amino acid sequence information complementary to that obtained from the product ion MS/MS spectra of the protonated peptide.

B. C-TERMINAL PEPTIDE MOLECULE DERIVATIZATIONS

Derivatization reactions specific to the C-terminal end of a peptide are reported less commonly than reactions targeted to the N-terminal end. Bennett and Day[52] reported that the C-terminal end of peptides could be derivatized to yield a positively charged quaternary amine via a reaction that produces the azlactone intermediate at that terminus. The reaction is shown in the scheme; the activated carbonyl reacts with an N-protected aminoalcohol to give the ester and reacted subsequently with iodomethane to provide the quaternary amine. The complete reaction scheme as described requires an overnight reaction. Yields were not reported, but for dipeptides at least, the positive ion mass spectra were improved in terms of the ratio of the intact cation of the derivatized peptide to the level of background signals from the glycerol liquid matrix.

Rose et al.[53] have described a method in which the C-terminal peptide of a protein subjected to enzymic digestion is singularly identified. This technique involves incorporation of ^{18}O into all α-carboxy groups liberated during the enzyme-catalyzed partial hydrolysis of the peptide in ^{18}O-labeled water. The first report[54] of this method involved the formation of volatile derivatives so that the final analysis could be completed with gas chromatography/mass spectrometry (GC/MS), and was therefore limited to relatively small peptide fragments. The second paper[56] describes the use of FAB mass spectrometry to monitor the extent of ^{18}O incorporation in larger C-terminal peptides; mixtures of the peptides are separated by HPLC. Data that indicate partial incorporation of the label must be interpreted very carefully because the occurrence of the hydrogen-addition reduction processes described in an earlier section confuses the measurement.

C. INTERNAL DERIVATIZATIONS

Peptides will react with a 2-fluoro-N-methyl pyridinium salt to form a pyridinium salt derivative with loss of hydrofluoric acid (HF) from hydroxyl, sulfhydryl, and primary amino groups.[50,55] These reactions can occur at any appropriate residue within the chain that is

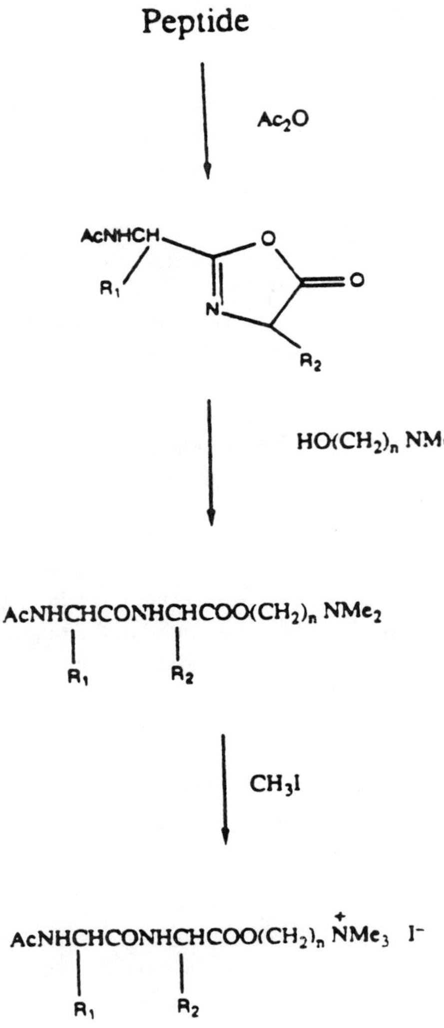

Peptide

↓ Ac$_2$O

AcNHCH$_{R_1}$ ... (structure: oxazolone ring with N, O, =O, R$_2$)

↓ HO(CH$_2$)$_n$ NMe$_2$

AcNHCHCONHCHCOO(CH$_2$)$_n$ NMe$_2$
| |
R$_1$ R$_2$

↓ CH$_3$I

AcNHCHCONHCHCOO(CH$_2$)$_n$ $\overset{+}{N}$Me$_3$ I$^-$
| |
R$_1$ R$_2$

Scheme 1

accessible to the reagent. Although the simple *N*-methyl pyridinium reagent used imparts a positive charge and distinct label to the peptide molecule, higher sensitivity can be expected for species that are surface active. Pyridinium salts have been developed that contain octyl group and pentafluorobenzyl group substituents. Derivatives formed from these three pyridinium compounds appear to exhibit increased surface activity in a number of commonly used FAB matrices.

The pyrylium salts described in the previous section will also react with internal residues that contain primary amine groups. Lysines will react to form an internal pyridinium salt. The double derivative (N-terminal and internal lysine) formed appears in the FAB and LSIMS mass spectrum as the singly charged ion formed by loss of one proton from the derivatized molecule.

D. THE IMPORTANCE OF DISULFIDE/SULFHYDRYL TRANSFORMATIONS

The oxidation state of cysteine-containing peptides is an important measurement in the assessment of the biological activity of a molecule. FAB and LSIMS can provide such a measurement, but only if sufficient attention is paid to the details of sample preparation so that unexpected oxidations and reductions are not encountered. The positive ion FAB or LSIMS mass spectrum of a cysteine-containing peptide can be characterized by changing ion abundances reflecting variations in the ratio of the oxidized to the reduced forms of the peptide. This ratio is affected by the liquid matrix selected, the primary beam conditions, and the presence of other components in the matrix or the sample mixture.

Buko and Fraser[26] provided a systematic FAB-based study of the extent of oxidations and reductions of cysteine-containing peptides. Earlier work had noted that reduction of cysteine linkages did indeed occur in FAB. Reduction of the disulfide bonds in insulin and somatostatin was noted by Buko et al.[56] In the case of the latter compound, the mass spectrum did not provide direct evidence for reduction of the peptide in the pattern of the isotopic envelope for the $(M+H)^+$ ion, but fragment ions that resulted from the reduction were noted in the spectrum. Confirmation of the reduction was obtained through the use of deuterated glycerol $[R(OD)_3]$, and observation of the appropriate mass shifts.

Williams et al.[17] noted that the positive ion FAB mass spectra of somatostatin, oxytocin, and vasopressin (all containing disulfide bridges) each contains an $(M+H)^+$ ion. However, because the spectra were not presented in the paper, the extent of reduction cannot be gauged in hindsight. Williams *did* note the presence of fragment ions corresponding to the reduction of the cysteine bridge. Morris et al.[57] noted that the positive ion FAB mass spectrum of peptides derived from human calcitonin exhibited ions corresponding to the reduced forms of the cysteine bridge, and further attributed the reduction to the use of thioglycerol as the liquid matrix. (Several papers in the literature claim incorrectly that this seminal paper of Morris reports no reduction.) Explicit reduction of the cysteine bridge followed by carboxymethylation shifted the masses of the two fragment peptides to the appropriate masses, confirming their participation in the bridge. Calcitonin itself was studied by FAB in the work of Larsen et al.[58] Synthetic salmon calcitonin was sputtered from dimethylsulfoxide and glycerol for positive ion FAB mass spectra and from glycerol, thioglycerol, and diethanolamine for negative ion mass spectra. Calcitonin in which the sulfhydryl groups were derivatized with iodoacetamide was also similarly analyzed. Larsen notes that the positive and negative ion mass spectra of the synthetic calcitonin were in concordance with respect to a 1:3 ratio of oxidized to reduced forms of the peptide (chosen to match the measured pattern of the isotopic envelope), and that this ratio did not change with the matrix used. Examination of the spectrum of the iodoacetamide derivatives showed both the original oxidized and underivatized forms of the calcitonin and peaks due to ions from the reduced forms. The sum of these observations confirms that the original sample contains a mixture of oxidized and reduced forms of the calcitonin. This result underscores the need to establish independently the purity of the biological sample before attribution of ions to artifactual reactions.

Mohara and Tanimoto[25] also studied the reduction of disulfide-containing peptides in LSIMS. In particular, this work was concerned with the extent of reduction in different matrices, and with changes in the additives in the matrix.

Knowing the mechanisms of cysteine reduction, several investigators have adopted specific strategies to identify disulfide-containing peptides. Sun and Smith[59] describe a performic acid oxidation that modifies certain residues in a peptide chain. Specifically, cystine, cysteine, methionine, and tryptophan residues undergo a characteristic change in molecular weight under the conditions used for the derivatization; each cysteinyl residue that reacts results in an increase in mass of 48 mass units. The performic acid oxidation approach has been used to show that the appropriate residues in a peptic digest of bovine insulin exhibit the appropriate mass shifts in the positive ion FAB mass spectrum, and the method was extended to elucidate the cysteinyl-containing residues in a tryptic digest of cyanogen bromide-treated ribonuclease-A. Treatment with cyanogen bromide prior to the performic acid oxidation removes ambiguity due to the reactivity of both methionyl and cysteinyl residues, because the former react with cyanogen bromide to produce the corresponding homoserine lactones.

Yazdanparast et al.[33] showed that prolonged irradiation of a disulfide-containing peptide in solution by a high-energy xenon beam results in a gradual reduction of the disulfide bond. Intrachain disulfide bonds are identified by an increase with time of the relative abundance of the protonated molecule of the oxidized form accompanied by a concomitant decrease in the abundance of the protonated molecule for the reduced form. A dithiothreitol/dithioerythritol matrix was used for part of this work. Samples were incubated in this matrix outside of the source of the mass spectrometer to show that the matrix itself could initiate a reduction of the disulfide bridge. The rate at which the reduction occurred, however, was much slower than that observed during the FAB analysis, implicating an acceleration of the reduction by the primary beam impact. In addition, the authors point out that the reactive species generated by primary beam impact thought to be involved in the reduction process are analogous to those thought to be formed by the UV light and electron irradiation of thiols and disulfides. Photoactivated reactions may therefore also play a role in the extent of reduction observed.

In further work, Yazdanparast et al.[60] described a protocol in which proteins are digested to mixtures of smaller peptides using any of a number of chemical or enzymatic methods that preserve the disulfide bonds present in the protein. The digested sample is then analyzed by FAB before and after explicit chemical reduction of the disulfide bridges by dithiothreitol and dithioerythritol in an alkaline solution. Reduction also occurs, as described earlier, with prolonged sputtering by the primary ion beam. Intra- as well as interchain disulfide bridges are identified based on the differences in the masses of the protonated molecules of the oxidized and the reduced mixtures.

Buko and Fraser[26] pointed out that small changes in peptide structure (for instance, the substitution of a single amino acid residue in a peptide with a total of six amino acids) can have drastic changes on the facility with which a disulfide bond in the peptide will reduce under standard conditions. Generally, interchain disulfide bonds are more susceptible to reduction than intrachain bridges, and FAB may be able to provide that differentiation. However, they also point out that samples of biological origin are likely to be completely uncharacterized with respect to the extent of oxidation and reduction of the disulfide bridges even prior to the FAB analysis. In addition, sample handling leads to additional changes outside of the sputtering process. For instance, some of the early work with insulin dissolved the sample in 50% formic acid, an excellent solvent, but also an excellent denaturant. One also notes that cysteine can be oxidized to cystine in dimethylsulfoxide to which HCl acid has been added. The cysteine/cystine transformation is indeed an important one, and FAB and LSIMS can provide useful information about the number and location of such residues in peptides. However, sample preparation and the spectral interpretation must always be completed carefully to avoid uncharacterized transformations.

Although it is not the purpose of this chapter to comment extensively on the use of MS/MS in the analysis of peptides and proteins, several groups have commented on the particular use of MS/MS in sequencing of cystine-bridged proteins and in the disulfide mapping of proteins. If a protonated molecule containing a disulfide bridge is fragmented in a collision-induced dissociation, the cleavage of the bridge bond between the sulfur atoms leads to one of the most abundant fragment ions in the spectrum, characterized by accompanying peaks 1 mass unit higher and lower. The triplet signal is especially easy to recognize in the product ion MS/MS spectrum.[61]

Most of this section has been concerned with the reduction of disulfide bonds in peptide molecules. As remarked earlier, Fujita et al.[34] noted that the molecular ions of other peptides also were observed to incorporate additional hydrogens. The relative level of this hydrogen reduction in some small peptides were determined by Musselman and Watson[62] for the solvents glycerol and thioglycerol. FAB analyses completed in glycerol solvent provided higher than expected intensities for ions at $(M+2)^+$, $(M+3)^+$, and $(M+4)^+$ where M is the mass of the neutral peptide molecule, but the mass spectra of the same compounds sputtered from a thioglycerol matrix provided isotopic intensities that match those predicted from the empirical formula. The use of a thioglycerol rather than a glycerol matrix was therefore recommended in the analysis of unknown peptides.

E. MONITORING OF DEGRADATION REACTIONS

Modern investigations in biochemistry and molecular biology deal routinely with peptides of molecular weights far in excess of the upper mass range exhibited by mass spectrometry. Certainly, significant experiments with electrospray ionization and laser desorption have pushed the upper mass limit in just the last few months to about 65,000 Da for the former, and about 150,000 Da for the latter.[63,64] FAB has been used almost exclusively with peptides of molecular weights of less than 10,000 Da. For sequence information, due to the lower probability that ions from larger peptides will dissociate into sequence-specific ions, the effective upper limit in peptide mass is about 4000 daltons. A common approach in FAB analysis of larger peptides is to degrade or digest the sample peptide into a mixture of smaller peptides more amenable to mass spectrometric sequencing. If a residue-specific cleavage reagent such as an enzyme is used, then the smaller peptides can be reassembled via sequence overlaps to provide the structure of the larger original peptide or protein. This experiment has been dubbed "FAB-mapping" by Morris and co-workers.[65,66] These investigators reduced the polypeptide hormone insulin with dithiothreitol, known to reduce the disulfide bonds (see Section D). The presence of both the A- and the B- chains of the original insulin was reflected in signals in the positive ion FAB mass spectrum for the molecular ions $(M+H)^+$ of both subunits. The same sample was subjected to a tryptic digestion, and by simple subtraction of the resulting fragment ions from the original molecular weight, a high proportion of the sequence information could be deduced.

FAB mapping depends on the specificity with which the various chemical or enzymic degradation reactions perform, because the reassembly of the original peptide depends on the removal of ambiguities in the determined sequence. Overlapping sequences are possible if enough sample is available to carry out several different digestions that cleave a large peptide at several different points, leading to smaller peptides with a substantial degree of confirmatory overlap. In some cases, careful interpretation of the FAB mass spectra measured after such digestions points out that unexpected rearrangements in peptide structure can occur.[72]

De Angelis[67] proposed a method in which acid hydrolysis is used for different lengths of time for separated aliquots of a peptide sample. For instance, 5 nmol portions of substance P are heated (90°C) with 2 µl of aqueous 5% HCl for 0.25, 0.5, 1, 2, 3, 4, and 6 h. Interpretation of the positive ion FAB mass spectra obtained from the seven discrete samples provides a series of overlapping peptides that are reassembled into the sequence of the original peptide.

IV. CONCLUSIONS

Means of interpretation of FAB and LSIMS mass spectra to deduce the amino acid sequence of a peptide have been studied in detail over the past 5 years. A standard nomenclature has been developed to provide a framework in which the systematic rules of peptide fragmentations can be discussed, and which has been extended to MS/MS methods for peptide sequencing as well.[68] Discordant observations with respect to the types of fragment ions observed in positive ion FAB and LSIMS mass spectra continue. For instance, changes in the relative dominance of a-, b-, and c-type ions (derived from reactions with charge retention at the N-terminus) and x-, y-, and z-type ions (charge retention at the C-terminus) have been reported from different research groups working with different groups of peptides. We assume a far too simple view of the sputtering and ionization process if we suppose that the subtle details of sample preparation and analysis in FAB will not ultimately play a role in determining some of these differences. New ions corresponding to side chain cleavages, rearrangement reactions, and multiple cleavages can sometimes be particularly abundant, and with increasing familiarity with the spectral interpretation of FAB spectra of peptides, the origins of these ions are being determined as well.[69]

Because the procedure of deducing a peptide sequence from its FAB mass spectrum is repetitive, a number of computer algorithms have been developed to perform this task. The first type of algorithm generates all permutations of amino acid residue combinations consistent with the molecular weight of the peptides determined from the FAB or LSIMS mass spectrum. The observed FAB sequence ions are then correlated with the predicted fragment ions.[70-72] A different kind of algorithm reported recently allows the peptide sequence to be determined stepwise from the mass differences in neighboring sequence ions appearing in the mass spectrum; this logic is similar to that used to manually sequence a peptide.[72] The authors claim that the advantage of this method is the ability to handle higher molecular weight peptides.

FAB and LSIMS are so widely available that they will probably remain the most commonly used methods in peptide analysis for the next 5 years. Most of this work will concentrate on studies of peptides that are under 4000 amu in molecular mass; if the samples to be studied are of higher mass, digestion or degradation techniques that bring the samples within this range will be used. Many mass spectrometrists will be forced to refine their ability to manipulate and react solution volumes that never exceed a few microliters, and to transfer those sample volumes without loss or contamination. An increasing percentage of this work will be reported in application areas, with a higher percentage of reports that do not include the mass spectrum itself. Many of the specialized journals now accept FAB data in the same way in which exact mass measurements are reported. For example, a new compound or peptide is reported, followed by analytical data such as "FAB mass spectrometry shows the (M+H)+ at m/z 1204." Neither the matrix nor the sample concentration is reported, nor is the spectrum tabulated or shown. We accept on faith that the peak reported is indeed a large signal in the mass spectrum, that the fragment ions are in concordance with the proposed structure, that there are no other signals present from compounds also in the mixture, that the isotope pattern expected is confirmed by that observed, and that no reduction, oxidation, complexation, or degradation of the sample is observed.

In the longer term, FAB and LSIMS must compete against the methods designed to allow the ionization of higher mass peptides, extending to m/z 20,000. At still higher masses, the competition evolves to a competition between mass spectrometry and nonspectroscopic methods of molecular mass determination, such as electrophoresis. Within the range of mass spectral accessibility, the relative merits of several different methods for sample ionization and several different instruments for mass analysis have been hotly debated. Time-of-flight (TOF) mass spectrometry, for instance, offers virtually unlimited mass range, but usually with a loss in resolution and with significant constraints on the methods of sample preparation that can be used and the methods of ionization that can be employed. Sector instruments provide nominal mass ranges to m/z 20,000, but in practice, transmission is such that only a few samples provide a sufficiently high ion flux so that these masses can be explored. Quadrupoles and ion traps are sufficient for molecules of masses of a few thousand; higher masses can be reached, but not yet routinely.

How then to increase the mass range of the instrument? It has been conceded for many years that examination of multiply charged ions was the most efficient way of increasing the mass range of an analyzer. However, doubly charged cations are relatively rare in FAB and LSIMS mass spectra, even for compounds that are stable as doubly positive ions in solution. A number of charge reduction processes, including electron reduction, complexation with an anion, and expulsion of a proton, conspire to produce singly charged ions. These bulk and selvedge reactions emphasize the nonchemical ionization facets of FAB mass spectrometry. Plasma desorption was the first ionization technique able to provide reproducible, albeit low, fluxes of multiply charged ions, providing several equivalent checks on the molecular mass of insulin. Electrospray and related means of ionization produce multiply charged ions with ease, and once sufficient numbers of the ionization sources are manufactured and brought into use, these ionization methods will provide the dominant competition for FAB and LSIMS for the analysis of peptides. Truly spectacular successes in the determination of molecular weights of peptides of up to m/z 20,000 have been reported already for these methods. A severe problem is the almost total lack of fragmentation for the sample molecules ionized by electrospray, limiting the interpreted information to molecular weight and not sequence information. MS/MS methods may provide some aid here, as they do in FAB and LSIMS, but the mechanisms for the collision-induced dissociations of multiply charged ions have not been established. Careful control of the instrumental and the sample parameters to cause dissociation by losses of characteristic fragments, and one charge at a time, are just now being investigated.[62] Any dissociation to fragment ions with significantly fewer charges will place the fragment ions outside of the mass range of the analyzer.

FAB and LSIMS, and the combination of these ionization methods with MS/MS, provide excellent information for peptide or peptide fragments of 10 to 15 residues in length. A two-step approach to very large peptide characterization can be foreseen in which the first step is electrospray ionization to determine the molecular weight, followed by one or several steps of digestion and cleavage and separation by liquid chromatography, with sequence determination by FAB and FAB MS/MS. An alternative approach to separation of a digest would be the use of capillary zone electrophoresis coupled with FAB mass spectrometry.[73]

The continued use of FAB in some form in peptide analysis seems assured. This overview has concentrated only on the aspects of sample derivatization and preparation. Aspects of spectral interpretation and reviews of applications are well covered elsewhere.[74,75] The successes of FAB and LSIMS have been all the more extraordinary for the disregard often given to the mechanics and fundamentals of sampling and sample preparation. Renewed attention to the complexities of the chemistry involved would be repaid amply with better quality mass spectra.

REFERENCES

1. **Hunt, D. F., Yates, J. R., III, Shabanowitz, J., Winston, S., and Hauer, C. R.,** Protein sequencing by tandem mass spectrometry, *Proc. Natl. Acad. Sci. U.S.A.,* 83, 6233, 1986.
2. **Williams, D. H.,** Bio-organic and biochemical applications of mass spectrometry: the study of peptides and proteins, *Adv. Mass Spectrom.,* 11A, 81, 1989.
3. **Burlingame, A. L. and Castagnoli, N., Jr., Eds.,** *Mass Spectrometry in the Health and Life Sciences,* Elsevier, New York, 1985.
4. **Hemling, M. E.,** Fast atom bombardment mass spectrometry and its application to the analysis of some peptides and proteins, *Pharm. Res.,* 4, 5, 1987.
5. **Biemann, K. and Scoble, H. A.,** Characterization by tandem mass spectrometry of structural modifications in proteins, *Science,* 237, 992, 1987.
6. **Gower, J. L.,** Matrix compounds for fast atom bombardment mass spectrometry, *Biomed. Mass Spectrom.,* 12, 191, 1985.
7. **De Pauw, E.,** Liquid matrices for secondary ion mass spectrometry, *Mass Spectrom. Rev.,* 5, 191, 1986.
8. **Cook, K. D., Todd, P. J., and Friar, D. H.,** Physical properties of matrices used for fast atom bombardment, *Biomed. Environ. Mass Spectrom.,* 18, 492, 1989.
9. **Allmaier, G. M.,** The influence of surface concentration in fast atom bombardment mass spectrometry of hydrophobic peptides, *Rapid Commun. Mass Spectrom.,* 2, 75, 1988.
10. **Gower, J. L.,** Matrix compounds for fast atom bombardment: a further review, *Adv. Mass Spectrom.,* 10B, 1537, 1986.
11. **Lehmann, W. D., Kessler, M., and Konig, W. A.,** Investigations on basic aspects of fast atom bombardment mass spectrometry: matrix effects, sample effects, sensitivity and quantification, *Biomed. Mass Spectrom.,* 11, 217, 1984.
12. **Sunner, J. A., Morales, A., and Kebarle, P.,** Dominance of gas-phase basicities over solution basicities in the competition for protons in fast atom bombardment mass spectrometry, *Anal. Chem.,* 59, 1378, 1987.
13. **Cooks, R. G. and Busch, K. L.,** Matrix effects, internal energies, and MS/MS spectra of molecular ions sputtered from surfaces, *Int. J. Mass Spectrom. Ion Phys.,* 53, 111, 1982.

14. **De Pauw, E.,** Matrix influence in LSIMS-FAB, *Adv. Mass Spectrom.,* 11A, 383, 1989.

15. **Sunner, J., Morales, A., and Kebarle, P.,** Temperature dependence of FAB spectra. Experiments and model calculations, *Adv. Mass Spectrom.,* 11A, 426, 1989.

16. **Schroder, E., Munster, H., and Budzikiewicz, H.,** Ionization by fast atom bombardment—a chemical ionization (matrix) process in the gas phase?, *Org. Mass Spectrom.,* 21, 707, 1986.

17. **Williams, D. H., Bradley, C. V., Santikarn, S., and Bojesen, G.,** Fast-atom bombardment mass spectrometry. A new technique for the determination of molecular weights and amino acid sequences of peptides, *Biochem. J.,* 201, 105, 1982.

18. **Seki, S.,** Addition of acid in SIMS, *Shitsuryo Bunseki,* 36, 17, 1988.

19. **De Pauw, E., Pelzer, G., and Marien, J.,** Matrix optimization for SIMS-FAB, *Adv. Mass Spectrom.,* 10B, 1563, 1986.

20. **Busch, K. L., Unger, S. E., Vincze, A., Cooks, R. G., and Keough, T.,** Desorption ionization mass spectrometry: sample preparation for secondary ion mass spectrometry, laser desorption, and field desorption, *J. Am. Chem. Soc.,* 104, 1507, 1982.

21. **Foti, S. and Saletti, R.,** Peptide sequencing by partial methanolysis and fast atom bombardment mass spectrometry, *Biomed. Environ. Mass Spectrom.,* 18, 168, 1989.

22. **Baczynskyj, L.,** New matrices for fast atom bombardment mass spectrometry, *Adv. Mass Spectrom.,* 10B, 1611, 1986.

23. **Tolun, E., Dass, C., and Desiderio, D. M.,** Trace level measurement of enkephalin peptides at the attomole/femtomole level by FAB-MS. Optimization of FAB matrix conditions, *Rapid. Commun. Mass Spectrom.,* 1, 77, 1987.

24. **Cleland, W. W.,** Dithiothreitol, a new protective reagent for SH groups, *Biochemistry,* 3, 480, 1964.

25. **Mohara, S. and Tanimoto, M.,** Reduction of disulfide-containing peptides in matrix on secondary ion mass spectrum measurement, *Mass Spectrosc.,* 35, 248, 1987.

26. **Buko, A. M. and Fraser, B. A.,** Peptide studies using a fast atom bombardment high field mass spectrometer and data system. IV. Disulfide-containing peptides, *Biomed. Mass Spectrom.,* 12, 577, 1985.

27. **Wirth, K. P., Junker, E., Rollgen, F. W., Fonrobert, P., and Przybylski, M.,** Reduction of azo-group-containing peptides in fast atom bombardment mass spectrometry, *J. Chem. Soc. Chem. Comm.,* 1387, 1987.

28. **Morgan, A. E., de Grefte, H. A. M., and Tolle, H. J.,** Effect of oxygen implantation on secondary ion yields, *J. Vac. Sci. Technol.,* 18, 164, 1981.

29. **Benninghoven, A., Rudenauer, F. F., and Werner, H. W.,** *Secondary Ion Mass Spectrometry. Basic Concepts, Instrumental Aspects, Applications and Trends,* Wiley-Interscience, New York, 1987.

30. **Burinsky, D. J., Dilliplane, R. L., DiDonato, G. C., and Busch, K. L.,** Reduction of methylene blue during the ionization process, *Org. Mass Spectrom.,* 23, 231, 1988.

31. **Collins, M. W.,** M.S. thesis, Indiana University, Bloomington, 1989.

32. **Hand, O. W., Detter, L. D., Lammert, S. A., Cooks, R. G., and Walton, R. A.,** Reduction induced by ion beams: hydrogenation of nitrogen-containing heterocycles and quinones in molecular secondary ion mass spectrometry, *J. Am. Chem. Soc.,* 111, 5577, 1989.

33. **Yazdanparast, Y., Andrews, P., Smith, D. L., and Dixon, J. E.,** A new approach for detection and assignment of disulfide bonds in peptides, *Anal. Biochem.,* 153, 348, 1986.

34. **Fujita, Y., Matsuo, T., Sakurai, T., Matsuda, H., and Katakuse, I.,** Mass distribution of peptide molecular ions in the secondary ionization process, *Int. J. Mass Spectrom. Ion Proc.,* 63, 231, 1985.

35. **Naylor, S., Moneti, G., and Guyan, S.,** Hydrophobic effects in the fast atom bombardment mass spectra of proteins and large peptides, *Biomed. Environ. Mass Spectrom.,* 17, 393, 1988.

36. **Bull, H. B. and Breese, K.,** Surface tension of amino acid solutions: a hydrophobicity scale of the amino acid residues, *Arch. Biochem. Biophys.,* 161, 665, 1974.

37. **Hopp, T. P. and Woods, K. R.,** Prediction of protein antigenic determinants from amino acid sequences, *Proc. Natl. Acad. Sci. U.S.A.,* 78, 3824, 1981.

38. **Kyte, J. and Doolittle, R. F.,** A simple method for displaying the hydropathic character of a protein, *J. Mol. Biol.,* 157, 105, 1982.

39. **Cornette, J. L., Cease, K. B., Margalit, H., Spouge, J. L., Berzofsky, J. A., and DeLisi, C.,** Hydrophobicity scales and computational techniques for detecting amphipathic structures in proteins, *J. Mol. Biol.,* 195, 659, 1987.

40. **Bursey, M. M. and Erickson, B. W.,** A search for evidence of secondary structure in the tandem mass spectra of peptides, paper ANYL-38, in the 197th National Meeting of the American Chemical Society, Dallas, April 9-14, 1989.

41. **Naylor, S., Findeis, A. F., Gibson, B. W., and Williams, D. H.,** An approach to the complete FAB analysis of enzymic digests of peptides and proteins, *J. Am. Chem. Soc.,* 108, 6359, 1986.

42. **Caprioli, R. M., Moore, W. T., and Fan, T.,** Improved detection of 'suppressed' peptides in enzymic digests analyzed by FAB mass spectrometry, *Rapid Commun. Mass Spectrom.,* 1, 15, 1987.

43. **Pucci, P., Ferranti, P., Marino, G., and Malorni, A.,** Characterization of abnormal human haemoglobins by fast atom bombardment mass spectrometry, *Biomed. Environ. Mass Spectrom.,* 18, 20, 1989.

44. **Akashi, S., Hirayama, K., Seino, T., Ozawa, S.-I., Fukuhara, K.-I., Oouchi, N., Murai, A., Arai, M., Murao, S., Tanaka, K., and Nojima, I.,** A determination of the positions of disulphide bonds in Paim I, α-amylase inhibitor from *Streptomyces corchorushii,* using fast atom bombardment mass spectrometry, *Biomed. Environ. Mass Spectrom.,* 15, 541, 1988.

45. **Naylor, S. and Moneti, G.,** Factors affecting the fragmentation of peptides in fast atom bombardment mass spectrometry, *Biomed. Environ. Mass Spectrom.,* 18, 405, 1989.

46. **van Breemen, R. B. and Le, J. C.,** Enhanced sensitivity of peptide analysis by fast-atom bombardment mass spectrometry using nitrocellulose as a substrate, *Rapid Commun. Mass Spectrom.,* 2, 20, 1989.

47. **Nielsen, P. F. and Roepstorff, P.,** Suppression effects in peptide mapping by plasma desorption mass spectrometry, *Biomed. Environ. Mass Spectrom.,* 18, 131, 1989.

48. **Kidwell, D. A., Ross, M. R., and Colton, R. J.,** Sequencing of peptides by secondary ion mass spectrometry, *J. Am. Chem. Soc.,* 106, 2219, 1984.

49. **O'Leary, M. H. and Samberg, G. A.,** Chemical modification of proteins by pyrylium salts, *J. Am. Chem. Soc.,* 93, 3530, 1971.

50. **Dunphy, J. C. and Busch, K. L.,** An overview of desorption ionization mass spectrometry in peptide and protein analysis, *Peptide Res.,* 1, 48, 1988.

51. **Duffin, K. L. and Busch, K. L.,** Derivatization of peptides with pyrylium salts for analysis by fast atom bombardment and secondary ion mass spectrometry, submitted for publication.

52. **Bennett, B. D. and Day, R. A.,** A novel C-terminus specific peptide derivatization designed to enhance FAB-MS sensitivity: chemical introduction of quaternary amine functionality specifically to C-terminus via azlactone intermediate, presented at the 35th ASMS Conference on Mass Spectrometry and Allied Topics, May 24-29, 1987, Denver, 568.

53. **Rose, K., Savoy, L.-A., Simona, M. G., Offord, R. E., and Wingfield, P.,** C-terminal peptide identification by fast atom bombardment mass spectrometry, *Biochem. J.,* 250, 253, 1988.

54. **Rose, K., Simona, M. G., Offord, R. E., Prior, C. P., Otto, B., and Thatcher, D. R.,** A new mass spectrometric C-terminal sequencing technique finds a similarity between γ-interferon and α2-interferon and identifies a proteolytically clipped γ-interferon that retains full antiviral activity, *Biochem. J.,* 215, 273, 1983.

55. **Brown, S. M. and Busch, K. L.,** A fluoromethylpyridinium salt reagent for functional-group-specific derivatization reactions in sample modifications for fast atom bombardment mass spectrometry, submitted for publication.

56. **Buko, A. M., Philips, L. R., and Fraser, B. A.,** Peptide studies using a fast atom bombardment high field mass spectrometer and data system. II. Characteristics of positive ionization spectra of peptides, m/z 858 to m/z 5729, *Biomed. Mass Spectrom.,* 10, 408, 1983.

57. **Morris, H. R., Panico, M., Etienne, T., Tippins, J., Girgis, S. I., and MacIntyre, I.,** Isolation and characterization of human calcitonin gene-related peptide, *Nature,* 308, 746, 1984.

58. **Larsen, B. S., Yergey, J. A., and Cotter, R. J.,** Evaluation of fast atom bombardment mass spectrometry for assessing the oxidation states of disulfide-containing peptides, *Biomed. Mass Spectrom.,* 12, 586, 1985.

59. **Sun, Y. and Smith, D. L.,** Identification of disulfide-containing peptides by performic acid oxidation and mass spectrometry, *Anal. Biochem.,* 172, 130, 1988.

60. **Yazdanparast, Y., Andrews, P. C., Smith, D. L., and Dixon, J. E.,** Assignment of disulfide bonds in proteins by fast atom bombardment mass spectrometry, *J. Biol. Chem.,* 262, 2507, 1987.

61. **Bean, M. F., Carr, S. A., Escher, E., and Moore, M.,** Solving difficult peptide structural problems by tandem mass spectrometry, paper WPA30 presented at the 37th Annual Conference on Mass Spectrometry and Allied Topics, Miami Beach, May 1989.

62. **Musselman, B. D. and Watson, J. T.,** Observation of solvent effects on abundance of polyhydrogen adducts (M+nH)$^+$ in fast atom bombardment mass spectrometry, *Biomed. Environ. Mass Spectrom.,* 14, 247, 1987.

63. **Barinaga, C. J., Edmonds, C. G., Udseth, H. R., and Smith, R. D.,** Sequence determination of multiply-charged peptide molecular ions by electrospray-ionization tandem mass spectrometry, *Rapid Commun. Mass Spectrom.,* 3, 160, 1989.

64. **Salephour, M., Perera, I., Kjellberg, J., Hedin, A., Islamian, M. A., Hakansson, P., and Sundqvist, B. U. R.,** Laser-induced desorption of proteins, *Rapid Commun. Mass Spectrom.,* 3, 259, 1989.

65. **Dell, A. and Morris, H. R.,** Fast atom bombardment — high field magnet mass spectrometry of 6000 dalton polypeptides, *Biochem. Biophys. Res. Commun.,* 106, 1456, 1982.

66. **Morris, H. R., Panico, M., Etienne, T., Tippins, J., Girgis, S. I., and MacIntyre, I.,** Isolation and characterization of human calcitonin gene-related peptide, *Nature,* 308, 746, 1984.

67. **De Angelis, F., Botta, M., and Nicoletti, R.,** Peptide sequencing by fast atom bombardment mass spectrometry: acid hydrolysis or tandem mass spectrometry?, *Biochem J. Lett.,* 245, 623, 1987.

68. **Roepstorff, P. and Fohlman, J.,** Proposal for a common nomenclature for sequence ions in mass spectra of peptides, *Biomed. Mass Spectrom.,* 11, 601, 1984.

69. **Stults, J. T. and Watson, J. T.,** Identification of a new type of fragment ion in the collisional activation spectra of peptides allows leucine/isoleucine differentiation, *Biomed. Environ. Mass Spectrom.,* 14, 583, 1987.

70. **Matsuo, T., Matsuda, H., and Katakuse, I.,** Computer program PAAS for the estimation of possible amino acid sequence of peptides, *Biomed. Mass Spectrom.,* 8, 137, 1981.

71. **Matsuo, T., Sakurai, H., Matsuda, H., Wollnik, H., and Katakuse, I.,** Improved PAAS. A computer program to determine possible amino acid sequences of peptides, *Biomed. Mass Spectrom.,* 10, 57, 1983.

72. **Sakurai, H., Matsuo, T., Matsuda, H., and Katakuse, I.,** PAAS 3: a computer program to determine probable sequence of peptides from mass spectrometric data, *Biomed. Mass Spectrom.,* 11, 396, 1984.

73. **Moseley, M. A., Deterding, L. J., Tomer, K. B., and Jorgensen, J. W.,** Capillary-zone electrophoresis/ fast atom bombardment mass spectrometry: design of an on-line coaxial continuous flow interface, *Rapid Commun. Mass Spectrom.,* 3, 87, 1989.

74. **Fenselau, C. and Cotter, R. J.,** Chemical aspects of fast atom bombardment, *Chem. Rev.,* 87, 501, 1987.

75. **Dell, A., Thomas-Oates, J. E., Rogers, M. E., and Tiller, P. R.,** Novel fast atom bombardment mass spectrometric procedures for glycoprotein analysis, *Biochimie,* 70, 1435, 1988.

Chapter 11

ON-LINE METHODS FOR PEPTIDE ANALYSIS BY CONTINUOUS-FLOW FABMS

William E. Seifert, Jr., William T. Moore, and Richard M. Caprioli

TABLE OF CONTENTS

I.   FABMS and Peptide Analysis 202

II.   Continuous-Flow FABMS 203
    A.   Introduction 203
    B.   Operation and Performance of Continuous-Flow FAB 203
        1.   Stability of CF-FAB Operation 203
        2.   Memory Effects in CF-FABMS 205
    C.   Performance Advantages of CF-FABMS in Peptide Analysis 205
    D.   Disadvantages of CF-FAB 207
    E.   Modes of CF-FAB Operation 207

III. Applications of CF-FABMS to Protein and Peptide Analyses 209
    A.   Protein Structure from On-Line Analysis of Enzymic Hydrolysates 209
    B.   LC/CF-FABMS 210
    C.   CZE/CF-FABMS 213

IV. Conclusion 217

References 219

I. FABMS AND PEPTIDE ANALYSIS

Since it was first introduced in 1981,[1] fast atom bombardment mass spectrometry (FABMS) has provided the analytical biochemist with a valuable tool for the amino acid sequence analysis of proteins and peptides. This "soft" ionization technique has allowed the direct mass spectrometric analysis of many biologically important molecules, including peptides and small proteins, without the necessity of chemical derivatization to improve volatility and stability of the molecule of interest. The use of FABMS has occurred not only in laboratories that specialize in instrumental analysis, but also in those using primarily liquid-based reaction chemistries such as enzymatic and Edman degradation techniques to sequence proteins and determine sites of modification. The success of FABMS and other desorption ionization techniques in the structural analysis of peptides has resulted in mass spectrometry being recognized as a vital component of an integrated approach toward the solution of structural problems in the biological sciences.

In the typical FABMS analysis, the peptide sample is first dissolved in the sample matrix, usually glycerol, thioglycerol, or some other suitable viscous compound. Several microliters of the sample matrix mixture are placed on a sample probe tip, and the sample is introduced into the ionization chamber of the mass spectrometer by means of a direct insertion probe. A beam of high-energy atoms (usually 6 to 8 kV Xe atoms) in the mass spectrometer source causes the molecules present on the surface of the sample-matrix droplet to be sputtered from the surface. Ions formed during this process are subsequently mass-analyzed.

Although this method of sample introduction is simple and easy to use, it has several shortcomings when applied to different analytical problems. First, the matrix in which the sample is dissolved is usually 80 to 95% glycerol (or another suitable liquid organic compound or viscous solution) in order for the droplet to survive introduction into the vacuum system. This matrix precludes the direct analysis of reactions that occur in primarily aqueous environments, such as enzymatic hydrolyses. In addition, the high concentration of organic components in the matrix causes a high background level, or chemical noise, to be produced on bombardment by the atom beam. The radiation damage to the organic matrix results in the presence of low-intensity peaks at every mass, which limits the sensitivity of the technique. Cluster ions formed from the organic matrix components often dominate the mass spectrum and can obscure the presence of analytes having the same nominal mass.[2] Cluster ions formed between matrix components and sample molecules decrease sensitivity by a dilution effect. Second, quantitation of components is difficult due to variations in ion intensities from sample to sample unless appropriate internal standards, known not to interfere with the analysis, are present. Third, standard FABMS is not easily applied to analyzing dynamic processes, where the concentration of reactants and products are changing rapidly. Following the course of reactions becomes a matter of taking as many isolated samples as possible with time and analyzing them as rapidly as they can be prepared.

FAB ionization is a surface phenomenon, and therefore the intensity of sample ions observed in the FAB mass spectrum is a function of the concentration of the molecules on the surface of the sample matrix droplet. A number of investigators[3-5] have shown that molecular ion signals from some peptides are not observed in the FABMS analysis, especially when present in mixtures with other peptides. This suppression effect has been studied in some detail by Naylor et al.,[5] and the hydrophilicity/hydrophobicity index of the peptide appears to play an important role in determining the concentration of the peptide at the surface of the droplet. Peptides that are hydrophilic in nature do not tend to migrate to the surface of the sample matrix droplet, especially in the presence of a more hydrophobic peptide, and therefore the $(M+H)^+$ ion of the hydrophilic peptide will be less intense than expected, and in some cases not observed at all. Thus, one cannot rely on the FAB mass spectrum of a peptide mixture to reflect the actual concentrations of peptides present in the solution.

II. CONTINUOUS-FLOW FABMS

A. INTRODUCTION

Initial efforts at the continuous introduction of liquid samples into a FAB source involved the use of a moving belt interface.[6,7] Fractions of a high-performance liquid chromatography (HPLC) eluent were deposited onto the belt as it was continuously fed into the source of the mass spectrometer where FAB occurred. Caprioli et al.[8] described a continuous-flow sample introduction probe that can operate at flow rates of 5 to 10 µl/min and achieves stable performance with glycerol concentrations in the carrier/mobile phase in the range 3 to 50%. The carrier solution is deposited directly onto the surface of a FAB probe tip via a fused-silica capillary tube that emerges through an orifice in the center of the tip. Fused-silica tubing is necessary to isolate the probe from the high voltage present at the probe tip in a magnetic sector mass spectrometer. Samples can be injected into the flow of carrier solution, included in the carrier solution itself, or the continuous-flow probe can be used as a liquid chromatographic interface to the mass spectrometer. Takeuchi and co-workers[9,10] coupled micro-HPLC and FABMS through an interface comprised of fused-silica capillary tubing and a porous metal frit, which served to disperse the mobile phase, concentrate the analytes and glycerol, and serve as the site of FAB ionization. The mobile phase, which consisted of water/acetonitrile or water/methanol solutions containing 10% glycerol, were introduced into the mass spectrometer ion source at flow rates of 0.5 to 1.0 µl/min.

B. OPERATION AND PERFORMANCE OF CONTINUOUS-FLOW FAB

The basic principle of successful operation of the continuous-flow FAB (CF-FAB) probe is the delivery of a primarily aqueous carrier solution to the target of the probe and the subsequent removal of the liquid at a similar rate to provide a thin fluid surface.[11] Instability of the ion beam generated in the bombardment process occurs if too much fluid is allowed to accumulate on the probe tip. Excessive accumulation results in surges in source pressure due to disruptive boiling of the liquid droplet, broad chromatographic peaks, and decreased sensitivity. Figure 1 shows a schematic diagram of a commercial CF-FAB probe. The removal of liquid from the surface of the probe tip is accomplished in this design by two methods: evaporation by gentle heating and removal by capillary action to a filter pad in contact with the probe tip. The standard pumping capacities of most commercial mass spectrometers can handle the evaporation of about 10 µl/min of water, acetonitrile or methanol, but by using filter pads to absorb the liquid from the probe tip, flow rates of up to 15 to 20 µl/min have been maintained. Excellent CF-FAB performance is achieved using this design with flow rates of 5 to 10 µl/min over 6 to 8 h of operation without removing the probe from the mass spectrometer.

1. Stability of CF-FAB Operation

A number of criteria is used to assess the stable operation of a CF-FAB probe. Caprioli[11] has used the following in the utilization of a CF-FAB probe on a Finnigan MAT-90 mass spectrometer:

1. A variation of less than ±10% when monitoring a background ion such as m/z 185 for glycerol or m/z 42 for acetonitrile
2. The lack of significant ripple in the peak tops of monitored background ions
3. Achieving a stable ionization gauge reading (2×10^{-4} torr)
4. Reproducible (±10%) peak areas for a standard compound (e.g., the peptide substance P)

Seifert et al.[12] used visual observation of a CF-FAB probe tip through the glass top of a Finnigan TSQ-70 triple-stage quadrupole mass spectrometer to adjust continuous-flow parameters such as flow rate and probe temperature to obtain optimal results. Although the CF-FAB inlet used

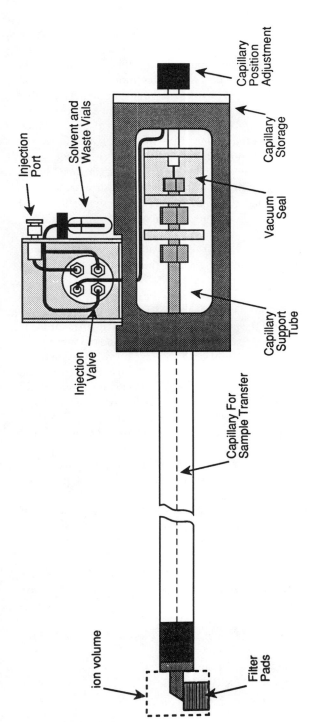

FIGURE 1. Schematic diagram of a commercial CF-FAB probe designed for a magnetic sector mass spectrometer. (Courtesy of Finnigan MAT Corporation.)

in these studies was of a different design (the fused-silica introduction tube is orthogonal to the probe tip), stable operation was achieved only when the probe tip was covered with just a thin film of carrier solution. Performance rapidly deteriorated when carrier solution was allowed to build up on the probe tip to form a droplet.

2. Memory Effects in CF-FABMS

Whenever samples are introduced into a mass spectrometer in a continuous fashion, there is always the possibility of carryover from one sample to the next. Reducing these memory effects in CF-FAB appears to be directly associated with attaining a stable mode of operation.[8,12] Conditions that cause either the buildup of carrier solution (too fast of a flow rate or decreased evaporation) or the accumulation of glycerol (restricted flow-rate or too high of a probe temperature) on the probe tip result in very broad peaks from, for example, a flow-injection analysis or the contamination of subsequent mass spectra with previous samples. Using proper operating conditions, the memory effects can be reduced to less than 0.1%.[11]

C. PERFORMANCE ADVANTAGES OF CF-FABMS IN PEPTIDE ANALYSIS

Reducing the amount of viscous organic matrix from 80 to 90% as normally used in standard FAB analyses to the 5 to 10% used in CF-FAB yields significant performance advantages. The first immediate effect is the decrease in the intensity of the chemical noise, the peak-at-every-mass background previously described. Structural features in the mass spectrum, which were previously obscured by this chemical noise, become discernable in the CF-FABMS spectrum.[13]

The decrease in nonspecific background ions also leads to an increase in sensitivity and the limit of detection of many peptides analyzed by CF-FAB. Caprioli and Fan[13] investigated the effect on sensitivity using several peptides in the 1000 to 2000 Da molecular weight range. Using standard FAB, a 5-pmol sample of the peptide substance P (mol wt = 1347 Da) gave a signal-to-background chemical noise ratio (S/N) of 3:1 when the quasi-molecular ion at m/z 1348 was monitored. The same amount of peptide gave a S/N ratio of about 100:1 by CF-FAB. A sample containing 100 fmol (136 pg) of substance P was indistinguishable from the chemical noise, whereas the same sample analyzed by CF-FAB yielded a spectrum in which the S/N ratio for the $(M+H)^+$ ion was measured to be approximately 4:1. The gain in sensitivity was estimated to be about 150-fold at this level. Further experiments with substance P and gramicidin S indicated that, in addition to the increase in sensitivity due to a decrease in the background noise, ion yields were also improved by approximately a factor of five due to the increase in the water content in the carrier solution.

Another advantage of CF-FAB is the observation of a decrease in the ion suppression effect.[14,15] As noted previously, standard FAB analyses of peptide mixtures often show the effects of ion suppression. In practice, this factor has meant that, in the FAB analysis of peptide mixtures such as those obtained by enzymic digestion of proteins, one can usually expect to observe only 50 to 80% of the total peptides in the mixture. Using CF-FAB to analyze peptide mixtures minimizes these suppression effects and enables the analyst to observe most, if not all, of the peptides in the mixture. An example of this phenomenon is shown in Figure 2 for a peptide present in the tryptic digest of 2 nmol of glucagon.[14] In the standard FAB analysis of the tryptic digest of glucagon, the peptide corresponding to amino acids 19 to 29, a relatively hydrophobic peptide having an $(M+H)^+$ at m/z 1352, is preferentially observed over that corresponding to residues 1 to 12 $(M+H)^+$ at m/z 1357), which is more hydrophilic.[3,5] This phenomenon is shown in Figure 2A, where the intensity for m/z 1357 is barely discernable above the background, whereas the peak at m/z 1352 is an intense signal. When the same amount of digest mixture is analyzed by CF-FAB, intense $(M+H)^+$ signals are observed for both peptides, as shown in Figure 2B.

Similar results were obtained when a mixture of seven synthetic, hydrophilic peptides were analyzed by both standard FAB and CF-FAB.[15] Each of the peptides in the mixture had the same two N-terminal and C–terminal amino acids. The hydrophilic nature of the peptides was varied

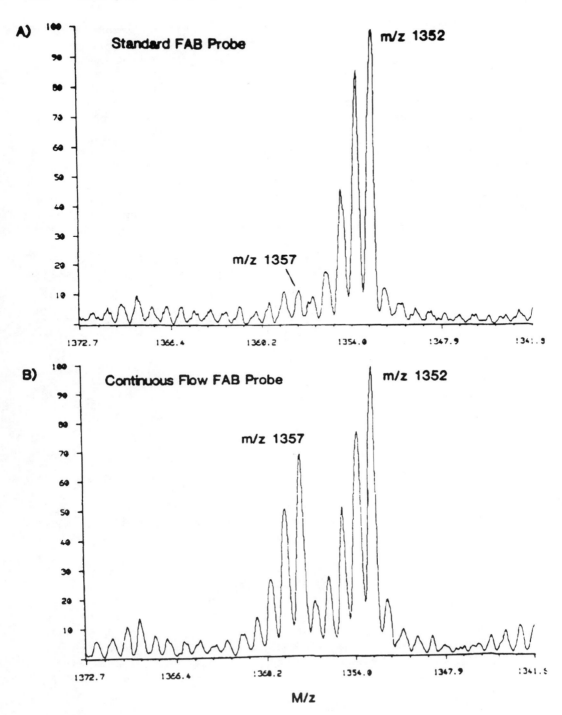

FIGURE 2.   Partial FAB mass spectra of the tryptic hydrolysate of glucagon over the quasi-molecular ion region for the fragments containing residues 19 to 29 (m/z 1352) and 1 to 12 (m/z 1357). (A) Standard FAB analysis; (B) continuous-flow FAB analysis. (From Caprioli, R. M., Moore, W. T., and Fan, T., *Rapid Commun. Mass Spectrom.*, 1, 15, 1987. With permission.)

by varying the middle three residues. Figure 3 compares the spectra obtained from the standard FAB and CF-FAB analyses of a mixture at pH 1 containing 300 pmol of each peptide. Significant differences in the intensities of the $(M+H)^+$ ions were observed between the two methods, with considerable suppression of the ion currents of peptides I, III, IV, and V occurring in standard FAB. Peptides that were suppressed the most in standard FAB had charges of +2 or +3 in the solution at pH 1, whereas the peptides with a +1 charge showed no suppression compared to CF-FAB. The extent to which ion suppression is observed in FAB may therefore, at least in the case of small hydrophilic peptides, depend significantly on the total number of positive charges of the peptide.

The decrease in the ion suppression observed in CF-FAB is believed to be due to two factors.[14] First, CF-FAB by its very nature is a dynamic process and the sample surface is constantly changing. The surface layers on the tip of the CF-FAB probe are constantly being refreshed with the flow of sample and carrier solution. The effective surface area is, therefore, greatly increased over that of the static FAB probe sample surface, and the concentration of analytes on the surface comes closer to reflecting the concentration in the solution. Second, because the carrier solution in CF-FAB is primarily aqueous in composition, a significant increase in sensitivity is observed over standard FAB, in which the matrix is primarily nonaqueous.

D. DISADVANTAGES OF CF-FAB

Although CF-FAB yields some important performance advantages, the dynamic nature of the technique presents some difficulties as well. One of these difficulties is the initial time required to attain a stable surface in the mass spectrometer source. This time is typically about 15 min, including the time to set up the pump and solvent flow. Because the sample surface undergoes constant fluctuation with the carrier flow, the FAB process gives rise to signals that have more variation than is normally observed with standard FAB. Averaging spectra over an injection peak has been found to be a valuable method for obtaining reproducible spectra from CF-FAB.[16]

Another disadvantage of CF-FAB is the higher source pressure produced in the mass spectrometer due to the introduction of the primarily aqueous carrier solution. Flow rates of 5 to 10 µl/min will produce a pressure in the source of approximately 2×10^{-4} torr. If a magnetic sector mass spectrometer is used, care must be taken to insure that proper electrical grounding is in effect so that high voltage arcs, which can occur at elevated source pressures, do not damage the electronics of the mass spectrometer. Also, because fused-silica capillary tubing is being used, solutions should be carefully filtered to remove any particulate matter, which could block the small orifices of the tubing or injection valves. For optimal results, care must also be taken to eliminate any dead volumes in connections.

E. MODES OF CF-FAB OPERATION

CF-FAB can be applied in a number of different ways to the analysis of peptides and peptide mixtures. Three basic modes of operation can be employed depending on the application and the specific information required: flow-injection analysis, constant-flow analysis, and an outgrowth of constant-flow analysis, chromatographic analysis. These modes of operation can be utilized in the batch sample processing of synthetic peptides, in the on-line monitoring of reactions to determine the change in concentration of reactants and/or products, and in the HPLC/CF-FAB or capillary zone electrophoresis (CZE)/CF-FAB analysis of peptide mixtures such as the enzymic hydrolysates of proteins.

Flow-injection analysis, in which the sample to be analyzed is injected into the carrier flow via a microinjection valve, provides a convenient and fast method for the repetitive injection of a sample every 2 to 3 min without cross contamination or for individual samples to take advantage of the increased sensitivity of CF-FAB. Reaction mixtures in which the concentrations of the reactants and products do not change rapidly may be analyzed by this method. An autosampling device can be added to the CF-FAB probe[17] to facilitate the measurement of kinetic parameters for the enzymic hydrolysis of large natural substrates. Samples are then

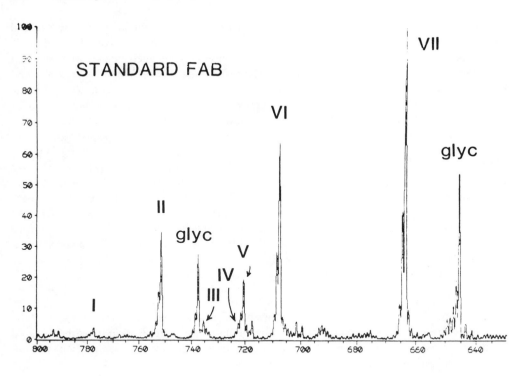

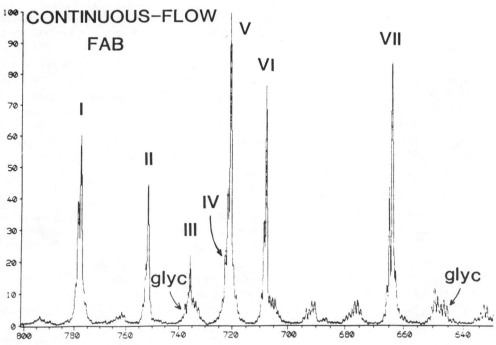

FIGURE 3. FAB mass spectra of a mixture of seven peptides. (Top) Standard FAB analysis. (Bottom) Continuous-flow FAB analysis. Peaks designated by "glyc" are ions from the glycerol matrix. (From Caprioli, R. M., Moore, W. T., Petrie, G., and Wilson, K., *Int. J. Mass Spectrom. Ion Proc.*, 86, 187, 1988. With permission from Elsevier Science Publishers, Physical Sciences and Engineering Division.)

injected every 2 min (or as rapidly as every 12 s) to monitor the increase in products or the decrease in reactants.

In cases where it is advantageous to constantly monitor the progress of a reaction, the constant-flow mode of operation can be used. In this mode, the reaction solution is continuously pumped into the mass spectrometer either by using atmospheric pressure or a syringe pump. A small concentration of glycerol (5%) is included in the reaction solution or is added via a make-up tee just prior to entering the mass spectrometer.[18]

CF-FAB can also be used as an interface between liquid chromatographic techniques and mass spectrometry.[19,20] If chromatographic separations are performed using capillary-bore HPLC or CZE, the entire eluent can be introduced into the ion source without splitting. Depending on the chromatographic requirements, glycerol can be included in the mobile phase or introduced after the chromatographic separation by a make-up tee.[18,21]

III. APPLICATIONS OF CF-FABMS TO PROTEIN AND PEPTIDE ANALYSES

A. PROTEIN STRUCTURE FROM ON-LINE ANALYSIS OF ENZYMIC HYDROLYSATES

CF-FABMS has been used on-line in a constant-flow mode to continuously monitor the enzymic degradation of peptides under reaction conditions compatible with the FAB process.[17] The progress of the reaction is directly monitored by flowing the reaction solution into the source, the flow being created by the pressure differential between the atmosphere and the source vacuum. In an experiment examining the degradation of α-MSH by pepsin, reaction progress profiles for substrate and products were generated by plotting selected ion chromatograms for the (M+H)$^+$ ions of the intact peptide and pepsin-generated fragments. From the decrease in the intensity of the quasi-molecular ion of the substrate and concomitant increase in those of the products, initial rate data were obtained.

It was also found that on-line continuous monitoring of an exopeptidase-catalyzed degradation would yield C-terminal sequencing information.[22] Because the buffer for this enzyme reaction was not directly compatible with the FAB process, FAB matrix was introduced with a make-up tee. By following the action of a mixture of carboxypeptidase Y and P on ribonuclease S peptide, the sequence of the last thirteen residues of the peptide was determined. Because these enzymes hydrolyze C-terminal residues from the peptide, new molecular species are produced for each hydrolysis step. An unambiguous deduction of the C-terminal sequence was obtained by observing the series of truncated molecular species generated that differ in mass by that of a single amino acid residue. These data are further corroborated by following the rise and fall of these species via their respective selected ion chromatograms with respect to time. Similarly, Ashcroft et al.[23] have demonstrated the utility of the technique to monitor the enzymic degradation of the peptide substance P by carboxypeptidase Y as a way to obtain C-terminal primary structure information. In this experiment, enzyme reaction solution and matrix solution were introduced into the source by atmospheric pressure, each flowing in through equal lengths of fused-silica capillary meeting in a tee. By selecting the appropriate i.d. and length of the capillaries, a 5 µl/min flow of the mixture could be obtained. By monitoring the fall of substrate and rise of product specific ion chromatograms, the authors could derive the sequence of the four C-terminal residues.

Gavard et al.[24] have recently described an on-line CF-FAB approach to obtaining C-terminal sequence information on large proteins that exceed the mass range of mass spectrometric detection for a particular instrument. A large protein is first treated by passing a solution of the protein through an immobilized endoprotease column. The resultant fragments leaving the column that are in the mass range of the instrument are identified. The exopeptidase carboxypeptidase is added to the solution of protein and incubated for 1 h. This solution is monitored after passage through the immobilized endoprotease column. Any disappearing or newly appearing fragments are deduced to be solely associated with the exoprotease-processing of the C-terminal fragment. In this way, the mass of the C-terminal fragment is identified. Monitoring the exoprotease digest with time can permit deduction of the sequence of the C-terminal fragment. The validity of this approach was demonstrated by a C-terminal sequence determination on a synthetic version of the first 29 residues of human growth hormone releasing factor (GRF), in which the sequence of the last four C-terminal residues was correctly deduced. The immobilized trypsin column alone generated a nine residue C-terminal fragment. After carboxypeptidases were added, this particular fragment disappeared and four new signals sequentially appeared allowing deduction of the sequence for the last four C-terminal residues. It should be noted that the method depends on the release of the C-terminal fragment having an $(M+H)^+$ within the mass range of the spectrometer.

B. LC/CF-FABMS

HPLC of peptides and proteins on the conventional[25] or the microbore scale[26] has become an indispensable tool for the protein chemist. The coupling of this technique to mass spectrometry, providing mass-specific detection, is currently of great interest for the analysis of enzymic digests of proteins, particularly those produced by recombinant DNA methods, and for the characterization of products from automated synthesis.[27,28] A major advantage of LC/MS is that chromatographically unresolved components can be effectively resolved by the mass selectivity of the MS detector.

One approach to coupling LC to FABMS is to use microbore (1 mm i.d.) or capillary (0.3 mm i.d.) HPLC columns and the CF-FAB probe as the MS interface.[19,29–31] Using microbore or capillary HPLC columns minimizes splitting ratios or eliminates them altogether, insuring high sensitivity through maximal transfer of analyte to the mass spectrometer. Microbore HPLC is performed using a system that has syringe pumps capable of accurately delivering low flow rates without pulsing, an important requirement for CF-FAB when used with capillary columns. A flow rate of 25 μl/min provides good chromatographic performance when a microbore reversed-phase (RP) column is used for peptide or protein separation. The effluent from the HPLC column developed at this rate is split 1:4 so that a flow of 5 μl/min is delivered to the CF-FAB probe. When acetonitrile gradient elution is used, a 9:1 precolumn splitter is installed before the sample injector in order to provide rapid gradient formation and still maintain a 25 μl/min column flow. The addition of glycerol into the chromatography solvents at a concentration of 5% does not significantly alter the chromatographic results for peptide separations. For example, the results obtained in Figure 4 show the LC/CF-FABMS analysis of 200 pmol of a tryptic digest of β-lactoglobulin A. Chromatographic separation was effected using a microbore RP column (C_8, 1 mm i.d. × 50 mm) and directing 20% of the eluate (representing 40 pmol of the digest) into the mass spectrometer. Figure 4A shows the total ion chromatogram of the fractionation of the tryptic peptides of β-lactoglobulin A. Figure 4B displays a "peptide chromatogram" created by overlaying individually normalized selected ion chromatograms of the peptides. In all, 19 peptides were identified that fell within the mass range scanned.

Using this approach, Caprioli and co-workers have shown that the CF-FABMS interface permits efficient chromatographic separation (using both microbore and capillary RP columns) and on-line mass analysis of mixtures of small molecular weight peptides, mixtures of closely related high molecular weight peptide hormones (insulins),[19] and complex mixtures of peptides found in the tryptic digests of proteins such as myoglobin,[19] β-lactoglobulin A,[28] cytochrome C,[26] apolipoprotein A,[27,28] and recombinant human growth hormone (rhGH).[28] The method has also been demonstrated to be remarkably sensitive in that a minor modification of a protein present at a 1% level can be detected and quantitatively estimated.[28]

Other workers have also reported applications involving the coupling of liquid chromatography and mass spectrometry with a CF-FAB interface. Ashcroft[32] has reported the advantage of using microbore RP-HPLC with CF-FABMS in the analysis of a mixture containing

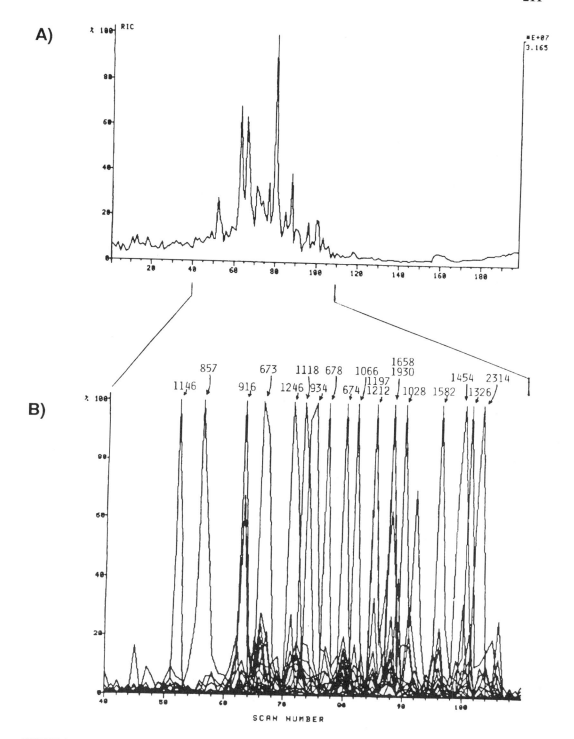

FIGURE 4. LC/CF-FABMS analysis of a tryptic digest of β-lactoglobulin A using a microbore C_8 column. (A) Total ion chromatogram; (B) "peptide chromatogram" generated by overlaying the selected ion chromatograms for the individual peptides identified.

picomoles of the peptides angiotensin II, angiotensin III, and met- and leu-enkephalin. Chromatographic separation was achieved on a 15-cm microbore ODS C_{18} column using a 30 min gradient of methanol from 0 to 60% keeping the concentrations of the mobile phase components trifluoroacetic acid (TFA) and glycerol constant at 0.1 and 5%, respectively. Because the RP-HPLC column was developed at 60 μl/min, a 1:11 post-column splitter was installed to reduce flow to 5 μl/min for CF-FABMS analysis. Generation of the specific ion chromatograms for each species indicated baseline separation of all four species, the order of elution being leu-enkephalin, met-enkephalin, angiotensin III, and angiotensin II.

Mock et al.[33] fractionated and FAB-mapped the tryptic peptides of pyridylethylated ribonuclease B to demonstrate the use of a continuous-flow probe as an interface to couple capillary RP-HPLC with FABMS. Chromatography was performed using a slurry-packed RP capillary column (0.31 mm i.d. × 250 mm, Rosil C_{18}) and developing the column at 10 μl/min flow-rate. A FAB matrix-containing acetonitrile gradient was formed at a 150 μl/min flow rate. Before introduction to the column, the flow was reduced 1:14 by a splitting tee. The resulting 10 μl/min flow of the column eluate was directed into the source of the mass spectrometer, which was equipped with a liquid nitrogen cold finger. With the source heater set to 65°C, the typical operating pressure was 1.5×10^{-5} torr. The authors were able to chromatographically resolve and mass-analyze 13 fragments derived from 100 pmol of the digest covering 70% of the primary structure.

Bourell et al.[34] used the FRIT-FAB interface to successfully FAB-map the tryptic peptides of recombinant growth hormone (rGH). The peptides were chromatographically separated by capillary RP-HPLC (0.32 × 150 mm, C_{18}) using a 65 min linear acetonitrile gradient (0 to 60%). The mobile phase contained 1% glycerol as a FAB matrix component. The flow rate was adjusted to approximately 5 μl/min by using a precolumn and injector splitter. In the analysis of 25 pmol of rGH digest, the authors found all the tryptic peptides expected in the mass range scanned (600 to 4000). There was excellent agreement between an on-line ultraviolet (UV)-monitored profile and the total ion chromatogram trace.

Cappiello et al.[35] used a FRIT-FAB continuous flow interface to couple capillary RP-HPLC with a four-sector tandem MS. In two-sector operation, effluent from the capillary column is directly introduced into the source for on-line FAB-mapping of a protein digest. On-line four-sector operation (LC/CF-FABMS/MS) is performed when chromatographic resolution is high and there is enough time between peaks to optimize MS-1 and MS-2 parameters for analysis of each peptide. For four-sector operation when chromatographic peaks overlap, the authors incorporated a "loop trapping" multiloop injection valve system that acts both as a fraction collector and an eventual sample dispenser. To demonstrate the features of the system, data were shown for the deduction of sequence information for the tryptic peptides of human hemoglobin. Product ion collisionally activated decomposition (CAD) spectra were obtained that permitted deduction of a peptide sequence from 50 pmol of myoglobin digest. The digest peptides were resolved into at least 11 separate peaks by a linear acetonitrile gradient. Fractions were collected with the multiloop system prior to its use as a sample dispenser for tandem MS analysis.

Deterding et al.[36] used a coaxial CF-FAB interface to mass analyze synthetic peptide mixtures and tryptic digests of bovine growth factor 1 to 29 and glucagon on a 4-sector tandem mass spectrometer. The peptides were fractionated by capillary RP-HPLC (0.32 mm i.d., C_{18}) employing acetonitrile gradient elution. The coaxial flow interface permits the introduction of matrix to the analyte at the FAB target area as it emerges from the capillary column exit, thus minimizing peak broadening and effects of glycerol on the chromatographic separation. The coaxial flow interface consists of a length of fused-silica capillary surrounded by a wider bore capillary that is the conduit for matrix delivery. Using the coaxial interface resulted in sharper peak definition and shorter elution times, as was demonstrated by comparing selected ion chromatograms for separations with and without glycerol in the mobile phase. These authors also demonstrated that they could obtain CAD spectra of FAB-desorbed ions in a quadrupole instrument. The amino acid sequence of a nonpeptide fragment derived from the tryptic digest of 41 pmol of bovine growth factor 1 to 29 was deduced in this way.

Other on-line LC/CF-FABMS studies were reported at the 11th International Mass Spectrometry Conference held in Bordeaux in late summer of 1988. McDowall et al.[37] used a CF-FAB probe interface to compare the coupling of on-line conventional scale HPLC and capillary dimension RP HPLC to FABMS for the analysis of small peptides. In order to accommodate the flow rates (on the order of 5 to 10 µl/min) typical of CF-FABMS systems, the conventional bore (4.6 mm) HPLC eluate was split. The same linear velocity, however, could be maintained with a smaller diameter (0.46 mm) RP capillary column, thus making splitting unnecessary. The capillary dimension RP HPLC was found to be the preferable approach for separating biomedical samples in which quantity was limited. Performing the separation in the capillary dimension allowed analysis of the maximal amount of sample, maximized the rate of delivery of the sample to the flow probe tip, and minimized the elution volume for the sample, thus maintaining the concentration of the analyte of interest high at the target area of the flow probe tip. From a comparative examination of a mixture of small peptides (M-R-F-A, bradykinin, leu-enkephalin, and gramicidin S), the authors found that the detection limits were approximately 100-fold lower for capillary dimension RP HPLC than those obtained with conventional-bore RP HPLC coupling using post-column splitting of the eluate.

Bateman and Jones[38] performed LC/CF-FABMS using a homemade capillary RP HPLC column (C_{18}, 0.32 mm × 10 cm) to resolve and mass-analyze a mixture of 100 pmol of closely related enkephalin pentapeptide analogs having $(M+H)^+$ ions ranging from m/z 554 to 573. A linear gradient of acetonitrile, from a concentration of 20 to 45% over a 15 min development time, permitted resolution of the four components within a total run time of 30 min. The mobile phase contained 5% glycerol and 0.1% TFA to insure stable ion production during the FAB process. The peak width at half height for the elution profile of leu-enkephalin was 12 s, demonstrating good chromatographic resolution for this system.

Lisek et al.[39] demonstrated the use of LC/CF-FABMS in a multidimensional quantitative assay for substance P, an 11-residue neuropeptide, in cerebrospinal fluid (CSF), a complex multi-component biological sample matrix. Quantitation was achieved via a stable isotope dilution assay using a deuterated analog of substance P as an internal standard. The CSF was acid extracted, and the substance P was further enriched by antisubstance P immunoaffinity chromatography and subsequent desalting of the immunoaffinity-column-eluted material over an RP C_4 disposable column. The desalted material was then analyzed by LC/CF-FABMS using a C_{18} RP fused-silica capillary column. The quasi-molecular ions for the endogenous substance P and the d_8-substance P internal standard were selectively monitored and compared to results obtained for a series of standards in order to quantitate the substance P in the CSF. The values obtained by this assay were comparable to those obtained by radioimmunoassay (RIA) and were free of the potential ambiguity that could be present in any RIA owing to antibody cross-reactivity.

C. CZE/CF-FABMS

CZE is a high-resolution microscale (attomole to picomole) separation system for charged molecules based on their differential rates of migration in an electric field, which is created by applying a high voltage (20 to 30kV) across an open tubular fused-silica capillary (10 to 100 µm i.d.) filled with a buffer solution.[40,41] Performing electrophoresis in a capillary column permits efficient cooling of the process so that Joule heating and associated convective disturbances are minimized, thereby reducing sample zone broadening and increasing separation efficiency. Another consequence of performing electrophoresis in silica capillaries is the creation of a phenomenon known as electro-osmotic flow,[42] which arises from the formation of a charged double layer at the inner surface of a silica capillary. The flow that is established under these circumstances is a smooth, unidirectional flow that has a nonparabolic or "plug flow" characteristic at a rate in the nl/min range. Electro-osmotic flow, therefore, significantly contributes to the high-resolution capability of CZE and can result in the number of theoretical plates in excess of 10^6. This affects eventual movement of all analytes toward the cathode under proper conditions whether they are cations, anions, or neutral molecules.

Since the publication of high-resolution CZE electropherograms of a tryptic digest of egg white lysozyme and a six component mixture of standard proteins in a paper by Jorgenson and Lukacs,[40] there has been a growing interest in exploiting capillary electrophoresis as an analytical tool in peptide and protein chemistry.[43-46] Presently, a high level of interest is evidenced by the number of presentations given at the First International Symposium on High Performance Capillary Electrophoresis held in Boston in April 1989. Approximately one half of the studies dealt with peptide and/or protein analysis. Several studies indicated the utility of CZE separations for the verification of structure and purity for rDNA protein products[47-55] and synthetic peptides derived by chemical methods.[56] Other presentations demonstrated the power of CZE for generating electropherogram "peptide maps" of enzymatically digested or chemically cleaved proteins.[57-61]

Coupling mass spectrometry to CZE for the study of protein structure adds an advantage unrivaled by that of other detection techniques such as UV monitoring. Characterizing a peptide by its molecular weight as well as its electrophoretic mobility can provide enough information for the rapid and unambiguous deduction of structure modification such as an amino acid deletion, substitution, or a post-translational modification.

CZE has been coupled to mass spectrometry using two different approaches. One involves the use of ionization techniques such as electrospray ionization (ES)[62-64] and atmospheric pressure ionization (API, also known as ionspray or pneumatically assisted electrospray).[65] The other approach of coupling CZE to mass spectrometry involves CF-FABMS.

The first report describing the coupling of CZE with CF-FABMS was by Minard et al.[66] An "open gap make-up" interface was used in an effort to preserve the separation efficiency of CZE, complete an electrophoretic circuit, and still permit the introduction of both FAB matrix solvent and CZE effluent flowing at disparate rates (5 μl/min vs. 100 nl/min, respectively). This interface consisted of one end of a CZE capillary (75 μm i.d. × 60 cm) and one end of a CF-FAB capillary (50 μm i.d. × 70 cm) epoxied in close proximity (a 20-μm gap) to a glass cover slip. This cover slip arrangement could then be immersed in a reservoir containing FAB matrix solution and the cathode. The other end of the CZE capillary was immersed in the anode compartment containing electrophoresis solution. The CF-FAB capillary was threaded through a commercial CF-FAB probe with a sample tip modified to enhance sensitivity and reduce peak tailing effects. The tip area was reduced approximately 80% from a 4-mm diameter to a diameter of 1.5 mm. The preliminary results obtained with this system encouragingly indicated that electrophoretic separation and mass analysis were achieved by CZE/CF-FABMS with a simple mixture of three synthetic peptides at low pmole concentrations. Specific ion chromatograms showed that substance P (mol wt 1347) was separated from the co-eluting mixture of met-enkephalin (mol wt 573) and leu-enkephalin (mol wt 555).

In a later experiment, Minard et al.[67] were able to effect the separation of met- and leu-enkephalin using an upgraded CZE/CF-FAB interface design and micellar electrokinetic capillary chromatography (MECC). This experiment is important in that it was the first indication that MECC could be coupled to mass spectrometry. MECC is an electrophoretic method that has been shown to be very effective for neutral or uncharged species. It involves the partitioning of these analytes between the electrophoretic solution and slowly moving micelles, which are usually formed of sodium dodecyl sulfate. Thus it was demonstrated that even "soap" solutions containing analyte after being passed through and diluted with a "make-up interface" can be introduced into the high vacuum FAB source with productive results.

Moseley and co-workers[68] have extended the coaxial "column within a column" capillary interface system that had been previously devised[18] to combine CZE and FABMS. The interface was designed to minimize the effect of the source vacuum on the flow of electrophoretic buffer in the CZE capillary and to minimize any disturbance of the electro-osmotic flow that would decrease the uniquely high-separation efficiency of CZE. In addition, it also allowed for electrophoretic transport to be preserved all the way to the FAB probe tip, thus eliminating the potential for any additional degradation of the separation efficiency due to pressure induced nonparabolic flow. The coaxial CF-FABMS interface consisted of an internal fused-silica CZE capillary (13 μm i.d.) and an outer surrounding capillary (150 μm i.d.). The FAB matrix (25% glycerol) is pumped into the outer sheath capillary through a 1/16 in. tee fitting mounted to the back end of the probe at a rate of 1 μl/min to maintain the 10^{-5} torr source operating pressure in the mass spectrometer. By pulling the internal CZE capillary slightly back (1 mm) from the end of the matrix sheath capillary, the vacuum-induced flow generated in the CZE capillary is minimized to an insignificant rate of 3 nl/min. The ground or cathodic end of the system is the probe tip or target, which floats at +8kV when inserted into the mass spectrometer source. A 30 kV setting from a high voltage power supply would result in a 22 kV voltage drop across the CZE capillary. Using the coaxial CZE/CF-FAB interface, the investigators were able to completely resolve mixtures of di-, tri-, and pentapeptides in amounts ranging from 30 to 600 fmol. Although there was a slight amount of peak trailing for the pentapeptide, very sharp peaks were obtained for the dipeptides demonstrating theoretical plate counts as high as 290,000. Recently, Moseley et al.[69] have reported the electrophoretic separation and mass-analysis of 20 fmol of the decapeptide angiotensin I (mol wt 1296) and 10 fmol of the *N*-acetylated derivative (mol wt 1338).

Caprioli and co-workers[55,70] designed a CZE/CF-FAB interface which is shown in Figure 5. This interface is multifunctional and consists of a combination of features that have been previously and separately employed by Minard et al.[20] and Moseley et al.[68,69] The design of the interface was influenced by practical considerations including analysis development and the types of samples to be analyzed. It includes a UV monitor, which aids in the off-line fine tuning of CZE conditions while simulating the vacuum of the mass spectrometer. The design was also influenced by the desire to scan over wide mass ranges during the analysis of complex mixtures of peptides such as those present in a protein digest. CZE was performed using an apparatus similar to that described by Jorgenson and Lukacs.[40] As shown in Figure 5, the effluent end of the CZE capillary (50 μm i.d. × 90 cm) and the inlet end of the CF-FAB capillary (75 μm i.d.) meet in a short segment of Teflon tubing (0.5 mm i.d.) placed in a plexiglass block consisting of two 90° intersecting passageways (1/16 in. i.d.). The other two connections to the block permit the introduction of a "flow-through" cathode and the CF-FAB matrix (5% glycerol, 3% acetonitrile). Enclosing the capillary juncture in a Teflon alignment-sleeve and the flow rate differential between that in the CF-FAB capillary (5 μl/min) and the CZE capillary (0.1 μl/min) permit efficient transfer of CZE eluate to the mass spectrometer, as the results in Figure 6 indicate. Figure 6A shows both the UV detector trace (upper) and the MS total ion chromatogram (lower) derived from a CZE/CF-FABMS analysis of a synthetic peptide preparation prior to purification of the desired product, a 15 residue peptide having the sequence MHRQETVDCLKKFNA(NH$_2$) (the major peak in the chromatograms). The UV cell was located near the cathode end of the capillary before the plexiglass block. Although direct comparison of the resolution recorded by the UV and MS detectors is difficult because of the differences in their physical characteristics and operating dynamics, if one regards the chromatograms as an indication of the "apparent" resolutions recorded by these detectors, then the profiles show that the apparent resolution is quite similar and that no major broadening has occurred as a consequence of transfer in the interface. A similar result is shown in Figure 6B for the CZE/CF-FABMS analysis of a tryptic digest of 40 pmol of rhGH. Again, the UV trace and the total ion chromatogram show very similar profiles. A further experiment to determine the sensitivity of the CZE/CF-FABMS system did, however, show evidence of zone broadening. Comparison of the UV traces and the reconstructed ion chromatograms for the analysis of 368

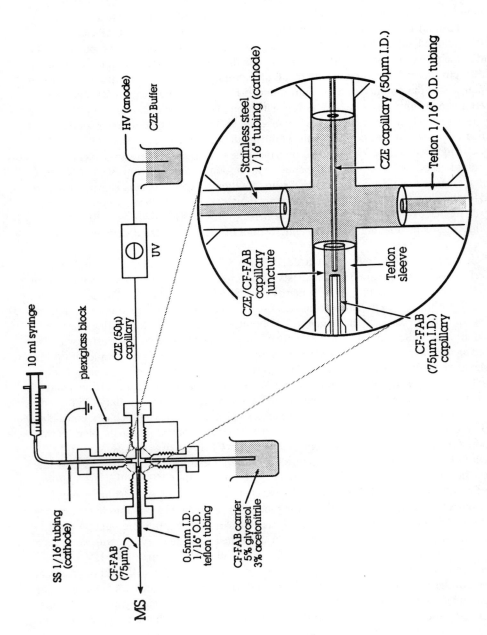

FIGURE 5. Schematic for coupling CZE with mass spectrometry via a CF-FAB interface. (From Caprioli, R. M., Moore, W. T., Martin, M., Dague, B. B., Wilson, K. B., and Moring, S., Coupling capillary zone electrophoresis and continuous-flow fast atom bombardment mass spectrometry for the analysis of peptide mixtures, *J. Chromatogr.*, 480, 247, 1989. With permission from Elsevier Science Publishers, Physical Sciences and Engineering Division.)

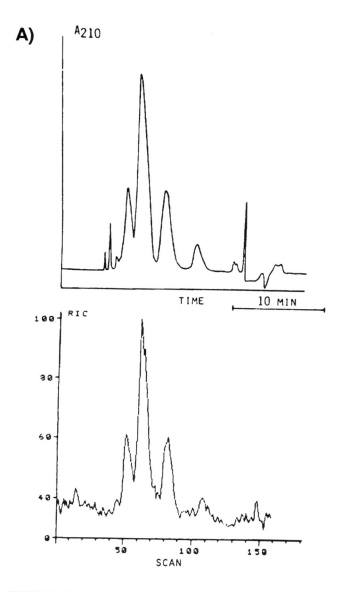

FIGURE 6. Comparison of the UV detector trace (upper) and the total ion chromatogram (lower) obtained from the CZE/CF-FABMS analysis of (A) a synthetic pentadecapeptide preparation and (B) the tryptic hydrolysis of 40 pmol of rhGH.

and 75 fmol of angiotensin II indicated a twofold zone broadening as a consequence of passage across the interface. However, even though there is some zone broadening, the results presented for the synthetic peptide crude mixture and the protein digest demonstrated that the degree is not severe enough to decrease the value of this approach as a viable alternative to HPLC/FABMS of synthetic peptide mixtures and enzymic digest of proteins.

IV. CONCLUSION

CF-FAB has become a valuable tool for the sequence determination of peptides and proteins, in addition to its uses in other areas of biochemistry and molecular biology. The method allows

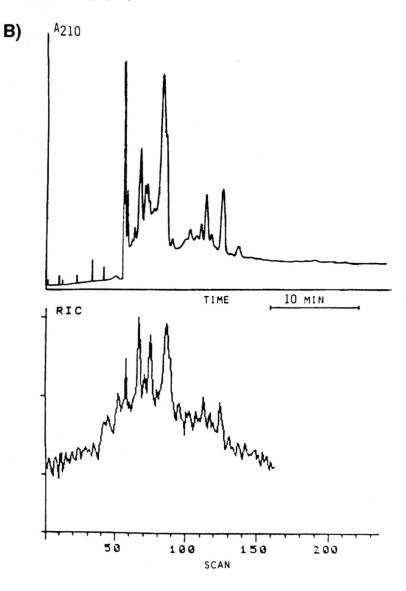

FIGURE 6 (continued).

for the direct and automated analysis of aqueous mixtures. By eliminating most of the organic matrix components normally used in standard FABMS, CF-FAB increases the sensitivity of the mass spectrometric analysis of peptides and also minimizes the suppression effect. CF-FAB has also been shown to be an effective interface for coupling mass spectrometry and liquid chromatographic techniques such as microbore LC or CZE, thereby adding mass specificity to the chromatographic separation. The technique has moved from the development stage, and is being successfully used to help solve analytical problems in a number of areas involving proteins and peptides.

REFERENCES

1. **Barber, M., Bordoli, R. S., Sedwick, R. D., and Tyler, A. N.,** Fast atom bombardment of solids (F.A.B.): a new ion source for mass spectrometry, *J. Chem. Soc. Chem. Commun.,* 325, 1981.
2. **Barber, M., Bordoli, R. S., Elliot, G., Sedgwick, R. D., and Tyler, A. N.,** Fast atom bombardment mass spectrometry, *Anal. Chem.,* 54, 645A, 1982.
3. **Clench, M. R., Garner, G. V., Gordon, D. B., and Barber, M.,** Surface effects in FAB mapping of proteins and peptides, *Biomed. Mass Spectrom.,* 12, 355, 1985.
4. **Lehmann, W. D., Kessler, M., and Koenig, W. A.,** Investigations on basic aspects of fast atom bombardment mass spectrometry: matrix effects, sample effects, sensitivity and quantification, *Biomed. Mass Spectrom.,* 11, 217, 1984.
5. **Naylor, S., Findeis, F., Gibson, B. W., and Williams, D. H.,** An approach toward the complete FAB analysis of enzymic digests of peptides and proteins, *J. Am. Chem. Soc.,* 108, 6359, 1986.
6. **Stroh, J. G., Cook, J. C., Milberg, R. M., Brayton, L., Kihara, T., Huang, Z., and Rinehart, K. L.,** On-line liquid chromatography/fast atom bombardment mass spectrometry, *Anal. Chem.,* 57, 985, 1985.
7. **Dobberstein, P., Korte, E., Meyerhoff, G., and Pesch, R.,** Investigation of an LC/MS interface for EI-, CI- and FAB-ionization, *Int. J. Mass Spectrom. Ion Phys.,* 46, 185, 1983.
8. **Caprioli, R. M., Fan, T., and Cottrell, J. S.,** Continuous-flow sample probe for fast atom bombardment mass spectrometry, *Anal. Chem.,* 58, 2949, 1986.
9. **Ito, Y., Takeuchi, T., Ishi, D., and Goto, M.,** Direct coupling of micro high-performance liquid chromatography with fast atom bombardment mass spectrometry, *J. Chromatogr.,* 346, 161, 1985.
10. **Takeuchi, T., Watanabe, S., Kondo, N., Ishii, D., and Goto, M.,** Improvement of the interface for coupling of fast atom bombardment mass spectrometry and micro high-performance liquid chromatography, *J. Chromatogr.,* 435, 482, 1988.
11. **Caprioli, R. M.,** Continuous-flow fast atom bombardment mass spectrometry, in *Biologically Active Molecules,* Schlunegger, U. P., Ed., Springer-Verlag, Berlin, 1989, 59.
12. **Seifert, W. E., Jr., Ballatore, A., and Caprioli, R. M.,** Direct analysis of drugs by continuous-flow fast-atom bombardment and tandem mass spectrometry, *Rapid Commun. Mass Spectrom.,* 3, 117, 1989.
13. **Caprioli, R. M. and Fan, T.,** High sensitivity mass spectrometric determination of peptides: direct analysis of aqueous solutions, *Biochem. Biophys. Res. Commun.,* 141, 1058, 1986.
14. **Caprioli, R. M., Moore, W. T., and Fan, T.,** Improved detection of "suppressed" peptides in enzymic digests analyzed by FAB mass spectrometry, *Rapid Commun. Mass Spectrom.,* 1, 15, 1987.
15. **Caprioli, R. M., Moore, W. T., Petrie, G., and Wilson, K.,** Analysis of mixtures of hydrophilic peptides by continuous-flow fast atom bombardment mass spectrometry, *Int. J. Mass Spectrom. Ion Proc.,* 86, 187, 1988.
16. **Caprioli, R. M.,** Continuous-flow fast atom bombardment mass spectrometry, *Trends Anal. Chem.,* 7, 328, 1988.
17. **Caprioli, R. M.,** Analysis of biochemical reactions with molecular specificity using fast atom bombardment mass spectrometry, *Biochemistry,* 27, 513, 1988.
18. **deWit, J. S. M., Deterding, L. J., Moseley, M. A., Tomer, K. B., and Jorgensen, J. W.,** Design of a coaxial continuous flow fast atom bombardment probe, *Rapid Commun. Mass Spectrom.,* 2, 100, 1988.
19. **Caprioli, R. M., DaGue, B., Fan, T., and Moore, W. T.,** Microbore HPLC/mass spectrometry for the analysis of peptide mixtures using a continuous flow interface, *Biochem. Biophys. Res. Commun.,* 146, 291, 1987.
20. **Minard, R. D., Chin-Fatt, D., Curry, P., Jr., and Ewing, A. G.,** Capillary electrophoresis/flow FAB MS, in Proc. 36th ASMS Conf. on Mass Spectrometry and Allied Topics, San Fransisco, June 5-10, 1988, 950, 1988.
21. **Tomer, K., Moseley, M., III, de Wit, J., Deterding, L., and Jorgenson, J.,** Evaluation and optimization of continuous flow fast atom bombardment interfaces between open tubular liquid chromatography and mass spectrometry, in *Advances in Mass Spectrometry,* Vol. II, Longevialle, P, Ed., Heyden & Son, London, 1989, 440.
22. **Caprioli, R. M.,** Enzymes and mass spectrometry: a dynamic combination, *Mass Spectrom. Rev.,* 6, 237, 1987.
23. **Ashcroft, A. E., Chapman, J. R., and Cotrell, J. S.,** Continuous-flow fast atom bombardment mass spectrometry, *J. Chromatogr.,* 394, 15, 1987.
24. **Gavard, P., Couderc, F., Courcelle, E., and Prome, J. C.,** Determination of the carboxyl-end sequence of proteins and peptides by combined enzymatic degradation and continuous flow FAB, in *Advances in Mass Spectrometry,* Vol II, Longevialle, P., Ed., Heyden & Son, London, 1989, 1426.
25. **Hearn, T. W., Regnier, F. E., and Wehr, C. T.,** HPLC of peptides and proteins, *Am. Lab.,* 14, 18, 1982.
26. **Simpson, R. J. and Nice, E. C.,** The role of microbore HPLC in the purification of subnanomole amounts of polypeptides and proteins for gas-phase sequence analysis, in *Methods in Protein Sequence Analysis,* Walsh, K. A., Ed., Humana Press, Clifton, NJ, 1986, 213.
27. **Covey, T., Lee, E., Bruin, A., and Henion, J.,** Liquid chromatography/mass spectrometry, *Anal. Chem.,* 58, 1451A, 1986.

28. **Canova-Davis, E., Chloupek, R. C., Baldonado, I. P., Battersby, J. E., Spellman, M. W., Basa, L. J., O'Connor, B., Pearlman, R., Quan, C., Chakel, J. A., Stults, J. T., and Hancock, W. S.,** Analysis by FAB-MS and LC of proteins produced by either biosynthetic or chemical techniques, *Am. Biotech. Lab.,* 6, 8, 1988.

29. **Caprioli, R. M., Moore, W. T., DaGue, B., and Martin, M.,** Microbore high-performance liquid chromatography-mass spectrometry for the analysis of proteolytic digests by continuous-flow fast-atom bombardment mass spectrometry, *J. Chromatogr.,* 443, 355, 1989.

30. **Caprioli, R. M., DaGue, B. B., and Wilson, K.,** Optimization of chromatographic conditions for combined microbore HPLC/continuous-flow fast-atom bombardment mass spectrometry, *J. Chromatogr. Sci.,* 26, 640, 1988.

31. **Caprioli, R. M.,** Coupling chromatographic techniques with FABMS for the structural analysis of biological compounds, in *Biologically Active Molecules,* Schlunegger, U. P., Ed., Springer-Verlag, Berlin, 1989, 79.

32. **Ashcroft, A. E.,** Micro- and microbore liquid chromatography-fast atom bombardment-mass spectrometry (LC-FAB-MS) analysis of peptides and antibiotics, *Org. Mass Spectrom.,* 22, 754, 1987.

33. **Mock, K., Firth, J., Evans, S., and Cotrell, J.,** Continuous flow FAB-MS for rapid structural comparison of analogous therapeutic proteins, *Proc. 37th ASMS Conf. on Mass Spectrometry and Allied Topics, Miami Beach, May 21–26, 1989,* 1989, 1000.

34. **Bourell, J. H., Henzel, W. J., and Stults, J. T.,** Analysis of enzymatically digested proteins by on line capillary HPLC/FABMS, *Proc. 37th ASMS Conf. on Mass Spectrometry and Allied Topics, Miami Beach, May 21–26, 1989,* 1989, 1015.

35. **Cappiello, A., Palma, P., Papayannopoulos, I. A., and Biemann, K.,** Efficient introduction of HPLC fractions into a high performance tandem mass spectrometer, *Proc. 37th ASMS Conf. on Mass Spectrometry and Allied Topics, Miami Beach, May 21–26, 1989,* 1989, 990.

36. **Deterding, L. J., Jorgenson, J. W., Moseley, M. A., Pleasance, S., Thibault, P., and Tomer, K. B.,** Analysis of proteolytic digests and peptide mixtures using coaxial continuous flow FAB mass spectrometry, *Proc. 37th ASMS Conf. on Mass Spectrometry and Allied Topics, Miami Beach, May 21-26, 1989,* 1989, 1006.

37. **McDowall, M. A., Gould, V., Slaughter, C., and Pramanik, B.,** On-line HPLC-MS with a continuous flow FAB interface, in *Advances in Mass Spectrometry,* Vol. 11, Longevialle, P., Ed., Heyden & Son, London, 1989, 1248.

38. **Bateman, R. H. and Jones, D. S.,** Mixture analysis by combined capillary column liquid chromatography - dynamic FAB mass spectrometry, in *Advances in Mass Spectrometry,* Vol. 11, Longevialle, P., Ed., Heyden & Son, London, 1989, 1234.

39. **Lisek, C. A., Bailey, J., Aimone, L. D., Benson, L., Yaksh, T. L., and Jardine, I.,** Quantitation of endogenous neuropeptides by combined microcolumn-LC/dynamic FAB-MS, in *Advances in Mass Spectrometry,* Vol. 11, Longevialle, P., Ed., Heyden & Son, London, 1989, 1214.

40. **Jorgenson, J. W. and Lukacs, K. D.,** Capillary zone electrophoresis, *Science,* 222, 266, 1983.

41. **Gordon, M. J., Huang, X., Pentoney, S. L., Jr., and Zare, R. N.,** Capillary electrophoresis, *Science,* 242, 224, 1988.

42. **Pretorius, V., Hopkins, B. J., and Schieke, J. D.,** Electro-osmosis, a new concept for high-speed liquid chromatography, *J. Chromatogr.,* 99, 23, 1974.

43. **Lauer, H. H. and McManigill, D.,** Capillary zone electrophoresis of proteins in untreated fused silica tubing, *Anal. Chem.,* 58, 166, 1986.

44. **Cohen, A. S. and Karger, B. L.,** High-performance sodium dodecyl sulfate polyacrylamide gel capillary electrophoresis of peptides and proteins, *J. Chromatogr.,* 397, 409, 1987.

45. **Grossman, P. D., Wilson, K. J., Petrie, G., and Lauer, H. H.,** Effect of buffer pH and peptide composition on the selectivity of peptide separations by capillary zone electrophoresis, *Anal. Biochem.,* 173, 265, 1988.

46. **McCormick, R.M.,** Capillary zone electrophoretic separation of peptides and proteins using low pH buffers in modified silica capillaries, *Anal. Chem.,* 60, 2322, 1988.

47. **Hancock, W. S.,** The application of capillary electrophoresis to the characterization of recombinant-DNA produced proteins, Abstr. W-L-1, presented at *1st Int. Symp. High Performance Capillary Electrophoresis, Boston, April 10-12, 1989.*

48. **Nielson, R. G., Riggin, R. M., and Rickard, E. C.,** Capillary electrophoresis of fragments produced by enzymatic digestion of biosynthetic human growth hormone, Abstr. W-L-2, presented at *1st Int. Symp. High Performance Capillary Electrophoresis, Boston, April 10-12, 1989.*

49. **Fazio, S. D., Vivilecchia, R. V., Sheridan, J. V., and LeSuer, L. F.,** Purity analysis of recombinant proteins by capillary zone electrophoresis in the pharmaceutical industry, Abstr. W-L-3, presented at *1st Int. Symp. High Performance Capillary Electrophoresis, Boston, April 10-12, 1989.*

50. **Gassmann, E., Kuhn, R., and Luedi, H.,** Industrial applications of high performance capillary electrophoresis, Abstr. T-L-3, presented at *1st Int. Symp. High Performance Capillary Electrophoresis, Boston, April 10-12, 1989.*

51. **Stover, F. S., Vinjamoori, D. V., and Dahl, W. E.,** Applications of capillary electrophoresis in an industrial analytical laboratory, Abstr. M-P-113, presented at *1st Int. Symp. High Performance Capillary Electrophoresis, Boston, April 10-12, 1989.*

52. **Nielsen, R. G., Sittampalam, G. S., and Rickard, E. C.,** Method optimization in capillary electrophoresis applied to protein and peptide separations, Abstr. M-P-123, presented at *1st Int. Symp. High Performance Capillary Electrophoresis, Boston, April 10-12, 1989.*

53. **Wu, S.-L., Frenz, J., and Hancock, W. S.,** Application of capillary electrophoresis to the characterization of human growth hormone, Abstr. M-P-127, presented at *1st Int. Symp. High Performance Capillary Electrophoresis, Boston, April 10-12, 1989.*

54. **Palmieri, R. and Ohms, J.,** Characterization of trypsin digested recombinant proteins by capillary electrophoresis, Abstr. T-P-114, presented at *1st Int. Symp. High Performance Capillary Electrophoresis, Boston, April 10-12, 1989.*

55. **Caprioli, R. M., Moore, W. T., Martin, M., Wilson, K. B., and Moring, S.,** Coupling capillary zone electrophoresis and continuous-flow fast atom bombardment mass spectrometry for the analysis of peptide mixtures, *J. Chromatogr.,* 480, 247, 1989.

56. **Yu, L. W. and Lane, P. A.,** Applications of capillary zone electrophoresis for the rapid analysis of pharmaceuticals and peptide fragments from automatic synthesizers, Abstr. W-L-6, presented at *1st Int. Symp. High Performance Capillary Electrophoresis, Boston, April 10-12, 1989.*

57. **Cohen, A. S., Guttman, A., Karger, B. L., Pearson, J., and McCroskey, M. C.,** Micropreparative collection of CNBr fragments from capillary gel columns followed by microsequencing, Abstr. M-P-120, presented at *1st Int. Symp. High Performance Capillary Electrophoresis, Boston, April 10-12, 1989.*

58. **Puma, P., Young, P., Karol, R. and Fuchs, M.,** Capillary electrophoresis as an adjunct to HPLC in the analysis of proteins and peptides, Abstr. M-P-127, presented at *1st Int. Symp. High Performance Capillary Electrophoresis, Boston, April 10-12, 1989.*

59. **Palmieri, R. and Rampal, J.,** Characterization of peptides and proteins by high performance capillary electrophoresis, Abstr. M-P-130, presented at *1st Int. Symp. High Performance Capillary Electrophoresis, Boston, April 10-12, 1989.*

60. **Rush, R. S., Cohen, A. S., and Karger, B. L.,** Studies in peptide mapping by high performance capillary electrophoresis, Abstr. M-P-137, presented at *1st Int. Symp. High Performance Capillary Electrophoresis, Boston, April 10-12, 1989.*

61. **Huang, E. and Henion, J.,** Capillary electrophoresis/mass spectrometry analysis of protein enzymatic digests with atmospheric pressure ionization, Abstr. T-P-124, presented at *1st Int. Symp. High Performance Capillary Electrophoresis, Boston, April 10-12, 1989.*

62. **Oliveras, J. A., Nguyen, N. T., Yonker, C. R., and Smith, R. D.,** On-line mass spectrometric detection for capillary electrophoresis, *Anal. Chem.,* 59 1232, 1987.

63. **Smith, R. D., Barinaga, C. J., and Udseth, H. R.,** Improved electrospray interface for capillary zone electrophoresis-mass spectrometry, *Anal. Chem.,* 60, 1949, 1988.

64. **Udseth, H. R., Loo, J. A., and Smith, R. D.,** Capillary isotachophoresis/mass spectrometry, *Anal. Chem.,* 61, 228, 1989.

65. **Lee, E. D., Henion, J., and Muck, W.,** On-line capillary zone electrophoresis/ion spray tandem mass spectrometry determination of small peptides, in *Proc. 36th ASMS Conf. on Mass Spectrometry and Allied Topics, San Francisco, June 5-10, 1988,* 1988, 1085.

66. **Minard, R. D., Chin-Fatt, D., Curry, P., Jr., and Ewing, A. G.,** Capillary electrophoresis/flow FAB MS, *Proc. 36th ASMS Conf. on Mass Spectrometry and Allied Topics, San Francisco, June 5-10, 1988,* 1988, 1085.

67. **Minard, R. D., Luckenbill, D., Curry, P., Jr., and Ewing, A. G.,** Capillary electrophoresis/flow FAB/MS, in *Advances in Mass Spectrometry,* Vol. 11, P. Longevialle, P., Ed., Heydon & Son Ltd, London, 1989, 436.

68. **Moseley, M. A., Deterding, L. J., Tomer, K. B., and Jorgenson, J. W.,** Capillary-zone electrophoresis/fast-atom bombardment mass spectrometry: design of an on-line coaxial continuous-flow interface, *Rapid Commun. Mass Spectrom.,* 3, 100, 1988.

69. **Moseley, M. A., Deterding, L. J., Tomer, K. B., Bragg, N., and Jorgenson, J. W.,** Capillary-zone electrophoresis and capillary liquid chromatography coupled with sector mass spectrometry using coaxial continuous flow FAB interfaces, *Proc. 37th ASMS Conf. on Mass Spectrometry and Allied Topics,* 1989, 114.

70. **Moore, W. T., Martin, M., DaGue, B., and Caprioli, R. M.,** Analysis of synthetic peptide mixtures and protein digests by CZE/CF-FABMS, *Proc. 37th ASMS Conf. on Mass Spectrometry and Allied Topics,* 1989, 106.

Chapter 12

INVESTIGATION OF AMINO ACID MUTATIONS BY HIGH RESOLUTION MASS SPECTROMETRY

Hisashi Matsuda, Takekiyo Matsuo, Itsuo Katakuse, and Yoshinao Wada

TABLE OF CONTENTS

I.   Principles of the Digit-Printing Method 224
      A.   Background 224
      B.   A Model Analysis of Sickle Cell Hemoglobin by Digit-Printing 225
      C.   Ambiguity in the Digit-Printing Method 230

II.   High Resolution Mass Spectrometer 231

III. Investigation of Hemoglobin Variants and Other Proteins 232
      A.   Mass Spectra of Tryptic Peptides of Normal Globin Chains 232
      B.   Hb F Izumi (γ6Glu → Gly) 248
      C.   Hb F Yamaguchi (γ80Asp → Asn) 248
      D.   Hb Providence (β82Lys → Asn → Asp) 248
      E.   Synthetic Variants (α42Tyr → Phe and α42Tyr → His) 251
      F.   Calmodulin 251
      G.   Prealbumin Variant 253

References 256

I. PRINCIPLES OF THE DIGIT-PRINTING METHOD

A. BACKGROUND

Protein variants that have alterations in the amino acid sequence are probably the specific underlying biochemical causes of many diseases called "molecular disease". The first important process to understand in molecular disease is the amino acid sequence of the variant. Mass spectrometry has offered a concrete strategy to characterize variants. We introduced this technique in 1983 and called it the "digit-printing method".[1] It is explained in this section.

The digit-printing method is defined as follows: If a peptide contains an amino acid substitution, the molecular weight of the peptide should change precisely according to the mass difference between the amino acids involved in the substitution. Stated conversely: if the mass difference of peptide is determined by mass spectrometry, the precise type and position of the substitution in the peptide would be determined. The strength of this strategy stems from the following four facts:

1. The number of natural amino acids is limited to 20;
2. The molecular weights of these amino acids are different from each other, except for Leu-Ile in the exact mass and Lys-Gln in the nominal mass;
3. Molecular weights of constituent peptides in a mixture can be determined precisely by mass spectrometry;
4. Finger-printing method.

Points one and two relate to the essential characters of amino acids, inherited since amino acids have been created. Point three was accomplished very recently, just after the introduction of soft ionization in the late 1970s. The mass spectrometric techniques that support the digit-printing method are explained in Section II. Point four refers to the background technique, which has been applied widely as the conventional method to characterize variants. Until a decade ago, the basic methodological approach for the structural analysis of hemoglobin variants and other protein variants consisted of the following steps:

1. Cleave the protein enzymatically (trypsin was most often used) or chemically into small peptide fragments.
2. Separate the peptides on a filter paper by a combination of electrophoresis and chromatography in two dimensions, and visualize by staining with ninhydrin. The procedure and the pattern of separated peptides are called "fingerprinting" and a "fingerprint", respectively.
3. Locate one or two peptides differing from normal peptides in their positions by comparing them with the pattern of normal protein.
4. Extract the abnormal peptides from the filter paper.
5. Analyze the amino acid composition of any abnormal peptide. This analysis may suffice to determine the exact type and position of the amino acid substitution.
6. Further analyze the abnormal tryptic peptide enzymatically or chemically (including by Edman degradation) if necessary to determine the precise position of the amino acid substitution.

High pressure liquid chromatography (HPLC) analysis of peptides has almost replaced the fingerprinting steps two, three and four. Without regard to the analytical methods used, the procedure for the separation of peptides and the resulting pattern are generally called "peptide-mapping" and a "peptide-map", respectively. In the late 1970s, the development of soft ionization techniques enabled mass spectrometry to determine the molecular weight of component peptides in peptide mixtures, and therefore available to the peptide-mapping. The name "digit-printing" was given because mass spectrometry characterizes peptides by the digital values of mass number and apparently from the analogy of fingerprinting as well.

TABLE 1
The 20 Common Protein Amino Acids and Related Residues

Amino acid	3-letter abbreviation of residue	Single-letter abbreviation	Formula of residue	Nominal residue weight	Exact residue weight
Glycine	Gly	G	C_2H_3NO	57	57.021464
Alanine	Ala	A	C_3H_5NO	71	71.037114
Serine	Ser	S	$C_3H_5NO_2$	87	87.032028
Proline	Pro	P	C_5H_7NO	97	97.052764
Valine	Val	V	C_5H_9NO	99	99.068414
Threonine	Thr	T	$C_4H_7NO_2$	101	101.047679
Cysteine	Cys	C	C_3H_5NOS	103	103.009185
Leucine	Leu	L	$C_6H_{11}NO$	113	113.084064
Isoleucine	Ile	I	$C_6H_{11}NO$	113	113.084064
Asparagine	Asn	N	$C_4H_6N_2O_2$	114	114.042927
Aspartic acid	Asp	D	$C_4H_5NO_3$	115	115.026943
Glutamine	Gln	Q	$C_5H_8N_2O_2$	128	128.058578
Lysine	Lys	K	$C_6H_{12}N_2O$	128	128.094963
Glutamic acid	Glu	E	$C_5H_7NO_3$	129	129.042593
Methionine	Met	M	C_5H_9NOS	131	131.040485
Histidine	His	H	$C_6H_7N_3O$	137	137.058912
Phenylalanine	Phe	F	C_9H_9NO	147	147.068414
Arginine	Arg	R	$C_6H_{12}N_4O$	156	156.101111
Tyrosine	Tyr	Y	$C_9H_9NO_2$	163	163.063329
Tryptophan	Trp	W	$C_{11}H_{10}N_2O$	186	186.079313
S-carboxymethylcysteine			$C_5H_7NO_3S$	161	161.014664
S-aminoethylcysteine			$C_5H_{10}N_2OS$	146	146.051384
Homoserine (C-terminal residue)			$C_4H_8NO_3$	118	118.050418
Homoserine lactone (C-terminal residue)			$C_4H_6NO_2$	100	100.039853
Proton			H^+	1	1.0072765
Water			H_2O	18	18.0105647

From Wada, Y., Matsuo, T., and Sakurai, T., *Mass Spectrom. Rev.*, 8, 379, 1989. With permission.

Residue weights of the amino acids are given in Table 1. The difference between residue weights and the number of different nucleotide bases in the corresponding codons are given in Table 2, from which the candidates of possible substitutions that fit the mass difference of peptides are deduced. The process of determining an amino acid substitution by digit-printing is summarized as follows:

1. Determine the molecular weight of intact protein.
2. Cleave the protein enzymatically or chemically into peptides.
3. Apply soft ionization mass spectrometry to the resulting peptide mixtures.
4. Identify the abnormal peptides by comparing the mass spectra of normal and abnormal proteins, and calculate the molecular weight difference between normal and abnormal peptides.
5. Choose the type of amino acid substitutions from Table 2.
6. Determine the type and the position of substitution.

B. A MODEL ANALYSIS OF SICKLE CELL HEMOGLOBIN BY DIGIT-PRINTING

The principle of digit-printing is explained here by the analysis of the abnormal hemoglobin, Hb-S (6β Glu → Val), in sickle cell anemia. (1) By tryptic digestion of β-chain, the sixth residue comes into peptide T1. (2) The mass spectrum of the tryptic mixture of Hb-S is shown in Figure

TABLE 2
Mass Difference between Amino Acid Residues and the Minimum Nucleotide Base Difference in the Corresponding Coding Triplets

Difference in nominal mass	Difference in exact mass	Residues	Min nucleotide base difference
0	0.000	Leu-Ile	1
0	0.036	Lys-Gln	1
1	0.948	Glu-Lys	1
1	0.959	Asn-Leu	2
1	0.959	Asn-Ile	1
1	0.984	Asp-Asn	1
1	0.984	Glu-Gln	1
2	1.943	Asp-Leu	2
2	1.943	Asp-Ile	2
2	1.962	Cys-Thr	2
2	1.979	Thr-Val	2
2	1.998	Met-Glu	2
2	2.016	Val-Pro	2
3	2.946	Met-Lys	1
3	2.982	Met-Gln	2
4	3.941	Cys-Val	2
4	3.995	Thr-Pro	1
6	5.996	Cys-Pro	2
6	6.018	His-Met	3
7	6.962	Tyr-Arg	2
8	8.016	His-Glu	2
9	8.964	His-Lys	2
9	9.000	His-Gln	1
9	9.033	Arg-Phe	2
10	10.010	Phe-His	2
10	10.021	Pro-Ser	1
10	10.075	Leu-Cys	2
10	10.075	Ile-Cys	2
11	11.034	Asn-Cys	2
12	12.018	Asp-Cys	2
12	12.036	Val-Ser	2
12	12.036	Leu-Thr	2
12	12.036	Ile-Thr	1
13	12.995	Asn-Thr	1
13	13.032	Gln-Asp	2
13	13.068	Lys-Asp	2
14	13.979	Asp-Thr	2
14	14.016	Ala-Gly	1
14	14.016	Thr-Ser	1
14	14.016	Leu-Val	1
14	14.016	Ile-Val	1
14	14.016	Gln-Asn	2
14	14.016	Glu-Asp	1
14	14.052	Lys-Asn	1
15	14.975	Asn-Val	2
15	14.975	Gln-Leu	1
15	14.975	Gln-Ile	2
15	15.000	Glu-Asn	2
15	15.011	Lys-Leu	2
15	15.011	Lys-Ile	1
16	15.959	Asp-Val	1
16	15.959	Glu-Leu	2

TABLE 2 (continued)
Mass Difference between Amino Acid Residues and the Minimum Nucleotide Base Difference in the Corresponding Coding Triplets

Difference in nominal mass	Difference in exact mass	Residues	Min nucleotide base difference
16	15.959	Glu-Ile	2
16	15.977	Cys-Ser	1
16	15.995	Ser-Ala	1
16	15.995	Tyr-Phe	1
16	16.014	Met-Asp	3
16	16.028	Phe-Met	2
16	16.031	Leu-Pro	1
16	16.031	Ile-Pro	2
17	16.990	Asn-Pro	2
17	16.998	Met-Asn	2
18	17.956	Met-Leu	1
18	17.956	Met-Ile	1
18	17.974	Asp-Pro	2
18	18.026	Phe-Glu	3
19	18.973	Phe-Lys	3
19	19.010	Phe-Gln	3
19	19.042	Arg-His	1
22	22.032	His-Asp	1
23	23.016	Trp-Tyr	2
23	23.016	His-Asn	1
24	23.975	His-Leu	1
24	23.975	His-Ile	2
25	25.049	Gln-Cys	3
25	25.061	Arg-Met	1
25	25.086	Lys-Cys	3
26	26.004	Tyr-His	1
26	26.016	Pro-Ala	1
26	26.033	Glu-Cys	3
26	26.052	Leu-Ser	2
26	26.052	Ile-Ser	2
27	27.011	Asn-Ser	1
27	27.011	Gln-Thr	2
27	27.047	Lys-Thr	1
27	27.059	Arg-Glu	2
28	27.995	Asp-Ser	2
28	27.995	Glu-Thr	2
28	28.006	Arg-Lys	1
28	28.031	Val-Ala	1
28	28.031	Met-Cys	3
28	28.043	Arg-Gln	1
29	28.990	Gln-Val	2
29	29.027	Lys-Val	2
30	29.974	Glu-Val	1
30	29.978	Trp-Arg	1
30	29.993	Met-Thr	1
30	30.011	Ser-Gly	1
30	30.011	Thr-Ala	1
31	31.006	Gln-Pro	1
31	31.042	Lys-Pro	2
32	31.972	Cys-Ala	2
32	31.972	Met-Val	1

TABLE 2 (continued)
Mass Difference between Amino Acid Residues and the Minimum Nucleotide Base
Difference in the Corresponding Coding Triplets

Difference in nominal mass	Difference in exact mass	Residues	Min nucleotide base difference
32	31.990	Glu-Pro	2
32	32.023	Tyr-Met	3
32	32.041	Phe-Asp	2
33	33.025	Phe-Asn	2
34	33.984	Phe-Leu	1
34	33.984	Phe-Ile	1
34	33.988	Met-Ero	2
34	34.021	Tyr-Glu	2
34	34.050	His-Cys	2
35	34.968	Tyr-Lys	2
35	35.005	Tyr-Gln	2
36	36.011	His-Thr	2
38	37.990	His-Val	2
39	39.011	Trp-Phe	2
40	40.006	His-Pro	1
40	40.031	Pro-Gly	2
41	41.027	Gln-Ser	2
41	41.063	Lys-Ser	2
41	41.074	Arg-Asp	2
42	42.011	Glu-Ser	2
42	42.047	Val-Gly	1
42	42.047	Leu-Ala	2
42	42.047	Ile-Ala	2
42	42.058	Arg-Asn	2
43	43.006	Asn-Ala	2
43	43.017	Arg-Leu	1
43	43.017	Arg-Ile	1
44	43.990	Asp-Ala	1
44	44.008	Met-Ser	2
44	44.026	Thr-Gly	2
44	44.059	Phe-Cys	1
46	45.988	Cys-Gly	1
46	46.021	Phe-Thr	2
48	48.000	Phe-Val	1
48	48.036	Tyr-Asp	1
49	49.020	Trp-His	3
49	49.020	Try-Asn	1
50	49.979	Tyr-Leu	2
50	49.979	Tyr-Ile	2
50	50.016	Phe-Pro	2
50	50.027	His-Ser	2
53	53.092	Arg-Cys	1
55	55.039	Trp-Met	2
55	55.053	Arg-Thr	1
56	56.063	Leu-Gly	2
56	56.063	Ile-Gly	2
57	57.021	Asn-Gly	2
57	57.021	Gln-Ala	2
57	57.033	Arg-Val	2
57	57.037	Trp-Glu	2
57	57.058	Lys-Ala	2

TABLE 2 (continued)
Mass Difference between Amino Acid Residues and the Minimum Nucleotide Base
Difference in the Corresponding Coding Triplets

Difference in nominal mass	Difference in exact mass	Residues	Min nucleotide base difference
58	57.984	Trp-Lys	2
58	58.005	Asp-Gly	1
58	58.005	Glu-Ala	1
58	58.021	Trp-Gln	2
59	59.048	Arg-Pro	1
60	60.003	Met-Ala	2
60	60.036	Phe-Ser	1
60	60.054	Tyr-Cys	1
62	62.016	Tyr-Thr	2
64	63.995	Tyr-Val	2
66	66.011	Tyr-Pro	2
66	66.022	His-Ala	2
69	69.069	Arg-Ser	1
71	71.037	Gln-Gly	2
71	71.052	Trp-Asp	3
71	71.074	Lys-Gly	2
72	72.021	Glu-Gly	1
72	72.036	Trp-Asn	3
73	72.995	Trp-Leu	1
73	72.995	Trp-Ile	3
74	74.019	Met-Gly	2
76	76.031	Phe-Ala	2
76	76.031	Tyr-Ser	1
80	80.037	His-Gly	2
83	83.070	Trp-Cys	1
85	85.032	Trp-Thr	2
85	85.064	Arg-Ala	2
87	87.011	Trp-Val	2
89	89.027	Trp-Pro	2
90	90.047	Phe-Gly	2
92	92.026	Tyr-Ala	2
99	99.047	Trp-Ser	1
99	99.080	Arg-Gly	1
106	106.042	Tyr-Gly	2
115	115.042	Trp-Ala	2
129	129.058	Trp-Gly	1

From Wada, Y., Maksuo, T., and Sakurai, T., *Mass Spectrom Rev.*, 8, 379, 1989.
With permission.

1b. (3) The normal peptide T1 at m/z 952 and the abnormal one at m/z 922 are detected. (4) In Table 2, five types of amino acid substitution: Glu → Val, Trp → Arg, Met → Thr, Ser → Gly, and Thr → Ala fit the decrease of 30 mass units, because only single amino acid substitution is considered. (5) Among them, the substitutions Trp → Arg, Met → Thr, and Ser → Gly are excluded because the resideues of Ser, Met, and Trp are not present in the normal sequence of T1. Accordingly, only two types. Glu → Val and Thr → Ala fit the molecular weight of the abnormal peptide. Then, as described later, electrophoresis of Hb S unequivocally indicates that the substitution accompanies a change of charge, which excludes the possibility of Thr → Ala substitution. As a result, Glu → Val is recognized as the only possible type of substitution.

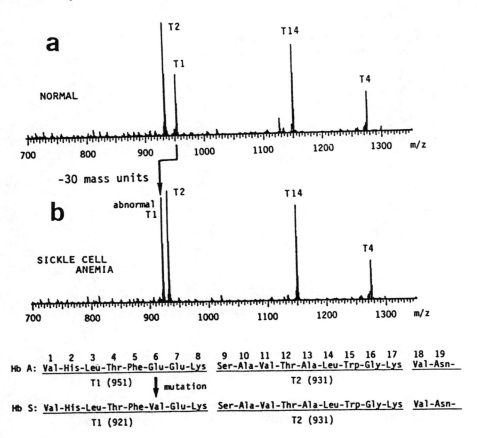

FIGURE 1.   Mass spectra of tryptic peptides of β-chains of normal hemoglobin (a) and sickle cell hemoglobin Hb S (b) in the region of m/z 920-960. N-terminal sequences are at the bottom. The molecular weight of the peptide is in parentheses. (From Wada, Y., Matsuo, T., and Sakurai, T., *Mass Spectrom. Rev.*, 8, 379, 1989. With permission.)

Though depending strongly on the amino acid sequence of the peptide fragment containing the substitution, very often only one mass spectrum can propose the most probable type and position of substitution.

C. AMBIGUITY IN THE DIGIT-PRINTING METHOD

Three types of ambiguity may remain after a single course of digit-printing in some cases.

1. More than two positions for a specific type of substitution are present in the affected peptide, as in the case of Hb-S, where peptide T1 contains two glutamic acid residues at the sixth and seventh position.
2. More than two types of substitution fit the mass difference, in the nominal mass or in the exact mass (Table 2).
3. The abnormal peptide fragment is too large to determine accurately the molecular weight. We encounter this problem particularly when the substitution involves Glu/Gln or Asp/ Asn, because the nominal mass difference between acid and amide residues is only one mass unit.

In order to overcome these ambiguities, the following procedures are used.

1. Use of another cleavage method
2. Chemical modification of specific amino acid residues
3. High mass resolution analysis
4. Referring to the known chemical properties of the variant
5. Single nucleotide base change
6. Amino acid sequencing by Edman degradation
7. MS/MS technique

Some of these methods were employed effectively in the studies described in Section III.

II. HIGH RESOLUTION MASS SPECTROMETER

A high performance mass spectrometer is indispensable for the analysis of proteins. The requirements for the instrument are high mass resolution, high ion transmission, high mass range, and high accuracy in mass calibration. A double-focusing mass spectrometer with electric and magnetic sectors is most suitable for this purpose.

The resolving power, R, of a mass spectrometer is given by:

$$R = A_\gamma / (A_x s + \Delta + d)$$

where s and d are the width of source and collector slits, respectively, and A_γ, A_x, and Δ are mass dispersion, image magnification, and total amount of aberrations, respectively. To achieve the most efficient detection, the width d is chosen to be $A_x s + \Delta$. A large numerator and a small denominator in this equation are both necessary to obtain high resolving power. Such high resolution is obtained by making $A_x s$ (as well as Δ) small. Because the ion transmission is restricted by the slit width, a small A_x is necessary to maintain high ion transmission. Therefore, a large value of A_γ / A_x is desirable for obtaining high resolution and high transmission simultaneously.

In order to reduce the image aberrations, Δ, the ion trajectories in a mass spectrometer should be calculated very accurately. A third-order calculation technique of ion trajectories has been studied, and the computer code third order ion optics (TRIO) has been developed at Osaka University.[4] This program is applicable to any ion optical system consisting of drift spaces, cylindrical or toroidal electric sectors, homogeneous or inhomogeneous magnetic sectors, and magnetic and electric quadrupole lenses and multipoles. The influence of the fringing fields is taken into consideration. The second- and third-order aberration coefficients of a mass spectrometer are calculated in a short time.

Two different types of mass spectrometers have been designed and constructed at Osaka University. One is the cylindrical electric sector, quadrupole, homogeneous magnetic sector (CQH) type shown in Figure 2 and the other the quadrupole, quadrupole, homogeneous magnetic sector, quadrupole, cylindrical electric sector (QQHQC) type shown in Figure 3.

The CQH type mass spectrometer, designed to achieve complete second-order focusing,[3] was constructed in 1975. The measurement of the second- and third-order aberration coefficients showed that the measured values were in good agreement with those calculated by TRIO.

The QQHQC type mass spectrometer was designed to achieve high mass analysis.[5] The distinguishing features of this mass spectrometer are high transmission through a narrow magnet gap, very small image magnification, small second- and third-order aberrations, and a very large magnet radius (1.25 m). Two electric quadrupole lenses are placed between the ion source and the magnetic sector and are used as the focusing mode in the vertical (y) direction and as the defocusing mode in the horizontal (x) direction. Therefore, by fitting lens strength, the ion beam profile in the magnet gap can be made very wide in the x-direction and narrow in the y-direction,

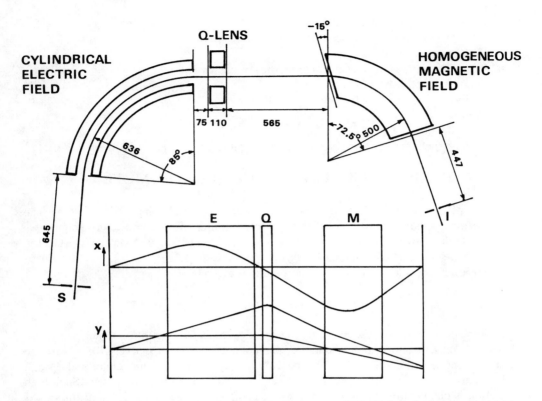

FIGURE 2. Schematic drawing of double-focusing mass spectrometer CQH. Horizontal (x) and vertical (y) ion trajectories are shown in the lower part. (From Matsuda, H., *Mass Spectrom. Rev.*, 2, 299, 1983. With permission.)

as shown in Figure 3. Image magnification of about 0.2 is obtained with this device. Consequently, high mass resolution and high beam transmission are obtained simultaneously. This mass spectrometer was constructed in 1987.[6] A mass resolution of 10,000 (90% valley) was obtained with an object slit of 250 μm. A maximum mass range is m/z = 25,000 at 10 kV acceleration. An example of a mass spectrum of CsI cluster ions is shown in Figure 4. Complete separation of clusters is observed up to m/z = 50,000.

III. INVESTIGATION OF HEMOGLOBIN VARIANTS AND OTHER PROTEINS

More detailed explanations and related references concerning the following items are given in Reference 2.

A. MASS SPECTRA OF TRYPTIC PEPTIDES OF NORMAL GLOBIN CHAINS

As indicated earlier in the description of the digit-printing method process (point three), the comparison of mass spectra between the normal and abnormal variant is very important. Both samples should be prepared under the same conditions. For this purpose, one must first obtain the standard mass spectra of tryptic peptides of normal globins. The procedure for sample preparation described here is a conventional one, because the purity acceptable for the conventional peptide mapping is sufficient for digit-printing. The molecule of human hemoglobin is composed of four globin polypeptide chains, which are α-, β-, γ-, and δ- chains. Amino acid sequences of tryptic fragments of four human globin chains are presented in Table 3, and the calculated molecular weights of the expected peptide fragments by various cleavages is presented in Table 4.

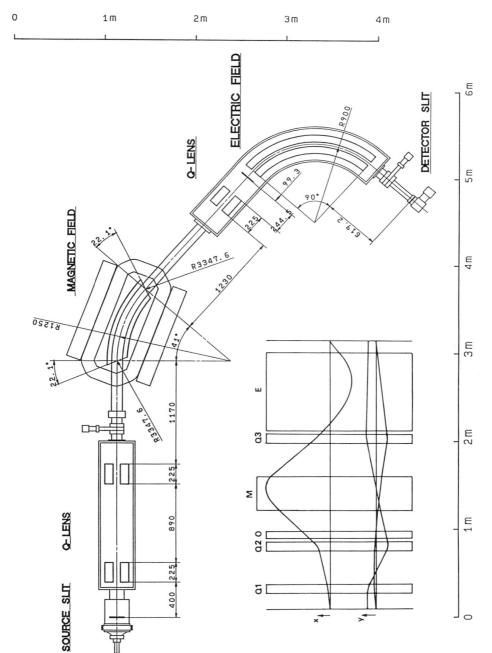

FIGURE 3. Schematic drawing of double-focusing mass spectrometer QQHQC. Horizontal (x) and vertical (y) ion trajectories show that the ion beam in the magnet gap is very wide in x and narrow in y directions.

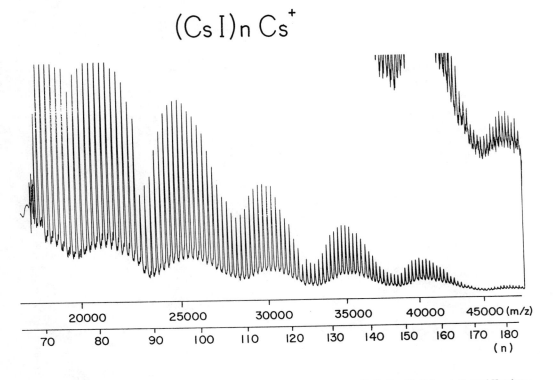

FIGURE 4.   SIMS mass spectrum of CsI clusters. (From Matsuda, H., Matsuo, T., Fujita, Y., Sakurai, T., and Katakuse, I., *Int. J. Mass Spectrom. Ion Proc.*, 91, 1, 1989. With permission.)

TABLE 3
Amino Acid Sequences and Mol wt of Tryptic Peptides

Tryptic Peptides of α-Globin Chain[a]

No.	Formula	Mol wt	Sequence
T1	$C_{32}H_{56}N_8O_{11}$	728.407	1 2 3 4 5 6 7 Val-Leu-Ser-Pro-Ala-Asp-Lys
T2	$C_{19}H_{36}N_6O_7$	460.265	8 9 10 11 Thr-Asn-Val-Lys
T3	$C_{25}H_{37}N_7O_6$	531.281	12 13 14 15 16 Ala-Ala-Trp-Gly-Lys
T4	$C_{65}H_{100}N_{20}O_{23}$	1528.727	17 18 19 20 21 22 23 24 25 26 Val-Gly-Ala-His-Ala-Gly-Glu-Tyr-Gly-Ala- 27 28 29 30 31 Glu-Ala-Leu-Glu-Arg
T5	$C_{51}H_{78}N_{10}O_{13}S$	1070.547	32 33 34 35 36 37 38 39 40 Met-Phe-Leu-Ser-Phe-Pro-Thr-Thr-Lys
T6	$C_{85}H_{120}N_{22}O_{24}$	1832.885	41 42 43 44 45 46 47 48 49 50 Thr-Tyr-Phe-Pro-His-Phe-Asp-Leu-Ser-His- 51 52 53 54 55 56 Gly-Ser-Ala-Gln-Val-Lys

TABLE 3 (continued)
Amino Acid Sequences and Mol wt of Tryptic Peptides

Tryptic Peptides of α-Globin Chain[a]

No.	Formula	Mol wt	Sequence
T7	$C_{16}H_{27}N_7O_5$	397.207	57 58 59 60 Gly-His-Gly-Lys
T8	$C_6H_{14}N_2O_2$	146.106	61 Lys
T9	$C_{128}H_{206}N_{38}O_{43}S$	2995.482	62 63 64 65 66 67 68 69 70 71 Val-Ala-Asp-Ala-Leu-Thr-Asn-Ala-Val-Ala- 72 73 74 75 76 77 78 79 80 81 His-Val-Asp-Asp-Met-Pro-Asn-Ala-Leu-Ser- 82 83 84 85 86 87 88 89 90 Ala-Leu-Ser-Asp-Leu-His-Ala-His-Lys
T10	$C_{12}H_{25}N_5O_3$	287.196	91 92 Leu-Arg
T11	$C_{38}H_{59}N_9O_{11}$	817.433	93 94 95 96 97 98 99 Val-Asp-Pro-Val-Asn-Phe-Lys
T12(CM)	$C_{138}H_{221}N_{35}O_{39}S$	3024.611	100 101 102 103 104 105 106 107 108 109 Leu-Leu-Ser-His-Cys-Leu-Leu-Val-Thr-Leu- 110 111 112 113 114 115 116 117 118 119 Ala-Ala-His-Leu-Pro-Ala-Glu-Phe-Thr-Pro- 120 121 122 123 124 125 126 127 Ala-Val-His-Ala-Ser-Leu-Asp-Lys
T13	$C_{57}H_{97}N_{13}O_{18}$	1251.707	128 129 130 131 132 133 134 135 136 137 Phe-Leu-Ala-Ser-Val-Ser-Thr-Val-Leu-Thr- 138 139 Ser-Lys
T14	$C_{15}H_{23}N_5O_4$	337.175	140 141 Tyr-Ars

Tryptic Peptides of β-Globin Chain[a]

No.	Formula	Mol wt	Sequence
T1	$C_{42}H_{69}N_{11}O_{14}$	951.503	1 2 3 4 5 6 7 8 Val-His-Leu-Thr-Pro-Glu-Glu-Lys
T2	$C_{43}H_{69}N_{11}O_{12}$	931.513	9 10 11 12 13 14 15 16 17 Ser-Ala-Val-Thr-Ala-Leu-Trp-Gly-Lys
T3	$C_{54}H_{91}N_{17}O_{21}$	1313.658	18 19 20 21 22 23 24 25 26 27 Val-Asn-Val-Asp-Glu-Val-Gly-Gly-Glu-Ala- 28 29 30 Leu-Gly-Arg
T4	$C_{62}H_{95}N_{15}O_{14}$	1273.718	31 32 33 34 35 36 37 38 39 40 Leu-Leu-Val-Val-Tyr-Pro-Trp-Thr-Gln-Arg

TABLE 3 (continued)
Amino Acid Sequences and Mol wt of Tryptic Peptides

Tryptic Peptides of β-Globin Chain[a]

No.	Formula	Mol wt	Sequence
T5	$C_{93}H_{135}N_{21}O_{30}S$	2057.940	41 42 43 44 45 46 47 48 49 50 Phe-Phe-Glu-Ser-Phe-Gly-Asp-Leu-Ser-Thr- 51 52 53 54 55 56 57 58 59 Pro-Asp-Ala-Val-Met-Gly-Asn-Pro-Lys
T6	$C_{11}H_{23}N_3O_3$	245.174	60 61 Val-Lys
T7	$C_{17}H_{29}N_7O_5$	411.223	62 63 64 65 Ala-His-Gly-Lys
T8	$C_6H_{14}N_2O_2$	146.106	66 Lys
T9	$C_{75}H_{120}N_{20}O_{23}$	1168.884	67 68 69 70 71 72 73 74 75 76 Val-Leu-Gly-Ala-Phe-Ser-Asp-Gly-Leu-Ala- 77 78 79 80 81 82 His-Leu-Asp-Asn-Leu-Lys
T10(CM)	$C_{63}H_{98}N_{16}O_{23}S$	1478.671	83 84 85 86 87 88 89 90 91 92 Gly-Thr-Phe-Ala-Thr-Leu-Ser-Glu-Leu-His- 93 94 95 Cys-Asp-Lys
T11	$C_{50}H_{75}N_{15}O_{15}$	1125.557	96 97 98 99 100 101 102 103 104 Leu-His-Val-Asp-Pro-Glu-Asn-Phe-Arg
T12(CM)	$C_{82}H_{132}N_{22}O_{20}S$	1776.971	105 106 107 108 109 110 111 112 113 114 Leu-Leu-Gly-Asn-Val-Leu-Val-Cys-Val-Leu- 115 116 117 118 119 120 Ala-His-His-Phe-Gly-Lys
T13	$C_{64}H_{95}N_{15}O_{19}$	1377.693	121 122 123 124 125 126 127 128 129 130 Glu-Phe-Thr-Pro-Pro-Val-Gln-Ala-Ala-Tyr- 131 132 Gln-Lys
T14	$C_{51}H_{88}N_{16}O_{14}$	1148.667	133 134 135 136 137 138 139 140 141 142 Val-Val-Ala-Gly-Val-Ala-Asn-Ala-Leu-Ala- 143 144 His-Lys
T15	$C_{15}H_{18}N_4O_4$	318.133	145 146 Try-His

Tryptic Peptides of γ-Globin Chain[a]

No.	Formula	Mol wt	Sequence
T1	$C_{41}H_{59}N_{11}O_{16}$	961.414	1 2 3 4 5 6 7 8 Gly-His-Phe-Thr-Glu-Glu-Asp-Lys
T2	$C_{45}H_{73}N_{11}O_{13}$	975.539	9 10 11 12 13 14 15 16 17 Ala-Thr-Ile-Thr-Ser-Leu-Trp-Gly-Lys

TABLE 3 (continued)
Amino Acid Sequences and Mol wt of Tryptic Peptides

Tryptic Peptides of γ-Globin Chain[a]

No.	Formula	Mol wt	Sequence
T3	$C_{53}H_{89}N_{17}O_{22}$	1315.637	18 19 20 21 22 23 24 25 26 27 Val-Asn-Val-Glu-Asp-Ala-Gly-Gly-Glu-Thr- 28 29 30 Leu-Gly-Arg
T4	$C_{62}H_{95}N_{15}O_{14}$	1273.718	31 32 33 34 35 36 37 38 39 40 Leu-Leu-Val-Val-Tyr-Pro-Trp-Thr-Gln-Arg
T5	$C_{89}H_{132}N_{22}O_{28}S$	1988.930	41 42 43 44 45 46 47 48 49 50 Phe-Phe-Asp-Ser-Phe-Gly-Asn-Leu-Ser-Ser- 51 52 53 54 55 56 57 58 59 Ala-Ser-Ala-Ile-Met-Gly-Asn-Pro-Lys
T6	$C_{11}H_{23}N_3O_3$	245.174	60 61 Val-Lys
T7	$C_{17}H_{29}N_7O_5$	411.223	62 63 64 65 Ala-His-Gly-Lys
T8	$C_6H_{14}N_2O_2$	146.106	66 Lys
T9	$C_{45}H_{81}N_{11}O_{15}$	1015.591	67 68 69 70 71 72 73 74 75 76 Val-Leu-Thr-Ser-Leu-Gly-Asp-Ala-Ile-Lys
T10	$C_{32}H_{53}N_9O_{11}$	739.386	77 78 79 80 81 82 His-Leu-Asp-Asp-Leu-Lys
T11(CM)	$C_{64}H_{99}N_{17}O_{23}S$	1505.682	83 84 85 86 87 88 89 90 91 92 Gly-Thr-Phe-Ala-Gln-Leu-Ser-Glu-Leu-His- 93 94 95 Cys-Asp-Lys
T12	$C_{50}H_{75}N_{13}O_{15}$	1097.551	96 97 98 99 100 101 102 103 104 Leu-His-Val-Asp-Pro-Glu-Asn-Phe-Lys
T13	$C_{81}H_{136}N_{20}O_{19}$	1693.029	105 106 107 108 109 110 111 112 113 114 Leu-Leu-Gly-Asn-Val-Leu-Val-Thr-Val-Leu- 115 116 117 118 119 120 Ala-Ile-His-Phe-Gly-Lys
T14	$C_{66}H_{96}N_{16}O_{21}$	1448.694	121 122 123 124 125 126 127 128 129 130 Glu-Phe-Thr-Pro-Glu-Val-Gln-Ala-Ser-Trp- 131 132 Gln-Lys
T15	$C_{48}H_{87}N_{15}O_{17}S$	1177.613	133 134 135 136 137 138 139 140 141 142 Met-Val-Thr-Gly-Val-Ala-Ser-Ala-Leu-Ser- 143 144 Ser-Arg
T16	$C_{15}H_{18}N_4O_4$	318.133	145 146 Tyr-His

TABLE 3 (continued)
Amino Acid Sequences and Mol wt of Tryptic Peptides

Tryptic Peptides of δ-Globin Chain[a]

No.	Formula	Mol wt	Sequence
T1	$C_{42}H_{69}N_{11}O_{14}$	951.503	1 2 3 4 5 6 7 8 Val-His-Leu-Thr-Pro-Glu-Glu-Lys
T2	$C_{44}H_{70}N_{12}O_{12}$	958.524	9 10 11 12 13 14 15 16 17 Thr-Ala-Val-Asn-Ala-Leu-Trp-Gly-Lys
T3	$C_{52}H_{89}N_{17}O_{19}$	1255.652	18 19 20 21 22 23 24 25 26 27 Val-Asn-Val-Asp-Ala-Val-Gly-Gly-Glu-Ala- 28 29 30 Leu-Gly-Arg
T4	$C_{62}H_{95}N_{15}O_{14}$	1273.718	31 32 33 34 35 36 37 38 39 40 Leu-Leu-Val-Val-Tyr-Pro-Trp-Thr-Gln-Arg
T5	$C_{92}H_{133}N_{21}O_{30}S$	2043.925	41 42 43 44 45 46 47 48 49 50 Phe-Phe-Glu-Ser-Phe-Gly-Asp-Leu-Ser-Ser- 51 52 53 54 55 56 57 58 59 Pro-Asp-Ala-Val-Met-Gly-Asn-Pro-Lys
T6	$C_{11}H_{23}N_3O_3$	245.174	60 61 Val-Lys
T7	$C_{17}H_{29}N_7O_5$	411.223	62 63 64 65 Ala-His-Gly-Lys
T8	$C_6H_{14}N_2O_2$	146.106	66 Lys
T9	$C_{75}H_{120}N_{20}O_{23}$	1668.884	67 68 69 70 71 72 73 74 75 76 Val-Leu-Gly-Ala-Phe-Ser-Asp-Gly-Leu-Ala- 77 78 79 80 81 82 His-Leu-Asp-Asn-Leu-Lys
T10(CM)	$C_{64}H_{99}N_{17}O_{24}S$	1521.677	83 84 85 86 87 88 89 90 91 92 Gly-Thr-Phe-Ser-Gln-Leu-Ser-Glu-Leu-His- 93 94 95 Cys-Asp-Lys
T11	$C_{50}H_{75}N_{15}O_{15}$	1125.557	96 97 98 99 100 101 102 103 104 Leu-His-Val-Asp-Pro-Glu-Asn-Phe-Arg
T12(CM)	$C_{59}H_{106}N_{16}O_{16}S$	1326.769	105 106 107 108 109 110 111 112 113 114 Leu-Leu-Gly-Asn-Val-Leu-Val-Cys-Val-Leu- 115 116 117 Ala-His-Arg
T13	$C_{21}H_{32}N_6O_6$	464.238	118 119 120 Asn-Gly-Lys
T14	$C_{64}H_{96}N_{16}O_{20}S$	1440.671	121 122 123 124 125 126 127 128 129 130 Glu-Phe-Thr-Pro-Gln-Met-Gln-Ala-Ala-Tyr- 131 132 Gln-Lys

TABLE 3 (continued)
Amino Acid Sequences and Mol wt of Tryptic Peptides

Tryptic Peptides of δ-Globin Chain[a]

No.	Formula	Mol wt	Sequence
T15	$C_{51}H_{88}N_{16}O_{14}$	1148.667	133 134 135 136 137 138 139 140 141 142 Val-Val-Ala-Gly-Val-Ala-Asn-Ala-Asn-Ala- 143 144 145 146 Leu-Ala-His-Lys
T16	$C_{15}H_{18}N_4O_4$	318.133	145 146 Tyr-His

Note: (CM) indicates that the cysteine residue in the peptide is S-carboxymethylated.

From Wada, Y., Matsuo, T., and Sakurai, T., *Mass Spectrom. Rev.*, 8, 379, 1989. With permission.

TABLE 4
Molecular Weight of the Peptides Generated by Cleavage with Trypsin, Lysyl Endopeptidase from *Achromobacter Lyticus Staphylococcus Aureus* V8 Protease, or Cyanogen Bromide

α-Globin Chain[a]

Trypsin

Peptide no.	Sequence no.	Nominal wt.	Exact wt.
T1	1—7	728	728.407
T2	8—11	460	460.265
T3	12—16	531	531.281
T4	17—31	1528	1528.727
T5	32—40	1070	1070.547
T6	41—56	1832	1832.885
T7	57—60	397	397.207
T8	61—61	146	146.106
T9	62—90	2994	2995.482
T10	91—92	287	287.196
T11	93—99	817	817.433
T12(CM)	100—127	3023	3024.611
T13	128—139	1251	1251.707
T14	140—141	337	337.175
T12a(AE)	100—104	614	614.321
T12b	105—127	2412	2413.337

Cyanogen bromide

Peptide no.	Sequence no.	Nominal wt.	Exact wt.
CB1	1—32	3294	3295.695
CB2	33—76	4752	4754.388
CB3(CM)	77—141	7097	7100.843

TABLE 4 (continued)
Molecular Weight of the Peptides Generated by Cleavage with Trypsin, Lysyl Endopeptidase from *Achromobacter Lyticus Staphylococcus Aureus* V8 Protease, or Cyanogen Bromide

α-Globin Chain[a]

Lysyl endopeptidase

Peptide no.	Sequence no.	Nominal wt.	Exact wt.
L1	1—7	728	728.407
L2	8—11	460	460.265
L3	12—16	531	531.281
L4	17—40	2580	2581.264
L5	41—56	1832	1832.885
L6	57—60	397	397.207
L7	61—61	146	146.106
L8	62—90	2994	2995.482
L9	91—99	1086	1086.619
L10(CM)	100—127	3023	3024.611
L11	128—139	1251	1251.707
L12	140—141	337	337.175
L10a(AE)	100—104	614	614.321
L10b	105—127	2412	2413.337

V8 protease

Peptide no.	Sequence no.	Nominal wt.	Exact wt.
S1	1—23	2304	2305.218
S2	24—27	438	438.175
S3	28—30	331	331.174
S4(CM)	31—116	9430	9434.885
S5	117—141	2736	2737.480

β-Globin Chain[a]

Trypsin

Peptide no.	Sequence no.	Nominal wt.	Exact wt.
T1	1—8	951	951.503
T2	9—17	931	931.513
T3	18—30	1313	1313.658
T4	31—40	1273	1273.718
T5	41—59	2057	2057.940
T6	60—61	245	245.174
T7	62—65	411	411.223
T8	66—66	146	146.106
T9	67—82	1668	1668.884
T10(CM)	83—95	1478	1478.671
T11	96—104	1125	1125.557
T12(CM)	105—120	1776	1776.971
T13	121—132	1377	1377.693
T14	133—144	1148	1148.667
T15	145—146	318	318.133
T10(AE)	83—95	1463	1463.708
T10a(AE)	83—93	1220	1220.586
T10b	94—95	261	261.132
T12a(AE)	105—112	872	872.515
T12b	113—120	907	907.503

TABLE 4 (continued)
Molecular Weight of the Peptides Generated by Cleavage with Trypsin, Lysyl Endopeptidase from *Achromobacter Lyticus Staphylococcus Aureus* V8 Protease, or Cyanogen Bromide

β-Globin Chain[a]

Cyanogen bromide

Peptide no.	Sequence no.	Nominal wt.	Exact wt.
CB1	1—55	6027	6030.084
CB2(CM)	56—146	9926	9931.194

Lysyl endopeptidase

Peptide no.	Sequence no.	Nominal wt.	Exact wt.
L1	1—8	951	951.503
L2	9—17	931	931.513
L3	18—59	4607	4609.295
L4	60—61	245	245.174
L5	62—65	411	411.223
L6	66—66	146	146.106
L7	67—82	1668	1668.884
L8(CM)	83—95	1478	1478.671
L9(CM)	96—120	2883	2884.517
L10	121—132	1377	1377.693
L11	133—144	1148	1148.667
L12	145—146	318	318.133
L8a(AE)	83—93	1220	1220.586
L8b	94—95	261	261.132
L9a(AE)	96—112	1979	1980.062
L9b	113—120	907	907.503

V8 protease

Peptide no.	Sequence no.	Nominal wt.	Exact wt.
S1	1—6	694	694.365
S2	7—7	147	147.053
S3	8—22	1615	1615.857
S4	23—26	360	360.164
S5	27—43	2093	2094.141
S6	44—90	4838	4840.486
S7(CM)	91—101	1362	1362.624
S8(CM)	102—121	2322	2323.226
S9	122—146	2678	2679.429

γ-Globin Chain[a]

Trypsin

Peptide no.	Sequence no.	Nominal wt.	Exact wt.
T1	1—8	961	961.141
T2	9—17	975	975.539
T3	18—30	1315	1315.637
T4	31—40	1273	1273.718

TABLE 4 (continued)
Molecular Weight of the Peptides Generated by Cleavage with Trypsin, Lysyl Endopeptidase from *Achromobacter Lyticus Staphylococcus Aureus* V8 Protease, or Cyanogen Bromide

γ-Globin Chain[a]

Trypsin

Peptide no.	Sequence no.	Nominal wt.	Exact wt.
T5	41—59	1988	1988.930
T6	60—61	245	245.174
T7	62—65	411	411.223
T8	66—66	146	146.106
T9	67—76	1015	1015.591
T10	77—82	739	739.386
T11(CM)	83—95	1505	1505.682
T12	96—104	1097	1097.551
T13	105—120	1692	1693.029
T14	121—132	1448	1448.694
T15(G)	133—144	1177	1177.613
T15(A)		1191	1191.628
T16	145—146	318	318.133
T11(AE)	83—95	1490	1490.719
T11a(AE)	83—93	1247	1247.597
T11b	94—95	261	261.132

Cyanogen bromide

Peptide no.	Sequence no.	Nominal wt.	Exact wt.
CB1	1—55	6014	6016.991
CB2(CM)	56—133	8651	8655.611
CB3(G)	134—146	1346	1346.694
CB3(A)		1360	1360.710

Lysyl endopeptidase

Peptide no.	Sequence no.	Nominal wt.	Exact wt.
L1	1—8	961	961.414
L2	9—17	975	975.539
L3	18—59	4540	4542.264
L4	60—61	245	245.174
L5	62—65	411	411.223
L6	66—66	146	146.106
L7	67—76	1015	1015.591
L8	77—82	739	739.386
L9(CM)	83—95	1505	1505.682
L10	96—104	1097	1097.551
L11	105—120	1692	1693.029
L12	121—132	1448	1448.694
L13(G)	133—146	1477	1477.735
L13(A)		1491	1491.750
L9a(AE)	83—93	1247	1247.597
L9b	94—95	261	261.132

TABLE 4 (continued)
Molecular Weight of the Peptides Generated by Cleavage with Trypsin, Lysyl Endopeptidase from *Achromobacter Lyticus Staphylococcus Aureus* V8 Protease, or Cyanogen Bromide

γ-Globin Chain[a]

V8 protease

Peptide no.	Sequence no.	Nominal wt.	Exact wt.
S1	1—5	589	589.250
S2	6—6	147	147.053
S3	7—21	1659	1659.883
S4	22—26	447	447.160
S5	27—90	6969	6972.712
S6(CM)	91—101	1362	1362.624
S7	102—121	2210	2211.278
S8	122—125	492	492.222
S9(G)	126—146	2304	2305.164
S9(A)		2318	2319.179

δ-Globin Chain[a]

Trypsin

Peptide no.	Sequence no.	Nominal wt.	Exact wt.
T1	1—8	951	951.503
T2	9—17	958	958.524
T3	18—30	1255	1255.652
T4	31—40	1273	1273.718
T5	41—59	2043	2043.925
T6	60—61	245	245.174
T7	62—65	411	411.223
T8	66—66	146	146.106
T9	67—82	1668	1668.884
T10(CM)	83—95	1521	1521.677
T11	96—104	1125	1125.557
T12(CM)	105—116	1326	1326.769
T13	117—120	464	464.238
T14	121—132	1440	1440.671
T15	133—144	1148	1148.667
T16	145—146	318	318.133
T10(AE)	83—95	1506	1506.714
T10a(AE)	83—93	1263	1263.592
T10b	94—95	261	261.132
T12a(AE)	105—112	872	872.515
T12b	113—116	457	457.301

Cyanogen bromide

Peptide no.	Sequence no.	Nominal wt.	Exact wt.
CB1	1—55	5982	5982.074
CB2(CM)	56—126	7879	7883.083
CB3	127—146	2137	2138.139

TABLE 4 (continued)
Molecular Weight of the Peptides Generated by Cleavage with Trypsin, Lysyl Endopeptidase from *Achromobacter Lyticus Staphylococcus Aureus* V8 Protease, or Cyanogen Bromide

δ-Globin Chain[a]

Lysyl endopeptidase

Peptide no.	Sequence no.	Nominal wt.	Exact wt.
L1	1—8	951	951.503
L2	9—17	958	958.524
L3	18—59	4535	4537.274
L4	60—61	245	245.174
L5	62—65	411	411.223
L6	66—66	146	146.106
L7	67—82	1668	1668.884
L8(CM)	83—95	1521	1521.677
L9(CM)	96—120	2879	2880.543
L10	121—132	1440	1440.671
L11	133—144	1148	1148.667
L12	145—146	318	318.133
L8a(AE)	83—93	1263	1263.592
L8b	94—95	261	261.132
L9a(AE)	96—112	1979	1980.062
L9b	113—120	903	903.529

V8 protease

Peptide no.	Sequence no.	Nominal wt.	Exact wt.
S1	1—6	694	694.365
S2	7—7	147	147.053
S3	8—26	1926	1927.016
S4	27—43	2093	2094.141
S5	44—90	4867	4869.476
S6(CM)	91—101	1362	1362.624
S7(CM	102—121	2318	2319.252
S8	122—146	2741	1742.406

Note: (CM) Cysteine residue is S-carboxymethylated. (AE) Cysteine residue is S-aminoethylated. A Peptides derived from Aγ. G Peptides derived from Gγ. Cleavage product at the S-aminoethylcysteinyl peptide bond is indicated by peptide number followed by "a" or "b".

From Wada, Y., Matsuo, T., and Sakurai, T., *Mass Spectrom. Rev.*, 8, 379, 1989. With permission.

The β-chain is used here as a model sample. The other chains were described in detail in Reference 2. A mass spectrum of a tryptic digest of the S-aminoethylated β-chain is shown in Figure 5. All tryptic peptides other than the small ones, T6 and T7, are identified. Other small peptides, T8 and T15, are detectable as combined peptides T8 + T9 and T14 + T15, respectively. Cysteine residues are present at positions 93 and 112 of the β-chain. S-aminoethylation allows the tryptic cleavage at the latter site and the release of two peptides from T12. On the other hand, the S-aminoethylcysteinyl bond at position 93 is resistant to the digestion, and thus an uncleaved

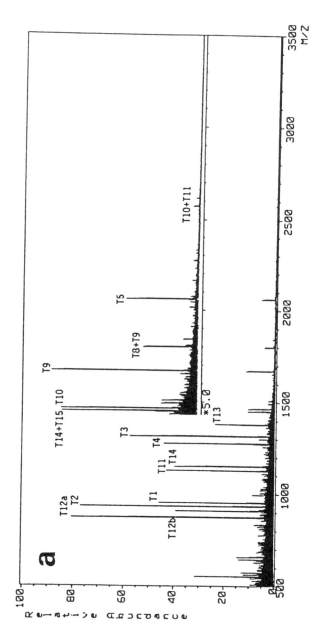

FIGURE 5. Mass spectra of tryptic peptides of S-aminoethylated (a) S-carboxymethylated (b) and unmodified (c) normal β-chains. (From Wada, Y., Matsuo, T., and Sakurai, T., *Mass Spectrom. Rev.*, 8, 379, 1989. With permission.)

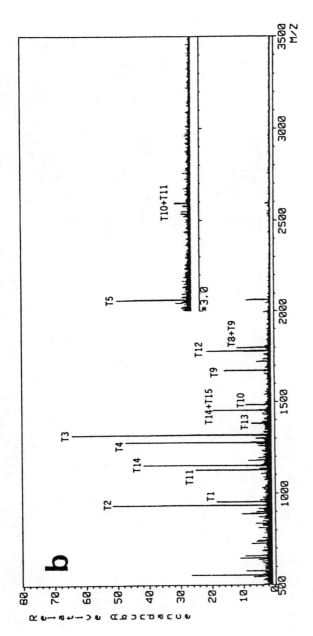

FIGURE 5 (continued).

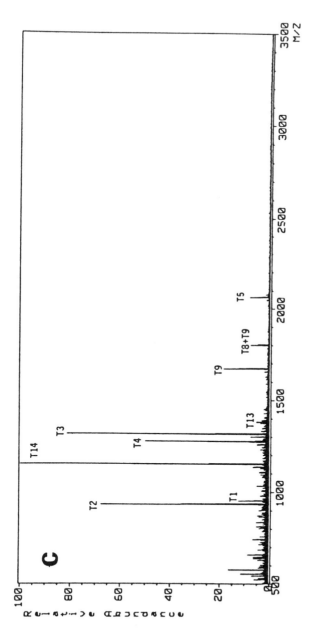

FIGURE 5 (continued).

peptide T10 is observed at m/z 1464.7. Another alkylation or S-carboxymethylation presents T10 and T12 at m/z 1478.7 and m/z 1777.0, respectively (Figure 5b). In the case of an unmodified chain, peptides T10, T11, and T12 are not observed (Figure 5c).

B. Hb F IZUMI (γ 6GLU →GLY)

This hemoglobin is the first variant characterized by the digit-printing using fast atom bombardment mass spectrometry (FABMS) or liquid secondary ion mass spectrometry (LSIMS). Moreover, the study typically presents the microanalysis by this strategy, because the γ-chain is a constituent of fetal hemoglobin, and the amount of material available is very limited.

This variant was identified at the cathodic side of normal γ-chain on electrophoresis, and thus an amino acid substitution of a neutral amino acid → basic, or an acidic → neutral was suggested. Approximately 1 mg of the abnormal γ-chain was obtained by ion-exchange chromatography.

The tryptic digest was analyzed by mass spectrometry. As shown in Figure 6, the (M+H)⁺ ion of peptide T1+T2 at m/z 1919 was missing, and a new ion appeared at m/z 1847. The decrease of 72 mass units indicates types of amino acid substitution of Glu → Gly or Trp → Asn from Table 2, and only the former fits the electrophoretic mobility. The normal peptide, however, contains two glutamic acid residues at the fifth and sixth positions. To determine the substituted position, an additional digestion with *Staphylococcus aureus* V8 protease was carried out on the tryptic digest. The peptides with the mol wt 646 and 1218 will appear if the substitution is at position 5, whereas the peptides with the mol wt 589 and 1275 will appear if the substitution is at position 6 (see the amino acid sequences in Figure 7). As shown in Figure 7, the ions at m/z 590 and m/z 1276 were identified in the mass spectrum of the abnormal chain, indicating the substitution at the 6th residue.

C. Hb F YAMAGUCHI (γ 80ASP→ ASN)

Digit-printing played an important role in the analysis of this γ-chain variant extracted from newborn blood dried onto a 10 mm disk of filter paper.

Electrophoresis indicated that the abnormal chain had a substitution of an acidic amino acid → neutral, or a basic → neutral. A very small amount of abnormal γ-chain was isolated from normal γ-chain, but not from normal β-chain, by chromatography. The mass spectrum of the tryptic digest presented in Figure 8 detected a shift of the peak for γ T10 from m/z 740 to m/z 739. The decrease of 1 mass unit indicates that either the 79th or 80th aspartic acid is replaced by asparagine. The position of substitution, however, remained undetermined.

The abnormal γ T10 was purified by reversed phase HPLC (RP HPLC); the identification of the abnormal peptide in the chromatogram was easy, because the abnormal peptide must appear a little later than the elution of the normal one (Figure 9). The substitution at the 80th residue was finally determined by a sequenator.

D. Hb PROVIDENCE (β82LYS→ ASN→ ASP)

This variant was characterized by a newly developed large-scale mass spectrometer.[6] The electrophoretic analysis of blood from a patient uniquely identified two β-chain variants, and they were isolated from the normal hemoglobin by chromatography. The purified chains were digested with trypsin and lysyl endopeptidase, then analyzed.

The mass spectra are shown in Figure 10. In either mass spectrum, peptide T9 at m/z 1669.9 and T10a at m/z 1221.6 are missing, and a new peak appears in the high mass region. These findings suggest that, in both cases, the mutation at lysine 82 prevents the cleavage at the peptide bond between peptides T9 and T10a. The large peptides must correspond to the residues 67 to 93, with an amino acid substitution at position 82. The precise mol wt 2857.4 and 2858.4, of the peptides found in the high mass region were determined. The molecular weights unequivocally indicated a Lys → Asn substitution for the former (Figure 10, top), and a Lys → Asp substitution

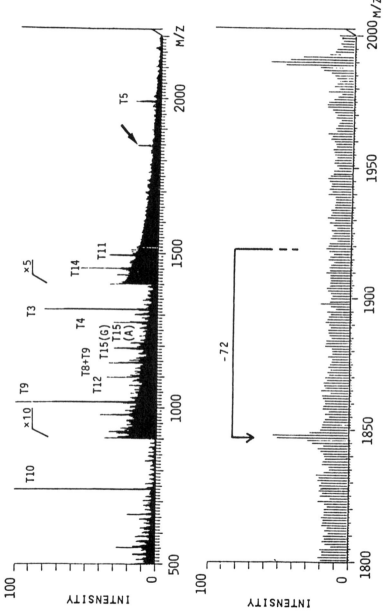

FIGURE 6. Mass spectrum of tryptic peptides of the γ-chain from Hb F Izumi. The arrow indicates an abnormal peak. (From Wada, Y., Matsuo, T., and Sakurai, T., *Mass Spectrom. Rev.*, 8, 379, 1989. With permission.)

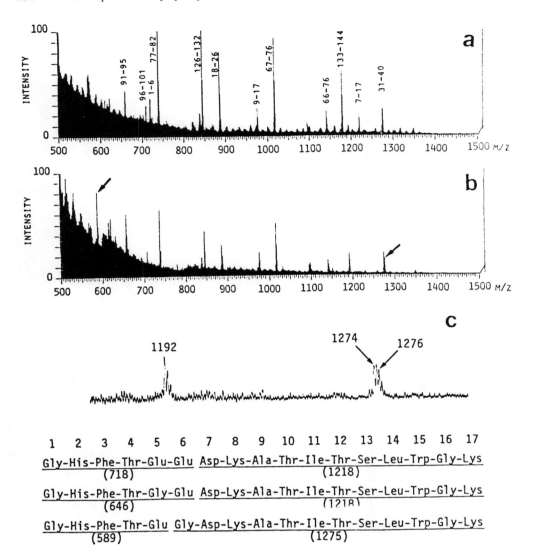

FIGURE 7.   Mass spectra of the digests with trypsin and V8 protease of normal γ-chain (a) and the γ-chain of Hb F Izumi (b and c). The arrows in spectrum b indicate abnormal peaks. The numbers given above the individual peaks in the upper mass spectrum are the sequence number. Amino acid sequences of normal peptide T1+2 (upper line), the peptide with 5Glu→Gly substitution (middle line), and the peptide with 6Glu→Gly substitution (lower line) are at the bottom. The spaces between residues indicate the site for the cleavage with V8 protease. (From Wada, Y., Matsuo, T., and Sakurai, T., *Mass Spectrom. Rev.*, 8, 379, 1989. With permission.)

for the latter (Figure 10, bottom). The presence of two substitutions in this case are explained by the deamidation process, which generates an aspartic acid residue from the substituted asparagine.

In general, the abundance of a protonated peptide in the mass spectrum of complex peptide mixtures is smaller than that given by the mass spectrum of a single isolated peptide. This fact limits the effective mass range and makes it difficult to determine the molecular weight of large peptides by using conventional mass spectrometers. In this study, a large mass spectrometer facilitated the precise measurement of the mass of the abnormal peptides.

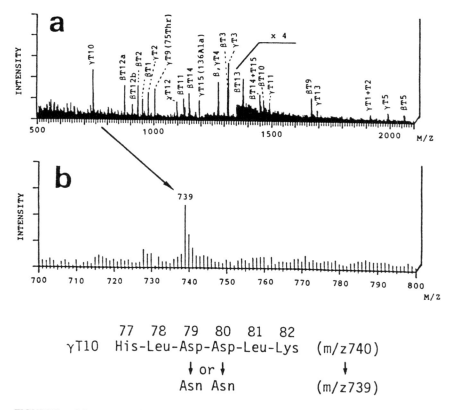

FIGURE 8.   Mass spectra of tryptic peptides of the non-α-chains (a mixture of β-and γ-chains) from the patient. The lower mass spectrum shows the region of m/z 700—800 of the upper mass spectrum. Amino acid sequence of γ T10 is given at the bottom. (From Wada, Y., Matsuo, T., and Sakurai, T., *Mass Spectrom. Rev.*,8, 379, 1989. With permission.)

E. SYNTHETIC VARIANTS (α42TYR→ PHE AND α42TYR→ HIS)

The development of genetic engineering enables us to synthetically make a mutation at a specific site in the globin chain. The mutants provide a great   deal of physicochemical information about the roles of the normal residues in the quaternary structure of hemoglobin molecule as well as about the effects of the substitutions on the molecular function. In the course of the investigation, the structure of the synthesized variant must be verified. For this purpose, mass spectrometry is very useful because of rapid characterization. For example, the mass spectra of tryptic digests of mutant globins in Figure 11 proved two new types of substitutions, α42Tyr→ Phe and α42Tyr→ His, created by Imai et al. at Osaka University.

F. CALMODULIN

The analysis of calmodulin, an intracellular protein composed of 148 amino acid residues, was carried out for the purpose of the assignment of amides in its sequence. The assignments, namely the determinations whether aspartic acid or asparagine and glutamic acid or glutamine, at positions 24, 60, 129, and 135, were not consistent among reports (see the amino acid sequences in Figure 12). As described previously, the difference in the molecular weight of acid and amide forms is 1 mass unit. In this study, we amplified the difference by glycinamidation[7] before tryptic digestion or cleavage with cyanogen bromide. The modification adds 56 mass units at every free carboxyl side group in the peptide.

For example, residues 129 and 135 are in the cyanogen bromide peptide CB8, which contains

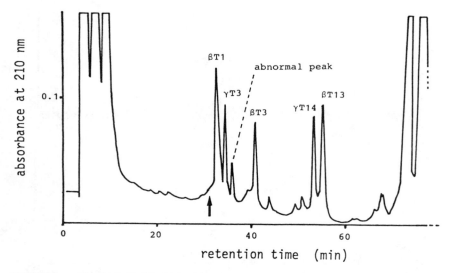

FIGURE 9. The first half of the high performance liquid chromatogram of tryptic peptides of non-α-chains from the patient. The normal γ T10 comes at the position indicated by an arrow.

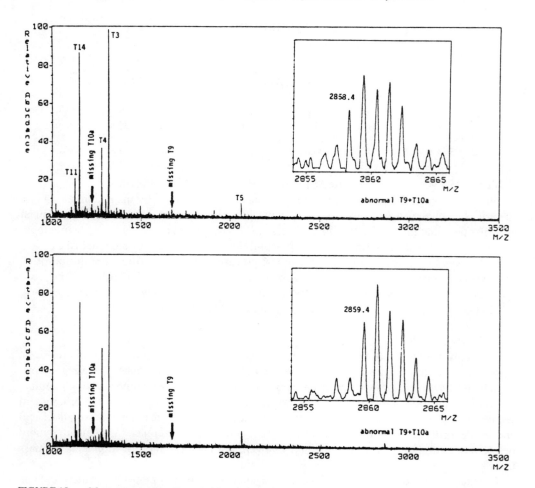

FIGURE 10. Mass spectra of the digests with trypsin and lysyl endopeptidase of the two abnormal β-chains from the patient. (From Wada, Y., Matsuo, T., and Sakurai, T., *Mass Spectrom. Rev.*, 8, 379, 1989. With permission.)

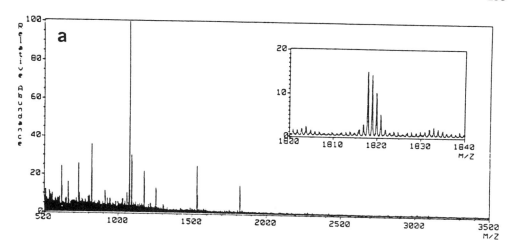

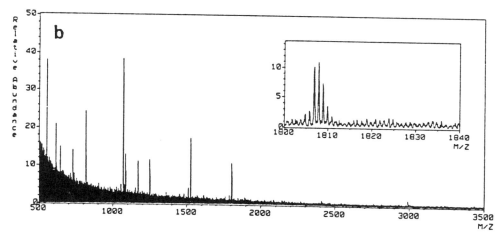

FIGURE 11.   Mass spectra of tryptic digests of synthetic variants α42Tyr→Phe (a) and α42Tyr→His (b). (From Wada, Y., Matsuo, T., and Sakurai, T., *Mass Spectrom. Rev.*, 8, 379, 1989. With permission.)

nine residues in acidic or amide form. The increase of 336 mass units by the glycinamidation of this peptide indicates that 6 of these 9 residues are in the acidic form (Figures 12c and 12d). Because 5 residues (127, 131, 133, 139, and 140) had been identified as being in the acidic form, it followed that either residue 129 or 135 must be in the amide form. The answer was provided by a combination of Edman degradation and mass spectrometry. After four degradation cycles, residues 129 to 144 were detected at m/z 1810 (Figure 13a). The subsequent cycle eliminated residue 129, and residues 130 to 144 were detected at m/z 1695 (Figure 13b). The decrease of 115 mass unit indicates that residue 129 is aspartic acid. Consequently, residue 135 is glutamine. Similar analyses on two other peptides (T3 and CB3) concluded the acid or amide forms at the residues in question.

G. PREALBUMIN VARIANT

Familial amyloidotic polyneuropathy (FAP) is an adult-onset, fatal neurologic disease of dominant inheritance. On the other hand, prealbumin is a serum protein with an unknown physiological role composed of 127 amino acid residues. A variant prealbumin, in which valine is replaced by methionine at position 30, coexists with the normal counterpart in the blood of FAP patients (see Figure 14, bottom). The molecular abnormality is detectable, even in the presymptomatic stage, and is diagnostic of FAP.

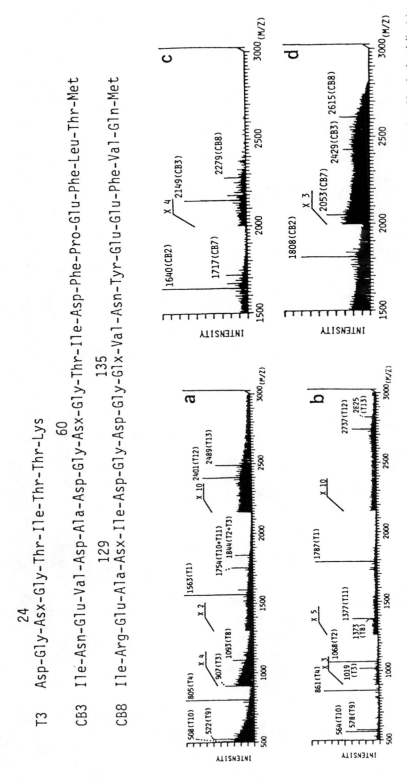

FIGURE 12. Mass spectra of peptide mixtures from bovine brain calmodulin. (a) Tryptic digest of unglycinamidated calmodulin. (b) Tryptic digest of glycinamidated calmodulin. (c) Tryptic digest of glycinamidated calmodulin. (d) Cyanogen bromide peptides of unglycinamidated calmodulin. (d) Cyanogen bromide peptides of glycinamidated calmodulin. The determined m/z values are given above individual peaks. The amino acid sequences shown at the top are those of the peptides containing the residues in dispute. "Asx" is aspartic acid or asparagine residue. "Glx" is glutamic acid or glutamine residue. (From Wada, Y., Matsuo, T., and Sakurai, T., *Mass Spectrom. Rev.*, 8, 379, 1989. With permission.)

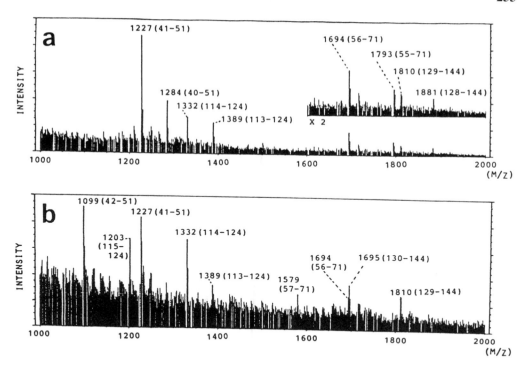

FIGURE 13.   Mass spectra of the residual peptides after four (a) and five (b) cycles of Edman degradation of cyanogen bromide peptides from bovine brain calmodulin. The m/z values are given above individual peaks, and the sequence numbers are in parentheses. (From Wada, Y., Matsuo, T., and Sakurai, T., *Mass Spectrom. Rev.*, 8, 379, 1989. With permission.)

The mass spectrum of the tryptic digest identified the peak at m/z 1398 near the normal peak of T4 at m/z 1366 (Figure 14a). The Val → Met substitution in this peptide was subsequently confirmed by a specific modification of methionine: treatment of a protein with performic acid changes methionine residues to methionine sulfone residues. As was expected, the peptide containing the substituted methionine gained 32 mass units, whereas the normal peptide containing no methionine residues was unchanged (Figure 14b). This analysis is a diagnostic use of digit-printing.

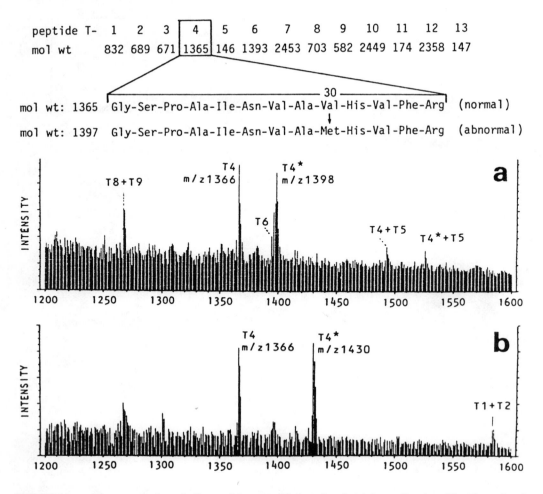

FIGURE 14.   Mass spectra of tryptic digests of the unmodified prealbumin (a) the prealbumin oxidized by performic acid (b) from the patient. * indicates the abnormal peptide T4. The molecular abnormality of the variant prealbumin is shown at the top. (From Wada, Y., Matsuo, T., and Sakurai, T., *Mass Spectrom. Rev.*, 8, 379, 1989. With permission.)

REFERENCES

1. **Wada, Y., Hayashi, A., Fujimura, M., Katakuse, I., Ichihara, T., Nakabushi, H., Matsuo, T., Sakurai, T., and Matsuda, H.,** Characterization of a new fetal hemoglobin variant, Hb F Izumi Aγ6Glu→Gly by molecular secondary ion mass spectrometry, *Biochim. Biophys. Acta,* 749, 244, 1983.
2. **Wada, Y., Matsuo, T., and Sakurai, T.,** Structure elucidation of hemoglobin variants and other proteins by digit-printing method, *Mass Spectrom. Rev.,* 8, 379, 1989.
3. **Matsuda, H.,** High-resolution, high-sensitivity mass spectrometers, *Mass Spectrom. Rev.,* 2, 299, 1983.
4. **Matsuo, T., Matsuda, H., Fujita, Y., and Wollnik, H.,** Computer program "TRIO" for third order calculation of ion trajectory, *Mass Spectrosc. Jpn.,* 24, 19, 1976.
5. **Matsuda, H.,** High-resolution, high transmission mass spectrometer, *Int. J. Mass Spectrom. Ion Proc.,* 66, 209, 1985.
6. **Matsuda, H., Matsuo, T., Fujita, Y., Sakurai, T., and Katakuse, I.,** A new mass spectrometer at Osaka University, *Int. J. Mass Spectrom. Ion Proc.,* 91, 1, 1989.
7. **Wada, Y., Hayashi, A., Katakuse, I., Matsuo, T., and Matsuda, H.,** Application of glycinamidation to the peptide mapping using secondary ion mass spectrometry, *Biomed. Mass Spectrom.,* 12, 122, 1985.

Chapter 13

THE MASS SPECTRAL ANALYSIS OF HEMOGLOBIN VARIANTS

Terry D. Lee and Samuel Rahbar

TABLE OF CONTENTS

I.   Introduction 258

II.   Hemoglobin Structure 258
      A.   Normal Hemoglobin 258
      B.   Abnormal Hemoglobin 259

III.   Analytical Methods 261

IV.   Mass Spectral Methods 263
      A.   Background 263
      B.   Analysis of Tryptic Digest Mixtures 263
      C.   Analysis of Isolated Peptides 269
      D.   Analysis of Intact Globin Chains 271

V.   Conclusions 272

References 273

I. INTRODUCTION

Probably no protein has been studied and characterized more thoroughly than hemoglobin. The well-defined function and availability of nearly unlimited quantities have made hemoglobin the ideal choice for structure-function studies. Even more importantly, hemoglobin is an essential component of human physiology. Abnormalities in hemoglobin structure occur frequently through mutations of the genes responsible for the coding and/or regulation of its synthesis. Worldwide, it is estimated that there are 200 million carriers of an abnormal globin gene. Many of these mutations result in clinical manifestations and hundreds of abnormal variants have been characterized so far. Whereas nothing can be done at this time to cure a patient, it is nevertheless important that the cause of a patient's symptoms be determined to avoid misdiagnosis and inappropriate treatment.

A number of techniques has been developed over the years to identify and characterize hemoglobin variants. Recent developments in mass spectral methods of analysis have provided a new and versatile set of tools for protein structural analysis. Mass spectrometry should now be considered a nearly indispensable tool for determining abnormal hemoglobin structures. Mass spectral techniques have the advantage of short analysis times, generally simple sample preparation, and provide a definitive physical measurement that correlates directly with a structural abnormality. Mass spectral methods are particularly useful for variants that are electrophoretically "silent", or difficult to separate by chromatographic methods.

As useful as mass spectrometry has become, suitable instrumentation is not available in most clinical environments. Physicians and protein chemists often have little knowledge of the nature of methods and instrumentation. At the same time, mass spectrometrists often have little knowledge or appreciation of the complexities of protein structural analysis. By itself, mass spectrometry is not sufficient to characterize fully an abnormal hemoglobin, but must be used in conjunction with other methods such as chromatographic separations, chemical and enzymatic degradations, derivatization, and protein sequence analysis. Consequently, a considerable gap must often be bridged between those researchers in need of mass spectral support and those able to provide it.

The main purpose of this chapter is to describe the recent progress that has been made in the analysis of hemoglobin variants by mass spectral methods, and to collect in one place basic information on various aspects of this field of research. Although the emphasis is on hemoglobin, the general techniques presented can be applied to the analysis of other variant proteins.

II. HEMOGLOBIN STRUCTURE

A. NORMAL HEMOGLOBIN

For a thorough treatment of the molecular, genetic, and clinical aspects of hemoglobin, the reader is referred to the excellent work of Bunn and Forget.[1] Only a brief summary of hemoglobin structure and different types of abnormalities is included here.

Hemoglobin is a globular protein having a molecular weight of approximately 62,000 Da. It consists of four subunits or chains of nearly equal size. Each subunit contains one heme group, which consists of a porphyrin ring with one atom of iron in the center. Although the heme group is the moiety that binds oxygen, it is the variations in the protein or globin part of the molecule that result in hemoglobin variants.

With the advent of DNA sequencing technology, the genetic basis for the variety of hemoglobin structures is well understood. There are two families of genes that code for the two types of chains found in the globin portion of the molecule. The α-chain family[2] found on chromosome 16 consists of a segment coding for the embryonic globin, zeta (ζ)-chain, two copies of the segment that codes for the α-chain, an incomplete α-chain gene referred to as pseudo-alpha (φα), and the recently described theta (θ)-chain gene.[3] The β-chain family is found on chromosome 11, and consists of single segments for the embryonic ε-gene, the β-gene, the δ-gene, a pseudo-beta (φβ) gene, and two different genes for the γ-chain. The expression of the various chains changes during the course of human development. The pseudo-genes are incomplete and not expressed. The θ-chain appears to be complete, but a functional protein has yet to be characterized.

The functional hemoglobin unit is a tetramer formed by two members of the α-chain family and two members from the β-chain family. The ζ- and ε-chains appear first, and together they form embryonic hemoglobin, which decreases generally to undetectable levels by the tenth week of gestation. As production of the ε- and ζ-chains decrease, they are replaced by the α- and γ-chains, which together form fetal hemoglobin (Hb F, $\alpha_2\gamma_2$). The production of the γ-chain starts to decrease before birth, to be replaced by the β-chain, which when combined with the α–chain, forms normal adult hemoglobin (Hb A, $\alpha_2\beta_2$). Fetal hemoglobin accounts for 60% of the total at birth, and falls to less than 1% during the second year after birth. Production of δ-chain begins shortly after birth, and together with α-chain forms Hb A2 ($\alpha_2\delta_2$). Levels of Hb A2 normally do not exceed a few percent. Consequently, Hb A accounts for >97% of the hemoglobin found in normal individuals, and most of the known variants arise by alteration of either the α- or β-chain. Several γ- and δ-chain variants have also been described recently.

The different hemoglobin chains are very similar in overall structure, differing primarily in the composition and sequence of the amino acid residues (Figure 1). Indeed, all chains are thought to be derived from the same primitive precursor gene. The tetrameric structure of hemoglobin is essential for its oxygen-carrying function. Each of the various hemoglobins has different properties. In the absence of α-chains, β- and γ-chains can form the necessary tetrameric structure (Hb H, β^4 and Hb Barts, γ^4); an ability that is totally lacking in the α-chain.

B. ABNORMAL HEMOGLOBIN

Abnormalities in hemoglobin, known as hemoglobinopathies, can occur as the result of mutations in the DNA sequences responsible for the structure and expression of the globin chains. A single base change in a globin gene can change the codon for one amino acid to that of another. Most often, the result is a single amino acid substitution in the synthesized protein molecule. Single amino acid replacements account for approximately 95% of the known hemoglobin variants. Although the change of 1 amino acid out of 141 (α-chain) or 146 (β-chain) may seem to be a small difference, it can have a profound effect on the function or stability of the hemoglobin molecule if it occurs in a region associated with heme binding or interaction between the chains. The first point mutation variant (which has a Glu replaced by Val (Glu > Val) in position 6 of the β-chain) to be determined was for sickle cell anemia.[4]

Other mutations have a more dramatic effect on the sequence of a chain. If the mutation occurs in one of the bases for the stop codon, which signals the termination of the polypeptide chain, it can be converted to an amino acid codon. The synthesized protein will have a C-terminal extension, as is the case with Hb Constant Spring[5] and several other highly unstable hemoglobin variants.[6] Insertions or deletions of one or more bases can result in shortened chains, elongated chains, or a shift in the reading frame over a portion of the sequence. Fusion subunits result from chromosomal crossover during meiosis. Known examples are members of the β-chain family, which are found close together on the same chromosome. The various Hb Lepore variants have abnormal chains with an N-terminal portion of the normal δ-chain combined with the C-terminal portion of normal β-chain.[7]

DNA or RNA changes that result in the reduction or absence of either the α- or β-chain are known as thalassemia syndromes. This change can occur due to alterations in transcription or processing of the globin messenger RNAs because of point mutations or outright deletions of portions of the globin genome. Deficient production of Hb A often results in greater than normal levels of Hb F and other minor hemoglobins. The exact cause of thalassemia is sometimes difficult to determine. Progress in DNA technology has provided techniques to identify different

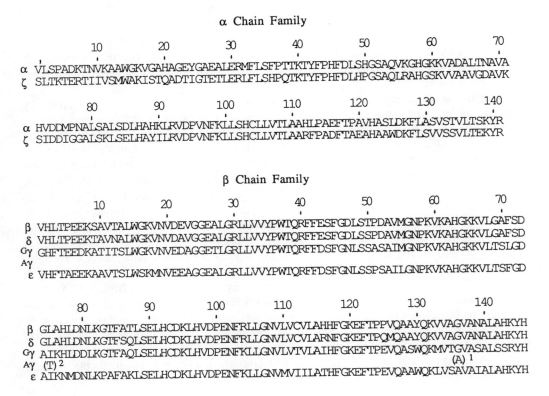

α Chain Family

```
         10        20        30        40        50        60        70
α  VLSPADKTNVKAAWGKVGAHAGEYGAEALERMFLSFPTTKTYFPHFDLSHGSAQVKGHGKKVADALTNAVA
ζ  SLTKTERTIIVSMWAKISTQADTIGTETLERLFLSHPQTKTYFPHFDLHPGSAQLRAHGSKVVAAVGDAVK

         80        90        100       110       120       130       140
α  HVDDMPNALSALSDLHAHKLRVDPVNFKLLSHCLLVTLAAHLPAEFTPAVHASLDKFLASVSTVLTSKYR
ζ  SIDDIGGALSKLSELHAYIILRVDPVNFKLLSHCLLVTLAARFPADFTAEAHAAWDKFLSVVSSVLTEKYR
```

β Chain Family

```
          10        20        30        40        50        60        70
β   VHLTPEEKSAVTALWGKVNVDEVGGEALGRLLVVYPWTQRFFESFGDLSTPDAVMGNPKVKAHGKKVLGAFSD
δ   VHLTPEEKTAVNALWGKVNVDAVGGEALGRLLVVYPWTQRFFESFGDLSSPDAVMGNPKVKAHGKKVLGAFSD
Gγ  GHFTEEDKATITSLWGKVNVEDAGGETLGRLLVVYPWTQRFFDSFGNLSSASAIMGNPKVKAHGKKVLTSLGD
Aγ
ε   VHFTAEEKAAVTSLWSKMNVEEAGGEALGRLLVVYPWTQRFFDSFGNLSSPSAILGNPKVKAHGKKVLTSFGD
```

```
          80        90        100       110       120       130       140
β   GLAHLDNLKGTFATLSELHCDKLHVDPENFRLLGNVLVCVLAHHFGKEFTPPVQAAYQKVVAGVANALAHKYH
δ   GLAHLDNLKGTFSQLSELHCDKLHVDPENFRLLGNVLVCVLARNFGKEFTPQMQAAYQKVVAGVANALAHKYH
Gγ  AIKHLDDLKGTFAQLSELHCDKLHVDPENFKLLGNVLVTVLAIHFGKEFTPEVQASWQKMVTGVASALSSRYH
Aγ  (T)²                                                           (A)¹
ε   AIKNMDNLKPAFAKLSELHCDKLHVDPENFKLLGNVMVIILATHFGKEFTPEVQAAWQKLVSAVAIALAHKYH
```

¹ Two genes code for the γ chain, one having Gly in position 136 (Gγ) and the other having Ala (Aγ).

² A γ chain variant of the chain with Ala 136 in 20-30% of newborns has Thr in position 75.

FIGURE 1.   Primary structure of human hemoglobin α-, ζ-, β-, δ-, γ-, and ε- chains.

thalassemias at the molecular level. More than 40 different point mutations and a number of deletions have been found responsible for different types of β-thalassemia in different populations of the world. A number of hemoglobin variants such as Hb Constant Spring, and Hb Knossus,[8] and Hb E have been reported to have the thalassemic phenotypes; namely, their abnormal chain is synthesized in a much lower amount than their normal counterparts.

Changes in hemoglobin structure can also result from chemical modifications of the normal globin chains. These modifications include glycosylation, carbamylation, and acetylation along with other covalent adducts. Modified globin chains are found in low levels in all individuals, but increased levels can be found when abnormal physiological conditions exist. The most noteworthy example is that of glycosylated hemoglobins, which become much more pronounced in individuals with diabetes mellitus.[9] A good example of a normal post-translational modification is acetylation of the amino terminus of γ-chains.

If an abnormality is inherited from only one parent, the individual is heterozygous with respect to that variation. If inherited from both parents, then the individual is homozygous. An important aspect to characterizing a hemoglobin variant from an individual is the characterization of the hemoglobins from close family members. Although rare, it is possible for an individual to inherit more than one abnormality. Two mutations, which by themselves have no clinical manifestations, can be quite serious when they occur together. The amount of abnormal hemoglobin present is generally less than the amount of normal hemoglobin. Normally, four copies of the α-chain gene (two on each chromosome) exist. If they are equally expressed and stable, the amount of abnormal hemoglobin present is 25%. For the β-chain, only two genes occur. The expected amount of a β-chain variant would be 50%. Actual amounts can be considerably less.

III. ANALYTICAL METHODS

In addition to the mass spectral methods covered below, a number of other analytical tools are very useful for characterization of hemoglobin variants, including electrophoresis, high performance liquid chromatography (HPLC), amino acid analysis, and automated microsequencing. Ever since Linus Pauling demonstrated in 1949 that the abnormal hemoglobin responsible for sickle cell anemia could be separated by electrophoresis, this technique has been used both as a routine screening method to detect abnormal hemoglobins and as a separation step to provide purified material for subsequent characterization. Protein mobilities during electrophoresis are dependant on the balance of positive and negative charges associated with the various ionizable functional groups. Amino acid substitutions that involve ionizable amino acid residues (Asp, Glu, Lys, Arg, Trp, His) will generally alter the electrophoretic mobility of the intact hemoglobin. On a practical level, variants can be divided into two general categories: those that have abnormal electrophoretic mobilities and those that are electrophoretically silent. Electrophoretic separations can be done in a variety of media but cellulose acetate is probably the most versatile.[10] Isoelectric focusing is a technique that separates components with different isoelectric points using gels with pH gradients formed by ampholyte mixtures. Immobilized pH gradients permit preparative scale electrophoresis over very narrow pH ranges. Although electrophoresis cannot determine the exact substitution involved, the data can be very useful for deciding between different possibilities for interpreting the mass spectral data.

Tests for hemoglobin instability are not only used to detect abnormalities, but can also serve as a means to provide preparations that are enriched in the relative amounts of abnormal components for subsequent characterization. The test for instability relies on precipitation using heat[11] or a chemical agent such as isopropyl alcohol.[12]

Separation of the abnormal globin chain from the normal chains, although not necessary, is often very desirable. Electrophoretic separations are possible on a preparative level, but it is easier to rely generally on ion-exchange chromatography[10] or HPLC methods. Using reversed-phase HPLC (RP HPLC), it is possible to separate all of the normal globin chains from hemolysates (Figure 2).[13] It is also possible to separate most abnormal chains that appear as extra peaks in the chromatogram. In addition, alterations in the relative amounts of the various chains are evident.

Although much of the mass spectral characterization of the hemoglobin variants can be done directly on the mixture created by enzymatic and chemical degradation (see later in chapter), HPLC purification of the abnormal fragments is essential for subsequent characterization by amino acid analysis and automated Edman microsequencing. Mass spectral analysis is also simplified if the peptide of interest is separated from the rest of the mixture. The separation of tryptic digest fragments by RP HPLC is generally straightforward (Figure 3).[14] HPLC-mapping of the variations in the digest mixture is often a major step in isolating fragments of a hemoglobin present in limited amounts. The full power of structural characterization methods can be brought to bear on small (<3500 amu), isolated peptides.

For proteins of known structure such as hemoglobin, amino acid analysis is often sufficient to define single amino acid substitutions in isolated peptides. With the rapid development of mass spectrometry and microsequencing, the trend in recent years has been to rely less and less on amino acid analysis. This is particularly true in those instances when sample amounts are very limited. Good microsequencing data can provide both composition and sequence information.

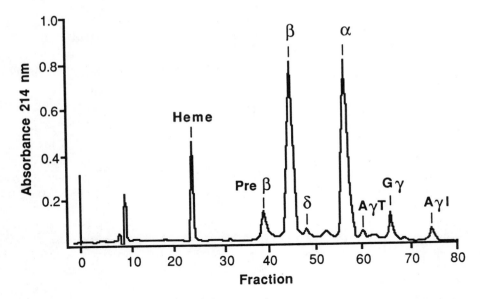

FIGURE 2.   RP HPLC separation of globin chains using a large pore Vydac C4 column after the method of Schroeder et al.[13]

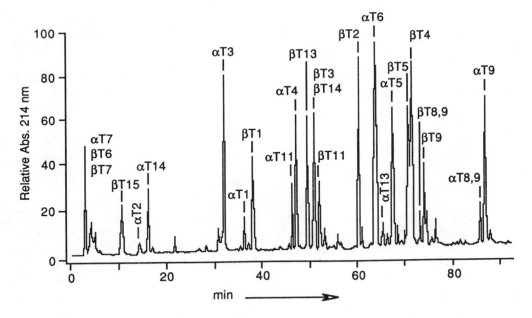

FIGURE 3.   RP HPLC separation of the tryptic digest mixture of whole globin from Hb A on a Vydac C18 after the method of Rahbar et al.[14]

However, microsequencing support may not always be available, and the possibility of solving the problem by alternative methods should not be rejected out-of-hand. The capacity of mass spectrometry to solve variant structures without assistance from microsequencing and amino acid analysis will increase. However, the complementary nature of mass spectral, Edman microsequencing, and amino acid analysis data cannot be overemphasized. Researchers who limit their choices of analytical tools run the risk of making their problems much more difficult to solve.

IV. MASS SPECTRAL METHODS

A. BACKGROUND

Pioneering work by Wada and co-workers[15,16] using field desorption (FD) mass spectrometry established the feasibility of identifying hemoglobin variants by mass spectral methods. The development of liquid secondary ion mass spectrometry (LSIMS), also known as fast atom bombardment (FAB), provided a much easier means of obtaining molecular weight and fragment ion information for peptides. The first example of the use of mass spectrometry to identify an unknown hemoglobin variant was the characterization of Hb F Izumi (A_γ 6 Glu > Gly).[17] This work established clearly the power of the technique, particularly with respect to the speed of the analysis and the small amount of sample required.

Mass spectral analysis is most applicable to those hemoglobinopathies where the sequence of one or more of the chains has been altered by a mutation in the gene. In those instances when the abnormality is due to a decreased level of chain synthesis, mass spectrometry can only confirm that the protein present has the expected sequence. The distinction between reduced levels of a particular chain caused by impaired synthesis and those caused by a rapidly degraded unstable variant is often difficult to make.

The power of mass spectral analysis for the efficient characterization of hemoglobin variants is most obvious in those cases where the abnormal chain or whole globin can be separated from the normal protein by electrophoresis or chromatography. The alteration of the normal structure has already been established and all that remains is to determine the exact nature of the structural difference. Most variants are the result of a single amino acid substitution in the peptide sequence. This change generally results in a mass difference (Table 1), except in those instances in which the two amino acids happen to have the same nominal mass value (i.e., Leu, Ile and Lys, Gln).

For the most part, analysis of mass spectral data involves comparisons of observed protonated molecular ion values with those predicted from fragments of the normal chains. These comparisons are made easier by computer programs that facilitate the various calculations needed. Computer programs that perform a variety of useful calculations are available from commercial[18] and other sources.[19-21]

B. ANALYSIS OF TRYPTIC DIGEST MIXTURES

The most common means of producing convenient-sized pieces of hemoglobin chains for structural analysis is digestion with the endoprotease trypsin. In favorable instances, a single tryptic digest is sufficient to solve the problem. A case in point is that of Hb Koln.[22] The mass spectrum of the tryptic digest mixture (Figure 4) shows a clear absence of the protonated molecular ion for tryptic fragment #11 (m/z 1127) and the appearance of a new peak at m/z 1159. Three single amino acid substitutions can account for the difference of 32 amu: Pro > Glu, Asp > Phe, and Val > Met. The first two possibilities can be excluded on the basis that both would require more than one base change in the DNA sequence, and that both would have different electrophoretic mobilities than normal hemoglobin. This variant is electrophoretically silent.

Trypsin cleaves generally after Lys and Arg. However, the presence of Pro on the C-terminal side usually prevents cleavage, and some sites are prone to incomplete digestion resulting in combined fragments. The fragments and corresponding mass values observed in the tryptic digest mixtures for the α-chain are given in Table 2, and those for the β-chain are given in Table 3. The relative intensities of the molecular ions in the mass spectrum are dependent on a number of factors other than concentration, and cannot be used to quantitate the amount of a given fragment. Generally speaking, the small fragments having only a few residues are difficult to detect, due in part to reduced sensitivity and interference from fragment ions from the larger peptides and from matrix ions. The appearance of the spectrum will vary greatly depending on the matrix used and the history of the sample, and considerable variation can exist from scan to

TABLE 1
Possible Amino Acid Substitutions Corresponding to Nominal Mass Differences between Normal and Variant Peptides

ΔM	Possible single amino acid substitutions	ΔM	Possible single amino acid substitutions
-129	W > G	-35	Y > Q Y > K
-115	W > A	-34	**F > L F > I** H > C Y > E M > P
-106	Y > G	-33	F > N
-99	**R > G** W > S	-32	**M > V** C > A F > D E > P Y > M
-92	Y > A	-31	**Q > P** K > P
-90	F > G	-30	**T > A** W > R M > T S > G E > V
-89	W > P	-29	Q > V K > V
-87	W > V	-28	**V > A R > Q** R > K D > S M > C E > T
-85	R > A W > T	-27	**N > S** K > T R > E Q > T
-83	**W > C**	-26	**P > A L > S I > S** Y > H E > C
-80	H > G	-25	**R > M** Q > C K > C
-76	**Y > S** F > A	-24	**H > L** H > I
-74	M > G	-23	**H > N** W > Y
-73	**W > L** W > I	-22	**H > D**
-72	**E > G** W > N	-19	**R > H** F > K F > Q
-71	W > D K > G Q > G	-18	**M > L M > I** D > P F > E
-69	**R > S**	-17	**N > P** M > N
-66	H > A Y > P	-16	**S > A D > V Y > F L > P C > S** I > P E > L E > I F > M M > D
-64	Y > V	-15	**K > I Q > L** Q > I K > L E > N N > V
-62	Y > T	-14	**A > G K > N L > V I > V E > D** T > S Q > N D > T
-60	**Y > C F > S** M > A	-13	**N > T** K > D Q > D
-59	**R > P**	-12	**I > T** D > C L > T V > S
-58	**E > A D > G** W > Q W > K	-11	**N > C**
-57	Q > A K > A R > V N > G W > E	-10	**P > S** L > C I > C F > H
-56	L > G I > G	-9	**H > Q** R > F H > K
-55	**R > T** W > M	-8	H > E
-53	**R > C**	-7	Y > R
-50	H > S Y > L Y > I F > P	-6	C > P H > M
-49	**Y > N** W > H	-4	T > P C > V
-48	**Y > D F > V**	-3	**M > K** M > Q
-46	**C > G** F > T	-2	**D > L** D > I C > T M > E V > P T > V
-44	**D > A F > C** T > G M > S	-1	**D > N** N > I E > Q E > K N > L
-43	**R > L R > I** N > A	0	**K = Q L = I**
-42	**V > G** L > A I > A R > N E > S		
-41	R > D Q > S K > S		
-40	**H > P** P > G		
-39	W > F		
-38	H > V		
-36	H > T		

Note: Substitutions in boldface type are possible with only one base change in the DNA code. The symbol ">" should be read as "replaced by".

scan in the relative intensities of the protonated molecular ions.[23] It is recommended that investigators begin with digests of normal hemoglobin to familiarize themselves with the characteristics of their particular system.

During trypsin digestion, a portion of a globin chain, known as the hydrophobic core, precipitates. As a consequence, the core peptide is generally not observed in a tryptic digest mixture. This can be prevented by derivitizing the cysteine residues. Changes in the mass values for the tryptic peptides induced by the more common Cys derivatization reagents are also given in Tables 2 and 3. Performic acid oxidation, which is very useful for converting Cys to the more

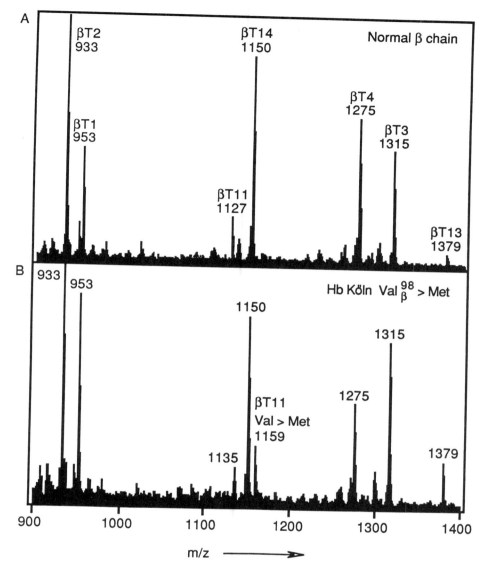

FIGURE 4.   Positive ion FAB spectrum of (A) normal and (B) abnormal (Hb Koln) β-chain tryptic digest mixture. Spectra were obtained from 2 ml of a 1-mg/ml solution of the tryptic digest mixture in a glycerol matrix using a JEOL HX100HF mass spectrometer. The sample was ionized with a beam of Xe atoms having 6 keV translational energy. Spectra were collected over the mass range of 200 to 3500 with a total cycle time of 45s.

hydrophilic cysteic acid group, degrades Trp, and converts Met to the methionine sulfone derivative. Care must also be taken to avoid chlorination of Tyr residues when using this reagent. One group has reported that commercial sources of trypsin, which are generally contaminated with small amounts of chymotrypsin, are useful for generating the fragments that include the core peptide for the β-chain.[24] More consistent results are likely to be achieved using trypsin that has been HPLC purified. It should also be noted that proteases will act on themselves. Peptides present in small amounts in digest mixtures may well be fragments of trypsin rather than a minor variant component.

In some cases, the change of a single amino acid can greatly alter the appearance of the FAB-map of the tryptic digest mixture. Such is the case for Hb Pasadena.[25] The mass spectrum of the

TABLE 2
Masses for Protonated Molecular Ions for Hemoglobin α-Chain Tryptic Fragments
(Including Oxidized, Aminoethylated, Pyridinylethylated, and
Carboxymethylated Forms)

Tryptic fragment no.	Position no.	Mass	Oxidized mass	Amino-ethylated mass	Pyridinyl-ethylated mass	Carboxy-methylated mass
8	61	147.11[a]				
10	91—92	288.20				
14	140—141	338.18[a]				
7	57—60	398.22				
2	8—11	461.27				
3	12—16	532.29	c			
1	1—7	729.41				
11	93—99	818.44				
5	32—40	1071.55	1103.54			
13	128—139	1252.72				
4	17—31	1529.73				
6	41—56	1833.89				
12	100—127	2967.61[b]	3015.60		3072.67	3025.62
12a	100—104			615.33[d]		
12b	105—127			2414.34[d]		
9	62—90	2996.49	3028.48			

Frequently Observed Peptides Resulting from Incomplete Digestion

7, 8	57—61	526.31				
10, 11	91—99	1087.63				
1, 2	1—11	1171.67				
8, 9	61—90'	3124.58				

[a]   Generally not observed due to interference from sample and matrix background.
[b]   Core peptide, not soluble under normal conditions.
[c]   Presence of Trp results in extensive degradation during oxidation.
[d]   Peptides resulting from cleavage at aminoethyl-Cys.

isolated abnormal β-chain digest mixture has 3 new ions (m/z 810, 921, and 1049 [Figure 5]). Mass differences between these new ions and the normal tryptic fragments do not equal any single amino acid substitutions except for the substitution of Pro for Lys in fragment T1. This substitution can be disregarded because it would require a minimum of two base changes in the DNA sequence and would remove the tryptic cleavage site. A computer search of the β-chain sequence matched the ion at m/z 810 and residues 76 to 82 corresponding to the C-terminal portion of fragment T9 (T9b in Figure 5), suggesting the creation of a new tryptic cleavage site. Substitution of Arg for Leu 75 yields the expected mass of 921 for the N-terminal portion of fragment T9 (T9a in Figure 5). The ion at 1049 is now readily explained by the addition of Lys to fragment T9a (T8,9a Figure 5), the result of incomplete digestion. With the proper computer help, analysis of the mass spectral data can be done quickly and the sequence of the abnormal chain is fully determined.

If the single amino acid substitution results in the elimination of a tryptic cleavage site, the spectrum of the digest mixture will show a new ion with a mass corresponding to the molecular weight of the two fragments minus a water molecule plus the mass difference for the substitution. The mass spectral analysis of such a variant has been described by Prome et al.[23]

Separation of the abnormal chain is not always necessary. The tryptic digest of oxidized whole globin from Hb Setif[26] (Figure 6) has an additional molecular ion at m/z 866. The intensity

TABLE 3
Masses for Protonated Molecular Ions for Hemoglobin β-Chain Tryptic Fragments (Including Oxidized, Aminoethylated, Pyridinylethylated, and Carboxymethylated Forms)

Tryptic fragment no.	Position no.	Mass	Oxidized mass	Amino-ethylated mass	Pyridinyl-ethylated mass	Carboxy-methylated mass
8	66	147.11[a]				
6	60—61	246.18[a]				
15	145—146	319.14[a]				
7	62—65	412.23				
2	9—17	932.52	c			
1	1—8	952.51				
11	96—104	1126.56				
14	133—144	1149.67				
4	31—40	1274.73	c			
3	18—30	1314.67				
13	121—132	1378.70				
10	83—95	1421.67	1469.66	1464.72[e]	1526.73	1479.68
10a	83—93			1221.59[d]		
10b	94—95			262.14[a,d]		
9	67—82	1669.89				
12	105—120	1719.97[b]	1767.96		1825.03	1777.98
12a	105—112			873.52[d]		
12b	113—120			908.51[d]		
5	41—59	2058.95	2090.94			

Frequently Observed Peptides Resulting from Incomplete Digestion

8, 9	66—82	1797.99				
10, 11	83—104	2529.22				
14, 15	133—146	1449.79				

Peptides Resulting from Cleavage at Other than Normal Tryptic Cleavage Sites.

14″	140—146	539.33				
14′	133—139	629.36				
12′	105—108	416.25[b]				
12″	109—120	1322.74[b]	1370.72		1427.80	1380.75
	109—112			476.29		

[a] Generally not observed due to interference from sample and matrix background.
[b] Core peptide, not soluble under normal conditions.
[c] Presence of Trp results in extensive degradation during oxidation.
[d] Peptides resulting from cleavage at aminoethyl-Cys.
[e] Intact peptide often observed despite the presence of an additional tryptic cleavage site.

of the ion is consistent with an α-chain variant, which would be expected to be present at levels less than 25%. Possible single amino acid substitutions obtained by computer analysis were Val > Phe and Asp > Tyr. The latter possibility was consistent with the observed mobility of the abnormal band on electrophoresis. In this instance, the abnormal tryptic peptide was readily purified by HPLC and the structural assignment confirmed by amino acid analysis.

In some instances, enzyme degradations can yield unexpected results. The mass spectrum of a tryptic digest mixture of an abnormal aminoethylated β-chain was normal except for a large decrease in the intensity of fragment βT4 and two new ions at m/z 606 and 677 (Figure 7). The other ions had previously been shown to be derived from the matrix and 2 pairs of fragments

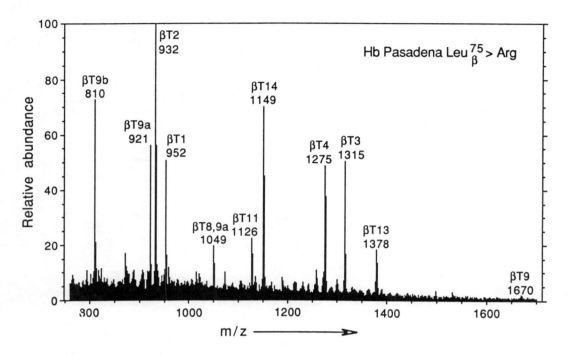

FIGURE 5.   Positive ion FAB mass spectrum of the tryptic digest mixture of the abnormal β-chain from Hb Pasadena (75 Leu > Arg). The substitution of Arg in Tryptic fragment 9 introduces a new cleavage site to generate 2 new fragments at m/z 810 and 921. A very small amount of normal βT9 is evident at m/z 1670. Analysis conditions the same as those for Figure 4.

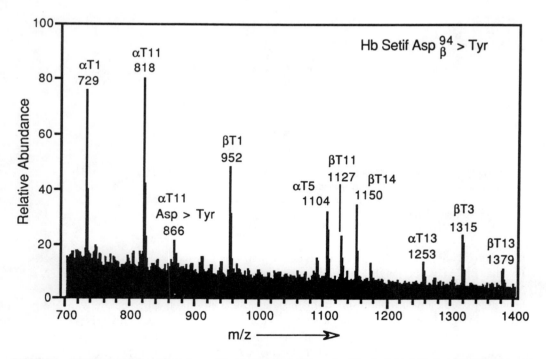

FIGURE 6.   Positive ion FAB mass spectrum of the tryptic digest mixture of oxidized whole globin from Hb Setif (β Asp94 > Tyr). Analysis conditions are the same as those for Figure 4.

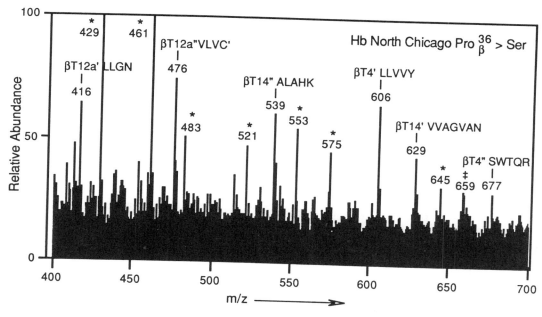

FIGURE 7.   Low mass region of the positive ion FAB mass spectrum of the tryptic digest mixture from aminoethylated β-chain from Hb North Chicago (36 Pro > Ser). Protonated molecular ions for peptides are indicated and the sequence indicated using the single letter amino acid code. C′ is used to designate aminoethylated Cys. Prominent ions derived from the glycerol matrix are marked with an *. The ion at m/z 659 is a fragment of trypsin. Conditions of the analysis are the same as those for Figure 4.

(βT12a′, βT12a″ and βT14′, βT14″) resulting from cleavage with trypsin after Asn. The ion at m/z 606 was consistent with the first five residues of βT4. The ion at m/z 677 would fit for the C-terminal portion of βT4 provided Ser was substituted for Pro. This assignment was confirmed by microsequence analysis of the purified peptide. The cleavage of βT4 after the Tyr residue was surprising because intact βT4 was not detected and no other chymotryptic fragments were observed.[27]

C. ANALYSIS OF ISOLATED PEPTIDES

Mass spectral analysis is not limited to the determination of molecular weights. If sufficient sample is available, the FAB mass spectrum of an isolated peptide will often contain sufficient information in the fragment ion series to determine the sequence. Fragmentation of any of the three types of bonds in the peptide backbone results in a number of ion series depending on the mechanism of the bond cleavage.

A good example is the characterization of an αT9 fragment from an abnormal hemoglobin. The mass difference of 22 between the abnormal peptide and normal αT9 is consistent only with the substitution of His for Asp. Unfortunately, 4 Asp residues occur in the sequence of αT9. The comparison of the mass values for ions in the spectrum with those predicted for each of the four possible structures is summarized in Figure 8. In the diagram, each of the large rectangles labeled with a position number is a map of the fragment ions for six of the more common ion series. The ions series are labeled across the top with the series designation of Roepstorff and Fohlman.[28] Numbers in the column labeled n correspond to the location of the cleavage site in the sequence starting at the N-terminus. Series members corresponding to ions observed in the spectrum are indicated with a black square. Positions labeled with an open circle had mass values that were below the mass range of the spectrum. The array is divided into five regions. The mass values for the fragments in regions I and V are common to all four possible structures. The limited

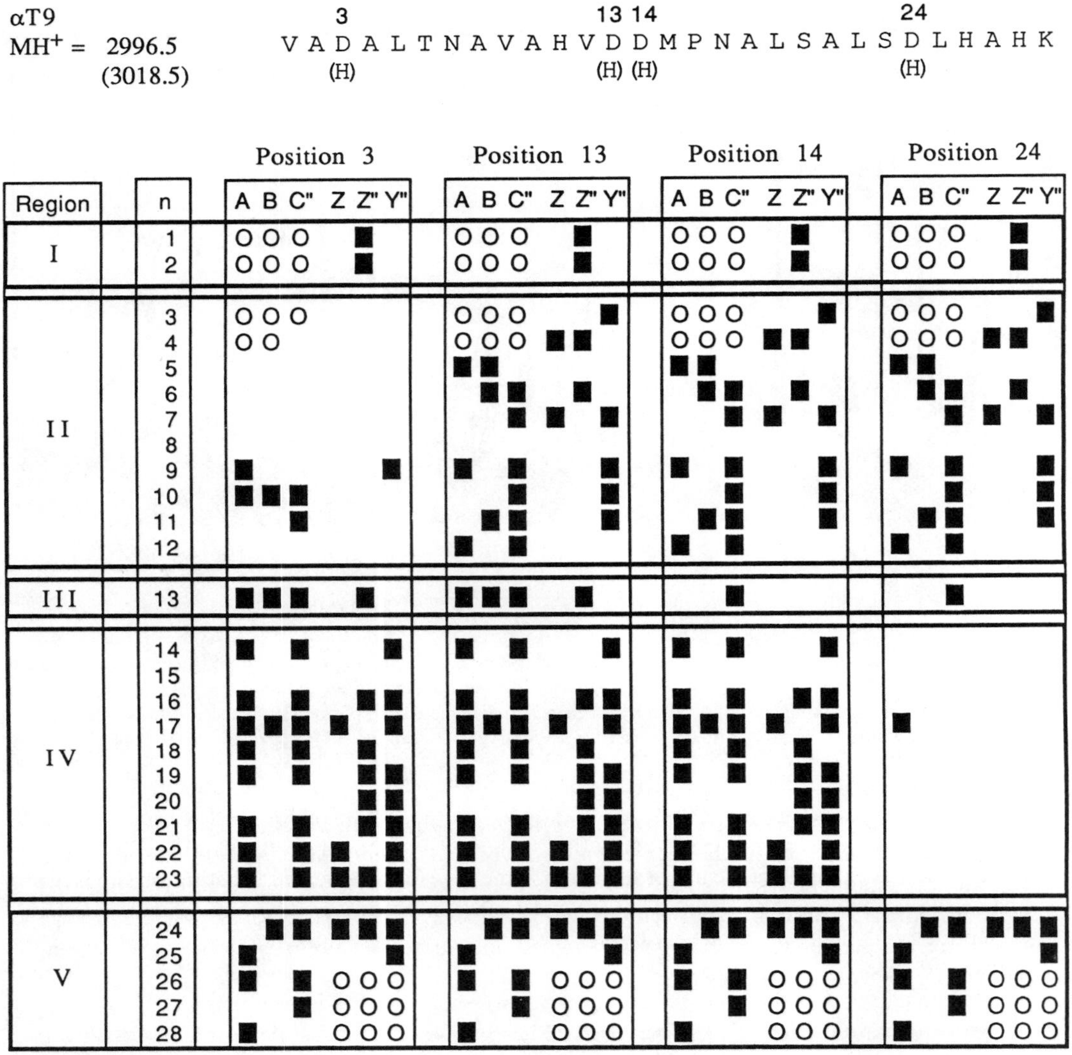

αT9
MH$^+$ = 2996.5
(3018.5)

	3		13 14		24
V A D A L T N A V A H V D D M P N A L S A L S D L H A H K					
	(H)		(H) (H)		(H)

		Position 3	Position 13	Position 14	Position 24
Region	**n**	A B C" Z Z" Y"	A B C" Z Z" Y"	A B C" Z Z" Y"	A B C" Z Z" Y"
I	1				
	2				
II	3				
	4				
	5				
	6				
	7				
	8				
	9				
	10				
	11				
	12				
III	13				
IV	14				
	15				
	16				
	17				
	18				
	19				
	20				
	21				
	22				
	23				
V	24				
	25				
	26				
	27				
	28				

FIGURE 8. Diagram used to correlate the fragment ions observed in the FAB mass spectrum of peptide αT9 from Hb Q Thailand (74 Asp > His) with those predicted for each of the possible substitution sites. Ion series are designated with the notation of Roepstorff and Fohlman[28] and series members are numbered from the N-terminus. Ions present in the mass spectrum corresponding to predicted fragments are represented with black squares. Open circles correspond to mass values below the mass range of the acquired spectrum.

number of observed ions in region II corresponding to substitution in position three eliminates this possibility. Similarly, the map of region IV eliminates the possibility of substitution at position 24. Only a single set of values for each of the 6 ion series distinguishes substitution at position 13 from that at position 14, and serves to make the point that the FAB spectra of peptides of very similar structure will differ very little. In this instance, the information in the spectrum is sufficient to locate the substitution at position 13. Thus, the variant was determined to be Hb Q-Thailand, α74 Asp > His.[29]

Instances occur when the fragment ions cannot be interpreted with certainty. A case in point is an abnormal αT12 peptide. The observed mass value for the protonated molecular ion was 3014.6, 1 amu less than normal αT9. Possible substitutions are Glu > Lys, Asp > Asn, and Glu > Gln. The first can be excluded because it would have created a new tryptic site. In addition,

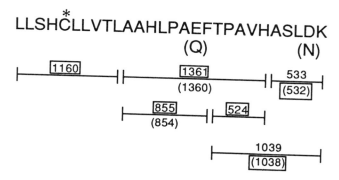

FIGURE 9. Diagram of chymotryptic fragments and m/z values for αT12. The mass values resulting from substitution of Gln for Glu or Asn for Asp are given in parentheses. Protonated molecular ions observed in the mass spectrum of the chymotryptic digest mixture from Hb Tarrant (126 Asp > Asn) are enclosed in boxes. The * indicates that the Cys residue has been converted to cysteic acid.

amino acid analysis was consistent for normal αT9. Asn and Gln are converted to Asp and Glu, respectively, during the acid hydrolysis step of amino acid analysis, so it was not possible to distinguish between the remaining two possibilities on this basis. Fragment ions in the spectrum were more consistent with the substitution of Asn for Asp, but the small 1 mass unit differences involved made it difficult to be sure because of the isotope distribution for high mass ions.

The problem was solved by further enzymatic digestion of the αT9 fragment using chymotrypsin. Chymotrypsin cleaves on the C-terminal side of hydrophobic and aromatic residues. The mass spectrum of the digest mixture contained protonated molecular ions consistent with substitution of Asn for Asp, whereas the corresponding ions for the other possibility were missing (Figure 9). Thus, the variant was assigned as Hb Tarrant α126 Asp > Asn.[30] Another example of the use of further enzyme digestion to locate a substitution site is the analysis of Hb Torino by Castagnola et al.[31] They used aminopeptidase M to generate a mixture of peptides with different numbers of amino acids removed from the N-terminus of the variant peptide.

With tandem mass spectrometry, it is not always necessary to isolate peptides in order to obtain sequence information. Using a variety of techniques,[32] it is possible to obtain product ion spectra from a selected parent ion. Thus, the mass spectrometer can be used to both separate and sequence the peptide. These techniques have not been widely used to study hemoglobin variants. Prome and co-workers[23] have concluded that collision-induced dissociation–mass-analyzed ion kinetic energy (CID-MIKE) spectra of tryptic fragments ending in Lys generally have good sequence ions, whereas those ending in Arg do not. In practice, if too many components occur in the mixture, the spectra cannot be interpreted unambiguously; thus, a certain amount of prior separation is necessary. If a chromatographic separation is necessary anyway, it is generally better to collect individual peaks and analyze only those not normally present in the chromatogram. New advances in coupling HPLC separations with mass spectral analysis will eventually make it possible to obtain sequence information on peptide components as peaks elute from the column.

D. ANALYSIS OF INTACT GLOBIN CHAINS

Until recently, the analysis of intact globin chains by mass spectrometry has not been possible. Now, new ionization methods are capable of routine analysis of proteins as large as the globin chains. A technique known as "electrospray"[33] or a closely related technique known as "ionspray"[34] would seem to have the necessary mass accuracy to distinguish mass differences

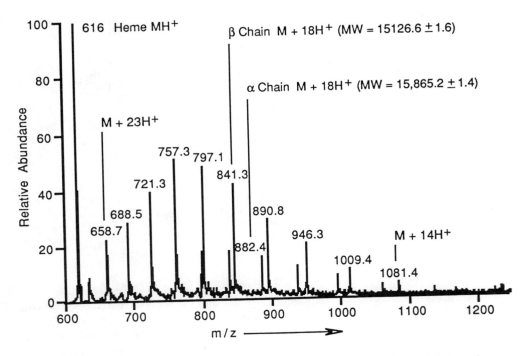

FIGURE 10.   Ionspray mass spectrum of Hb A showing ions derived from α-chain, β-chain, and Heme. Spectrum obtained from a 5 ml or a 0.1 mg/ml solution of hemoglobin in 3 mM TFA in 50% aqueous methanol, using a Sciex API III triple quadrupole mass spectrometer.

greater than a few Daltons. The method uses an electrospray process to produce multiply charged molecular ions by evaporation of solvent droplets. The analysis of Hb A yields the protonated molecular ion for the heme group and series of multiply charged molecular ions for both the α- and β-chains (Figure 10). The reason for the difference in sensitivity between the two chains is not known. Each multiply charged molecular ion provides an independent measurement of the molecular weight. Such an analysis could serve as a useful screening technique, much like electrophoresis but with the added advantage that it would also work on silent variants with a sufficiently large mass difference and define which chain it involves. For complete characterization, it would still be necessary to degrade the globin chain into smaller pieces using enzymatic or chemical methods.

V. CONCLUSIONS

Mass spectral methods and instrumentation provide a multiplicity of strategies for detecting and determining abnormal hemoglobin variants. Spectra of intact globin chains from simple hemolysates provide rapid means to screen for structural abnormalities. Spectra of whole globin enzyme digest mixtures provide confirmation of the major portion of the expected structure and may reveal unusual peptides. The mass difference of the peptide alone may be sufficient information to define the single amino acid substitution, or at the very least point the way for further investigation. Complete characterization often depends on isolation of the abnormal peptide component through chromatographic or other means. Sequence information from the fragment ions in a spectrum has been shown to be very useful both on isolated peptides and enzyme digest mixtures using tandem mass spectrometry. The future prospects for mass spectral analysis of protien variants are very exciting. It should eventually be possible to provide full characterization of many variants using a single process that combines on-column enzymatic degradation, HPLC separation, and tandem mass spectral analysis with computerized interpretation of the result. Even without future advances, mass spectrometry should be considered an essential component of variant protein characterization.

REFERENCES

1. **Bunn, H. F. and Forget, B. G.**, *Hemoglobin: Molecular, Genetic and Clinical Aspects*, W. B. Saunders, Philadelphia, 1986.
2. **Higgs, D. R., Vickers, M. A., Wilkie, A. O. M., Pretorius, I.-M., Jarman, A. P., and Weatherall, D. J.**, A review of the molecular genetics of the human α-globin gene cluster, *Blood*, 73, 1081, 1989.
3. **Marks, J., Shaw, J. P., and Shen, C. K. J.**, Structure and genomic complexity of primate θ1-globin gene, a novel α-globin-like gene, *Nature*, 321, 785, 1986.
4. **Ingram, V. M.**, Abnormal human haemoglobins. III. The chemical difference between normal and sickle cell haemoglobins, *Biochim. Biophys. Acta*, 36, 402, 1959.
5. **Milner, P. F., Clegg, J. B., and Weatherall, D. J.**, Hemoglobin H disease due to a unique hemoglobin variant with an elongated α-chain, *Lancet*, April 10, 729, 1971.
6. **Huisman, T. H. J.**, International Hemoglobin Information Center, IHIC variants list, *Hemoglobin*, 13, 223, 1989.
7. **Baglioni, C.**, The fusion of two peptide chains in Hemoglobin Lepore and its interpretation as a genetic deletion, *Proc. Natl Acad. Sci., U.S.A.*, 48, 1880, 1962.
8. **Arous, N., Galacteros, F., Fessas, P., Loukopoulos, D., Blouquit, Y., Komis, G., Sellaye, M., Boussiou, M., and Rosa, J.**, Structural study of hemoglobin Knossus, β22 (B9) Ala > Ser. A new abnormal hemoglobin present as a silent β-thalassemia, *FEBS Lett.*, 147, 247, 1982.
9. **Rahbar, S., Blumenfeld, O., and Ranney, H. M.**, Studies of an unusual hemoglobin in patients with diabetes mellitus, *Biochem. Biophys. Res. Commun.*, 36, 838, 1969.
10. **Huisman, T. H. J.**, The hemoglobinopathies, in *Methods in Hematology*, Churchill Livingstone, Edinburgh, 1986.
11. **Lehmann, H. and Huntsman, R. G.**, *Man's Haemoglobins*, North-Holland, Amsterdam, 1974.
12. **Carrell, R. W. and Kay, R.**, A simple method for the detection of unstable haemoglobins, *Br. J. Haematol.*, 23, 615, 1972.
13. **Shelton, J. B., Shelton, J. R., and Schroeder, W. A.**, High performance liquid chromatographic separations of globin chains on a large-pore C-4 column, *J. Liquid Chromatogr.*, 7, 1969, 1984.
14. **Rahbar, S., Lee, T. D., Baker, J. A., Rabinowitz, L. T., Asmeron, Y., Legesse, K., and Ranney, H. M.**, Reverse phase high performance liquid chromatography and secondary ion mass spectrometry. A strategy for identification of ten hemoglobin variants, *Hemoglobin*, 10, 379, 1986.
15. **Matsui, T., Matsuda, H., Katakuse, I., Wada, Y., Fujita, T., and Hayashi, A.**, Field desorption mass spectra of tryptic peptides of human hemoglobin chains, *Biomed. Mass Spectrom.*, 8, 25, 1981.
16. **Wada, Y., Hayashi, A., Fujita, T., Matsuo, T., Katakuse, I., and Matsuda, H.**, Structural analysis of human hemoglobin variants with field desorption mass spectrometry, *Biochim. Biophys. Acta*, 667, 233, 1981.
17. **Wada, Y., Hayashi, A., Masanori, F., Katakuse, I., Ichihara, T., Nakabushi, H., Matsuo, T., Sakurai, T., and Matsuda, H.**, Characterization of a new fetal hemoglobin variant, Hb F Izumi Ag 6 Glu > Gly, by molecular secondary ion mass spectrometry, *Biochim. Biophys. Acta*, 749, 244, 1983.
18. **Intelligenetics**, *PC Gene*, 5.17, for the IBM PC, Mountain View, CA , 1988.
19. **Andrews, P. C.**, *PROCOMP*, 1.0, for the IBM PC, Biochemical Department, Purdue University, West Lafayette, IN, 1989.
20. **Lee, T. D. and Vemuri, S.**, *MacProMass*, 1.0, for the MacIntosh, Division of Immunology, City of Hope, Duarte CA, 1989.
21. **Hojrup, P.**, *GPMA*, 4.2A, for the IBM PC, Department of Molecular Biology, Odense University, Odense, Denmark, 1988.
22. **Carrell, R. W., Lehmann, H., and Hutchinson, H. E.**, Haemoglobin Koln (β98 Val > Met): an unstable protein causing inclusion-body anaemia, *Nature*, 210, 915, 1966.
23. **Prome, D., Prome, J. C., Blouquit, Y., Lacombe, C., Rosa, J., and Robinson, J. D.**, FAB mapping of proteins: detection of mutation sites in abnormal human hemoglobins, *Spectrosc. Int. J.*, 5, 157, 1987.
24. **Pucci, P., Ferranti, P., Marino, G., and Malorni, A.**, Characterization of abnormal human haemoglobins by fast atom bombardment mass spectrometry, *Biomed. Environ. Mass Spectrom.*, 18, 20, 1989.

25. **Johnson, C. S., Moyes, D., Schroeder, W. A., Shelton, J. B., Shelton, J. R., and Beutler, E.**, Hemoglobin Pasadena, $\alpha2\beta2$ 75(E19) Leu > Arg: identification by high performance liquid chromatography of a new unstable variant with increased oxygen affinity, *Biochim. Biophys. Acta,* 623, 360, 1980.

26. **Wajcman, H., Belkhodja, O., and Labie, D.**, Hb Setif: G1(94) α Asp > Tyr. A new hemoglobin variant with substituion of the residue involved in a hydrogen bond between unlike subunits, *FEBS Lett.,* 27, 298, 1972.

27. **Rahbar, S., Louis, J., Lee, T. D., and Asmerom, Y.**, Hemoglobin North Chicago ($\beta36$ [C2] Proline > Serine): a new high affinity hemoglobin, *Hemoglobin,* 9, 559, 1985.

28. **Roepstorff, P. and Fohlman, J.**, Proposal for a common nomenclature for sequence ions in mass spectra of peptides, *Biomed. Mass Spectrom.,* 11, 601, 1984.

29. **Lorkin, P. A., Charlesworth, D., Lehmann, H., Rahbar, S., Tuchinda, S., and Lie-Injo, L. E.**, Two haemoglobins Q, $\alpha74$ (EF3) and $\alpha75$ (EF4) aspartic acid > histidine, *Br. J. Haemotol.,* 19, 117, 1970.

30. **Moo-Penn, W. F., Johnson, M. H., Wilson, S. M., Therrell, B. L., and Schmidt, R. M.**, Hemoglobin Tarrant: $\alpha126$ (h9) Asp > Asn. A new hemoglobin variant in the $\alpha_1\beta_1$ contact region showing high oxygen affinity and reduced cooperativity, *Biochim. Biophys. Acta,* 490, 443, 1977.

31. **Castagnola, M., Dobosz, M., Landolfi, R., Pascali, V. L., DeAngelis, F., Vettore, L., and Perona, G.**, Determination of neutral haemoglobin variants by immobilized pH gradient, reversed-phase high-performance liquid chromatography and fast-atom bombardment mass spectrometry: the case of Hb Torino $\alpha43$ (CE1) Phe > Val, *Biol. Chem. Hoppe-Seyler,* 369, 241, 1988.

32. **McLafferty, F. W.**, *Tandem Mass Spectrometry,* John Wiley & Sons, New York, 1983.

33. **Mann, M., Meng, C. K., and Fenn, J. B.**, Interpreting mass spectra of multiply charged ions, *Anal. Chem.,* 61, 1702, 1989.

34. **Covey, T. R., Bonner, R. F., Shushan, B. I., and Henion, J. D.**, The determination of protein, oligonucleotide and peptide molecular weights by ion-spray mass spectrometry, *Rapid Commun. Mass Spectrom.,* 2, 249, 1988.

Chapter 14

DETECTION AND LOCATION OF DISULFIDE BONDS IN PROTEINS BY MASS SPECTROMETRY

David L. Smith and Yiping Sun

TABLE OF CONTENTS

I.   Introduction 276

II.   Methods for Locating Disulfide Bonds in Proteins 276

III. Detection of Disulfide-Containing Peptides 277

IV.   Procedure for Locating Disulfide Bonds in Proteins 281

V.   Disulfide-Related Cross Linkages 285

VI.   Summary 285

Acknowledgments 286

References 286

I. INTRODUCTION

Disulfide bonds, which are formed in proteins by the oxidation of cysteinyl thiols, are among the most frequently encountered post-translational modifications. These cross linkages play an important role in stabilizing the three-dimensional structure of extracellular proteins by placing constraints on the locations of cysteinyl residues. Disulfide-containing enzymes usually have full activity only when the half-cystinyl residues are joined correctly. The correlation between activity and disulfide connectivities extends to small neuropeptides such as oxytocin and vasopressin. Although intracellular reduction potentials usually limit the concentration of disulfide bonds in the cytoplasm, there are circumstances under which proteins do form disulfide bonds inside cells. For example, there is considerable evidence indicating that crystallins, structural proteins found in the lens of the eye, undergo disulfide bonding to form aggregates that scatter light. This process is believed to be one of the causes of cataracts.

Because of the relation between protein/peptide activities and disulfide bond connectivity, it is important to have reliable methods for locating disulfide bonds in proteins. In addition to providing unambiguous results, the ideal method should not be limited by the amino acid sequence or size of the protein, and it should be highly sensitive. Much of the interest in developing improved methods for locating disulfide bonds in proteins stems from recent advances in biotechnology that have led to the facile production of peptide and protein drugs, many of which contain disulfide bonds. Because of the potential toxicity of related substances that have incorrectly formed disulfide connectivities, it is important to determine with certainty the disulfide connectivities in these products. The purpose of this chapter is to describe the role played by fast atom bombardment mass spectrometry (FABMS) in locating disulfide and related bonds in proteins.

II. METHODS FOR LOCATING DISULFIDE BONDS IN PROTEINS

A variety of methods has been used to locate disulfide bonds in proteins. The method[1-3] used most frequently requires cleaving a protein between half-cystinyl residues to give peptides that contain a single disulfide bond. These peptides, which may contain intra- or intermolecular disulfide bonds, are isolated and identified subsequently by either their amino acid composition or their sequence. Although this approach is well established and has been used with considerable success, it is limited in many respects. For example, disulfide-containing peptides must be isolated if they are to be identified by only their amino acid composition or sequence. Isolation to homogeneity may be particularly difficult if a relatively nonspecific cleavage reaction is used to fragment the protein, or if the protein is large. Although highly specific cleavage reagents, such as trypsin, are preferred, their use is limited to proteins in which the half-cystinyl residues are separated by certain amino acids that are susceptible to cleavage.

An alternate approach is based on the observation that disulfide bonds in a protein can often be reduced sequentially to thiols.[4-6] If this stepwise reduction is the case, the protein is reduced partially such that only the most labile disulfide bond is cleaved. The thiols so formed are derivatized with a reagent, such as radiolabeled iodoacetic acid. The protein is then reduced completely, and subsequently cleaved enzymatically. The location of the labile disulfide bond is determined by isolating and identifying the radiolabeled peptides. Although this approach has been very useful for identifying disulfide bonds that are easily cleaved, it is often accompanied by disulfide rearrangements. Diagonal electrophoresis has been used extensively to identify disulfide-containing peptides.[4]

FABMS, which is now used extensively for structural characterization of peptides and proteins, has proved useful for locating disulfide bonds in proteins. Morris and Pucci[7] used FABMS to locate the disulfide bonds in insulin. In doing so, they were the first to demonstrate the potential of FABMS as a tool for determining disulfide bond connectivities in proteins.

Locating disulfide bonds in insulin is a considerable challenge because insulin has two adjacent cysteinyl residues that are not readily cleaved with specific endopeptidases. As a result, disulfide-containing peptides diagnostic for the connectivities of these half-cystinyl residues cannot be formed directly. However, the authors were successful in producing the necessary disulfide-containing peptides by using a combination of peptic digestion, treatment with carboxypeptidase A, and multiple Edman degradations. The disulfide-containing peptides were identified by their molecular weights, which were determined by FABMS analysis of the digests. It is important to note that this very significant accomplishment was achieved without isolating the peptides, as would be required for their identification by conventional methods.

Several recent studies have illustrated that FABMS is a useful method for locating disulfide bonds in proteins. Takao et al.[8] located two disulfide bonds in hen egg-white lysozyme, and Yazdanparast et al.[9] demonstrated the generality of the method by locating all of the disulfide bonds in hen egg-white lysozyme and ribonuclease A. Akashi et al.[10] have used FABMS to locate the disulfide bonds in Paim I, an α-amylase inhibitor isolated from the culture filtrate of *Streptomyces corchorushii*. Raschdorf et al.[11] have since used a similar approach to verify disulfide connectivities in a biosynthetic insulin-like growth factor. It is interesting to note that they also isolated a related enzyme that had been produced with incorrect disulfide linkages. The potential of FABMS for identifying and locating a variety of protein cross linkages related to disulfide bonds is enormous. For example, Toren et al.[12] reported evidence for lanthionine and sulfoxide cross linkages in insulin-related substances.

The primary advantage of FABMS rests with the fact that peptides can be identified when present in mixtures. This approach substantially reduces the requirement for chromatographic separations, and is particularly important for identifying disulfide-containing peptides derived from large proteins, or peptides produced by low-specificity cleavage reactions.[7,13-15] It is also important to note that disulfide-bonded peptides are identified by their molecular weight. Because molecular weight is unrelated to chromatographic properties of the peptide or its constituent amino acids, information based on FABMS is complementary to information derived from amino acid composition or sequence. Despite the strength of FABMS for identifying disulfide-containing peptides, it is obviously prudent to utilize all information and methods that may be available to determine the disulfide connectivities in a protein. Burman et al.[16] have demonstrated the power of combining mass spectrometry with amino acid composition and sequence data by locating the seven disulfide bonds in bovine neurophysin.

III. DETECTION OF DISULFIDE-CONTAINING PEPTIDES

Many of the samples to be analyzed by FABMS contain a mixture of peptides, only some of which are disulfide bonded. Because it is these peptides that will be used to determine the disulfide connectivities, it is important to have methods for rapidly distinguishing disulfide-containing peptides from peptides that do not contain disulfide bonds. One approach is to reduce the disulfide bonds with a reducing agent, such as dithiothreitol or mercaptoethanol[17] and to analyze the sample again by FABMS. Peaks due to disulfide-containing peptides will be replaced by new peaks corresponding to the reduced products. If a peptide contains an intramolecular disulfide bond, a new peak will appear 2 mass units higher. If a peptide contains an intermolecular disulfide, new peaks corresponding to the molecular ions of the two constituent peptides will usually be found after reduction of the disulfide bond. This approach is illustrated in Figures 1A and B, which are the FAB mass spectra of the disulfide linked peptide produced by trypsinolysis of somatostatin (SST)-14 taken before and after reduction of the intermolecular disulfide bond with a 3:1 mixture of dithiothreitol and dithioerythritol. Reactants and products of this reaction are given in Equation 1. The FAB mass spectrum taken prior to chemical reduction has prominent peaks for the molecular ions (M+H)$^+$ of the disulfide-containing peptide (m/z 933), the peptide cleaved from the interior of SST-14 (m/z 741), and a

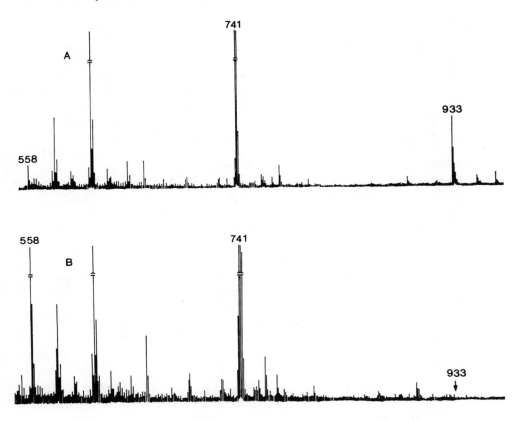

FIGURE 1.   FAB mass spectra of the disulfide linked peptide produced by trypsinolysis of SST-14 that was digested with trypsin. Spectra taken before (A) and after (B) reduction of the disulfide bond with a 3:1 mixture of dithiothreitol and dithioerythritol. The molecular ion (m/z 933) of the disulfide-containing peptide (AlaGlyCysLys/ThrPheThrSer Cys) is not present after reduction. The molecular ion (m/z 558) of one of the constituent peptides (ThrPheThrSerCys) is present before and after reduction. The large peak at m/z 741 is the molecular ion of a peptide cleaved from the interior of SST-14. Other peaks are due to the matrix.

small peak for one of the constituent peptides (m/z 558). After chemical reduction (Figure 1B), the peak for the disulfide-containing peptide (m/z 933) is no longer present, and the peak corresponding to the constituent peptide (m/z 558) is substantially stronger. A signal corresponding to the other constituent peptide (m/z 378) could not be distinguished from the matrix peaks.

Although reduction of disulfide bonds is a well-established method for denoting disulfide-containing peptides, it cannot be used to identify thiol-containing peptides. This deficiency is significant, because identifying the constituent peptides (i.e., thiol-containing peptides) is normally the first step in identifying disulfide-containing peptides. Sun et al.[18] have demonstrated that performic acid can be used to conveniently induce a predictable shift in the mass of peptides that contain disulfide bonds as well as thiols. These reactions are illustrated here for peptides (P) with intra- (Equation 2) or inter- (Equation 3) molecular disulfide bonds. Performic acid efficiently oxidizes half-cystinyl groups to cysteic acid. Peptides that contain an inter-molecular disulfide bond are converted into their constituent peptides (Equation 3) with a concomitant increase in molecular weight of 49 mass units for each half-cystinyl residue. The molecular weight of peptides that contain an intramolecular disulfide increase by 98 mass units upon oxidation (Equation 3). It is noted that Met and Trp residues are also oxidized by performic acid under similar conditions.[18]

$$\underset{|}{\overset{S\text{———}S}{|}} \qquad \xrightarrow{\text{OXIDATION}} \qquad \underset{|}{SO_3H} \qquad \underset{|}{SO_3H} \qquad (2)$$

$$P\text{-}S\text{-}S\text{-}P \qquad \xrightarrow{\text{OXIDATION}} \qquad PSO_3H \qquad PSO_3H \qquad (3)$$

The FAB mass spectra presented in Figure 2 illustrate the use of performic acid with [Arg^8]-vasopressin, a peptide that contains nine amino acids and one intramolecular disulfide bond. The spectrum in Figure 2A was obtained with 500 pmol of peptide dissolved in approximately 1 µl of a 50:50 mixture of glycerol and 1-thioglycerol. Spectrum B was obtained after reacting 0.5 µl of performic acid with another 500 pmol aliquot of the peptide for 5 min on the probe tip. The oxidized peptide was analyzed by adding approximately 1 µl of matrix and analyzing as before. The peak at m/z 1084 in Figure 2A is due to the MH$^+$ ion of [Arg^8]-vasopressin. This peak, which is not present in the FAB mass spectrum of the oxidized vasopressin, is replaced with a new peak (m/z 1182), which is 98 mass units higher. The increase in mass is due to the addition of six oxygens and two hydrogens to the half-cystinyl residues comprising the disulfide bond, as illustrated in Equation 2.

In addition to these chemical methods, a dual electrode electrochemical high performance liquid chromatography (HPLC) detector may be used to denote peptides that contain disulfide bonds.[19-22] This highly specific detector is based on the oxidation reaction (Equation 4) that occurs at a mercury amalgam electrode biased +0.15 V relative to a Ag/AgCl reference electrode

$$\text{Peptide-SH} + \text{Hg}^\circ \rightarrow (\text{Peptide-S})_2\text{Hg} \qquad (4)$$

Disulfide-containing peptides are detected when a second mercury electrode (-1.0 V relative to Ag/AgCl) is placed directly upstream from the thiol-detecting electrode. This second electrode reduces a small portion of the disulfides to give thiols, which are detected subsequently at the downstream electrode. An example of the reduction reaction that occurs at the upstream electrode is given in Equation 1.

The selectivity of the electrochemical detector for identifying disulfide-containing peptides may be illustrated with a tryptic digest of cyanogen bromide (CNBr)-treated hen egg-white lysozyme, which is a small protein with 129 amino acids and a mol wt of 14,600. This protein has four disulfide bonds, and has often been used as a model for the development of new methods for locating disulfide bonds.[2,8,9] The structure of hen egg-white lysozyme is given in Figure 3. Chromatograms obtained with ultraviolet (UV) (230 nm) and EC detectors operating in series are given in Figure 4. The chromatograms presented in Figures 4B and C were obtained by operating the EC detector in the two- and one-electrode modes, respectively. When operated in

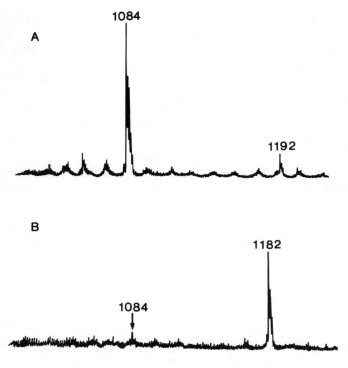

FIGURE 2.   FAB mass spectra of [Arg⁸]-vasopressin taken before (A) and after (B) on-probe oxidation of the disulfide bond with performic acid. The peak at m/z 1192 in (A) is the thioglycerol adduct of reduced [Arg⁸]-vasopressin.

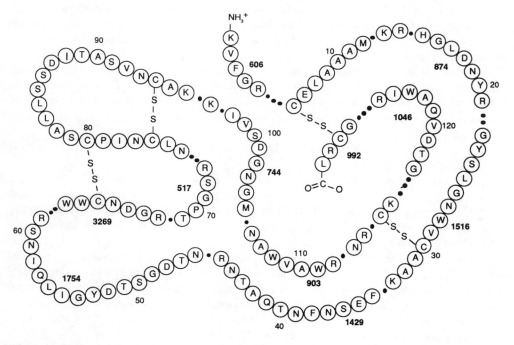

FIGURE 3.   The structure of hen egg-white lysozyme and the MH⁺ of the peptides found in the FAB mass spectrum of the CNBr/tryptic digest.

the two-electrode mode (Figure 4B), the EC detector responds to disulfide- and thiol-containing peptides. When operated in the one-electrode mode (Figure 4C), it responds only to thiol-containing peptides. Hence, the difference in the chromatograms obtained for the two modes of operation indicates those peptides that contain disulfide bonds.

The results presented in Figure 4 show that this digest contains many peptides that have disulfide bonds, and essentially no thiol-containing peptides. It is important to note that only a small portion of the disulfide bonds are reduced. As a result, the chromatographic fractions that contain peptides with disulfide bonds can be collected and analyzed by FABMS. Portions of the FAB mass spectrum of the fraction indicated by the arrow in Figure 4B are given in Figure 5. The peak at m/z 3269 is due to the Asn 74–Lys 96 segment joined to the Trp 62–Arg 68 segment by a disulfide bond (see Figure 3). Because disulfide bonds are reduced partially during FABMS analysis, peaks corresponding to the molecular ion of the constituent peptides (Asn 74–Lys 96, m/z 2335, and Trp 62–Arg 68, m/z 936) are also present. The assignment of the peak at m/z 3269 as a disulfide-containing peptide was confirmed by its absence in the mass spectrum of an aliquot in which the disulfide bonds had been reduced.

IV. PROCEDURE FOR LOCATING DISULFIDE BONDS IN PROTEINS

The general approach by which FABMS is used to locate disulfide bonds in proteins can be illustrated with hen egg-white lysozyme[8,9] (Figure 3). Disulfide-containing peptides suitable for analysis by FABMS were produced by treating lysozyme with CNBr and trypsin. CNBr cleaves on the C-terminus side of Met, causing the protein to unfold enough so that trypsin can cleave after Arg and Lys. An aliquot of the digest was analyzed by FABMS to determine the molecular weights of the peptides formed by cleavage of the original protein. Results of this analysis are presented in Table 1, which lists the molecular weights and assignments of peptides found by FABMS.

The molecular weights of peptides found in the digest were related to specific segments of the native protein by determining which peaks in the FAB mass spectrum were due to disulfide-containing peptides. This assignment was accomplished by reducing the disulfide bonds with chemical reagents, such as dithiothreitol or mercaptoethanol, and analyzing again by FABMS. Peaks that were no longer present in the FAB mass spectrum were designated as disulfide-containing peptides. Assignment to specific segments of the native protein was made by removing the N-terminal amino acid by a one-step Edman degradation reaction and analyzing again by FABMS. The assignments of peptides to unique segments of the protein was confirmed by correlating the shift in the molecular weights of peptides with the loss of specific amino acids.

It has now become apparent that computer-assisted analysis can facilitate assignment of the molecular weights of peptides to specific segments of a protein.[14,15] For example, a systematic search of the amino acid sequence of lysozyme for peptides that could give a molecular ion at m/z 874 (see Table 1) shows that there are four possible segments. This prediction is based on the premise that none of the peptides has been modified, and that cleavage could occur at any peptide linkage. The four tentative assignments are given in Table 2. Because the peptides present in this digest were formed by cleaving on the C-terminal side of Met, Arg, and Lys, it follows that the N-terminal residue of a peptide will most likely be preceded by one of these residues. Likewise, the C-terminal residues of the correct assignment should also be one of these amino acids. The N- and C-terminal peptides of the protein, as well as peptides formed by nonspecific cleavage are not so constrained. It is apparent from Table 2 that the only candidate segment consistent with these restrictions is the peptide composed of residues 15 to 21. It follows that the peptide with molecular ion at m/z 874 is assigned uniquely to the 15 to 21 segment. A similar approach can be used to assign uniquely all but the last three peptides (m/z 3269, 1516, and 992) listed in Table 1. Because these peptides cannot be assigned to specific segments of

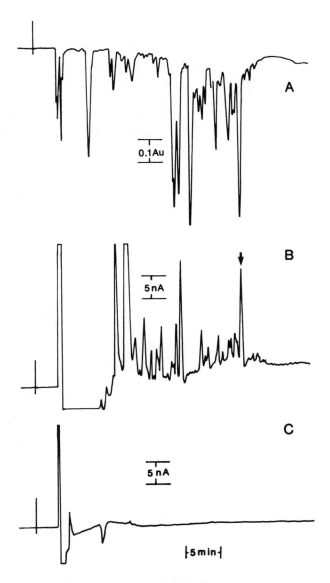

FIGURE 4.   HPLC chromatograms of a tryptic digest of hen egg-white
lysozyme obtained with UV (A) and two- (B) and one- (C) electrode
electrochemical detection of peptides. Chromatograms A and B were
obtained simultaneously, and chromatogram C was obtained after injecting
a second aliquot of the digest.

the protein, they are likely pairs of segments that are joined by disulfide bonds. Peaks in the FAB
mass spectrum that cannot be assigned to any segment of the protein may also be due to post-
translational modifications to the protein or to impurities in the sample.

Whether a particular peptide contains a disulfide bond can often be determined from the
presence in the FAB mass spectrum of related peptides that are formed by reduction of the
disulfide bond during FABMS analysis. This *in situ* reduction of disulfide bonds was reported
first by Yazdanparast et al.[23] As a result, protonated molecular ions corresponding to the
constituent peptides of a pair of segments joined by a disulfide bond are usually present in the
FAB mass spectrum. An application of this approach may be illustrated with results presented

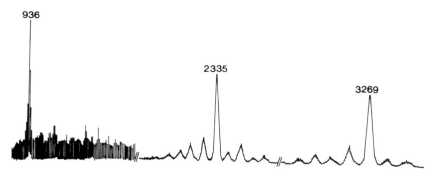

FIGURE 5.   Portions of the FAB mass spectrum (resolution 1000) of the chromatographic fraction indicated by the arrow in Figure 4B. The monoisotopic masses 2335 and 3269 correspond to average masses of 2339 and 3273, respectively. See Table 1 and Figure 3 for their identification.

TABLE 1
Peptides Found in the Tryptic
Digest of CNBr-Treated Hen Egg-White Lysozyme

Residues	Mass of MH[+ a]
1—7	606
2—5	478
6—12[b]	660
15—21	874
22—33[b]	1269
34—45	1429
46—61	1754
62—68[b]	936
69—73	517
74—96[b]	2335
98—105	744
106—112	903
113—114	289
117—125	1046
126—128[b]	335
62—68 / 74—96[c]	3269
22—33 / 115—116[c]	1516
6—12 / 126—128[c]	992

[a]   Monoisotopic mass of MH[+].
[b]   Peptides containing cysteine.
[c]   " / " indicates peptide segments joined by a disulfide bond.

in Table 1. For example, the molecular ion at m/z 3269 corresponds to a peptide composed of the 2 segments 74 to 96 (m/z 2335) and 62 to 68 (m/z 936) joined by a disulfide bond. The presence of both constituent peptides in the FAB mass spectrum substantiates the assignment of the disulfide-containing peptide. The molecular ion at m/z 992 can likewise be assigned uniquely to segments 6 to 12 (m/z 660) and 126 to 128 (m/z 335), which are joined by a disulfide bond. The remaining unidentified peptide (m/z 1516) is assigned to the segments 22 to 33 (m/z 1269) and 115 to 116 (m/z 250). The molecular ion of the 115 to 116 segment (m/z 250) could not be distinguished from the background ions in this low-mass region. The molecular ion of peptides that contain an intramolecular disulfide bond appear 2 mu higher than the original peaks after the disulfide bond is reduced.

The data presented in Tables 1 and 2 demonstrate that the location of two of the disulfide

TABLE 2
Summary of All Possible Peptides with MH⁺ 874
that Could be Obtained by Cleaving
Hen Egg-White Lysozyme at Any Peptide Linkage

Residues	Sequence	Mass of MH⁺
14—20	RHGLDNY	874
15—21	HGLENYR	874
18—24	DNYRGYS	874
92—99	VNCAKKIV	874

TABLE 3
Disulfide-Containing Peptides Found in the Partial Acid Hydrolysate
of the 62-68/74-96 Peptide Isolated from the
Tryptic Digest of CNBr-Treated Hen Egg-White Lysozyme

Residues	Mass of MH⁺	Linkage
62-64/78-87	1497	Cys-64/Cys-80
62-64/78-86	1382	Cys-64/Cys-80
74-77/88-96	1368	Cys-76/Cys-94

bonds (Cys–6/Cys–127 and Cys–30/Cys–115) in hen egg-white lysozyme can be determined from molecular weight information derived from a single FAB mass spectrum of the digested protein. The unique assignment of peptide molecular weights to specific segments of the native protein is facilitated by computer-assisted analysis. This approach is useful because it lists all of the possible assignments. A unique assignment can be made with this information when only one of the tentative assignments is consistent with the specificity of the cleavage reactions used to produce the peptides. The confidence with which tentative assignments are disallowed depends on the specificity of the cleavage reaction. For example, trypsin frequently displays chymotryptic behavior, cleaving after tyrosine, phenylalanine, and tryptophan.

Disulfide bond connectivities can be determined only if peptides that contain a single disulfide bond are produced from the original protein. The location of the remaining 2 disulfide bonds (Cys–64/Cys–80, Cys–76/Cys–94) could not be determined from these data because the 62–68/74–96 peptide (m/z 3269) contained both disulfide bonds. These disulfide bonds form a disulfide knot, which is representative of one of the more difficult problems that one encounters in locating disulfide bonds in protein. The resolution of this problem rests with finding a method for cleaving the protein between Cys–76 and Cys–80. This segment (–Cys–Asn–Ile–Pro–Cys–) is not amenable to enzymatic cleavage, both because of its amino acid composition and because it is likely buried in the knot. Peptides appropriate for locating these disulfide bonds were, however, formed by isolating the 62–68/74–96 peptide by reversed-phase HPLC (RP HPLC) and subjecting it to partial acid hydrolysis. This cleavage method is relatively unselective and usually cleaves peptides at many different sites. The masses and assignments of three disulfide-containing peptides found in this hydrolysate are given in Table 3.

The power of FABMS for identifying peptides in mixtures is demonstrated by the fact that peptides corresponding to nearly the entire sequence of lysozyme were found and identified in the digest without any chromatographic fractionation. Frequently, the complete assignment is not possible because some peptides suppress the FABMS signals from other peptides that are not concentrated along the surface of the matrix. For example, a similar analysis of the tryptic digest of CNBr-treated ribonuclease A failed to detect any disulfide-containing peptides. However, disulfide-bonded peptides suitable for locating all of the disulfide bonds in ribonuclease A were detected after the digest had been fractionated by RP HPLC.[9]

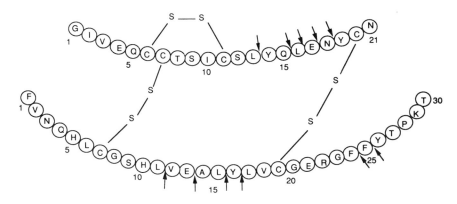

FIGURE 6.   Structure of human insulin and the sites where it is cleaved by pepsin.

V. DISULFIDE-RELATED CROSS LINKAGES

In the future, FABMS will likely play a major role in identifying and locating new or unexpected protein cross linkages. Because most previous attempts to identify new cross linkages have relied on acid hydrolysis to remove the cross linkage from the protein, attention has focused on acid-stable cross linkages. As a result, very little information about acid-labile cross linkages has been reported.

Results reported by Toren et al.[12] illustrate the potential of FABMS for identifying and locating disulfide-related cross linkages. The authors used methods similar to those described previously for lysozyme to investigate disulfide-related cross linkages in insulin B-chain dimers isolated from an A-chain/B-chain combination reaction that has been used to produce biosynthetic insulin. The dimers were fragmented with pepsin and analyzed by a combination of RP HPLC and FABMS. A preliminary investigation designed to determine likely sites for peptic cleavage was performed with human insulin. The structure and sites at which pepsin cleaved human insulin are given in Figure 6. This finding was an important step because pepsin is not a highly selective protease and the cleavage sites could not be predicted with certainty.

Mass spectral analysis of a peptic digest of the B-chain dimers suggested that several different types of cross linkages had been formed. For example, protonated molecular ions corresponding to the disulfide-containing peptides B5–11/B5–13, B5–13/B5–13, B1–11/B5–13, B1–13/B5–11, and B1–13/B5–13 were found in the FAB mass spectra of the HPLC–fractionated digest. These peptides clearly place a disulfide bond between Cys–7 and Cys–7 of two B-chains. Another disulfide-containing peptide, B1–11/B16–25, was found, indicating that some of the B-chain dimers were joined by a disulfide bond between Cys–7 and Cys–19.

Of particular interest was a large peak in the FAB mass spectrum corresponding to the B1–11/B16–25 peptide, less 32 mass units. Because Nashef et al.[24] found lanthionine (see Figure 7) in the acid hydrolysate of bovine insulin, which had been exposed to similar alkaline conditions, this ion was assigned tentatively to the lanthionine-bonded peptide pair B1–11/B16–25. Although lanthionine is stable to acid hydrolysis, this is the first report of an attempt to locate a lanthionine cross linkage in a protein. Evidence was also presented for the formation of a sulfoxide linkage (see Figure 7) joining the B1–11 and B16–25 segments. It is significant that these unexpected cross linkages were identified and located without isolating the different B-chain dimers.

VI. SUMMARY

As the need for identifying and locating cross linkages in proteins increases, FABMS will play an increasingly important role in this complex analytical challenge. A variety of chemical

B1-11/B16-25
MH^+ 2409
Lanthionine Linked Dimer

B1-11/B16-25
MH^+ 2425
Sulfoxide Linked Dimer

FIGURE 7. Structures of the lanthionine and sulfoxide cross linkages found in B-chain dimers isolated from an A-chain/B-chain combination reaction mixture.[12]

and enzymatic methods has been used to produce peptides that contain cross linkages. These peptides are identified by molecular weight, and related to specific segments of the original protein by computer-assisted analysis. In the case of disulfide cross linkages, which are the most common type of protein cross linkage, reduction, oxidation, and electrochemical HPLC detection have proved useful for identifying disulfide-containing peptides. Identification and location of other types of protein cross linkages, such as lanthoinine and sulfoxide, will be facilitated by FABMS and will constitute an even greater challenge because the methods for distinguishing peptides that contain these cross linkages are not highly developed.

ACKNOWLEDGMENTS

This work was supported by grants GM 40384 and EY 07609 from the National Institutes of Health.

REFERENCES

1. **Staswick, P. E., Hermodson, M. A., and Nielsen, N. C.,** Identification of the cystines which link the acidic and basic components of the glycinin subunits, *J. Biol. Chem.*, 259, 13431, 1984.
2. **Wetlaufer, D. B., Parente, L., and Haeffner-Gormley, L.,** Use of proline-specific endopeptidase in the isolation of all four "native" disulfides of hen egg white lysozyme, *Int. J. Pept. Protein Res.*, 26, 83, 1985.
3. **Thannhauser, T. W., McWherter, C. A., and Scheraga, H. A.,** Peptide mapping of bovine pancreatic ribonuclease A by reverse-phase high-performance liquid chromatography, *Anal. Biochem.*, 149, 322, 1985.
4. **Creighton, T. E.,** Disulfide bond formation in proteins, in *Methods in Enzymology*, Vol. 107, Wold, F. and Moldave, K., Eds., Academic Press, New York, 1984, 305.
5. **Gray, W. R., Luque, F. A., Galyean, R., Atherton, E., Sheppard, R. C., Stone, B. L., Reyes, A., Alford, J., McIntosh, M., Olivera, B. M., Cruz, L. J., and Rivier, J.,** Conotoxin GI: disulfide bridges, synthesis, and preparation of iodinated derivatives, *Biochemistry*, 23, 2796, 1984.
6. **Reeve, J. R., Jr. and Pierce, J. G.,** Disulfide bonds of glycoprotein hormones, *Int. J. Pept. Protein Res.* 18, 70, 1981.
7. **Morris, H. R. and Pucci, P.,** A new method for rapid assignment of S-S bridges in proteins, *Biochem. Biophys. Res. Commun.*, 126, 1122, 1985.
8. **Takao, T., Yoshida, M., Hong, Y., Aimoto, S., and Shimonishi, Y.,** Fast atom bombardment (FAB) mass spectra of protein digests: hen and duck egg-white lysozymes, *Biomed. Mass Spectrom.*, 11, 549, 1984.

9. **Yazdanparast, R., Andrews, P. C., Smith, D. L., and Dixon, J. E.,** Assignment of disulfide bonds in proteins by fast atom bombardment mass spectrometry, *J. Biol. Chem.*, 262, 2507, 1987.

10. **Akashi, S., Hirayama, K., Seino, T., Ozawa, S., Fukuhara, K., Oouchi, N., and Murai, A.,** A determination of the positions of disulphide bonds in Paim I, α-Amylase inhibitor form *Streptomyces corchorushii*, using fast atom bombardment mass spectrometry, *Biomed. Environ. Mass Spectrom.*, 15, 541, 1988.

11. **Raschdorf, F., Dahinden, R., Maerki, W., Richter, W. J., and Merryweather, J. P.,** Location of disulphide bonds in human insulin-like growth factors (IFGs) synthesized by recombinant DNA technology, *Biomed. Environ. Mass Spectrom.*, 16, 3, 1988.

12. **Toren, P., Smith, D., Chance, R., and Hoffman, J.,** Determination of interchain crosslinkages in insulin β-chain dimers by fast atom bombardment mass spectrometry, *Anal. Biochem.*, 169, 287, 1988.

13. **Sun, Y., Zhou, Z., and Smith, D. L.,** Location of disulfide bonds in proteins by partial acid hydrolysis and mass spectrometry, in *Techniques in Protein Chemistry*, Hugli, T., Ed., Academic Press, New York, 1989, 178.

14. **Smith, D. L. and Zhou, Z.,** Strategies for locating disulfide bonds in proteins, in *Methods in Enzymology*, Vol. 193, McCloskey, J. A., Ed., Academic Press, New York, 1990.

15. **Zhou, Z. and Smith, D. L.,** Assignment of disulfide bonds in proteins by partial acid hydrolysis and mass spectrometry, *J. Protein Chem.*, 9, 1990.

16. **Burman, S., Wellner, D., Chait, B., Chaudhary, T., and Breslow, E.,** Complete assignment of neurophysin disulfides indicates pairing in two separate domains, *Proc. Natl. Acad. Sci. U.S.A.*, 86, 429, 1989.

17. **Konigsberg, W.,** Reduction of disulfide bonds in proteins with dithiothreitol, in *Methods in Enzymology*, Vol. 25, Hirs, C. H. W. and Timosheff, S. N., Eds., Academic Press, New York, 1972, 185.

18. **Sun, Y. and Smith, D. L.,** Identification of disulfide-containing peptides by performic acid oxidation and mass spectrometry, *Anal. Biochem.*, 172, 130, 1988.

19. **Sun, Y., Andrews, P. C., and Smith, D. L.,** *J. Protein Chem.*, 2, 1990.

20. **Allison, L. A. and Shoup, R.E.,** Dual electrode liquid chromatography detector for thiols and disulfides, *Anal. Chem.*, 55, 8, 1983.

21. **Lazure, C., Rochemont, J., Seidah, N. G., and Chretien, M.,** Novel approach to rapid and sensitive localization of protein disulfide bridges by high-performance liquid chromatography and electrochemical detection, *J. Chromatogr.*, 326, 339, 1985.

22. **Garvie, C. T., Straub, K. M., and Lynn, R. K.,** Quantitative liquid chromatographic determination of disulfide-containing peptide analogues of vasopressin with dual Hg/Au electrochemical detection, *J. Chromatogr.*, 413, 43, 1987.

23. **Yazdanparast, R., Andrews, P. C., Smith, D. L., and Dixon, J. E.,** A new approach for detection and assignment of disulfide bonds in peptides, *Anal. Biochem.*, 153, 348, 1986.

24. **Nashef, A. S., Osuga, D. T., Lee, H. S., Ahmed, A. I. Whitaker, J. R., and Feeney, R. E.,** *J. Agric. Food Chem.*, 25, 245, 1977.

Chapter 15

TANDEM MASS SPECTROMETRY FOR DETERMINING THE AMINO ACID SEQUENCE OF CYCLIC PEPTIDES AND FOR ASSESSING INTERACTIONS OF PEPTIDES AND METAL IONS

Ronald L. Cerny and Michael L. Gross

TABLE OF CONTENTS

I. Introduction 290

II. Amino Acid Sequence Determination of Cyclic Peptides 291
  A. Method Development with Synthetic Peptides 291
  B. Applications 296
  C. Unified Approach to Determining Amino Acid Sequence
     of Cyclic Peptides 298

III. Metal-Ion Cationized Peptide Decompositions 306
  A. Introduction 306
  B. Alkali Metal Cationized Peptides 307
  C. Dipeptides Complexed with Monovalent Cations 310
  D. MS/MS for Determining Sites of Metal Ion Interaction 311

IV. Future Prospects 311
  A. High Mass 311
  B. High Mass Resolution 312
  C. High Sensitivity 312

References 313

I. INTRODUCTION

There is little need to "sing the praises" of tandem mass spectrometry (MS/MS) to mass spectrometrists. It is necessary, however, to ensure that the message is conveyed to structural chemists. The acceptance of MS/MS as an important tool for structure determination can be gauged easily by the growing number of literature studies in which the technique is used and by the commitment of instrument manufacturers to the design and marketing of tandem mass spectrometers. This growth is precipitated certainly by the advent of fast atom bombardment (FAB) in 1981.[1] The use of this "soft" ionization technique opened up a wealth of opportunities for mass spectrometry in the analysis of biomolecules. Probably the largest contributor to the popularity of the combined technique of FAB and MS/MS is the relative ease with which MS/MS can be merged with FAB once a suitable instrument is incorporated into a laboratory. Recent reviews on the application of MS/MS to biomolecules,[2] including peptides[3] and lipids,[4,5] attest to the acceptance and utility of FAB and MS/MS.

By combining MS/MS with FAB, the strong points of FAB (soft ionization, abundant molecular species, ease of employment) can be seized upon whereas the weak points (background-peak-at-every-mass, lack of structurally informative fragment ions, discrimination effects) can be minimized. This combination also utilizes the greatest asset of MS/MS: the ability to relate fragment ions to a specific precursor ion. The structural determination of biomolecules is aided by the propensity of these species to fragment upon CA into their basic building blocks. With peptides, for example, the dominant fragmentation processes involve cleavages along the peptide backbone. Similarly, oligonucleotides and oligosaccharides tend to cleave at the phosphate and glycosidic linkages, respectively, to form smaller nucleotides or saccharides. By studying these types of fragmentations and the interrelationship among these building blocks, a major part if not all of the structure of the biomolecule can be determined.

The purpose of this chapter is to provide insight into the development of the combination of FAB with MS/MS, particularly in its use for sequencing cyclic peptides and in studying the interaction of peptides with metals. The amino acid sequence determination of linear peptides has been reviewed recently by Biemann,[3] and is not covered in this chapter. The sequence determination of cyclic peptides has been an area of interest at the University of Nebraska laboratory for a number of years. Over that time, there has been a continuing evolution of a methodogy for using FAB combined with MS/MS in sequencing cyclic peptides. This chapter presents a chronological view of this evolution by describing some examples from various stages along the way. Our current approach is demonstrated with the structure proofs of some novel cyclic peptides, which are isolated from natural products and may be potent anticancer agents.

The development of the ability to determine the sequence of these biomolecules stimulates interest in how these molecules work in a biological system and the relationship between structure and activity. Of particular interest to us is the complexation of cationic species by peptides, proteins, and enzymes. Complexation reactions are important in a wide range of biological systems. A few examples are in siderochromes for the transport of iron,[6] and as catalysts for reactions of either the complexing peptide itself[7] or as a co-factor in enzyme activity.[8] By using MS/MS, it may be possible to determine the site of complexation within the gas phase peptide and thereby decipher intrinsic interactions between the peptide and the metal ion. Our initial studies have concentrated on the interaction of small peptides with alkali metal cations.

Finally, we discuss briefly the future prospects for MS/MS, and what roles ionization, activation, and detection may play in further refinement of the technique.

II. AMINO ACID SEQUENCE DETERMINATION OF CYCLIC PEPTIDES

A. METHOD DEVELOPMENT WITH SYNTHETIC PEPTIDES

Cyclic peptides are difficult to sequence by classical methods such as Edman degradation because of the problems encountered in trying to hydrolyze specific peptide bonds and to form a single linear species for the determination. Electron ionization (EI) of cyclic peptides results in a variety of competitive fragmentation processes as well as large abundances of ions formed by rearrangement processes.[9,10] Nevertheless, Munson and co-workers have used EI to study the correlation of cyclic peptide ring-opening with peptide bond nonplanarity.[11] In contrast to EI, cyclic peptides form abundant $(M+H)^+$ ions upon FAB, making them quite amenable to FAB and MS/MS. The first obvious test of the usefulness of the combined technique is to evaluate its ability to determine the structure of known cyclic peptides.

Protonation of an amide nitrogen in a cyclic peptide followed by ring cleavage at the N-acyl bond gives a linear acylium ion, as shown in Equation 1. The major fragment ions formed either by metastable decompositions or upon collisional activation result from losses of amino acid residues from the C-terminus. If one amide nitrogen in the cyclic peptide is considerably more basic than any of the others, one linear acylium form of $(M+H)^+$ is formed predominantly; in that case, fragment ions in the MS/MS spectrum are related easily to the sequence of the original cyclic peptide.

$$H_2NCH^1RCONHCH^2RCONHCH^3RCO \rightarrow$$

$$H_2NCH^1RCONHCH^2RCONHCH^3RCO^+ \rightarrow \qquad (1)$$

$$H_2NCH^1RCONHCH^2RCO^+ + CO + CH^3R=NH$$

As an example, the FAB MS/MS spectrum of cyclo-(Leu-Pro-Gly)$_4$ is shown in Figure 1; Table 1 summarizes the fragment ion assignments.[12] The fragmentation process can be described on the basis of protonation at the amide nitrogen of Pro, Ala, or Gly followed by ring opening. Elimination of amino acid residues from the acylium ions results in the three series of fragment ions a_1-k_1, a_2-k_2, and a_3-k_3. (The original publication predates the now widely accepted nomenclature proposed by Roepstorff.[13] According to this latter nomenclature, the ions constitute a series of B_n ions formed as shown in Equation 1. Ions referred to as "primes" (d_1, e_1 ...) for cyclic peptides are, in Roepstorff terminology, a-type fragment ions formed by loss of a carbonyl group from the acylium ions.) Fragmentations initiated with protonation at proline are the major processes observed. The prolyl nitrogen is much more basic than the amide nitrogens of the aliphatic glycine and leucine. This increased basicity directs the protonation and the fragmentation. The other two series of ions that arise by protonation at primary amides are also present, but of low abundance. The dominance of the ion series from protonation at proline simplifies the spectrum and makes the sequence determination relatively straightforward.

If, however, the amide basicities are more equal, or if two or more strongly basic sites are present, several different molecular acylium ions are produced. The $(M+H)^+$ ions of the cyclic peptide that are selected for MS/MS are actually a *mixture* of different ring-opened acylium ions of the same mass. The MS/MS spectrum, therefore, is a composite of the individual spectra of each component. Interpretation becomes difficult because the interrelationship of the fragment ions is difficult to decipher. Such is the case for cyclo-(Lys-Pro-D-Ala)$_2$. Its FAB MS/MS spectrum (Figure 2) contains abundant fragment ions resulting from protonation at both Pro and Lys amide nitrogens; the ion series resulting from protonation at the less basic alanine amide is

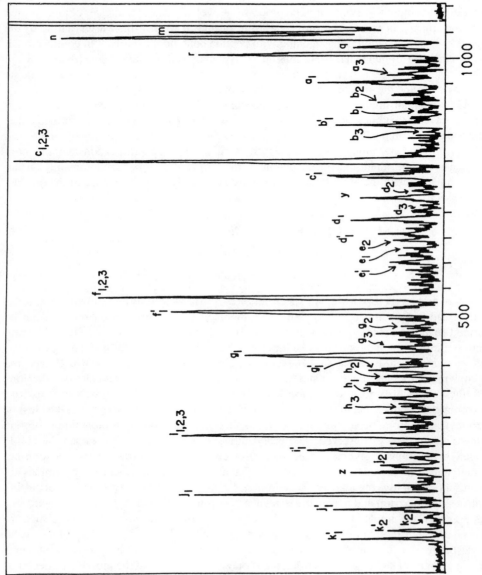

FIGURE 1.   FAB/MS/MS spectrum of the (M+H)⁺ ion, m/z 1069, of *cyclo*-(Leu-Pro-Gly)₄. (From Tomer, K. B., Crow, F. W., Gross, M. L., and Kopple, K. D., *Anal. Chem.*, 56, 880, 1984. Copyright 1984 American Chemical Society. With permission.)

TABLE 1
MS/MS Fragmentation of *cyclo*-(Leu-Pro-Gly)$_4$

Protonation on proline (1)		Protonation on leucine (2)		Protonation on glycine (3)		Protonation on sidechains	
Ion	Loss of[a]	Ion	Loss of[a]	Ion	Loss of[a]	Ion	Loss of[d]
a_1	Leu	a_2	Gly	a_3	Pro	m	NH_3
b_1	Gly	b_2	Pro	b_3	Leu	n	CO[e]
c_1[b]	Pro	c_2	Leu	c_3	Gly	q	C_3H_7
d_1	Leu	d_2	Gly	d_3	Pro	r	C_4H_9
e_1	Gly	e_2	Pro	e_3	Leu		
f_1	Pro	f_2	Leu	f_3	Gly		
g_1	Leu	g_2	Gly	g_3	Pro		
h_1	Gly	h_2	Pro	h_3	Leu		
i_1[c]	Pro	i_2	Leu	i_3	Gly		
j_1	Leu	j_2	Gly	j_3	Pro		
k_1	Gly	k_2	Pro	k_3	Leu		

[a] The designated loss is of the specified residue and all other residues listed above it. For example, ion c_1 refers to loss of Leu, Gly, and Pro although it has not been established that the losses are sequential.

[b] Ion c_1 also loses CO and C_3H_7 to yield ion y.

[c] Ion i_1 also loses CO and C_3H_7 to yield ion z.

[d] All ions arise from loss of the indicated fragment from the protonated molecular ions.

[e] This loss can arise from any of the ring-opened $(M + H)^+$ ions.

From Tomer, K. B., Crow, F. W., Gross, M. L., and Kopple, K. D., *Anal. Chem.*, 56, 880, 1984. Copyright 1984 American Chemical Society. With permission.

of much lower abundance. As is the case for method development studies involving known compounds, the fragment ions can be assigned on the basis of the known sequence, although this assignment would be difficult for an unknown of this type. The fragments are labeled in the figure to identify those fragment ions resulting from protonation at either proline or lysine.

As one goes to cyclic peptides of higher mass and thus containing more amino acids, the probability increases of having several amide sites of comparable basicity. The MS/MS spectrum becomes increasingly complex owing to its composite nature. In a sense, the spectrum is actually "information rich" and "information redundant" because all the fragment ions result from some form of expected fragmentation processes. Interpretation, however, is likely to be hindered by the exuberance of information. Furthermore, sensitivity is reduced because the total fragment ion current is dispersed over many species.

Other problems can arise in interpretation because of competitive fragmentations involving the N-terminus. Loss of an amino acid residue from the N-terminus of a ring-opened species in a Y-type process (Equation 2) may add confusion for the proper assignment of ions. When this loss occurs, the successive losses of amino acid residues from the C-terminus of the ring-opened $(M+H)^+$ ion will not account for all the fragment ions observed. This ambiguity makes it difficult to define unequivocally a sequence.

$$H_2NCH^1RCONHCH^2RCO(NHCHRCO)_nNHCHRCO^+ \rightarrow$$

$$H_2NCH^2RCO(NHCHRCO)_nNHCHRCO^+ + NH=CH^1R + CO \quad (2)$$

One possible approach to overcoming this problem is to determine the presence of ions containing the components of a dipeptide (e.g., $H_2NCH^1RCONHCH^2RCO^+$) and for the elimination of these species from the parent species in either the metastable ion or the CA spectrum of $(M+H)^+$.[14] If two different dipeptide fragments can be defined (e.g., AB and CD), then two possible sequences emerge: the so-called retro-isomers [i.e., (c(A-B-C-D) or c(D-C-

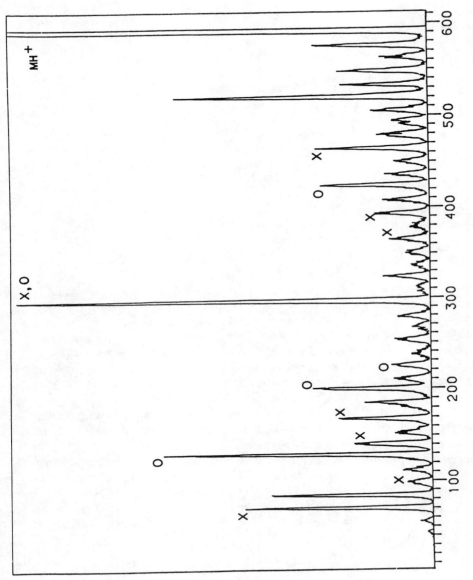

FIGURE 2.   FAB/MS/MS spectra of the (M+H)⁺ ion, m/z 593, of *cyclo*-(Lys-Pro-D-Ala)₂. Fragments resulting from protonation at Pro are labeled by X, those arising from protonation at Lys are indicated by O. (From Tomer, K. B., Crow, F. W., Gross, M. L., and Kopple, K. D., *Anal. Chem.*, 56, 880, 1984. Copyright 1984 American Chemical Society. With permission.)

B-A)], can be distinguished if either the N-terminus or the C-terminus of the ring-opened linear acylium ions can be identified. The N-terminus of the acylium ion can be ascertained if the X_{n-1} type ion, formed by elimination of $^1RCH=NH$ (Equation 3), is observed. This N-terminus specific process is, unfortunately, not always observed.

$$H_2NCH^1RCONHCH^2RCO(NHCHRCO)_nNHCHRCO^+ \xrightarrow{-^1RCH=NH}$$

$$HCONHCH^2RCO(NHCHRCO)_nNHCHRCO^+ \qquad\qquad (3)$$

Another approach for determining the termini involves examining the CA spectrum of the (M+H-CO)$^+$ immonium ion, A_n, which can fragment via two different processes as shown in Equation 4 for a tripeptide. The competitive reactions involve the two different termini of the A_n ion. The N-terminus amino acid can be eliminated to form an A_n that is one residue smaller (this ion is not an A_{n-1} ion, because the loss is from the N-terminus and not the C-terminus, as the A_{n-1} nomenclature would describe). Alternatively, the imine on the C-terminus can be lost, resulting in the B_{n-1} ion. The latter process, which is C-terminus specific, appears to be a favored fragmentation pathway for many ions of this nature.

$$
\begin{array}{c}
\qquad\qquad \xrightarrow{-^1RCHNHCO} H_2NCH^2RCONH=CHR^3 \\
H_2NCH^1RCONHCH^2RCONH=CH^3R \xrightarrow{\ \sim H\ } \\
\qquad\qquad \xrightarrow{-NH=CH^3R} H_2NCH^1RCONHCH^2RCO^+
\end{array}
\qquad (4)
$$

The sequence determination of an unknown cyclic peptide serves as an example of the application of this methodology. A synthetic pentapeptide was known to contain the amino acids Ser, Leu, Phe, and two Gly. The ions observed in the metastable and CA spectra of the protonated species of m/z 462 are summarized in Table 2. The dipeptides Gly-Ser, Ser-Leu, Leu-Phe, and Gly-Phe are observed. Only two possible acylium ions and only two corresponding sequences are consistent with these results (see Scheme 2). Differentiation of the two possibilities is accomplished by means of the metastable ion spectrum of the immonium ion of m/z 321 [M+H-Leu-CO]$^+$ (see Table 3). The most abundant fragmentation results from elimination of C_8H_9N or NH=CH-CH$_2$Ph, indicating that the structure of the m/z 321 ion must be Ser-Gly-Gly-NH=CH-CH$_2$Ph and not Phe-Gly-Gly-NH=CH-CH$_2$OH. This reasoning eliminates sequence **3**, leaving **1** as the correct acylium ion and **2** as the correct sequence of the cyclic pentapeptide (see below).

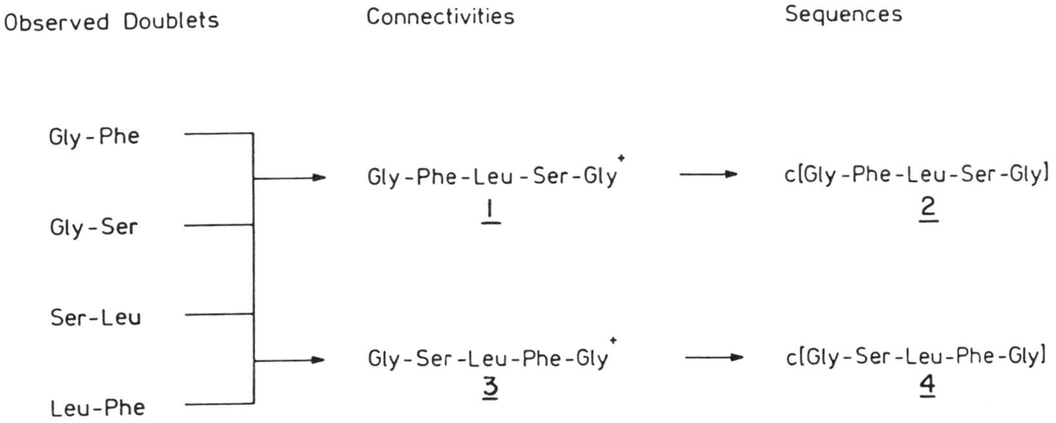

Observed Doublets	Connectivities	Sequences

Scheme 1

TABLE 2
Unimolecular and Collisionally Activated Decomposition (CAD) of m/z 462[a]

| m/z | Relative intensities | | Assignment |
	MI	CA	
444	100	100	$(M + H)^+ -H_2O$
434	42	82	$(M + H)^+ -CO$
405	1.0	8.0	$(M + H)^+ -Gly$
387	0.6	3.2	$(M + H)^+ -(Gly+H_2O)$
375	1.9	8.8	$(M + H)^+ -Ser$
349	11	52	$(M + H)^+ -Leu$
321		16	$(M + H)^+ -(Leu+CO)$
318	1.1	b	$(M + H)^+ -(Gly+Ser)$
301	0.8	5.2	$(M + H)^+ -(Gly+Ser+NH_3)$
279	1.0	b	$(GlyGlyPheNH_3)^+$
262	0.9	13	$(M + H)^+ -(Ser+Leu)$
258	0.9	b	$(M + H)^+ -(Gly+Phe)$
202	1.4	22	$(M + H)^+ -(Leu+Phe)$
145		11	$(Gly+Ser)^+$
120	0.3	64	$(Phe-CO)^+$
86		43	$(Leu-CO)^+$
60		10	$(Ser-CO)^+$
30		5.6	$(Gly-CO)^+$

[a]   The signal-to-noise ratio was >5:1.
[b]   Overlapping of broadened peaks prevents exact intensity measurements.

From Eckart, K., Schwartz, H., Tomer, K. B., and Gross, M. L., *J. Am. Chem. Soc.*, 107, 6765, 1985. Copyright 1985 American Chemical Society. With permission.

TABLE 3
Unimolecular Decomposition of m/z 321 [Protonated "Unknown"-(Leu+CO)]

m/z	Relative intensity	Assignment
304	88	m/z 321-NH_3
286	11	m/z 321-(NH_3+H_2O)
234	6	m/z 321-(Ser)
202	100	m/z 321-(C_8H_9N)
177	23	m/z 321-(Gly+Ser)
145	4	$(GlySer)^+$
120	70	$(Phe-CO)^+$

From Eckart, K., Schwartz, H., Tomer, K. B., and Gross, M. L., *J. Am. Chem. Soc.*, 107, 6765, 1985. Copyright 1985 American Chemical Society. With permission.

B. APPLICATIONS

The first use of FAB combined with MS/MS for the sequence determination of a peptide or cyclic peptide was reported in a study of HC-toxin.[15] This host-specific toxin, which attacks corn, was known to be a cyclic tetrapeptide made up of 2 Ala, 1 Pro, and 1 novel amino acid residue, 2-amino-9,10-epoxy-8-oxodecanoic acid (Aeo). On the basis of EI fragmentations, the toxin had been proposed incorrectly as c(Pro-Aeo-Ala-Ala).[16]

Upon FAB, the toxin produces an abundant $(M+H)^+$ of m/z 437. The CA spectrum of $(M+H)^+$ (see Figure 3) contains an abundant fragment ion of m/z 240 resulting from elimination of Aeo from a linear acylium ion, formed presumably by protonation at the prolyl amide nitrogen. This ion of m/z 240 contains the 3 remaining amino acid residues and, therefore, defines the sequence of the peptide. CA of source-produced ions of m/z 240 results in fragment ions of m/z 212 (–CO),

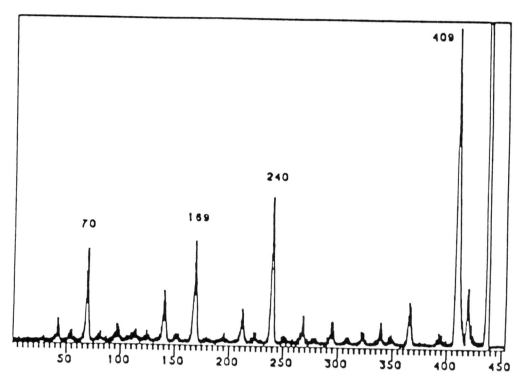

FIGURE 3. CID spectrum of the (M+H)$^+$ ion, m/z 437, of HC toxin. (From Gross, M. L., McCrery, D. A., Crow, F. W., Tomer, K. B., Pepe, M. R., Ciuffetti, L. M., Knoche, H. W., Daly, J. M., and Dunkle, L. D., *Tetrahedron Lett.*, 23, 5381, 1982. Copyright 1982 Pergamon Press. With permission.)

169 (-Ala), 141-(Ala + CO), and 70-(2Ala + CO), establishing the sequence of the m/z 240 ion as Pro-Ala-Ala. To insure that a rearrangement does not occur within the FAB source and does not obfuscate the interpretation, the MS/MS spectrum of source-produced ions of m/z 169 was compared to that obtained for ions of m/z 169 arising specifically from the ion of m/z 240. This critical discrimination was accomplished by an MS/MS/MS experiment. That is, all source-produced ions were activated in a collision cell located outside the source. The m/z 169 ions produced uniquely from m/z 240 ions were selected by employing the first analyzer of the triple sector instrument used in this study. Collisional activation of the m/z 169 ion was then accomplished in the second collision cell, located after the first stage mass analyzer.[17] The two spectra are identical, thus establishing that the amino acid losses are sequential. The correct structure of the HC-toxin was determined to be c(Pro-Ala-Ala-Aeo).

The structure of an HC-toxin analog was also determined by using FAB and MS/MS.[18] The analog differs from the originally identified toxin in that one of the Ala residues is replaced by a Gly residue. As with HC-toxin, the acylium ion formed by protonation at the prolyl nitrogen followed by ring-opening sequentially eliminates, upon CA, Aeo, Gly, and Ala. The structure of the HC-toxin analog, therefore, is c(Pro-Ala-Gly-Aeo).

A more challenging sequence problem was the cyclic peptide scytonemin A isolated from a blue-green alga.[19] From amino acid analysis of a 6N HCl hydrolyzate and detailed NMR studies, the peptide amino acid content was determined. Accurate mass measurement of the (M+H)$^+$ ion confirmed an empirical formula consistent with this content.

MS/MS of (M+H)$^+$ provided insufficient information to sequence the peptide. From nuclear magnetic resonance (NMR) studies, it was thought that the cyclic peptide contained a peptide side chain. The FAB mass spectrum and the CA spectrum of (M+H)$^+$ both contained an abundant fragment ion of m/z 1332, a loss of 131u from the protonated molecular species, consistent with elimination of N-acetylated Ala. No other significant information can be obtained from the CA spectra of either the $(M+H)^+$ (of m/z 1463) or the fragment of m/z 1332.

When the peptide was hydrolyzed under mild conditions, two major acyclic peptide fractions were separated and isolated. After acetylation and esterification, the first fraction produced upon FAB an $(M+H)^+$ of m/z 781. CA of this ion produced a number of abundant fragment ions. These ions can be rationalized by the fragmentation scheme outlined in Figure 4a. A portion of the first fraction was subjected to methanolysis. The FAB mass spectrum of the resultant material showed that a mass shift of $(M+H)^+$ to m/z 813 had been produced; its linked-field scan CA spectrum is shown in Figure 4b. Both CA spectra are consistent with a sequence of Ser-Gly-HyMePro-HyMePro-Leu-Hse (HyMePro is 4-hydroxy-3-methylproline and Hse is homoserine).

The FAB mass spectrometry of a second fraction (again acetylated and esterified) produced $(M+H)^+$ of m/z 1022. The CA spectrum of this ion showed that sequential losses of N-acetyl-Phe (to give an ion of m/z 833), Gly (to give the m/z 776 ion), and O-acetyl-HyLeu (to give the m/z 605 ion) from the N-terminus. The sequence of this portion was identified as Phe-Gly-HyLeu-MePro-Adha. (HyLeu is threo-3-hydroxyleucine, MePro is 3-methylproline, and Adha is 3-amino-2,5,9-trihydroxy-10-phenyldecanoic acid).

From different nuclear Overhauser effect (NOE) studies by NMR, the interlinkages Adha-Ser and Hse-Phe were established. These data defined the sequence of the cyclic portion of the peptide. Further NMR studies were used to determine the location of the N-acetyl Ala group, and the final structure was established to be that of **5** (see opposite page).

The previous example provides an important point for structural chemists. That is, in most cases, other types of critical supplemental information besides the mass spectral or the NMR data can be obtained readily. Just as significant advances in mass spectrometry have occurred over the past few years in the areas of ionization methods, mass range, and sensitivity, the NMR field has also progressed rapidly. High field NMR, homonuclear and heteronuclear correlation spectroscopy (COSY) experiments[20] and the use of coherence and heteronuclear bond correlations[21] to determine amino acid sequences have increased the power of NMR as a tool for structure determinations for many compounds including peptides and cyclic peptides. Although the sensitivity of NMR is a drawback, sufficient material may be isolated in studies of natural products so that these highly informative experiments can be performed. When the sample quantity is limited, however, mass spectrometry, and in particular MS/MS, will be the major source of structural information.

Information on amino acid content from only picomole quantities of the peptide can be obtained by classical methods after complete acid hydrolysis. Derivatization of small amounts of material followed by a second different mass spectral analysis can lead to additional insight. By comparing mass shifts in the spectra of the derivatized and nonderivatized samples, functionalities and their locations can sometimes be determined. Another possibility is the partial degradation of the sample to smaller subunits that then can be analyzed further by mass spectrometry.

C. UNIFIED APPROACH TO DETERMINING AMINO ACID SEQUENCE OF CYCLIC PEPTIDES

Building on our past experiences, we have evolved a concerted approach to sequence determinations, making use of other possible sources of information as well as obtaining MS/MS data. A mass spectral methodology has proven effective in sequencing peptides isolated from natural products. The steps are as follows:

1. Obtain FAB mass spectrum and verify molecular species.
2. Determine exact mass of molecular species.

3. Obtain MS/MS spectrum (both metastable ion and collisionally induced) of (M+H)⁺.
4. Compare the FAB and the MS/MS mass spectra.
5. Obtain MS/MS spectra of source-produced fragment ions that are significant fragments observed in the MS/MS spectrum of (M+H)⁺.
6. Determine (as necessary) exact masses of fragment ions to verify empirical formulae.

Some of these points appear rather obvious. Nevertheless, the significant information available from MS/MS spectra and exact mass measurements of source-produced fragment ions should not be overlooked, as has often been the case in other structure proofs. The utility of this approach will be demonstrated with a discussion of two examples, addressing each step in the methodology.

Normally, the identification of the molecular species is straightforward. Peptides, and in particular cyclic peptides, usually produce abundant (M+H)⁺ ions. Care must be taken to avoid mistakes when dealing with the exceptions. As an example, the FAB mass spectrum of dolastatin 13, a cyclic peptide isolated from the Indian sea hare *Dolabella auricularia*,[22] is shown in Figure 5. Using a protonating matrix such as glycerol or dithioerythritol/dithiothreitol (DTE/DTT), the sample produces an apparent (M+H)⁺ of m/z 888. Addition of a Li⁺ salt to the FAB matrix, however (see insert), results in the formation of (M+Li)⁺ of m/z 912. The ion of m/z 888 results from elimination of H₂O from the protonated molecular species.

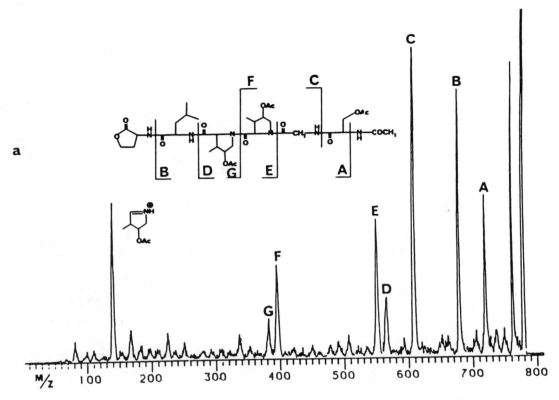

FIGURE 4.   (a) CAD mass spectrum of the (M+H)+ ion (m/z 781) of compound **5** showing fragmentations from the acetylated amino terminus [ions A (m/z 721), C (610), E (553), G (384)] and lactonized carboxyl terminus [ions B (m/z 280), D (567), F (398)]. (b) Linked scan mass spectrum of (M+H)+ ion (m/z 813) of compound **6** showing 31 amu increase in mass (due to addition of methanol to the carboxyl terminus) of fragment ions resulting from losses from the amino terminus. Ions A, C, and E are found at m/z 752, 642, and 585, respectively, whereas ions B, D, and F are found at the same positions, viz. m/z 680, 567, and 398, respectively. (From Helms, G. L., Moore, R. E., Niemczura, W. P., Patterson, G. M. L., Tomer, K. B., and Gross, M. L., *J. Org. Chem.*, 53, 1298, 1988. Copyright 1988 American Chemical Society. With permission.).

Exact mass measurements can be quite useful in verifying molecular formula. Their utility for *determining* a molecular species, however, decreases dramatically as the molecular weight of the peptide increases. Consider, for example, a peptide with a mass of ~500u. A usual error limit for exact mass determination is ±5ppm. Even within these limits, however, a large number of possible elemental compositions exists. The list of 20 to 30 possible combinations of C, H, N, O, and S can be narrowed usually to 5 to 10 possibilities that would be compatible for a peptide. The number of reasonable formulae, however, goes up by a factor of 5 to 10 for a peptide of mass 1000. Another way of looking at this relationship is to compare the ability to distinguish between two formulae differing by N_2(28.0061) and CO(27.9949) (a difference of 0.0112). At mass 200, the two formulae differ by 5.6 ppm, and can be distinguished routinely from each other. At mass 1000, however, the relative difference is 1.1 ppm. This small difference is within the experimental precision (and error) of accurate mass measurements; it would be difficult to rule out either formula on the basis of exact mass data alone.

Even having established an exact mass and empirical formula, one should keep in mind that this information, by itself, is of limited use. An example is an ion of 518.2336u. Considering only the 20 "natural" amino acids, we note that there are *37 combinations* of amino acids that have this exact mass and empirical formula. Once you take into account sequences, there are 5136 *different* peptides with the same exact mass! (Furthermore, this numerical analysis considers only normal linear peptides!)

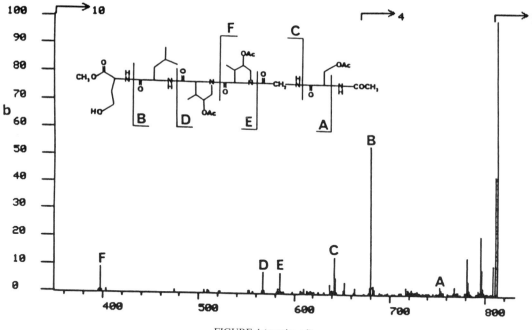

FIGURE 4 (continued).

A large amount of information can be obtained by comparing the FAB mass spectrum with the MS/MS spectrum obtained for (M+H)⁺. It is often stated that, compared to the MS/MS spectrum, few high mass ions are observed in the FAB mass spectrum. This statement is, however, somewhat misleading because multiplication factors of 25 to 200 are normally used to display fragment ions in the MS/MS spectra. If comparable gain factors are applied to the normal FAB spectrum, fragment ions can usually be observed. Most of the fragment ions seen in the MS/MS spectrum will also appear in the FAB mass spectrum. The usual limiting factor for identifying these fragments is the level of matrix background interference. Problems also arise if the sample is a mixture or if the sample contains a significant amount of impurities. In many cases, however, purification and separation techniques such as high pressure liquid chromatography (HPLC) can result in a relatively pure, single component sample.

Problems can sometimes arise in the mass assignments of ions in the MS/MS spectrum due to translational energy losses incurred by the fragment ions upon CA. When an electrostatic analyzer functions as MS-II, these energy losses result in ions being detected at lower voltages (and, hence, lower mass) than expected.[23] The degree of translational energy loss is compound- and fragmentation process-dependent and difficult to predict, but usually causes no more than a 3u shift. Comparing the masses of ions observed in the FAB spectrum to those assigned to fragment ions produced upon CA provides a check of those assignments. (It should be noted that the problem of translational energy loss can usually be compensated for if MS-II is a double-focusing mass spectrometer).

The MS/MS spectra of *source-produced* fragment ions can provide important information for determining the correct sequence of a peptide. From an amino acid analysis, a cyclic peptide hymenistatin 1 (isolated from a South Pacific Ocean sponge)[24] was found to contain 3 proline residues. As was discussed previously, protonation and the resultant ring-opening is likely to occur at any one of these three prolyl amide nitrogens. The result is the formation of three *different* isobaric linear acylium ions. The MS/MS spectrum of (M+H)⁺ is, therefore, a composite of the individual MS/MS spectra of these different species. The FAB mass spectrum and the MS/MS spectrum of (M+H)⁺ (Figures 6 and 7, respectively) both contain fragment ions of m/z 780, 794, and 796. These ions can result from losses of Ile, Pro, and Val, respectively,

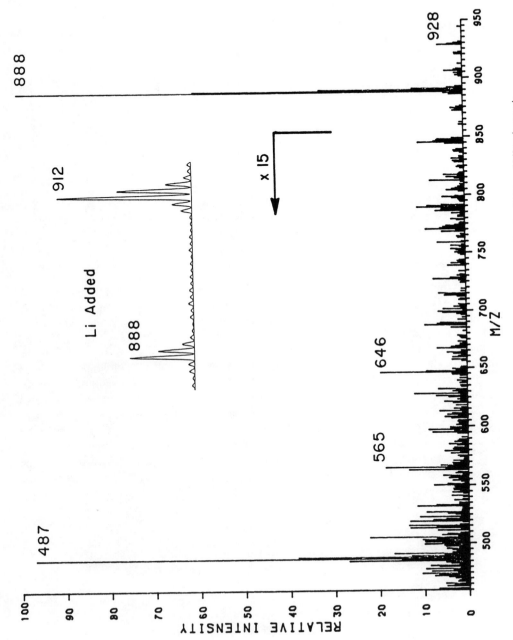

FIGURE 5. FAB mass spectrum of dolastatin 13; insert, FAB spectrum after the addition of LiI to the matrix.

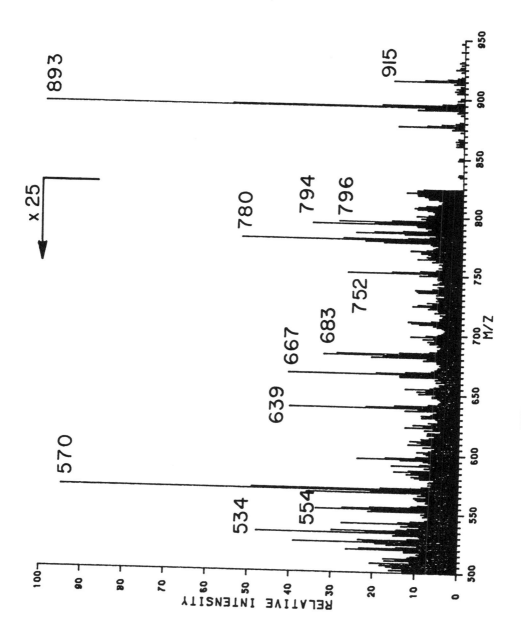

FIGURE 6.   FAB mass spectrum of hymenistatin 1.

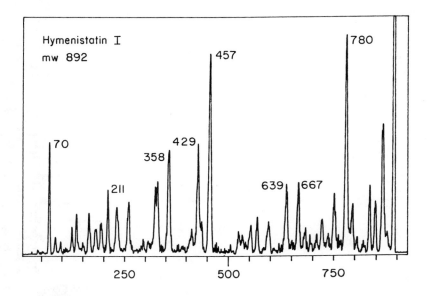

FIGURE 7.   FAB/MS/MS spectrum of (M+H)⁺ ion, m/z 893, of hymenistatin 1.

from the C-terminus of the three different acylium ions. These fragments are, therefore, characteristic of the individual acylium ions, and the MS/MS spectra of these three different *source-produced* fragment ions are no longer composite in nature. This system provides a means to resolve the isobaric mixture. The interrelationship of fragment ions described in Scheme 2 allows determination of the cyclic peptide sequence as c(Pro-Pro-Tyr-Val-Pro-Leu-Ile-Ile).

MS/MS spectra of source-produced fragment ions can also be useful for confirming that amino acid losses are sequential. The MS/MS spectrum of an (M+H)⁺ produced by FAB of dolastatin 13[22] is shown in Figure 8. Upon magnification, several abundant fragment ions are observed. A preliminary assumption may be that the fragment ions of m/z 789, 646, and 487 are formed by sequential losses. By comparing MS/MS spectra of source-produced fragment ions of these masses (Figure 9), it is clearly evident that these three fragment ions are not related and are produced by competitive fragmentation processes. The complexity of the peptide structure is apparent from these MS/MS spectra. Even though the three ions do not result from sequential eliminations, their subsequent fragmentations converge to produce common product ions, reflecting those portions of their structures that are common.

The structure of dolastatin was determined to be **6** (see opposite page, bottom); a fragmentation pathway identifying the ions observed in the MS/MS spectrum of (M+H)⁺ is shown in Scheme 3. The basic structure consists of a six amino acid cyclic portion with a two amino acid side chain. The observed propensity for the loss of water is due to the presence of a unique hemiacetal that is derived presumably from a Phe-Glu dipeptide precursor. Cyclization of the peptide occurs through the threonine oxygen. The fragment ions result from cleavage of the side chain bond as well as by a number of different ring-opening processes and eliminations of various amino acid residues.

An objection to obtaining MS/MS spectra of source-produced fragment ions may be raised on the basis of limited sample quantity. The purpose of these additional experiments is either to confirm that losses are sequential or to determine the interrelationship of fragment ions observed in the MS/MS spectrum of (M+H)⁺. This can be accomplished by obtaining fewer than normal scans (or an equivalent number at a faster scan rate). Often, the most useful information is the presence or absence of abundant fragment ions, and there is less concern about lower abundance ions. S/N, therefore, is not as important as it would be if these data were the only ones being

cyclo [Pro^a - Pro^b - Tyr - Val - Pro^c - Leu - Ile - Ile]

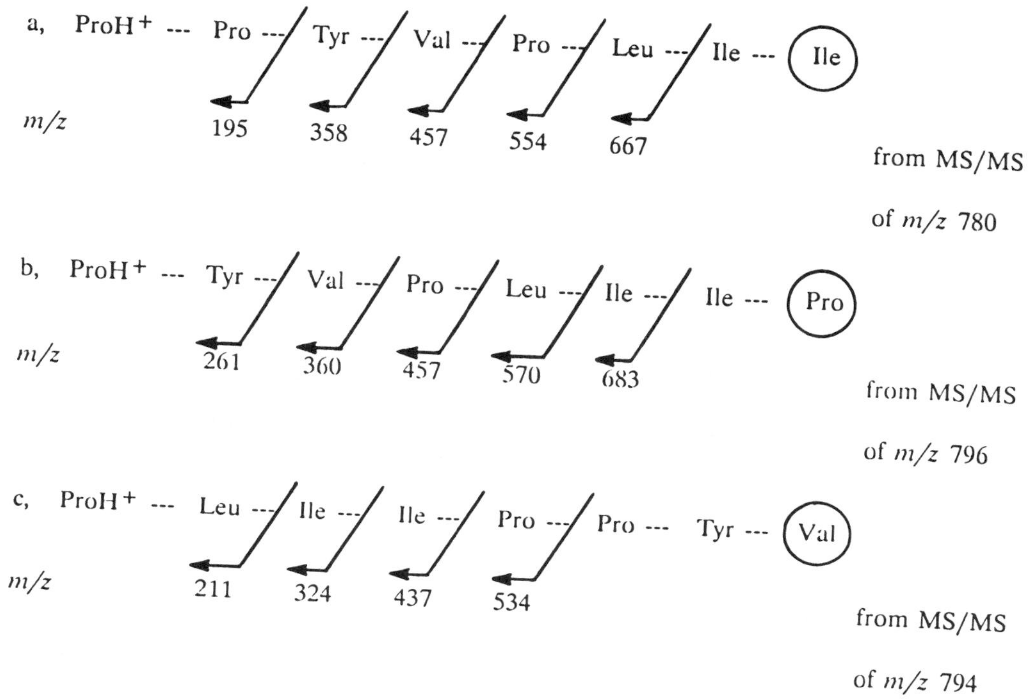

a, ProH$^+$ --- Pro ---/ Tyr ---/ Val ---/ Pro ---/ Leu ---/ Ile --- ⟨Ile⟩

m/z   195   358   457   554   667

from MS/MS

of *m/z* 780

b, ProH$^+$ --- Tyr ---/ Val ---/ Pro ---/ Leu ---/ Ile ---/ Ile --- ⟨Pro⟩

m/z   261   360   457   570   683

from MS/MS

of *m/z* 796

c, ProH$^+$ --- Leu ---/ Ile ---/ Ile ---/ Pro ---/ Pro --- Tyr --- ⟨Val⟩

m/z   211   324   437   534

from MS/MS

of *m/z* 794

Scheme 2

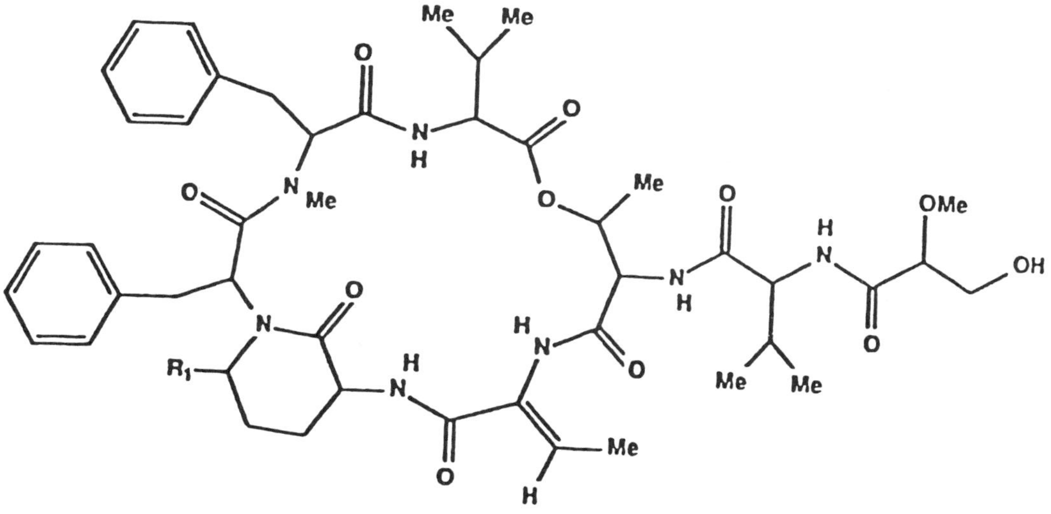

R$_1$ = OH

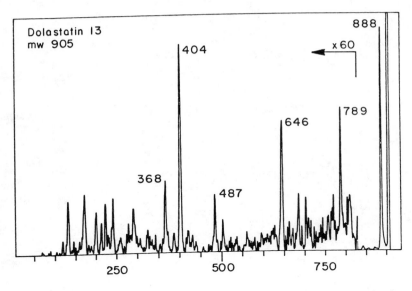

FIGURE 8.   FAB/MS/MS spectrum of (M+H)⁺ ion, m/z 906, of dolastatin 13.

obtained. Several of these confirmatory MS/MS scans can be obtained with a single sample loading.

The final point in our methodology checklist is exact mass measurement of fragment ions. From the earlier discussion on exact masses, the main purpose for these experiments is to verify empirical formula proposed for fragment ions. Instances occur, however, where key information can be obtained from exact mass measurements of lower mass ions, for which empirical formula can be *determined* with greater reliability. By determining both the formula and relationship (from MS/MS spectra) of a few lower mass fragment ions, an alternative approach of building up the peptide sequence rather than degrading it from (M+H)⁺ can be employed.

The test of any methodology is its successful application to determinations of unknown structure. MS/MS coupled with FAB ion production has been used in our laboratory for six structure proofs.[15,18,19,22,24,25] Although not all steps in the checklist were used, we find that the complete methodology is particularly important in collaborative projects with Pettit and the Cancer Research Institute at Arizona State University. The peptides, which are isolated from marine animals and plant life, are of reasonably high molecular weight, contain unusual or modified amino acids and may have complex side chains. This complexity often requires application of all the measurements cited in the checklist.

III. METAL-ION CATIONIZED PEPTIDE DECOMPOSITIONS

A. INTRODUCTION

Most research and applications of mass spectrometry to peptides are devoted to sequencing. Peptides are now sequenced by means of MS/MS experiments of (M+H)⁺ ions. As was pointed out in the discussion of the cyclic peptide research, these (M+H)⁺ precursors often comprise a mixture of species protonated at different sites, and the MS/MS spectra of (M+H)⁺ are composite in nature.

Two reasons exist for studying metal ion cationized peptides; the first is analytical in nature and attempts either to overcome deficiencies or to complement current methodology. It would be desirable for sequencing to form a single charge-localized ion species. One approach, which was proposed by others,[26] is through chemical modification or derivatization. An alternate and simpler approach is cationization to form (M+Cat)⁺. Cationization may be thought of as a simple

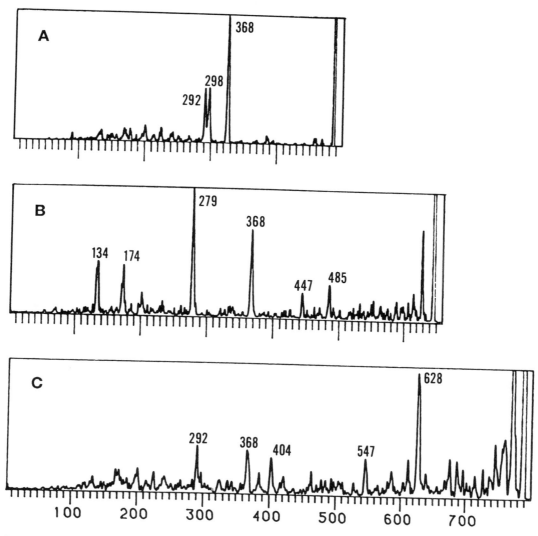

FIGURE 9. CA spectra of source-produced fragment ions of dolastatin 13; (A) of m/z 487, (B) of m/z 646, (C) of m/z 789.

derivatization that can be carried out *in situ* in the FAB matrix. The hope is that the cationization process is more selective, or that it occurs at different (relative to protonation) sites in the peptide. Precedents for the success of cationization are well established for structure studies of sugars,[27,28] bile salts,[29] fatty acids and alcohols,[30-32] nucleotides, and nucleosides,[33,34] and the structural information is often complementary to that of MS/MS experiments of (M+H)+.[35,36]

The second reason is found in the study of bioactivity involving cationization of peptides and proteins. In this area, it is important to ascertain where the metal is located. The MS/MS study of cationized peptides should provide information on these sites of interaction. The gas phase interactions reflect intrinsic properties of the metal-peptide complex, free from solvation effects.

B. ALKALI METAL CATIONIZED PEPTIDES

The first reported study involving cationization of peptides was of a tripeptide with Na+. Russell and co-workers[37,38] suggested that the site of interaction in the peptide is at highly basic sites such as the N-terminus or at side chains containing an amine or amide functionality. Others have proposed alternate interactions. Renner and Spiteller[39] suggested that the alkali metal

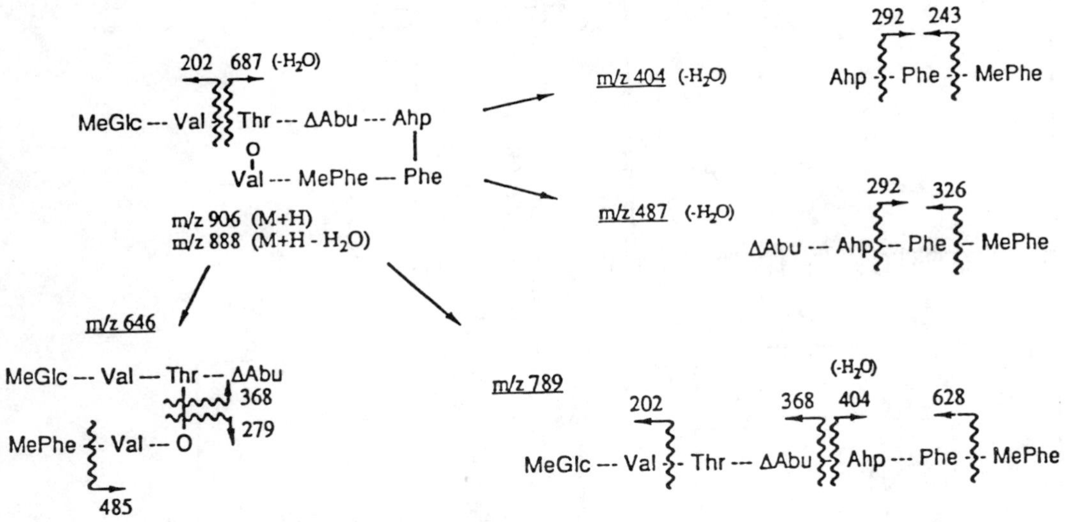

202   687 (-H₂O)

MeGlc --- Val {{ Thr --- ΔAbu --- Ahp
        O
        Val --- MePhe — Phe

m/z 906 (M+H)
m/z 888 (M+H - H₂O)

m/z 646

292   243
Ahp }- Phe }- MePhe

m/z 404 (-H₂O)

292   326
ΔAbu --- Ahp}- Phe }- MePhe

m/z 487 (-H₂O)

MeGlc --- Val — Thr --- ΔAbu
                    368
MePhe }- Val --- O   279

485

m/z 789

(-H₂O)
202   368   404   628
MeGlc --- Val }- Thr --- ΔAbu }} Ahp --- Phe }- MePhe

Scheme 3

interacts with the C-terminus carboxylate. They found that the major fragmentation process of the three cationized peptides studied involved elimination of the C-terminal amino acid residue. Tang et al.[40] observed a similar process for cationized leucine- and methionine-enkephalin in a reflecting time-of-flight (TOF) mass spectrometer. Our experience with over 60 cationized di- to nonapeptides[41] suggests that C-terminal amino acid elimination is general, establishing a C-terminus interaction and providing a method for determining the C-terminal amino acid.

The CA spectra of (M+H)⁺ and (M+Li)⁺ of methionine-enkephalin offer a comparison of the information that can be obtained from these two different species (see Figure 10). The CA spectrum of the (M+H)⁺ gives abundant information for identifying the peptide sequence. N-terminus and C-terminus-containing ions are both produced. In contrast, the CA spectrum of the (M+Li)⁺ contains principally only higher mass ions; the most abundant is generated by the elimination of Met from the C-terminus. In the Roepstorff terminology,[13] this production is termed to be (B₄+Li+OH)⁺. The other major fragment ion of m/z 403 is the Li⁺ analog of the A₄ ion observed in the MS/MS spectrum of (M+H)⁺, and is discussed later.

The mechanism for the elimination of the C-terminal amino acid is shown in Scheme 4. The peptide is expected to be in a zwitterionic form, and the initial interaction with the metal ion involves the deprotonated C-terminus. The complexation with Li⁺ or Na⁺ polarizes the carbonyl bond adjacent to the terminal amino acid residue. Nucleophilic attack of the negatively charged oxygen on this site results in elimination of CO and an imine. The product ion is another peptide less the C-terminal amino acid.

Several pieces of evidence support this mechanism:

1. The elimination of the amino acid at the C-terminus occurs in a consecutive way. The CA spectrum of the m/z 449 ion produced exclusively from elimination of the C-terminal Leu from (M+Li)⁺ upon CA of leucine enkephalin eliminates preferentially the next amino acid Phe to form an ion of m/z 302.
2. Collisional activation of an isotopically labeled peptide gives neutral losses consistent with the loss of CO and an imine from the cationized species.
3. Collisional activation of tripeptide analogs demonstrates that both a free acid at the C-terminus and a carbonyl functionality on the adjacent residue are required for the fragmentation process to occur.

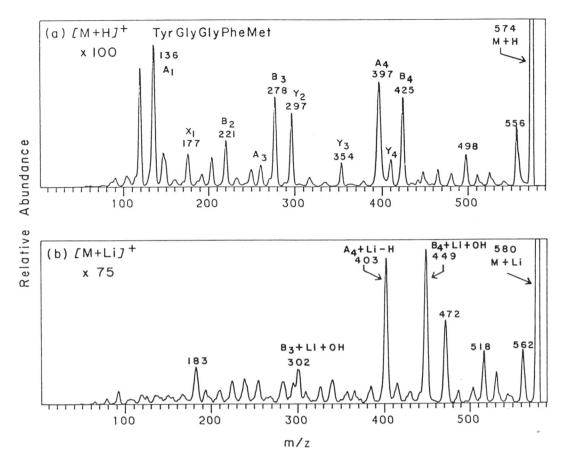

FIGURE 10.   CA mass spectra of methionine-enkephalin: (A) (M+H)⁺, m/z 574; (B) (M+Li)⁺, m/z 580. (From Grese, R. P., Cerny, R. L., and Gross, M. L., *J. Am. Chem. Soc.*, 111, 2835, 1989. Copyright 1989 American Chemical Society. With permission.)

Scheme 4

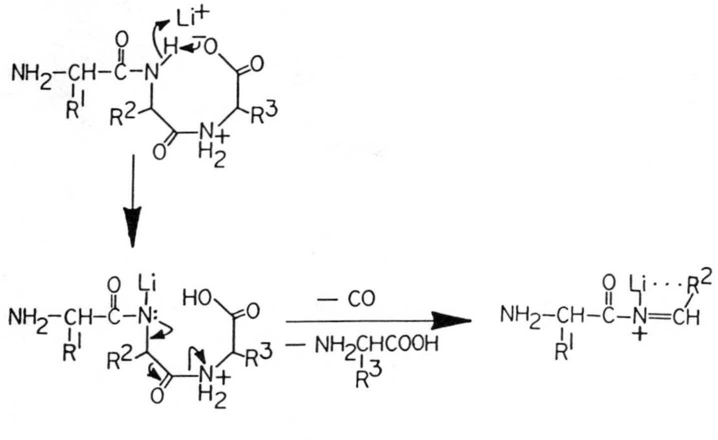

Scheme 5

Although the metastable C-terminus elimination is general and normally the major fragmentation process, there is one exception. If an Arg residue adjoins the C-terminus, a facile loss of H_2O occurs from the product $[B_{n-1}+Li+OH]^+$. This example is the only case thus far where the general trend does not hold.

Upon CA of cationized peptides, a fragmentation process that competes with the C-terminal amino acid elimination results in the formation of A-type ions with the metal ion substituted for a hydrogen, as proposed in Scheme 5. The $(A_n+Li-H)^+$ ions are most abundant when their formation takes place at the site of amino acids whose side chains provide stabilization for the binding of a cation at the deprotonated amide site. His offers that opportunity; the pyridine nitrogen of the side chain imidazole ring of His is involved in the formation of a stable, six-membered ring chelate of the metal ion.

Even for peptides containing His, the metastable ion spectra are dominated by the $(B_n+Li+OH)^+$ ions. The alternative process that results in the formation of $(A_n+Li-H)^+$ becomes more competitive upon CA. There is an increase in the amount of $(A_n+Li-H)^+$ formed; the abundance of $(B_n+Li+OH)^+$ remains relatively constant in both metastable ion and CA experiments. The fragmentation to give $(A_n+Li-H)^+$ has a slightly higher energy barrier than that for production of $(B_{n-1}+Cat+OH)^+$, accounting for the favored formation of the latter in metastable ion decompositions. The former process, however, is favored kinetically when the $(M+Li)^+$ is energized by CA. Isomerization to a deprotonated amide and subsequent decomposition to the former species occur rapidly. This is particularly true when His is available to lower the transition state energy for $(A_n+Li-H)^+$ production.

C. DIPEPTIDES COMPLEXED WITH MONOVALENT CATIONS

Dipeptides constitute another special case in the cationization of peptides. The $(M+Cat)^+$ ions decompose metastably via two different competitive processes (Equation 5). The first process leads again to the formation of $(B_n+Li+OH)^+$ ions, analogous to the process discussed previously for larger peptides.

$$H_3NCH^1RCONHCH^2RCOOLi \longrightarrow \begin{cases} H_3NCH^1RCOOLi + CO + NH=CH^2R \\ H_3NCH^2RCOOLi + CO + NH=CH^1R \end{cases} \qquad (5)$$

The relative abundance of the two fragment types is dependent on the basicity of the two amino acids; if one of the constituent amino acids contains a side chain capable of accepting a proton, the ion containing that amino acid is favored. For example, the relative amounts of the two products are approximately equal for aliphatic dipeptides containing Ala, Leu, and Val. If one amino acid contains a high proton affinity side chain (e.g., His, Arg, Lys), the fragmentation process involving the formation of that amino acid as the charged species will be dominant. Other amino acids (e.g., Phe, Met, Ser) are intermediate in their behavior. The overall behavior of alkali metal-complexed dipeptides is consonant with an interaction located at the C-terminus. Li^+, as expected, is the most strongly complexed.

The presence of Gly upsets the trend; the formation of the cationized Gly amino acid dominates, regardless of position, when placed in competition with the other aliphatic amino acids. This deviation is presumably because of the relative instability of imine formed in the elimination of that portion of glycine ($NH=CH_2$) as compared to the more substituted imines formed from the aliphatic side chains of the other amino acids.

The most abundant fragment produced upon CA is $(Y+Li+H)^+$. The basic amino acids, however, are still favored in the fragmentation, irrespective of their location in the dipeptide. This occurrence can hinder the sequencing of these peptides by this method.

D. MS/MS FOR DETERMINING SITES OF METAL ION INTERACTION

It has been proposed that certain sequences of amino acids within a protein provide environments for the specific complexation of certain metals. The presence of these "motifs" in a protein can be indicative of their biological function or at least of their role in complexation reactions. One example suggesting the existence of such motifs is the binding of calcium by proteins in the white muscle of fish and amphibia.[42] The structures of various parvalbumin proteins that are present in tissue provide a pattern or sequence of amino acids in the regions where the calcium ions are believed to be bound. An example can be made by comparing these amino acid sequences with those found in troponin C, the subunit of the muscle regulatory protein that binds calcium in mammalian systems. The sequences in the two different systems are strikingly similar. The sequences can be reduced to an easily recognizable pattern consisting of regions of several invariant amino acids (which must be in specific positions) and other sites where the amino acids are interchangeable. The presence of this "motif" in a protein may indicate specific activity for calcium.

Most of the studies that have investigated the sites of complexation of metals by peptides have relied on techniques such as circular dichroism and NMR. Conclusions are based on *changes* in measurements with and without the metal ion. By using MS/MS combined with FAB, it may be possible to determine the site of metal ion complexation directly by observing the fragmentation processes of the cationized peptides. Work is currently underway to determine the feasibility of establishing the existence of specialized "motifs" for various cationic species by using FAB and MS/MS.

IV. FUTURE PROSPECTS

The foregoing discussion illustrates the utility of desorption ionization (DI) combined with MS/MS in two of many areas in which the combination has been applied effectively. In the case of cyclic peptides and metal ion peptide complexes, the approach reveals both structural information (i.e., sequence) and the nature of metal ion interactions with peptides. Thus, there are strong incentives to develop further the capability of DI and MS/MS.

There are three areas that, when addressed appropriately, will lead to improvements in the capability of DI and MS/MS to deal with structural problems: (1) high mass, (2) high mass resolution, and (3) greater sensitivity.

A. HIGH MASS

The 3-sector tandem mass spectrometer[43] used for most of the studies reported here is perhaps the first to be applied extensively to problems of structural determination of biomolecules. It was conceived in 1978 before the requirement for high mass had been delineated. Its upper mass limit is 1800u at full accelerating potential (8 kV). Today, 11 years later, sector instruments having upper mass limits of 15,000u are reasonably routine because of advances in magnet technology (e.g., larger radius magnets of inhomogeneous fields or having non-normal entry faces).[44]

Although TOF and possibly Fourier transform mass spectrometers achieve higher upper mass limits, the current limit of sector instruments far exceeds our ability to induce fragmentation in large molecules. The upper mass limit for inducing fragmentation by collision with neutral gases is ca. 3000u, an upper mass limit that is accessible to both sector and quadrupole instruments.

B. HIGH MASS RESOLUTION

The spectra reported in this chapter show poor mass resolution because the second stage of the tandem mass spectrometer is only an electrostatic analyzer. It is clear today that a routine improvement in resolution of ca. 20 times can be made by using a double-focusing mass spectrometer as the second stage analyzer.[45] The first instrument for analytical chemistry that incorporated two double-focusing spectrometers in tandem was that of McLafferty and students.[46] For fundamental studies and complex mixture analysis, high mass resolution is also needed in the first stage. The 3-sector tandem instrument at the University of Nebraska was the first to offer that capability.[43] High "front-end" resolution, however, has not yet become important for biomolecule structural studies.

The more elegant approach to high resolution mass analysis in the second stage is Fourier transform mass spectrometry (FTMS).[47] Although the high resolution of FTMS has been demonstrated convincingly for small molecule systems, this achievement is considerably more difficult for large molecules, especially biomolecules. It is conceivable that fragments of biomolecules as large as m/z 5000 can be produced *routinely* and mass-analyzed at a resolution of 50,000 by using FT technology. Encouragement is offered by the recent demonstration of Wilkins and co-workers,[48] in which ions of m/z 6000 from a synthetic polymer were mass-analyzed at a resolution of 60,000.

C. HIGH SENSITIVITY

The three areas of instrumentation research that need to be emphasized in order to increase the sensitivity of DI and MS/MS for biomolecule studies are ionization, detection, and activation. Current detection limits in favorable cases are in the 1 to 10 pmol range. It is necessary to make these detection limits routine and to improve the ultimate detection limit by a factor of 100. These levels of sensitivity would make mass spectrometry competitive with more classical biochemical techniques such as the Edman degradation or DNA sequencing for peptide and protein structure determination.

Sector instruments are plagued by the requirement to use DI and liquid matrices (e.g., FAB, liquid secondary ion mass spectrometry [SIMS]). The liquid matrix introduces chemical noise that limits the amount of detectable analyte signal. One approach that allows the use of continuous ion beams that are needed usually for sector instruments is continuous-flow (CF) FAB.[49] Here, the replacement of the organic liquid matrix by water, for example, leads to a significant reduction of chemical noise. This technique is covered in detail in a chapter by Caprioli. So far, however, there have been only a few demonstrations of CF FAB and MS/MS.

Another approach is to use a form of DI that does not require a liquid matrix. Examples are field desorption, laser desorption, and pulsed SIMS. Impressive success has been demonstrated recently for laser desorption.[50] Unfortunately, these ionization modes are not compatible with sector or hybrid tandem mass spectrometers because the ion beams are short-lived, highly fluctuating, or even purposefully pulsed. The utilization of multichannel plates may improve the marriage between MS/MS and DI from the solid state.

The other approach is to merge FT or TOF mass spectrometry and matrixless DI, taking advantage of these spectrometers' inherent "multichannel advantage". Some progress has been made in the FTMS area.[51]

The improvement in detector technology has been mentioned already. The addition of multichannel plates to sector instruments will not only allow these instruments to be used with ionization methods that produce fluctuating ion beams, but also permit simultaneous recording of portions of the fragment ion spectrum. The first demonstration of the use of a multichannel plate on a contemporary mass spectrometer was by scientists at Kratos.[52] Biemann and students[53] have extended this technology in an elegant way to a preexisting four-sector tandem mass spectrometer. We have been interested in constructing a multichannel tandem sector mass spectrometer since 1985, and have sought sufficient funds and a commitment by an instrument company to build a new generation spectrometer having significant multichannel capability in its original design. This instrument was recently completed by VG Analytical of Manchester, U.K. The instrument incorporates a special Mattauch-Herzog type design for the second stage.

The final area in which improvements will be found is ion activation. To date, collisional activation has been the general means of ion activation. The limitation of ever-decreasing energy input as the molecular mass of the ion increases is now well known. Other methods of ion activation are needed if MS/MS is to be applied to analytes at subpicomole levels or of high molecular weight (>3000u). Two approaches appear promising: photoactivation with pulsed ultraviolet (UV) lasers and surface activation.[54] Both methods have the potential of overcoming the energy transfer limitations of collisional activation with small molecule target gases. Incorporation of these techniques into tandem sector or FT mass spectrometers will require creative design and development. Progress with photoactivation of biomolecules in both types of instruments was reported recently.[55,56]

REFERENCES

1. **Barber, M., Bordoli, R. S., Sedgwick, R. D., and Tyler, A. N.,** *J. Chem. Soc. Chem. Commun.*, 325, 1981.
2. **Tomer, K. B.,** *Mass Spectrom. Rev.*, 8, 445, 1989.
3. **Biemann, K. and Martin, S.,** *Mass Spectrom. Rev.*, 6, 75, 1987.
4. **Jensen, N. J. and Gross, M. L.,** *Mass Spectrom. Rev.*, 6, 497, 1987.
5. **Adams, J.,** *Mass Spectrom. Rev.*, 9, 141, 1990.
6. **Neilands, J. B.,** Microbial iron transport compounds (siderochromes), in *Inorganic Biochemistry*, Vol. 1, Eichhorn, G. L., Ed., Elsevier, Amsterdam, 1973.
7. **Jones, M. M. and Hix, J. E., Jr.,** Metal induced ligant reactions involving small molecules, in *Inorganic Biochemistry*, Vol. 1, Eichhorn, G. L., Ed., Elsevier, Amsterdam, 1973.
8. **Scratton, M. C.,** Metal enzymes, in *Inorganic Biochemistry*, Vol. 1, Eichhorn, G. L., Ed., Elsevier, Amsterdam, 1973.
9. **Anderegg, R. J., Biemann, K., Buchi, G., and Cushman, M.,** *J. Am. Chem. Soc.*, 98, 3365, 1976.
10. **Tuinman, A. A.,** in *Mycotoxins and Phycotoxins*, Steyn, P. S. and Vlegaar, R., Eds., Elsevier, Amsterdam, 1985.
11. **Banlak, E. L., II, Ma, Y.-C., Gierasch, L. M., and Munson, B.,** *J. Am. Chem. Soc.*, 111, 5487, 1989.
12. **Tomer, K. B., Crow, F. W., Gross, M. L., and Kopple, K. D.,** *Anal. Chem.*, 56, 880, 1984.
13. **Roepstorff, P. and Fohlman, J.,** *Biomed. Mass Spectrom.*, 11, 601, 1984.
14. **Eckart, K., Schwarz, H., Tomer, K. B., and Gross, M. L.,** *J. Am. Chem. Soc.*, 107, 6765, 1985.
15. **Gross, M. L., McCrery, D. A., Crow, F. W., Tomer, K. B., Pope, M. R., Ciuffetti, L. M., Knoche, H. W., Daly, J. M., and Dunkle, L. D.,** *Tetrahedron Lett.*, 23, 5381, 1982.
16. **Liesch, J. M., Sweeley, C. C., Stahfeld, G. D., Anderson, M. S., Weber, D. J., and Scheffer, R. P.,** *Tetrahedron*, 38, 45, 1982.
17. **Burinsky, D. J., Cooks, R. G., Chess, E. K., and Gross, M. L.,** *Anal. Chem.*, 54, 295, 1982.
18. **Kim, S.-D., Knoche, H. W., Dunkle, L. D., McCrery, D. A., and Tomer, K. B.,** *Tetrahedron Lett.*, 26, 969, 1985.
19. **Helms, G. L., Moore, R. E., Niemczura, W. P., Patterson, G. M. L., Tomer, K. B., and Gross, M. L.,** *J. Org. Chem.*, 53, 1298, 1988.

20. **Bax, A.,** *Two-Dimensional Nuclear Magnetic Resonance in Liquids*; Delft University Press, Delft, Holland, 1982.

21. **Bax, A. and Drobny, G. J.,** *Magn. Reson.*, 61, 306, 1985.

22. **Pettit, G. R., Kamano, Y., Herald, C. L., Dufresne, C., Cerny, R. L., Herald, D. L., Schmidt, J. M., and Kizu, H.,** *J. Am. Chem. Soc.*, 111, 5015, 1989.

23. **Neumann, G. M., Sheil, M. M., and Derrick, P. J.,** *Z. Naturforsch.*, 39a, 584, 1984.

24. **Pettit, G. R., Clewlow, P. J., Dufresne, C., Doubeck, D. L., Cerny, R. L., and Rutzler, K.,** *Canadian J. Chem.*, in press.

25. **Pettit, G. R., Kamano, Y., Herald, C. L., Dufresne, C., Bates, R. B., Schmidt, J. M., Cerny, R. L., and Kizu, H.,** submitted, *J. Org. Chem.*, 55, 2989, 1990.

26. **Kidwell, D. A., Ross, M. M., and Colton, R. J. J.,** *J. Am. Chem. Soc.*, 106, 2219, 1984.

27. **Röllgen, F. W., Giessman, U., Borchers, F., and Levsen, K.,** *Org. Mass Spectrom.*, 13, 459, 1978.

28. **Puzo, G., Fournie, J.-J., and Prome, J.-C.,** *Anal. Chem.*, 57, 892, 1985.

29. **Liehr, J. G., Kingston, E. E., and Beynon, J. H.,** *Biomed. Mass Spectrom.*, 12, 95, 1985.

30. **Adams, J. and Gross, M. L.,** *J. Am. Chem. Soc.*, 108, 6915, 1986.

31. **Adams, J. and Gross, M. L.,** *Anal. Chem.*, 59, 1576, 1987.

32. **Adams, J. and Gross, M. L.,** *Org. Mass Spectrom.*, 23, 307, 1988.

33. **Tomer, K. B., Gross, M. L., and Deinzer, M. L.,** *Anal. Chem.*, 58, 2527, 1986.

34. **Mallis, L. M., Raushel, F. M., and Russell, D. H.,** *Anal. Chem.*, 59, 980, 1987.

35. **Cerny, R. L., Tomer, K. B., and Gross, M. L.,** *Org. Mass Spectrom.*, 21, 655, 1986.

36. **Gross, M. L., Tomer, K. B., Cerny, R. L., and Giblin, D. E.,** in *Mass Spectrometry in the Analysis of Large Molecules*; McNeal, C. J., Ed., John Wiley & Sons, New York, 1986, 171.

37. **Mallis, L. M. and Russell, D. H.,** *Anal. Chem.*, 58, 1076, 1986.

38. **Russell, D. H., McGlohon, E. S., and Mallis, L. M.,** *Anal. Chem.*, 60, 1818, 1988.

39. **Renner, D. and Spiteller, G.,** *Biomed. Mass Spectrom.*, 15, 75, 1988.

40. **Tang, X., Ens, W., Standing, K. G., and Westmore, J. B.,** *Anal. Chem.*, 60, 1791, 1988.

41. **Grese, R. P., Cerny, R. L., and Gross, M. L.,** *J. Am. Chem. Soc.*, 111, 2835, 1989.

42. **Coffee, C. J., Bradshaw, R. A., and Kretsinger, R. H.,** *Adv. Exp. Med. Biol.*, 48, 211, 1973.

43. **Gross, M. L., Chess, E. K., Lyon, P. A., Crow, F. W., Evans, S., and Tudge, H.,** *Int. J. Mass Spectrom. Ion Phys.*, 42, 243, 1982.

44. **Cottrell, J. S. and Greathead, R. J.,** *Mass Spectrom. Rev.*, 5, 215, 1986.

45. **Sato, K., Asada, T., Ishihara, M., Kunihiro, F., Kammei, Y., Kubota, E., Costello, C. E., Martin, S. E., Scoble, H. A., and Biemann, K.,** *Anal. Chem.*, 59, 1652, 1987.

46. **McLafferty, F. W., Todd, P. J., McGilvery, O. C., and Baldwin, M. A.,** *J. Am. Chem. Soc.*, 102, 3360, 1980.

47. **Laude, D. A., Jr., Johlman, C. L., Brown, R. S., Weil, D. A., and Wilkins, C. L.,** *Mass Spectrom. Rev.*, 5, 107, 1986.

48. **Ijames, C. F. and Wilkins, C. L.,** *J. Am. Chem. Soc.*, 110, 2687, 1988.

49. **Caprioli, R. M.,** *Mass Spectrom. Rev.*, 6, 237, 1987.

50. **Karas, M. and Hillenkamp, F.,** *Anal. Chem.*, 60, 2299, 1988.

51. **Wilkins, C. L., Chowdhury, A. K., Nuwaysir, L. M., and Coates, M. L.,** *Mass Spectrom. Rev.*, 8, 67, 1989.

52. **Cottrell, J. S. and Evans, S.,** *Anal. Chem.*, 59, 1990, 1987.

53. **Hill, J. A., Biller, J. E., Martin, S. A., Biemann, K., Yoshidome, K., and Sato, K.,** *Int. J. Mass Spectrom. Ion Proc.*, 92, 211, 1989.

54. **Mabud, Md. A., Dekrey, M. J., and Cooks, R. G.,** *Int. J. Mass Spectrom. Ion Proc.*, 67, 285, 1985.

55. **Martin, S. A., Hill, J. A., Kittrell, C., and Biemann, K.,** *J. Am. Soc. Mass Spectrom.*, 1, 107, 1990.

56. **Hunt, D. F., Shabanowitz, J., and Yates, J. R.,** *J. Chem. Soc. Chem. Commun.*, 8, 548, 1987.

Chapter 16

ANALYSIS OF PERMETHYLATED PEPTIDES BY MASS SPECTROMETRY

Keith Rose

TABLE OF CONTENTS

I.   Introduction 316

II. Chemistry 316
     A.   Sample Preparation 316
     B.   The Permethylation Reaction 317
     C.   Acylation of Amino Groups 317
     D.   Formation of Derivatives of Cysteine 318
     E.   Formation of Derivatives of Arginine 318

III. Mass Spectrometry of Permethylated Peptides 319
     A.   Fragmentation 319
     B.   Sample Introduction 320

IV. Applications 320
     A.   Blocked N-Termini 320
     B.   Carboxyterminal Peptide Identification 322
     C.   Other Applications 322

V.   Conclusions 322
     A.   Advantages and Disadvantages of the Permethylation Approach 322
     B.   When to Use the Permethylation of Peptides 323

References 323

I. INTRODUCTION

Since the early days of organic mass spectrometry, this powerful and general analytical technique has been applied to the study of peptides. The history of the application of mass spectrometry to the characterization of polypeptides and proteins represents the struggle to overcome two great problems, which shall be referred to as the "separation" problem and the "volatility" problem. The separation problem arises from the fact that a protein is generally characterized through the structural analysis of a fairly large set of its corresponding peptides, which must be purified (separated) for the purpose. The volatility problem stems from the zwitterionic, hydrogen-bonded nature of peptides. Of course, the separation and volatility problems are linked: the larger the peptide accessible to structural analysis (i.e., the better the volatility problem is resolved), the smaller the number of (in principle overlapping) peptides that need be isolated to obtain the structure of the sample protein (i.e., the separation problem becomes less acute).

The volatility problem, which was a formidable one for the earlier methods of sample introduction and ionization, was solved by forming chemical derivatives. Formation of chemical derivatives has the disadvantages of the time needed to prepare them and the possibility of sample loss, but may confer the advantages of improved mass spectrometric fragmentation characteristics and the possibility of using gas-liquid chromatography (and liquid chromatography) for sample introduction. Gas chromatography is a means of separating very small quantities of material and can provide resolution comparable with reversed-phase high pressure liquid chromatography (RP HPLC) separation methods; it therefore offers a solution to the separation problem. Formation of chemical derivatives followed by analysis by combined gas chromatography-mass spectrometry (GC-MS) is thus a very attractive approach to the characterization of polypeptides and small proteins because it overcomes both the separation problem and the volatility problem and can assist structural analysis by directing mass spectrometric fragmentation predominantly to the peptide bonds.

Many chemical derivatives have been examined, and methods for preparing the most used of these have improved considerably over the last 20 to 30 years. The principal derivatives[1-5] are acyl-trimethylsilyl-polyaminoalcohols, acyl-permethyl derivatives, acyl-esters, trimethylsilyl derivatives, and Schiff-base derivatives. Of these derivatives, polyaminoalcohols and permethyl derivatives are the most used to solve real world protein problems, although acyl-esters have been found to be useful in work with peptide antibiotics. With the advent of ionization techniques applicable to large underivatized peptides — field desorption (FD), plasma desorption (PD), fast atom bombardment (FAB), secondary ion mass spectrometry (SIMS), laser desorption/ionization, thermospray, etc. (see other chapters) — some decline in interest in the formation of chemical derivatives would have been anticipated. In fact, derivatization is still actively investigated (see, e.g.,[6]) as a means of gaining selectivity and absolute sensitivity of detection with some of these newer ionization techniques.

The reader is referred to previous reviews[1-5] for the historical development of the formation of derivatives of peptides for mass spectrometric analysis. This review focuses on the permethylation of peptides as it is practiced today and gives some indications as to future developments.

II. CHEMISTRY

A. SAMPLE PREPARATION

As is true of most mass spectrometric techniques, samples must be pure in ways that surprise many biochemists. To someone who has worked hard to produce a single band on an electrophoretic gel, it is disappointing to learn that the sample is contaminated with inorganic salts, sealing film, plasticizers, detergents such as Triton, etc. For the permethylation reaction, only traces of inorganic salts can be tolerated, and better results are obtained if the sample is free from the contaminants mentioned here; significant amounts of sodium dodecyl sulfate can be tolerated, however.[7] Often, a simple clean-up procedure (RP HPLC or gel filtration) is adequate, but it may be risky when only very small quantities of sample are available.

B. THE PERMETHYLATION REACTION

The permethylation reaction, as its name implies, involves the substitution by a methyl group for every hydrogen atom attached to a heteroatom (an atom other than carbon). Powerful reagents (nowadays dimethylsulfinyl sodium and methyl iodide) are employed, necessitating prior acylation of amino groups and prior derivatization of the disulfide bond of cystine residues and the guanido group of arginine residues if peptides containing such groups are to form suitable derivatives upon permethylation.

Full experimental details of reagent preparation and permethylation reaction conditions have been published[7] and therefore are not repeated here. Strict adherence to the protocol is important. For example, if dimethylsulfinyl sodium is left in contact with the acylated peptide for more than the prescribed time, then nucleophilic attack by the base at the carbonyl group of the X-Pro bond becomes appreciable, particularly where X = trifluoroacetyl (i.e., where Pro- is N-terminal). Such an attack leads to cleavage of the X-Pro bond. Three improvements to the published procedure are worthy of note. Two have been mentioned previously.[8] The first is that higher yields are obtained if the glass sample tubes are either new tubes that have been rinsed with distilled water and pyrolyzed at 500°C for about 60 min, or are tubes that have been silanized. This change presumably reduces adsorption of peptide to the tubes, which were originally acid-washed. The second improvement concerns the replacement of triphenylmethane from Fluka with material from British Drug Houses for the preparation of dry DMSO and of dimethylsulfinyl sodium; use of the Fluka material appears to give rise to volatile, colored by-products. The third improvement consists of using, for the preparation of dimethylsulfinyl sodium, a flask having a male cone with a short protrusion beyond the cone. This alteration has the effect of preventing possible contamination of the preparation with silicone grease brought down from the (previously female) cone by condensation of the mist of DMSO produced during the reaction with sodium hydride; such contamination used to occur from time to time and led to the rejection of those batches of dimethylsulfinyl sodium.

Using these techniques good results may be generally obtained using 2 to 10 nmol peptide. Results[7] obtained with partial hydrolysates of the standard substrate glucagon represent a significant improvement over work published at the same time in which a less optimized permethylation procedure was used.[9]

The permethylation reaction is not applicable to sample amounts below about 2 nmol, and the reasons for this limit are unclear. Attempts to use amino acids or peptides as carriers met with little success.[10] On the hypothesis that adsorption was to blame, we have investigated the use of solvent-leached polypropylene tubes, but could not reduce the presence of extractables to an acceptable level. If the problem of permethylating smaller amounts could be solved, permethylation would find wider application because it is possible to chromatograph and detect subpicomole quantities of peptides that have been permethylated in amounts greater than 2 nmol. It should be noted that, even with the optimized permethylation procedure described here, C-methylation can occur, especially at glycyl residues adjacent to prolyl residues. Use of trideuteriomethyl iodide in the permethylation reaction is thus useful because it permits a distinction to be made between an artefactual C-methyl-glycyl (alanyl) residue and a naturally occurring one. However, the extent of C-methylation encountered under optimized conditions[7] is relatively small.

C. ACYLATION OF AMINO GROUPS

Prior acylation of amino groups is necessary to prevent their quaternization during the subsequent permethylation reaction. The acetyl group is useful for this purpose. It is of low mass, stable, and introduced easily by volatile reagents. We prefer aqueous pyridine,[7] because it is a powerful solvent for peptides and because the commonly used alternative, methanol, can lead to esterification and consequent problems upon permethylation (β-ketosulfoxide formation, alkylation, and elimination to give M+10 signals[11]). An important application involving N-terminal labeling with a deuterated reagent is discussed below. N-terminal prolyl residues have been found to be prone to a side reaction if permethylated after acetylation,[12] but such a side reaction cannot occur if the prolyl residue carries the trifluoroacetyl group. The trifluoroacetyl group offers advantages over the acetyl group of higher volatility[13,14] and improved mass spectral characteristics.[7,13] It is conveniently introduced using volatile reagents,[7] although when following this protocol, care must be taken to pierce holes of sufficient diameter in the aluminum foil used to cover the sample tubes. This permits removal of phenol in the first pumping step; otherwise, an excessive amount of base is required for the subsequent permethylation.

D. FORMATION OF DERIVATIVES OF CYSTEINE

Although the thiol group of cysteine (nondisulfide bonded) may be permethylated to give S-methylcysteine, modification of the disulfide bond of cystine residues prior to the permethylation reaction is essential if volatile derivatives are to be formed. This necessity does not usually present a problem, because one of the first steps of primary structure determination involves the cleavage of any disulfide bonds and the blocking of all cysteine residues. The purpose of the cleavage is twofold: to disrupt the native conformation in order to give proteolytic agents freer access to the polypeptide backbone, and to stabilize the potentially reactive thiols and disulfides, preferably as water-soluble derivatives. Several methods are currently in use and give rise to modified peptides with quite different properties. Oxidation with performic acid yields cysteic acid ($-SO_3H$, acidic), oxidative sulfitolysis produces S-sulfocysteine ($-SSO_3H$, acidic), and reduction followed by alkylation gives S-methylcysteine ($-SCH_3$, neutral), S-carboxymethylcysteine ($-SCH_2CO_2H$, weakly acidic), S-carboxamidomethylcysteine ($-SCH_2CONH_2$, neutral), S-aminoethylcysteine ($-SCH_2CH_2NH_2$, basic), or S(4-pyridylethyl)cysteine ($-SCH_2CH_2C_5H_4N$, weakly basic), depending on the reagent used for the alkylation.

When analysis is to be performed by mass spectrometry following the permethylation reaction, it is the S-methyl derivative that is the most suitable. The S-methyl derivative, which is conveniently introduced,[15] is sufficiently stable to survive intact the permethylation reaction, gives rise to easily interpretable mass spectra, and, by virtue of its small neutral nature, leads to derivatives that are relatively volatile. The S-carboxymethyl (and probably S-carboxamidomethyl) derivative is prone to undergo both C-methylation (on the carbon flanked by both sulfur and a carbonyl group) and a β-elimination reaction during permethylation, leading to formation of a dehydroalanyl peptide. Morris[16-18] has reported that such an elimination is quantitative under his experimental conditions. Although the products of these side reactions may often be resolved by gas chromatography, their presence is an unnecessary complication. S-aminoethyl derivatives are compatible with permethylation (when acylated along with other amino groups), but are considerably less volatile than S-methyl ones and can give rise to additional peaks in the mass spectrum.[11] S-pyridylethyl derivatives have not been permethylated, but would be expected to be prone to alkylation on the pyridine nitrogen leading to an involatile quaternized salt. Cysteic acid derivatives are not suitable because they cannot be methylated. S-sulfo derivatives are little used and rather unstable.

E. FORMATION OF DERIVATIVES OF ARGININE

Residues of arginine, like residues of cystine (but not cysteine), do not form suitable derivatives upon direct acylation and permethylation. As with cystine, if no appropriate action is taken prior to permethylation, then peptides containing those residues will not generally give results either by normal direct insertion or by GC-MS methods. It is important to realize that other peptides not containing these residues but present at the same time will be derivatized normally and will give good results.

One way of dealing with the problem of arginine is to remove that polar residue by conventional methods (Edman degradation or treatment with carboxypeptidase B). Such a tactic is possible only when the residue to be removed is close to the N- or C-terminus and forfeits information if applied to a mixture of peptides because the residues released can seldom be assigned to particular peptides. Some prior knowledge of the sequence is implied, but this information may be available in the case of tryptic peptides.

When arginine is present but not close to a terminal position, it becomes necessary to modify the guanido group in a separate reaction prior to the permethylation reaction. As with the acylation reaction (described previously), we demand that the reaction product should, besides being volatile and possessing good mass spectral characteristics, be formed in good yield on a small scale without forfeiting sequence information. In addition, in order to avoid complications, the reaction should affect only the residue that it is designed to modify. In these respects, the two most useful modifications of a residue of arginine are hydrazinolysis to a residue of ornithine, and condensation with pentane-2,4-dione (acetylacetone) to yield a residue of dimethylpyrimidylornithine.

Hydrazinolysis requires use of a very powerful nucleophile under rather vigorous conditions.[17] Backbone cleavage has been reported[16] as has formation of side chain hydrazides.[16] Hydrazinolysis is suitable for small peptides, but the side reactions become more troublesome with increasing peptide length. Where possible, the hydrazinolysis reaction should be monitored by RP HPLC[19] and the modified peptide recovered by this means in a preparative run.

The preferred modification for rendering arginine residues compatible with the permethylation reaction is condensation with pentane-2,4-dione. This reagent has been used under strongly acidic conditions,[20,21] but basic conditions are preferable.[22,23] Very mild conditions have been described for modification of proteins, but solubility problems may be encountered.[24] We have adapted the mild conditions by including urea[25] (not guanidine, which would be expected to react with the reagent) as follows. Protein or peptide is dissolved in a reagent solution (50 µl solution for nmol amounts of sample, or up to 10 mg/ml for larger quantities of sample) made by adding pentane-2,4-dione (0.5 ml) to a saturated solution of $NaHCO_3$ in deionized 8 M urea (4.5 ml) and adjusting the pH to 8.5 by dropwise addition of a solution (100 mg/ml) of Na_2CO_3 in deionized 8 M urea. After 48 h at 37°C, the reaction mixture may either be acidified carefully with pure acetic acid and sample recovered by gel filtration in dilute acetic or formic acid, or, after acidification followed by extraction with diethylether, the aqueous phase may be applied to RP HPLC for recovery of sample. Peptides containing dimethylpyrimidylornithine in place of arginine elute later than their unmodified counterparts. Besides the much milder conditions, condensation with pentane-2,4-dione has another advantage over hydrazinolysis: the reaction product has a characteristic ultraviolet (uv) absorption band at about 310 nm ($\varepsilon = 5600$ at pH 1).[24] Ornithine is formed quantitatively on hydrolysis under standard conditions for amino acid analysis.[24]

There are other methods (see, for example, Reference 26) for the modification of residues of arginine prior to permethylation, but they are mainly of historical interest.

III. MASS SPECTROMETRY OF PERMETHYLATED PEPTIDES

A. FRAGMENTATION

Space does not permit a full presentation of the modes of fragmentation of acetyl and trifluoroacetyl permethyl peptides. The general fragmentation of permethylated peptides under electron impact ionization has been well described.[3] Further useful information is given by Dell and Morris.[11] The N–C cleavage undergone by Asp, Asn, and the aromatic residues has been found to occur also with S-alkyl Cys residues,[7] and unusual fragmentations at Pro and Trp have been described.[12] However, a glance at some representative spectra[3,7] shows that spectral interpretation is usually straightforward.

B. SAMPLE INTRODUCTION

Most studies of permethylated peptides have involved use of either the direct insertion probe or GC-MS for sample introduction.[3,5] Both techniques are applicable to mixtures of peptides. The direct insertion probe has permitted peptides up to 16 residues in length to yield sequence information, although 10 residues is more common. When using the direct insertion probe, peptides of up to about 10 residues may be analyzed with a good chance of obtaining most if not all of the amino acid sequence information, and a few components of differing volatility may be separated by fractional distillation. When using GC-MS, peptide size is limited to about five residues for practical purposes (i.e., when several of the heavier residues are present) but considerable sample complexity may be tolerated (digest of a small protein).

Several authors have studied liquid chromatography-mass spectrometry (LC-MS) as a means of separating permethylated peptides and introducing them into the mass spectrometer, thereby in principle removing the volatility restriction of GC-MS. It should nevertheless be pointed out that the largest derivatives chromatographed, the enkephalins, are in fact amenable to GC-MS.[27] Normal phase LC was used in one study[28] and RP LC in a more extended study.[29] It was shown[29] that permethylated peptides give good peak shapes under practical RP HPLC conditions and that good on-line mass spectra may be obtained. Indeed, the results obtained are better than they appear because the spectrum of Phe–Val–Gln–Trp was misinterpreted. (In their Figure 4b, the authors thought they were dealing with Phe–Val–Glu–Trp; in fact, the spectrum is clearly that of Phe–Val–Gln–Trp: m/z 487 is due to the sequence ion Phe–Val–Gln+, m/z 417 and 530 are due to C-terminal ammonium ions +H2–Gln–Trp–OMe and +H2–Val–Gln–Trp–OMe, respectively, and m/z 372 is probably due to H++PCA*–Trp–OMe; *PCA = pyrrolidone carboxylic acid.) Unfortunately, neither group[28,29] used trifluoroacetylation or the latest[7] conditions for the permethylation reaction, so some chemical problems were encountered (mainly C-methylation) and large amounts of sample were used.

Permethylated peptides have been examined by FAB mass spectrometry (FABMS) using the standard FAB direct introduction probe.[30] Figure 1 shows the positive ion FAB mass spectrum of the chemotactic tetrapeptide Val–Gly–Ser–Glu as its N^α-trifluoroacetyl-N,O-permethyl derivative. Except in the molecular ion region, the spectrum is similar to that obtained by GC-MS with electron impact (EI) ionization[27] and consists of a dominant series of acylium sequence ions. In particular, and in contrast to FAB mass spectra of underivatized peptides (see other chapters), sequence ions are more prominent than the protonated or sodium cationized molecular ions (note the scale factor in Figure 1). Sequence ions are also prominent in the PD spectra of permethylated peptides, in marked contrast to the dominant molecular ion species observed when underivatized peptides are studied by this ionization technique.[47] Incidentally, the mass spectrum in Figure 1 testifies to the stability of permethylated peptide derivatives: the sample had been permethylated 7 years previously and stored at room temperature in the dark. Future applications of LC-MS and FABMS of permethylated peptides are discussed later in this chapter.

IV. APPLICATIONS

A. BLOCKED N-TERMINI

A major application of mass spectrometry of permethylated peptides is the characterization of the N-terminal regions of proteins and polypeptides that are N^α-blocked and thereby not directly amenable to the Edman degradation. In this application, due to Gray and del Valle,[31] a deuterated reagent is used to block amino groups on a protein (blocking groups present naturally are not displaced in this procedure). After partial hydrolysis of the protein with an enzyme to produce fragments of a size suitable for mass spectrometry, the permethylation reaction is used both to create a volatile derivative of the N-terminal fragment and in the same process to quaternize free amino groups generated on non-N-terminal fragments by the partial hydrolysis

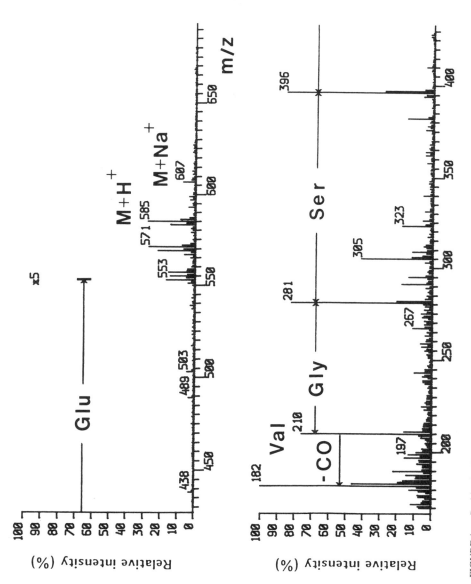

FIGURE 1. Positive ion FAB mass spectrum of Val–Gly–Ser–Glu as its N^{α}-trifluoroacetyl-N,O-permethyl derivative. The protonated molecular ion appears at m/z 585. The spectrum was obtained using a Kratos MS 50 S mass spectrometer fitted with an M-Scan FAB retrofit. Sample was bombarded with 9 kV xenon ions in a matrix of dithiothreitol-dithioerythritol-acetic acid. The author is grateful to Luc-Alain Savoy for providing this spectrum.

step. Thus, the chloroform extraction that follows the permethylation reaction serves to separate the hydrophobic derivative of the N-terminal peptide from the more hydrophilic quaternized salts of the non-N-terminal peptides. Numerous applications of this procedure have been cited in previous reviews.[3,5,32] More recent examples include acyl-CoA-binding protein (which was found to be acetylated) from cow liver,[33] fructose-1,6-bisphosphate aldolase (also found to be acetylated) from *Drosophila*,[34] and pancreatic spasmolytic polypeptide (confirmed to be blocked by a pyroglutamyl residue) from pig pancreas.[35] The method is particularly useful for viral proteins, which are often found to be acetylated and which may be analyzed successfully even in the presence of nucleic acid.[36]

B. CARBOXYTERMINAL PEPTIDE IDENTIFICATION

A technique in which mass spectrometry is used and that permits the identification of the C-terminal peptide of a protein has been described.[37] It involves the incorporation of ^{18}O into all α-carboxyl groups liberated during enzyme-catalyzed partial hydrolysis of the protein, followed by mass spectrometry to identify as the C-terminal peptide the only peptide that does not incorporate any ^{18}O. In earlier applications of the technique, peptide separation and identification was obtained by GC-MS analysis of permethylated derivatives.[37-41] More recently, HPLC has been used to separate underivatized peptides, and identification of the C-terminal peptide has been made by FABMS.[42-44] The HPLC-based method has advantages over the former method: it is more convenient (no derivatives to form) and the C-terminal peptide is identified as a free peptide, not a derivative, so non-mass spectrometric methods may be used to assist characterization. In view of these advantages, the HPLC-based method has replaced the previous method based on permethylation.

C. OTHER APPLICATIONS

Previous reviews[1-3,5,32] have cited the application of the permethylation reaction to the analysis of γ-carboxyglutamic acid in proteins and of glycopeptides. It should be mentioned that the permethylation reaction, employed originally in the carbohydrate field, is still used extensively for carbohydrate analysis by mass spectrometry (see, for example, Reference 45). The permethylation reaction has been used[46] to form a derivative of an acylhomoserine lactone cleaved by cyanogen bromide (CNBr) from a protein having N-terminal acetyl-Met-. Although successful in the case cited, where large amounts of material were available (70 nmol), the use of dimethylsulfinyl sodium cannot be recommended generally when ester (lactone) groups are present because β-ketosulfoxide formation and a subsequent elimination reaction may occur.[11] An alternative analytical procedure, which avoids permethylation, is available for acylhomoserine lactones.[8]

V. CONCLUSIONS

A. ADVANTAGES AND DISADVANTAGES OF THE PERMETHYLATION APPROACH

The principal advantages of the permethylation-mass spectrometric approach to the determination of the structure of peptides and proteins are (1) the applicability to mixtures of peptides through on-line separation of derivatives by fractional distillation, GC-MS, or LC-MS, (2) the formation of very stable derivatives with good mass spectrometric properties for amino acid sequence determination, and (3) the applicability to N^α-blocked proteins. Disadvantages include (1) the necessity to form prior derivatives of Arg and Cys (a delicate synthesis with small amounts of material), (2) the failure to obtain useful information with sample sizes of less than about 2 nmol, and (3) the harsh reaction conditions, which may dislodge or transform a fragile post-translational modification. A further disadvantage, which may become a problem if derivatives of very large peptides are introduced by LC-MS, FABMS, or other techniques, is the small but nevertheless significant degree of C-methylation that may occur at some positions (particularly Gly adjacent to Pro): nonquantitative reaction will tend to complicate spectral interpretation at high mass.

B. WHEN TO USE THE PERMETHYLATION OF PEPTIDES

Permethylation is applied most advantageously (in terms of effort and sample expended per results obtained) to blocked N-termini, particularly for viral proteins, where the method works even in the presence of some nucleic acid and without the need for isolation of the N-terminal peptide prior to analysis. If preliminary results are confirmed, permethylation may permit amino acid sequence information to be obtained by FABMS or other high-mass ionization techniques. If so, then efforts to reduce sample requirement for the reaction would be worthwhile and essential if the permethylation approach is to be as successful as current tandem mass spectrometry (MS/MS) techniques are with underivatized peptides. One would wish to perform the derivatization reaction on a protein digest and separate the large peptide derivatives by HPLC or supercritical fluid chromatography, ideally on-line to mass spectrometry. Such an approach, if successful, would be preferable to attempting fractional distillation from a probe of large peptides or to attempting the analysis by GC-MS of very complex mixtures of smaller peptides as is done at present.

REFERENCES

1. **Priddle, J. D.,** Mass spectrometry in the determination of primary structure, in *Chemistry of Macromolecules,* Part IIA, Vol. 24, Offord, R. E., Ed., University Park Press, Baltimore, 1979, 1.
2. **Falter, H.,** Amino acid sequence determination by mass spectrometry, in *Advanced Methods in Protein Sequence Determination,* Molecular Biology, Biochemistry and Biophysics Series, Vol. 25, Needleman, S. B., Ed., Springer-Verlag, Berlin, 1977, 123.
3. **Biemann, K.,** Amino acid sequence in oligopeptides and proteins, in *Biochemical Applications of Mass Spectrometry,* Vol. 1 (Suppl.), Waller, G. R. and Dermer, O. C., Eds., John Wiley & Sons, New York, 1980, 469.
4. **Krutzsch, H. C.,** Polypeptide sequencing with dipeptidyl peptidases, *Methods Enzymol.,* 91, 511, 1983.
5. **Biemann, K. and Martin, S. A.,** Mass spectrometric determination of the amino acid sequence of peptides and proteins, *Mass Spectrom. Rev.,* 6, 1, 1987.
6. **Vath, J. E., Zollinger, M., and Biemann, K.,** Method for the derivatization of organic compounds at the sub-nanomole level with reagent vapor, *Fresenius Z. Anal. Chem.,* 331, 248, 1988.
7. **Rose, K., Simona, M. G., and Offord, R. E.,** Amino acid sequence determination by g.l.c.-mass spectrometry of permethylated peptides: optimization of the formation of chemical derivatives at the 2-10 nmol level, *Biochem. J.,* 215, 261, 1983.
8. **Rose, K., Savoy, L.-A., Simona, M. G., Offord, R. E., Wingfield, P. T., Mattaliano, R. J., and Thatcher, D. R.,** The state of the N-terminus of recombinant proteins: determination of N-terminal methionine (formylated, acetylated, or free), *Anal. Biochem.,* 165, 59, 1987.
9. **Pedersen, C. T. and Roepstorff, P.,** Comparison of different derivatives for GC/MS analysis of peptides, *Int. J. Mass Spectrom. Ion Phys.,* 48, 193, 1983.
10. **Auffret, A. D. and Williams, D. H.,** A relatively sensitive method for the mass spectrometric sequencing of peptides, *J. Chem. Soc. Chem. Commun.,* 692, 1979.
11. **Dell, A. and Morris, H. R.,** Primary structure of a chloramphenicol acetyltransferase: mass spectrometric studies, *Biomed. Mass Spectrom.,* 8, 128, 1981.
12. **Auffret, A. D., Blake, T. J., and Williams, D. H.,** Mass spectrometric sequence studies of a superoxide dismutase from *Bacillus stearothermophilus, Eur. J. Biochem.,* 113, 333, 1981.
13. **Nau, H.,** Gas chromatography-mass spectrometry of permethylated peptides and their reduced and trimethylsilylated derivatives, *J. Chromatogr.,* 121, 376, 1976.
14. **Priddle, J. D., Rose, K., and Offord, R. E.,** The separation and sequencing of permethylated peptides by mass spectrometry directly coupled to gas-liquid chromatography, *Biochem. J.,* 157, 777, 1976.

15. **Heinrikson, R. L.,** The selective S-methylation of sulfhydryl groups in proteins and peptides with methyl-*p*-nitrobenzenesulfonate, *J. Biol. Chem.,* 246, 4090, 1971.
16. **Morris, H. R., Williams, D. H., Midwinter, G. G., and Hartley, B. S.,** A mass-spectrometric sequence study of the enzyme ribitol dehydrogenase from *Klebsiella aerogenes, Biochem. J.,* 141, 701, 1974.
17. **Morris, H. R., Dickinson, R. J., and Williams, D. H.,** Studies towards the complete sequence determination of proteins by mass spectrometry. Derivatization of methionine, cysteine, and arginine containing peptides, *Biochem. Biophys. Res. Commun.,* 51, 247, 1973.
18. **Morris, H. R., Dell, A., Petersen, T. E., Sottrup-Jensen, L., and Magnusson, S.,** Mass-spectrometric identification and sequence location of the ten residues of the new amino acid (γ-carboxyglutamic acid) in the N-terminal region of prothrombin, *Biochem. J.,* 153, 663, 1976.
19. **Honegger, A., Hughes, G. J., and Wilson, K. J.,** Chemical modification of peptides by hydrazine, *Biochem. J.,* 199, 53, 1981.
20. **Shemyakin, M. M.,** Primary structure determination of peptides and proteins by mass spectrometry, *Pure Appl. Chem.,* 17, 313, 1968.
21. **Shemyakin, M. M., Ovchinnikov, Y. A., Vinogradova, E. I., Feigina, M. Y., Kiryushkin, A. A., Aldanova, N. A., Alakhov, Y. B., Lipkin, V. M., and Rosinov, B. V.,** Mass spectrometric determination of the amino acid sequence in arginine-containing peptides, *Experientia,* 23, 428, 1967.
22. **Leclercq, P. A., Smith, L. C., and Desiderio, D. M.,** Modification, permethylation, and mass spectrometry of arginine-containing oligopeptides at the 100 nanomolar level, *Biochem. Biophys. Res. Commun.,* 45, 937, 1971.
23. **Vetter-Diechtl, H., Vetter, W., Richter, W., and Biemann, K.,** Ein für Massenspektrometrie und Gaschromatographie geeignetes Argininderivat, *Experientia,* 24, 340, 1968.
24. **Gilbert, H. F. and O'Leary, M. H.,** Modification of arginine and lysine in proteins with 2,4-pentanedione, *Biochemistry,* 14, 5194, 1970.
25. **Wallace, C. J. A. and Rose, K.,** The semisynthesis of analogues of cytochrome *c*: modifications of arginine residues 38 and 91, *Biochem. J.,* 215, 651, 1983.
26. **Rose, K., Priddle, J. D., and Offord, R. E.,** Affinity technique for the isolation of polypeptides containing arginine modified with cyclohexane-1,2-dione, and their analysis by combined gas-liquid chromatography-mass spectrometry, *J. Chromatogr.,* 210, 301, 1981.
27. **Rose, K., Bairoch, A., and Offord, R. E.,** Amino acid sequence determination by gas chromatography-mass spectrometry of permethylated peptides: the application of capillary columns, *J. Chromatogr.,* 268, 197, 1983.
28. **Roepstorff, P., McDowall, M. A., Games, M. P. L., and Games, D. E.,** Peptide sequencing by combined liquid chromatography/mass spectrometry, *Int. J. Mass Spectrom. Ion Phys.,* 48, 197, 1983.
29. **Yu, T. J., Schwartz, H. A., Cohen, S. A., Vouros, P., and Karger, B. L.,** Sequence analysis of derivatized peptides by high-performance liquid chromatography-mass spectrometry, *J. Chromatogr.,* 301, 425, 1984.
30. **Rose, K.,** Analysis of peptides and proteins by fast atom bombardment mass spectrometry (FAB-MS), *Spectrosc. Int. J.,* 7, 39, 1989.
31. **Gray, W. R. and del Valle, U. E.,** Application of mass spectrometry to protein chemistry. I. Method for amino-terminal sequence analysis of proteins, *Biochemistry,* 9, 2134, 1970.
32. **Carr, S. A. and Biemann, K.,** Identification of posttranslationally modified amino acids in proteins by mass spectrometry, *Methods Enzymol.,* 106, 29, 1984.
33. **Mikkelsen, J., Hojrup, P., Nielsen, P. F., Roepstorff, P., and Knudsen, J.,** Amino acid sequence of acyl-CoA-binding protein from cow liver, *Biochem. J.,* 245, 857, 1987.
34. **Malek, A. A., Hy, M., Honegger, A., Rose, K., and Brenner-Holzach, O.,** Fructose-1,6-bisphosphate aldolase from *Drosophila melanogaster*: primary structure analysis, secondary structure prediction, and comparison with vertebrate aldolases, *Arch. Biochem. Biophys.,* 266, 10, 1988.
35. **Rose, K., Savoy, L.-A., Thim, L., Christensen, M., and Jørgensen, K. H.,** Revised amino acid sequence of pancreatic spasmolytic polypeptide exhibits greater similarity with an inducible pS2 peptide found in a human breast cancer cell line, *Biochim. Biophys. Acta,* 998, 297, 1989.
36. **Rose, K., Kocher, H. P., Blumberg, B. M., and Kolakofsky, D.,** An improved procedure, involving mass spectrometry, for N-terminal amino acid sequence determination of proteins which are Na-blocked, *Biochem. J.,* 217, 253, 1984.
37. **Rose, K., Simona, M. G., Offord, R. E., Prior, C. P., Otto, B., and Thatcher, D. R.,** A new mass-spectrometric C-terminal sequencing technique finds a similarity between γ-interferon and α₂-interferon and identifies a proteolytically clipped γ-interferon that retains full antiviral activity, *Biochem. J.,* 215, 273, 1983.
38. **Blumberg, B. M., Rose, K., Simona, M. G., Roux, L., Giorgi, C., and Kolakofsky, D.,** Analysis of the Sendai virus M gene and protein, *J. Virol.,* 52, 656, 1984.
39. **Liang, S.-M., Allet, B., Rose, K., Hirschi, M., Liang, C.-M., and Thatcher, D. R.,** Characterization of human interleukin 2 derived from *Escherichia coli, Biochem. J.,* 229, 429, 1985.
40. **Wingfield, P., Payton, M., Tavernier, J., Barnes, M., Shaw, A., Rose, K., Simona, M. G., Demczuk, S., Williamson, K., and Dayer, J.-M.,** Purification and characterization of human interleukin-1β expressed in recombinant *Escherichia coli, Eur. J. Biochem.,* 160, 491, 1986.

41. **Wingfield, P., Payton, M., Graber, P., Rose, K., Dayer, J.-M., Shaw, A. R., and Schmeissner, U.,** Purification and characterization of human interleukin-1α produced in *Escherichia coli, Eur. J. Biochem.,* 165, 537, 1987.
42. **Rose, K., Savoy, L.-A., Simona, M. G., Offord, R. E., and Wingfield, P.,** C-terminal peptide identification by fast atom bombardment mass spectrometry, *Biochem. J.,* 250, 253, 1988.
43. **Rose, K., Savoy, L.-A., and Offord, R. E.,** The use of mass spectrometry to identify carboxyl terminal peptides, *J. Protein Chem.,* 7, 281, 1988.
44. **Wingfield, P., Graber, P., Turcatti, G., Movva, N. R., Pelletier, M., Craig, S., Rose, K., and Miller, C. G.,** Purification and characterization of a methionine-specific aminopeptidase from *Salmonella typhimurium, Eur. J. Biochem.,* 180, 23, 1989.
45. **Dell, A.,** F.A.B.-mass spectrometry of carbohydrates, *Adv. Carbohydr. Chem. Biochem.,* 45, 19, 1987.
46. **Hemling, M. E., Carr, S. A., Capiau, C., and Petre, J.,** Structural characterization of recombinant hepatitis B surface antigen protein by mass spectrometry, *Biochemistry,* 27, 699, 1988.
47. **Roepstorff, P.,** personal communication.

Chapter 17

APPLICATIONS OF MASS SPECTROMETRY FOR CHARACTERIZATION OF NEUROPEPTIDES

Chhabil Dass

TABLE OF CONTENTS

I.   Introduction 328

II.   FAB Mass Spectrometry 329

III.   Fragmentation Characteristics 330
        A.   FAB Mass Spectrometry 330
              1.   Types of Fragment Ions and Their Nomenclature 330
              2.   Normal Magnet-Scan FAB Mass Spectra 331
                    a.   The Influence of FAB Irradiation Time and Concentration of
                          Peptide on the Fragment Ion Yields 334
              3.   FAB-*B/E*-CAD Mass Spectra 334
                    a.   Low-Mass Discrimination 338
        B.   EI-CI Mass Spectra 338
        C.   Field Desorption Mass Spectra 338
        D.   Laser Desorption/Laser Ionization 338
        E.   Metal Ion-Peptide Interaction 339

IV.   Sequence Ion Information 339
        A.   Magnet-Scan FAB Mass Spectra 339
        B.   *B/E*-CAD Data 340

V.   Solvent-Solute Interactions 340

VI.   Quantification of Peptides 341

VII.   Conclusions 341

References 344

I. INTRODUCTION

The discovery in 1975 of two pentapeptides, methionine enkephalin (ME=YGGFM) and leucine enkephalin (LE=YGGFL),[1] as endogenous opioids prompted intensive research into brain neurochemistry. Since then, an increasing number of polypeptides with opioid activity has been isolated and characterized in the central nervous system (CNS) and periphery.[2] All endogenous opioid peptides belong to one of the three peptide families, namely, endorphins, enkephalins, and dynorphins, each deriving from a distinct gene-product precursor. Proopiomelanocortin (pOMC) is the precursor of β-endorphin, ACTH, and α-MSH-related peptides; preproenkephalin A is the precursor for ME, LE, and several larger enkephalin peptides; and preproenkephalin B is another precursor for LE and for dynorphin- and α-neoendorphin-related peptides.[2] In addition, there is pharmacological evidence for multiple opioid receptors (μ, δ, κ, σ, ε, etc.).[3]

Opioid peptides play several important roles in various biochemical and neurochemical processes.[4] Their concentration level in body fluids can be correlated with various psychiatric and metabolic disorders.[5] A clear knowledge of the molecular events in any biological process demands that various peptides involved must be quantified in body tissues and fluids with a high level of molecular specificity.[6] Quantification methods based on mass spectrometry/mass spectrometry (MS/MS) readily meet those rigorous criteria of molecular specificity. Quantification of opioid peptides is discussed in detail by Desiderio in this book. It is obvious that before undertaking analytical measurements of endogenous peptides a firm understanding of their mass spectrometric behavior must be provided from model studies.

This chapter discusses mass spectral fragmentation characteristics of several biologically important synthetic neuropeptides. A major emphasis is on the use of the fast atom bombardment (FAB)[7] technique. An objective comparison of the data obtained in our laboratory using the magnet-scan[8] and the B/E linked-field-collisionally activated dissociation (CAD)[9] techniques is presented. With the knowledge of fragmentation behavior, it is possible to deduce the amino acid sequence of a peptide. Furthermore, that knowledge also helps in selecting specific fragment ion(s) suitable for quantification of the endogenous peptides in biological tissues using the selected reaction monitoring (SRM) technique.[10] These two aspects are discussed further in the latter part of this chapter. A brief description on the literature data for opioid peptides obtained by other MS techniques is also included.

Development of MS techniques to sequence and quantify peptides is the focus of current research.[8-15] In the past, the potential of MS in this field had remained dormant because the use of conventional electron- (EI) and chemical ionization (CI) MS methods requires extensive chemical derivatization of peptides to increase their volatility. The advent of soft ionization techniques based on desorption of an analyte has greatly advanced the applications of MS in the analysis of highly polar and thermally labile compounds such as peptides. Underivatized peptides are now routinely analyzed by FAB,[7] secondary ion mass spectrometry (SIMS),[16] field desorption (FD),[17] laser desorption (LD),[18] and ^{252}Cf-plasma desorption (Cf-PD)[19] techniques. The applications of some of these techniques are discussed in detail by other authors in this book. Of these techniques, FAB has gained the most widespread acceptance in the analysis of biologically important compounds. Much of the success of FAB is attributed to its simplicity, reproducibility, ease of adaptability, and extended mass range. An appropriate liquid matrix is vital to the success of FAB analysis. The use of a liquid matrix in FAB, however, often complicates a mass spectrum because amino acid sequence-determining ions are overwhelmed in many cases by matrix ions. In addition, solvent-solute interactions are frequently observed.[20] In order to overcome these potential pitfalls of FAB, efforts are being made to develop methods utilizing tandem mass spectrometry, also referred to as MS/MS.[21] MS/MS is uniquely a powerful tool for structure analysis. This technique makes it possible, in the context of FAB, to isolate by MS-I an ion of interest, the protonated molecular ion (M+H)$^+$, from the chemical noise and other interferences. The decomposition products of that $(M+H)^+$ ion are mass-analyzed by MS-II to provide the sequence of the peptide. Excitation of the mass-analyzed ion by collisional activation (CA) increases its fragmentation. This overall approach requires the use of a reversed-geometry (magnet, B precedes electric sector, E) double-focusing, multi-sector, multi-quadrupole, or hybrid mass spectrometer.

Thus, state-of-the-art MS/MS can be a useful peptide-sequencing technique. Currently, however, the upper mass range for obtaining sequence ion information of oligopeptides by using MS/MS is limited to approximately 3000 Da.[11] This limitation is primarily attributed to the reduced probability of desorption of the bulkier molecule, the large number of possible fragmentation channels, and a decrease in the energy available for bond dissociation due to an increased number of degrees of freedom.[11,12,22] As a consequence, the information obtained in the case of high-mass peptides is not complete, and thus of limited use.

Although tandem mass spectrometry is an ideal MS/MS technique for sequencing oligopeptides (see other chapters in this book), the linked-field scan at constant B/E has also been applied to sequence peptides.[9,23] This chapter discusses further applications of this alternative mode of a MS/MS technique. Using a double-focusing instrument of forward geometry (B follows E), mass-selection of an $(M+H)^+$ ion is accomplished by linking B and E, and fragmentation occurs in the first field-free region. A scan of B and E, such that the ratio B/E is maintained constant throughout the scan, furnishes a mass spectrum of product ions obtained exclusively from a mass-selected $(M+H)^+$ ion.[24]

The conclusions drawn here regarding mass spectral fragmentation characteristics are based on the data obtained from several synthetic opioid peptides that contain five to ten amino acid residues and are related to the enkephalin sequence in the molecular mass (M_r) range of 400 to 1235u. Included in this list are ME, LE, ME–K, ME–KK, ME–KR, ME–RF, ME–RGL, LE–R, dynorphin A fragments 1 to 7, 1 to 8, 1 to 9, and 1 to 10 (LE–RRIRP), β- (LE–RKYP) and α-neoendorphins (LE–RKYPK), β-casomorphin (YPFPGPI), and proctolin (RYLPT). Most biologically active peptides are bracketed in their corresponding precursor molecule with pairs of the two basic amino acids arginine and lysine. Thus, after a trypsin-like enzymatic cleavage step, peptides containing arginine or lysine at the C-terminus are produced. Therefore, the study of those C-terminally extended enkephalin peptides is quite important.

LE and ME are the two opioid peptides most extensively studied by using MS. The first structural evidence of the existence of endogenous ME and LE was obtained by MS.[1] The acetylated and permethylated derivative of ME was sequenced by Yu et al.[25] by the combination of high pressure liquid chromatography (HPLC)-MS and isobutane CI. Hunt et al.[26] applied CI-MS and tandem quadrupole MS for sequence determination of [²H₀]/[²H₃]-N-acetyl-N,O-permethylated ME. LE and ME were used as examples by Barber et al.[27] to demonstrate the utility of FABMS in characterizing underivatized peptides. A recent report discussed the application of a hybrid instrument for analysis of LE.[28] We have studied in detail a set of C-terminally extended enkephalin peptides by FABMS[8] and B/E linked-field scan techniques.[9] A 4-sector tandem mass spectrometer was also employed for analyzing dynorphin A fragments 1 to 7 and 1 to 13.[29] Tandem mass spectrometry coupled with CAD was used by Gross and co-workers for LE variants[30] and for endorphin and ACTH fragments.[12] Laser ionization was recently evaluated for sequence analysis of LE and ME.[31] Tang et al.[32] reported SIMS of LE, ME, and their metal ion adducts with Na⁺, K⁺, and Ag⁺ ions. SIMS was also applied by Seki et al.[16] for sequencing of LE, ME, α-neoendorphin, and β-neoendorphin.

II. FAB MASS SPECTROMETRY

FABMS has made a dramatic contribution in the field of biomedical science. Although FABMS is well established, a brief description of the technique is in order. In the FAB mode of ionization, an analyte is dissolved or suspended in a liquid matrix, and bombarded by a kiloelectronvolt energy primary atom beam.[7] Ionization is believed to occur by desorption of the preformed ions present on the surface of the matrix or by gas phase ion-molecule reactions occurring in the selvedge region. FAB produces energetically stable $(M+H)^+$ and $(M-H)^-$ as molecular ion species in positive and negative modes of analysis, respectively. A liquid matrix is the essence of the FAB technique because it helps to maintain constant and steady ion signal. Prolonged ion current is essential for the successful use of FAB in sector mass spectrometers and for high-resolution and MS/MS experiments. Fluidity, low volatility, potent solvent properties, chemical inertness, and acidity or basicity are the attributes of a successful matrix. Glycerol, thioglycerol, a mixture of dithiothreitol (DTT) and dithioerythritol (DTE), diethanolamine, and sulfolane are some of the more commonly used matrices.[33-35] Manipulation of the composition of the FAB matrix by addition of acid, alkali, surfactants, etc. is also helpful in controlling ion current yields.[36] For further reading on the FAB technique, the reader is referred to two excellent reviews on this subject[37,38] and to a chapter by Busch in this book.

The data reported here were acquired on a VG 7070E-HF (Manchester, U.K.) double-focusing forward-geometry mass spectrometer outfitted with a Digital PDP 11/24 minicomputer-based VG 11-250 M+ data system. The ion source was a standard VG FAB system equipped with an Ion Tech B11NF saddle-field atom gun. Xe atoms of approximately 7 keV impact energy and emission current corresponding to 1 mA were used as the ionizing beam. FAB-desorbed ions were accelerated to a potential of 6 kV. CAD was accomplished by using helium as a collision gas. The linked-field scan data were acquired in the multichannel analyzer (MCA) mode.

An acidified (1 μl of 0.1 N HCl per 0.5 μl of the FAB matrix) mixture of DTT and DTE (3:1, w/w) or glycerol was used as a FAB matrix. Generally, 1 to 2 nmol of peptide was dissolved in these matrices.

III. FRAGMENTATION CHARACTERISTICS

A. FAB MASS SPECTROMETRY

Peptides are very complex molecules consisting of 20 naturally occurring amino acid residues. In solution, peptides are known to fold to the state of lowest free energy, and to exist in several conformations. A particular conformer may predominate, depending upon the chemical environment in the solution and the nature (hydrophobicity, charge, size, etc.) of amino acid residues in the peptide chain. For example, the β-turn structure is less prevalent in an aqueous solution of ME than of LE.[39] It is our belief that the formation of an intact $(M+H)^+$ or $(M-H)^-$ ion of a peptide and its subsequent fragmentation by FABMS will be controlled by the specific conformation of the peptide in the FAB matrix, the lattice forces associated with solute-solvent interaction, and the charge distribution in the peptide. In view of these complexities, it is not easy to predict *a priori* the fragmentation behavior of a peptide and occurrence of all specific amino acid sequence-determining fragment ions in a mass spectrum. As a result, it is difficult, if not impossible, to make any generalization about the fragmentation characteristics for peptides, a limitation noted by us[8] and others.[11]

1. Types of Fragment Ions and Their Nomenclature

Fragmentation of peptides subjected to FAB was first discussed by Williams and co-workers.[40] This study and several subsequent studies[8,11,16] have established that amino acid sequence-determining ions are formed by cleavage of three types of bonds in the backbone of peptides (Scheme 1), viz., (a) the bond between the carbonyl group and α-carbon (–CHR–CO–), (b) the peptide amide bond (–CO–NH–), and (c) the amino-alkyl bond (–NH–CHR–), with charge retention by either N- or C-terminus fragments. In addition, ions due to internal fragments (i.e., the ions formed by cleavage of two bonds in the peptide chain) and to certain immonium ions are also commonly observed. Also, cleavage of the side chain of certain amino acids yields

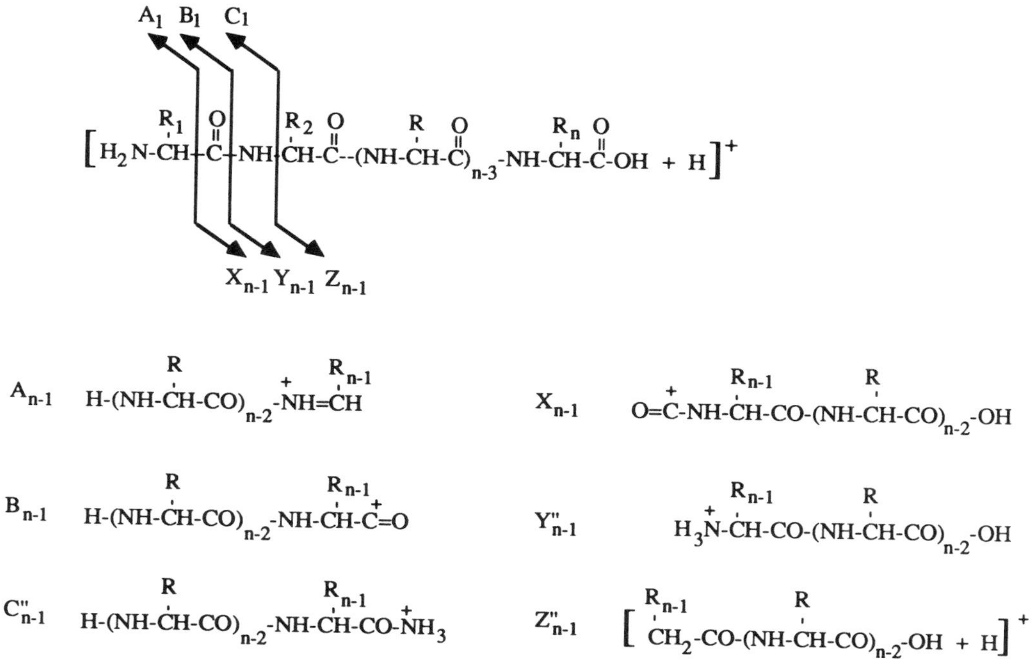

A_1 B_1 C_1

$$\left[H_2N-\overset{R_1}{\underset{|}{CH}}-\overset{O}{\underset{\|}{C}}-NH-\overset{R_2}{\underset{|}{CH}}-\overset{O}{\underset{\|}{C}}-(NH-\overset{R}{\underset{|}{CH}}-\overset{O}{\underset{\|}{C}})_{n-3}-NH-\overset{R_n}{\underset{|}{CH}}-\overset{O}{\underset{\|}{C}}-OH + H \right]^+$$

X_{n-1} Y_{n-1} Z_{n-1}

$$A_{n-1} \quad H-(NH-\overset{R}{\underset{|}{CH}}-CO)_{n-2}-NH\overset{+}{=}\overset{R_{n-1}}{\underset{|}{CH}}$$

$$X_{n-1} \quad O=\overset{+}{C}-NH-\overset{R_{n-1}}{\underset{|}{CH}}-CO-(NH-\overset{R}{\underset{|}{CH}}-CO)_{n-2}-OH$$

$$B_{n-1} \quad H-(NH-\overset{R}{\underset{|}{CH}}-CO)_{n-2}-NH-\overset{R_{n-1}}{\underset{|}{CH}}-\overset{+}{C}=O$$

$$Y''_{n-1} \quad \overset{+}{H_3}N-\overset{R_{n-1}}{\underset{|}{CH}}-CO-(NH-\overset{R}{\underset{|}{CH}}-CO)_{n-2}-OH$$

$$C''_{n-1} \quad H-(NH-\overset{R}{\underset{|}{CH}}-CO)_{n-2}-NH-\overset{R_{n-1}}{\underset{|}{CH}}-CO-\overset{+}{N}H_3$$

$$Z''_{n-1} \quad \left[\overset{R_{n-1}}{\underset{|}{CH_2}}-CO-(NH-\overset{R}{\underset{|}{CH}}-CO)_{n-2}-OH + H \right]^+$$

Scheme 1

sequence-specific ions. In essence, six different fragmentation processes, apart from the internal fragments and the side chain losses, are involved in formation of sequence ions. Discordancy in the fragmentation behavior, however, exists for different peptides. For example, not every peptide yields all expected sequence ion-series and furthermore, their relative abundances also vary significantly, perhaps due to the factors discussed here.

The nomenclature discussed by Roepstorff and Fohlman[41] is a widely accepted proposal to designate different sequence ions (Scheme 1), although Biemann[42] has suggested recently certain modifications to that scheme. According to Roepstorff and Fohlman, N-terminus sequence ions formed by the aforementioned three processes are designated as A, B, and C, respectively, and corresponding C-terminus sequence ions are denoted as X, Y, and Z. Addition of one or two H-atoms is observed generally in the formation of C-, Y-, and Z-sequence ions, and is represented by the number of apostrophes attached to the right of that letter.

2. Normal Magnet-Scan FAB Mass Spectra

A detailed study on the FABMS behavior of selected enkephalin peptides was reported by us.[8] The representative examples of the magnet-scan mass spectra are presented here for ME-RF and ME-RGL in Figures 1 and 2, respectively. In spite of the fact that the first four amino acids are common to all opioid peptides derived from POMC, preproenkephalin A, and preproenkephalin B, a clear generalization of fragmentation reactions for these peptides cannot be made. The following salient features, however, are derived from the data:

1. FAB of the neuropeptides of interest here yields a relatively abundant $(M+H)^+$ ion, the most important characteristic in determining the M_r of a peptide.
2. The principal mode of sequence ion formation in the pentapeptides ME and LE is cleavage of the peptide amide bond to form B- and Y″-type sequence ion series.
3. In contrast, the foremost fragmentation process of the C-terminally extended enkephalin peptides is fission of the amino-alkyl bond to yield Z″-type sequence ions (see Figures 1

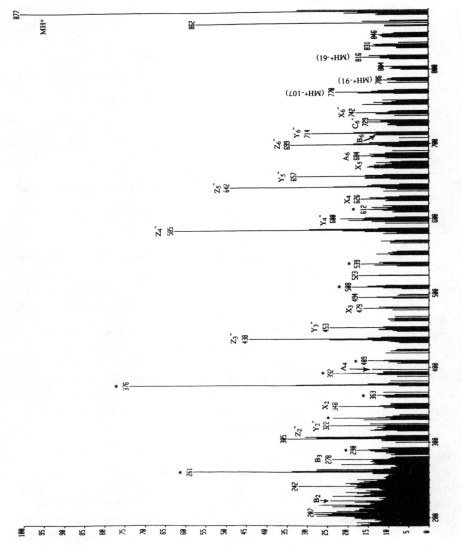

FIGURE 1. FAB mass spectrum of ME-RF. The peaks labeled by * are internal fragments. ([M+H]⁺ – 107), ([M+H]⁺ – 91), and ([M+H]⁺ – 61) represent the losses of tyrosine, phenylalanine, and methionine side chains, respectively. For the sequence of ME-RF, see Figure 5.

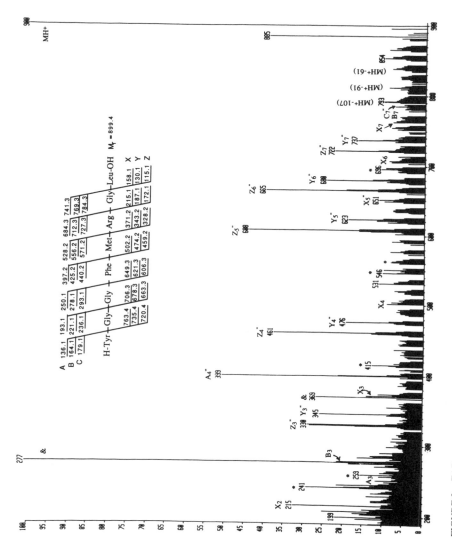

FIGURE 2. FAB mass spectrum of ME-RGL. For peaks labeled by *, ([M+H]⁺ − 107), ([M+H]⁺ − 91), and ([M+H]⁺ − 61), see Figure 1 legend. Peaks labeled by & are due to matrix ions.

and 2). The Y″ ions appear as satellite peaks in these spectra, and can be used as additional corroborative evidence for deriving the sequence.

4. In general, the N-terminus ions are less abundant than the C-terminus ions in the spectra of the C-terminally extended enkephalins. One of the factors responsible for the predominance of the C-terminus ions is the presence of basic amino acid residues (arginine or lysine) in the C-terminus extended portion of the enkephalin peptides. The high basicity of arginine or lysine favors formation of the C-terminus ions by directing protonation and fragmentation. The FAB spectrum of proctolin, a pentapeptide of RYLPT sequence, supports further this rationale. Because of the presence of N-terminus arginine, the N-terminus ions are more abundant than the C-terminus ions in that spectrum.

5. It is of interest that certain amino acid residues have overwhelming influence on the yield of sequence ions. For example, the sequence ions terminating at phenylalanine, tyrosine, or proline residues are overrepresented in a specific sequence ion-series. The preponderance of $Z_4″$ and $Z_5″$ ions among the Z ion-series for ME–RF (Figure 1) and ME–RGL (Figure 2), respectively, illustrates this point. Similarly, the immonium ions A_1, A_4, and m/z 120 are always present in high abundance in the spectra of all of the enkephalin-containing peptides studied here. Charge-stabilization afforded by the aromatic ring of phenylalanine and tyrosine residues facilitates formation of these sequence ions.[8]

6. Another unique feature of the opioid peptides is the loss of side chain moieties resulting from homolytic cleavage of the β,γ-bond. The ($[M+H]^+ - 107$), ($[M+H]^+ - 91$), and ($[M+H]^+ - 61$) ions are formed by the loss of tyrosine, phenylalanine, and methionine side chains, respectively (see Figures 1 and 2). These fragment ions may be used to indicate the presence of the respective amino acid residues.

7. Several internal fragments are also present in the FAB mass spectra of these opioid peptides, and provide additional sequence ion information.

8. The low-mass sequence ions predominate in the FAB mass spectra (not shown in Figures 1 and 2), but sometimes a clear distinction of these sequence specific ions is difficult because the matrix ions are overwhelmingly abundant in the low-mass region of the FAB mass spectra.

a. The Influence of FAB Irradiation Time and Concentration of Peptide on the Fragment Ion Yields

An interesting observation was made during the course of acquisition of FAB mass spectra of peptides.[8] Fragment ion intensities were found to increase with time of FAB irradiation, as illustrated in Figure 3, for dynorphin$_{1-7}$. This phenomenon is attributed to an increase in internal energy of the molecular ion. With time, the FAB solvent is depleted faster. Thus, the analyte-solvent clusters desorbed in the gas phase contain decreased numbers of solvent molecules. Under these circumstances, less energy is dissipated in desolvating those fewer solvent molecules, causing the molecular ions to be energy-rich. The data presented in Figure 4, which demonstrates the effect of concentration of dynorphin$_{1-7}$ on fragment ion yields, provide further credence to this rationale.

3. FAB-*B/E*-CAD Mass Spectra

As mentioned earlier in this chapter, a FAB mass spectrum is often dominated by ions formed due to matrix ionization or to matrix interaction with the sample. These interfering ions are present almost at every mass, cannot be reduced by signal averaging, and make sequence ions less readily recognizable. Furthermore, in many instances, a FAB mass spectrum exhibits little fragmentation, thus limiting its utility for complete sequencing of a peptide. These difficulties are further compounded by the presence of interfering impurities when dealing with biological samples. MS/MS, coupled with CAD of the molecular ion, has gained an increasing number of applications to sequence biologically important peptides.[12-15] The most important attributes of

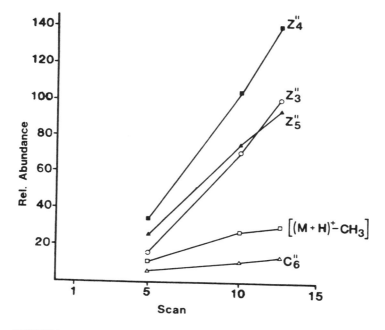

FIGURE 3. Increase in fragmentation of the $(M+H)^+$ ion of dynorphin$_{1-7}$ with time of irradiation. (From Dass, C. and Desiderio, D. M., *Anal. Biochem.*, 163, 52, 1987. With permission.)

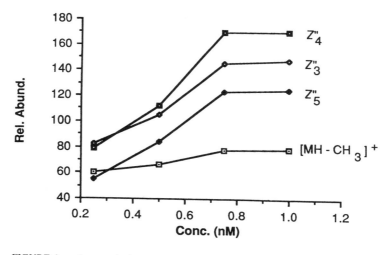

FIGURE 4. Increase in fragment ion intensities with concentration of dynorphin$_{1-7}$. (From Dass, C., and Desiderio, D. M., *Anal. Biochem.*, 163, 52, 1987. With permission.)

the MS/MS technique are enhanced fragmentation and signal-to-noise ratio, and elimination of interference from the matrix as well as from the sample. The use of MS/MS increases confidence in assigning the amino acid sequence to an unknown peptide because all the fragment ions in the spectrum must originate from the mass-selected precursor (i.e., the $[M+H]^+$ or $[M–H]^-$ ion of that peptide). We have used the linked-field MS/MS at constant B/E on a forward-geometry mass spectrometer as a tool to sequence peptides.[9] Figure 5 is a typical example, showing the B/E-CAD spectrum of ME-RF. The data obtained for several opioid peptides are summarized as follows:

1. The most important process in the formation of sequence ions of the collisionally activated peptides is cleavage of the peptide amide bond to yield the B- and Y″-type sequence ions. These two ion series are most complete. This finding is in contrast to the magnet-scan FAB mass spectra of the enkephalin peptides discussed previously, where it was observed that the dominant fragmentation is cleavage of the amino-alkyl bond to yield Z″-type sequence ions. The Z-type ion series, on the other hand, is very weak in the *B/E*-CAD spectra. Another contrasting observation of the two types of mass spectra is that only one H-atom is added to the Z-type C-terminus ions in the *B/E*-CAD mass spectra, whereas the corresponding ions in the magnet-scan spectra in most cases add two H-atoms.

2. A-type ions constitute another important ion series from the viewpoint of sequence analysis. Most of the peptides studied yield complete A-type ion series.

3. Fragmentation processes leading to the formation of C- and X-type sequence ions are less important. Only the C_{n-1} sequence ion is significant.

4. The losses of neutral moieties from specific amino acid side chains that were observed in the magnet-scan FAB mass spectra are also facile for the collisionally activated (M+H)⁺ ions of the peptides. For example, all tyrosine-containing peptides exhibit an abundant ion at ([M+H]⁺– 107). In fact, this fragment ion forms the base peak in the spectra of many of the opioid peptides (see Figure 5). Because tyrosine constitutes the N-terminus of all opioid peptides, this unique loss of 107u may be of analytical significance in screening these opioid peptides in biological samples by using a constant neutral loss linked-field scan. Leucine, isoleucine, arginine, lysine, phenylalanine, and methionine residues also exhibit a characteristic loss of neutral moieties.

5. The influence of basic amino acid residues on the yield of sequence ions is also apparent. Most C-terminally extended enkephalin peptides yield a complete set of the Y-type sequence ions. The effect, however, is not as pronounced in the *B/E*-CAD spectra as is noted in the magnet-scan spectra of the opioid peptides. Unlike the normal FAB mass spectra, A- and B-type N-terminus sequence ions are also present in high abundance in the *B/E*-CAD mass spectra.

6. Substantial evidence supporting the influence of certain amino acids on fragmentation of peptide bonds, also noted earlier in the magnet-scan spectra, is obtained in the CAD spectra. When cleavage takes place in the backbone of the peptides at phenylalanine and proline residues, the resulting sequence ions are overrepresented. For example, Y_2-, Y_3-, Y_4- (see Figure 5), Y_5-, Y_6-, and Y_7-ions in the penta-, hexa-, hepta-, octa-, nona-, and decapeptides, respectively, containing the enkephalin sequence are conspicuous in the Y-type ion series. Likewise, relative abundances of Y_2-, Y_4-, Y_5-, and Y_6-ions in β-casomorphin, and Y_2-ion in proctolin are a testament to that conclusion.[9] The B_4-acylium ion in the spectra of ME, LE–NH₂, and [D–Ala²]-LE is the most abundant of the B-type ions.[9] Further evidence of the generality of the influence of a phenylalanine residue on the yield of sequence ions derives from the dominance of A_4 and m/z 120 immonium ions, which are formed by cleavage of the bond at the carboxylic end of phenylalanine. Also, the A_8-ion in the spectra of the two neoendorphins is the most abundant in the A-type sequence ion series. The A_1 ions are also of significant abundance in these spectra, and point to a similar role of the aromatic ring of a tyrosine residue.

7. Like the magnet-scan spectra, the *B/E*-CAD spectra also contain internal fragments.

In sum, it can be stated that conventional magnet-scan and *B/E*-CAD FABMS techniques yield broadly similar mass spectra, but the pattern of the sequence ions in the two techniques is not identical. These distinguishing features are attributed to differences in energetics of the FAB-induced vs. collisionally activated (M+H)⁺ ions of these enkephalin peptides. One of the plausible explanations for different internal energies of the ions is that, in contrast to the collisionally activated ions, the FAB-desorbed ions are associated with solvent molecules.[20]

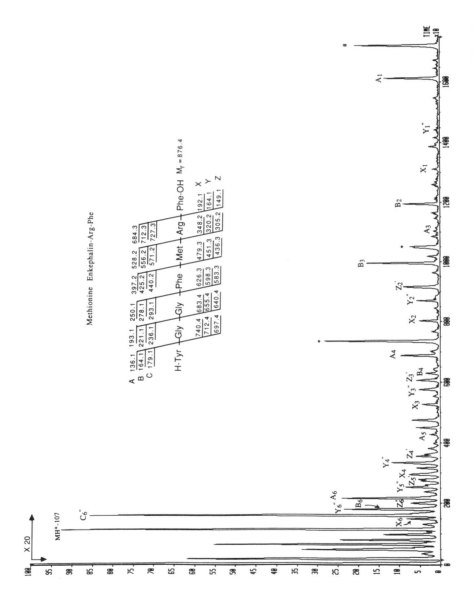

FIGURE 5. *B/E*-CAD mass spectrum of ME-RF. For peaks labeled by * and ([M+H]⁺ − 107), see Figure 1 legend. The peak labeled by # is an immonium ion of m/z 120 ($H_2NCHCH_2C_6H_5$). (From Dass, C. and Desiderio, D. M., *Int. J. Mass Spectrom. Ion Proc.*, 92, 267, 1989. With permission from Elsevier Science Publishers.)

This phenomenon, as discussed in the previous section, will influence the internal energy of the ions.

a. Low-Mass Discrimination

Low-mass discrimination for collisionally-activated ions was reported recently by Tomer et al.[12] and discussed at great length by Derrick and co-workers.[43] In the B/E-CAD spectra being discussed here, the relative abundance of sequence ions in a series also decrease monotonically with a decrease in the mass of the fragment ions, unless intercepted by a more favorable site of fragmentation (discussed earlier). The trend of the relative abundance of the A- and Y-sequence ion series (see Figure 5) clearly supports this contention. As a consequence of this mass discrimination effect, an A_1-immonium ion, which is very abundant in the spectra of the low-mass enkephalins, is almost negligible in the corresponding spectra of the high-mass peptides.

B. EI-CI MASS SPECTRA

The first structural evidence of endogenous ME and LE was obtained from the EI-MS data of a porcine brain extract.[1] Analysis of peptides by EI- or CI-MS, however, requires their conversion to sufficiently volatile derivatives, including trimethylsilylation[44] and acetylation followed by permethylation[45] of the peptide. Isobutane-CI in combination with LC-MS has been used to sequence the N-acetylated-N,O-permethylated derivatives of LE[46] and ME.[25] The dominant fragmentation in these spectra was cleavage of the peptide amide bond to produce B- and Y-type sequence ions. Hunt et al.[26] demonstrated the utility of CI-MS and MS/MS in determining the sequence of $[^2H_0]/[^2H_3]$-N-acetyl-N,O-permethylated ME. Both of these techniques produced B- and Y-type sequence ions. The $(M+H)^+$ and the B-type N-terminus ions were recognized readily in the CI mass spectrum by the isotopic doublets separated by 3u due to the deuterated isomer.

C. FIELD DESORPTION MASS SPECTRA

The discovery of FD significantly advanced the use of MS for analysis of nonvolatile and thermally labile compounds.[17] This soft ionization technique, by providing a means to transport involatile molecules to gas-phase ions, obviates the need for chemical derivatization. Abundant molecular ions such as $(M+H)^+$, $(M+Na)^+$, $(M+K)^+$, etc. are produced. Desiderio and co-workers[47-49] demonstrated the applications of FD in characterizing and quantifying LE. Although ions observed by FD-MS were related to the molecular ion only,[48] the sequence information was obtained in conjunction with the B/E-CAD technique.[49] Like the CAD data of the FAB-produced ions discussed previously, the principle fragmentation was cleavage of the peptide-amide bond to yield B- and Y-type sequence ions. Collisional activation also produced the A_4 N-terminus sequence ion series. Cleavage of the β,γ-bond, which resulted in the losses of leucine, phenylalanine, and tyrosine side chains, was also observed.

D. LASER DESORPTION/LASER IONIZATION

LD provides another useful method to ionize nonvolatile and fragile molecules of biomedical relevance. This mode of ionization is most suited to pulse mass spectrometers such as a Fourier-transform MS[50] or time-of-flight (TOF) instrument.[31] The use of multiphoton ionization (MPI) advances the possibility of selective ionization of a particular analyte. Furthermore, the degree of fragmentation can be controlled by tuning the intensity of photons. Recently, Grotemeyer and Schlag[31] used MPI with photons of 272.16 nm wavelength, corresponding to the π-π^* transition in the aromatic ring of a tyrosine residue, to obtain mass spectra of LE and ME. The dominant ion in these spectra, naturally, was the hydroxytropylium ion (m/z 107) resulting from the tyrosine residue. The sequence ions arise as a result of cleavage of the peptide amide bond with charge retention, mainly by the N-terminus ions, to yield the B-type ions. The loss of CO from these acylium ions resulted in A-type sequence ions. The disadvantage of this experimental protocol, however, is that only fragments containing aromatic amino acids such as tyrosine and phenylalanine have a greater propensity for ionization. Consequently, only those C-terminus ions that contain these aromatic amino acid residues were observed in the mass spectra of LE and ME. This limitation could pose difficulties in complete sequencing of larger peptides.

E. METAL ION-PEPTIDE INTERACTION

Derivatization of peptides as a strategy to provide sequence-specific ions is being increasingly explored.[32,51-54] Addition of alkali metal ion salt to the peptide-matrix mixture forms $(M+Cat)^+$ species, the yield of which is related directly to the strength of the peptide-metal ion bond. The bond energies decrease in the order Li, Na, K, etc. Collisional activation of the metal ion adducts with LE and ME (e.g., formed with Li and Ag salts) yields sequence-specific fragment ions in the product ion spectrum, the most prominent being the $(B+Cat+OH)^+$-type ions.[32,53,54] The $(A_4+Li-H)^+$ ion was also prominent in those spectra. Collisional activation of high translational energy ions also produced losses of side chain moieties.[53] The mechanism of the loss of the C-terminus amino acid to yield B-type of ions, however, is at variance. Tang et al.[32] proposed chelation of the metal ion, transfer of OH of the terminal carboxylic group to the amide group of the adjacent amino acid, and loss of CO and an imine (Scheme 2), whereas Grese et al.[53] provided evidence supporting the mechanism suggested by Renner and Spiteller,[52] in which metal binds to the C-terminal and fragmentation is triggered by nucleophilic attack of the negative oxygen (Scheme 3).

IV. SEQUENCE ION INFORMATION

The Edman degradation is the most popular chemical method for sequence determination of peptides. It relies on stepwise cleavage of a peptide from the N-terminus followed by identification of the cleaved amino acids. This method, however, suffers from limitations: blocked N-terminus peptides cannot be sequenced and certain modified amino acids are difficult to detect. FABMS offers new avenues for sequence ion determination of oligopeptides.[11] FABMS is the only alternative when a peptide has a blocked N-terminus.[55-57]

A. MAGNET-SCAN FAB MASS SPECTRA

Establishing the molecular ion is of foremost importance in sequence determination of an unknown peptide. The most intense ion at the high-mass end of a spectrum of a peptide is generally due to the molecular ion. The alkali metal ion adducts[27] and the $([M+H]^+ + 12)$ ion adduct[20] discussed later can aid in the recognition of the molecular ion. Next in importance is identification of both of the terminal amino acid residues. All opioid peptides studied here contain a tyrosine at the N-terminus, which is readily recognized by the presence of an abundant immonium ion of m/z 136 (A_1 ion) (not shown in Figures 1 and 2). Further identification of the N-terminus is obtained by considering the Y''_{n-1} and Z''_{n-1} sequence ions, which are formed by the loss of a tyrosine residue from the $(M+H)^+$ ion. Identification of the C-terminus amino acid residue, however, is not straightforward because of the interference of matrix ions at the low-mass end of the spectrum. Fortunately, the C''_{n-1} sequence ion is of reasonable abundance in the FAB mass spectra of these opioid peptides, and may be used to identify the C-terminus residue. In most cases, the A_{n-1} and B_{n-1} ions are also helpful. Once the terminal amino acids have been established, the remaining amino acid residues are recognized by considering the Z-type sequence ion series. The positive ion FABMS data cannot be used for bidirectional sequencing of the C-terminally extended ME and LE peptides because no complete N-terminus sequence ion series is observed. However, it was noted by us in a previous publication[8] that the use of negative ion FAB can provide bidirectional sequencing. The utility of positive ion FABMS for sequence analysis of peptides is demonstrated in Figures 1 and 2 for the heptapeptide ME-RF and octapeptide ME-RGL, respectively.

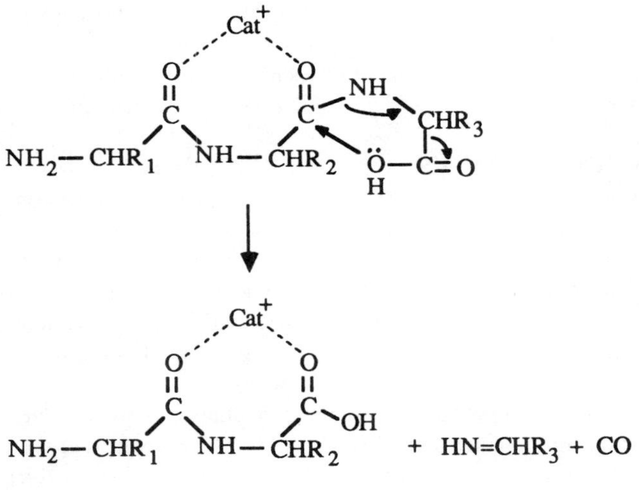

$$Cat^+$$

$$NH_2-CHR_1 \quad NH-CHR_2 \quad \overset{\cdot\cdot}{O}-C{=}O$$

Scheme 2

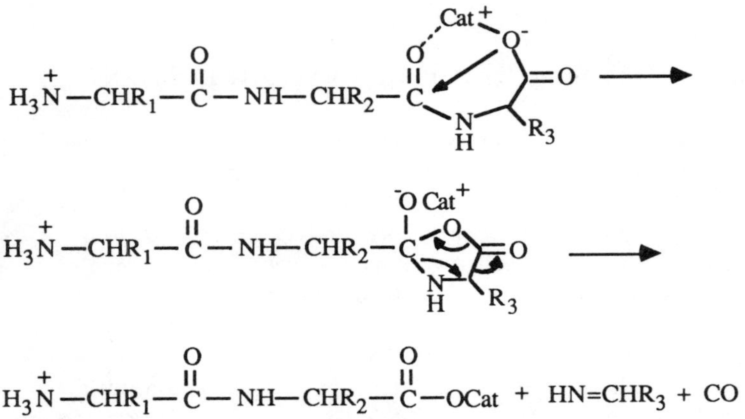

$$NH_2-CHR_1 \quad NH-CHR_2 \quad + \quad HN{=}CHR_3 + CO$$

Scheme 2

$$H_3\overset{+}{N}-CHR_1-\overset{O}{\overset{\|}{C}}-NH-CHR_2-C \quad \longrightarrow$$

$$H_3\overset{+}{N}-CHR_1-\overset{O}{\overset{\|}{C}}-NH-CHR_2-C \quad \longrightarrow$$

$$H_3\overset{+}{N}-CHR_1-\overset{O}{\overset{\|}{C}}-NH-CHR_2-\overset{O}{\overset{\|}{C}}-OCat \quad + \quad HN{=}CHR_3 + CO$$

Scheme 3

B. *B/E*-CAD DATA

Figure 5 illustrates the utility of FAB-CAD in sequencing neuropeptides. As is evident, the low-mass ions in a *B/E*-CAD spectra are free from matrix interference, and therefore, the amino acid residues at the N- and C-termini are readily recognized by considering Y''_1 and A_1 sequence ions, respectively. Further corroboration of the terminal amino acid residues is obtained from the A_{n-1}, B_{n-1}, C''_{n-1}, and Y''_{n-1} sequence ions, because all of them are abundant in those spectra. Another advantage of the *B/E*-CAD data is the possibility of bidirectional sequencing, because for most enkephalin-related peptides, the sequence ions include a complete Y-type and nearly complete A- or B-type ion series. Thus, utilizing complementary information afforded by these N- and C-terminal sequence ion series, it is possible to obtain a complete amino acid sequence of nmol amounts of an oligopeptide by use of the linked-field scan technique at constant *B/E*.

V. SOLVENT-SOLUTE INTERACTIONS

In a recent study,[20] we demonstrated that irradiation by a beam of fast atoms induces adduct ion formation between peptides and certain liquid matrices such as glycerol, ethylene glycol,

1,4-butanediol, and 1,5-pentanediol. The most dominant of the adducts, when glycerol is used as a FAB matrix, is the $([M+H]^+ + 12)$ ion (Figure 6). Other ions corresponding to the addition of 24, 42 (4), 56, 74, 86, 104, 116, and 134u to the $(M+H)^+$ ion of a peptide are also significant. These adduct ions are absent when α-thioglycerol or DTT/DTE are used as matrices. The adducts are also absent in an acidic solution of glycerol. The B/E linked-field MS/MS spectrum of the $([M+H]^+ + 12)$ ion established the free N-terminus amino group or the primary amine group of lysine and arginine as the site of glycerol attachment. A mechanism consistent with the addition of ionized glycerol or its oligomers to a peptide and subsequent fragmentation of the transitory intermediate adduct to form the adduct ions mentioned previously was proposed (Schemes 4 and 5).[20]

VI. QUANTIFICATION OF PEPTIDES

Measurement of tissue levels of the opioid peptides can provide information leading to an understanding of their role in various neurological and pathophysiological events. In view of complexities of biological samples, the methods used to quantify endogenous peptides must be highly structure-specific. Monitoring a specific sequence ion(s) arising by fragmentation of a mass-resolved $(M+H)^+$ ion of a peptide in the field-free regions of a mass spectrometer forms the basis of a high level of molecular specificity achieved by MS/MS techniques.[10] For successful use of this technique, which is known as SRM, the following three criteria must be satisfied: (1) the fragment ion selected for quantifying a peptide must be unique to that peptide; (2) that fragment ion should be sufficiently abundant so that highly-sensitive measurements compatible with tissue and cerebrospinal fluid (CSF) levels of that peptide can be made; and (3) an internal standard must be available in which an appropriate stable isotope is located at a site that furnishes a corresponding labeled fragment ion.

The B/E-CAD data suggest that selection of a specific fragment ion(s) for quantification studies is not straightforward and depends upon the peptide in question. In general, the Y_{n-1}, Y_{n-2}, A_4, A_{n-1}, A_{n-2}, B_{n-1}, and B_{n-2} sequence ions and the $([M+H]^+ - 107)$ ion are the predominant structurally diagnostic ions for the opioid peptides, and therefore, may be used for quantitative measurements. Of these, the Y-type C-terminus ions are more appropriate because monitoring the A- or B-type N-terminus sequence ions may not provide the necessary highest level of molecular specificity due to the fact that the first four N-terminus amino acid residues are common to all opioid peptides. Furthermore, incorporation of ^{18}O-atoms into the C-terminus carboxyl group is a simple method to synthesize an internal standard.[58] The molecular specificity of the SRM method can be increased further by monitoring several metastable reactions. The Y_{n-1}, Y_{n-2}, $((M+H)^+ - 107)$, and a few A- and B-type sequence ions qualify for this type of analytical measurement.

VII. CONCLUSIONS

In conclusion, it can be stated that FAB, alone or combined with CAD, provides an effective means of characterizing biologically active peptides. High performance four-sector instruments (see Chapter 7) are ideally suited to sequence peptides and proteins. These instruments are able to mass-select the precursor and mass-measure the resulting product ions at increased mass resolution, thus avoiding any ambiguity in the precursor-product relation. Mass selection of a precursor ion, on the other hand, is generally poor in the linked-field technique using a two-sector forward-geometry (EB) instrument, and may lead to spurious signals and increased noise level in a B/E spectrum. The data thus obtained may be of inferior quality compared to the corresponding data obtained by mass-analyzed ion-kinetic energy (MIKE) or high-performance MS/MS. Of course, one advantage of linked-field data over that obtained by MIKE using BE or EBE instruments is increased mass resolution of the product ions in the MS/MS spectrum. The

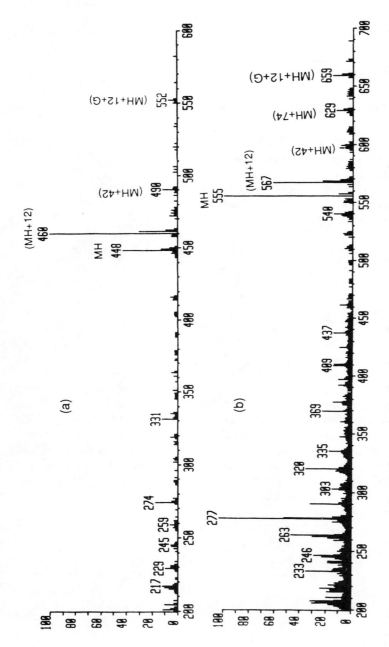

FIGURE 6.   Positive ion FAB mass spectra of (a) PFGK and (b) leucine enkephalinamide, demonstrating adduct ion formation when glycerol is used as the FAB matrix. (From Dass, C. and Desiderio, D. M., *Anal. Chem.*, 60, 2723, 1988. Copyright 1988 American Chemical Society. With permission.)

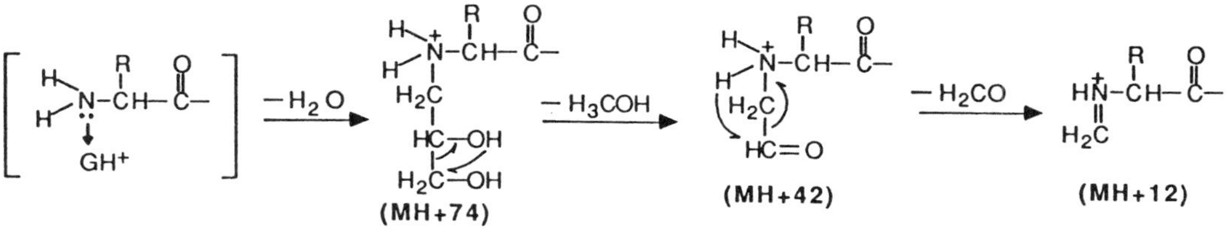

$$\left[\begin{array}{c} H \\ H \end{array} \overset{R}{\underset{GH^+}{N-CH-C-}} \overset{O}{\parallel} \right] \xrightarrow{-H_2O} \overset{H}{\underset{H_2C}{\overset{+}{N}-CH-C-}} \cdots \xrightarrow{-H_3COH} \cdots \xrightarrow{-H_2CO} \cdots$$

Scheme 4 (From *Anal. Chem.*, 60, 2723, 1988. With permission.)

$$[(M+G_2H)]^+ \xrightarrow[-CH_3OH]{-H_2O} (M+CH_2CHO+G)^+ \xrightarrow{-CH_2O} (MH-2H+CH_2+G)^+ \xrightarrow{-H_2O} (MH-3H+CH_2+CH_2CHOH\,CH_2OH)^+$$

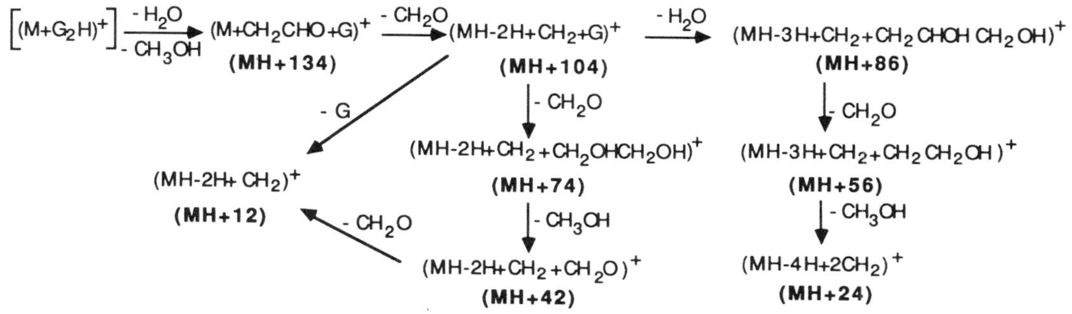

Scheme 5 (From *Anal. Chem.*, 60, 2723, 1988. With permission.)

discussion presented here demonstrates that the *B*/*E*-CAD technique using modest instrumentation can also provide meaningful sequence ion information.

The following salient points also emerge from the discussion presented in this chapter:

1. In the magnet-scan FAB mass spectra, the Z-type C-terminus ion series is complete for the enkephalin-related peptides, and must be considered for deducing the amino acid sequence. On the other hand, the B-type N-terminus and Y-type C-terminus ions provide complementary sequence ion information in the *B*/*E*-CAD spectra. These distinct features of the two types of mass spectra are attributed to differences in the internal energy of the $(M+H)^+$ parent ions.

2. The presence of phenylalanine, tyrosine, and proline has a marked influence on the abundance of sequence ions in both magnet-scan and CAD mass spectra.

3. The location of a basic amino acid also controls the yield of fragment ions in both types of mass spectra.

4. Fragmentation increases with increasing time of FAB irradiation and also with increasing concentration of a peptide.

5. A discrimination against low-mass ion exists in the CAD spectra.

6. For most peptides studied here, bidirectional sequencing is possible using the linked-field scan at constant *B*/*E*.

7. Irradiation by FAB induces adduct formation between peptides and glycerol.

REFERENCES

1. **Hughes, J., Smith, T. W., Kosterlitz, H. W., Fothergill, L. A., Morgan, B. A., and Morris, H. R.,** identification of two related pentapeptides from the brain with potent opiate agonist activity, *Nature*, 258, 577, 1975.
2. **Udenfriend, S. and Meienhofer, J., Eds.,** *The Peptides*, Vol. 6, Academic Press, New York, 1984.
3. **Kosterlitz, H. W. and Peterson, S. J.,** Types of opioid receptors: relation to antinociception, *Philos. Trans. R. Soc. London Ser. B*, 308, 291, 1985.
4. **Martin, J. B., Brownstein, M. J., and Krieger, D. T., Eds.,** *Brain Peptides Update*, Wiley-Interscience, New York, 1987, 261.
5. **Martin, J. B. and Barchas, J. D.,** *Neuropeptides in Neurological and Psychiatric Diseases*, Raven Press, New York, 1986.
6. **Desiderio, D. M.,** *Analysis of Neuropeptides by Liquid Chromatography and Mass Spectrometry*, Elsevier, Amsterdam, 1984.
7. **Barber, M., Bordoli, R. S., Sedgwick, R. D., and Tyler, A. N.,** Fast atom bombardment of solids (FAB): a new ion source for mass spectrometry, *J. Chem. Soc. Chem. Commun.*, 325, 1981.
8. **Dass, C. and Desiderio, D. M.,** Fast atom bombardment mass spectrometry analysis of opioid peptides, *Anal. Biochem.*, 163, 52, 1987.
9. **Dass, C. and Desiderio, D. M.,** Characterization of neuropeptides by fast atom bombardment and B/E linked-field scan techniques, *Int. J. Mass Spectrom. Ion Proc.*, 92, 267, 1989.
10. **Desiderio, D. M. and Dass, C.,** The measurement of leucine enkephalin at the femtomole level by mass spectrometry/mass spectrometry methods, *Anal. Lett.*, 19, 1963, 1986.
11. **Biemann, K. and Martin, S.,** Mass spectrometric determination of the amino acid sequence of peptides and proteins, *Mass Spectrom. Rev.*, 6, 1, 1987.
12. **Tomer, K. B., Gross, M. L., Zappey, H., Fokkens, R. H., and Nibbering, N. M. M.,** FAB and tandem mass spectrometry for endorphin- and ACTH- peptides of molecular weight to 2000, *Biomed. Environ. Mass Spectrom.*, 15, 649, 1988.
13. **Hunt, D. F., Shabanowitz, J., Yates, J. R., Zhu, N.-Z., Russell, D. H., and Castro, M. E.,** Tandem quadrupole Fourier-transform mass spectrometry of oligopeptides and small proteins, *Proc. Natl. Acad. Sci. U.S.A.*, 84, 620, 1987.
14. **Hunt, D. F., Yates, J. R., Shabanowitz, J., Winston S., and Hauer, C. R.,** Protein sequencing by tandem mass spectrometry, *Proc. Natl. Acad. Sci. U.S.A.*, 83, 6233, 1986.
15. **Biemann, K. and Scoble, H. A.,** Characterization by tandem mass spectrometry of structural modifications in proteins, *Science*, 237, 992, 1987.
16. **Seki, S., Kambara, H., and Naoki, H.,** Sequence analysis of an unknown peptide by molecular secondary ion mass spectrometry, *Org. Mass Spectrom.*, 20, 18, 1985.
17. **Beckey, H. D.,** *Principles of Field Desorption Mass Spectrometry*, Pergamon Press, New York, 1977.
18. **Wilkins, C. L. and Yang, C. L. C.,** Laser desorption Fourier transform mass spectra and mass measurement accuracy for bradykinins, *Int. J. Mass Spectrom. Ion Proc.*, 72, 195, 1986.
19. **Sundquist, B. and Macfarlane, R. D.,** ^{252}Cf-Plasma desorption mass spectrometry, *Mass Spectrom. Rev.*, 4, 421, 1985.
20. **Dass, C. and Desiderio, D. M.,** Particle beam induced interaction between peptides and liquid matrices, *Anal. Chem.*, 60, 2723, 1988.
21. **McLafferty, F. W.,** *Tandem Mass Spectrometry*, Wiley-Interscience, New York, 1983.
22. **Dass, C., Fridland, G. H., Tinsley, P. W., Killmar, J. T., and Desiderio, D. M.,** Characterization of β-endorphin in human pituitary by fast atom bombardment mass spectrometry of trypsin-generated fragments, *Int. J. Pept. Protein Res.*, 34, 81, 1989.
23. **Katakuse, I. and Desiderio, D. M.,** Positive and negative fast atom bombardment-collisionally-activated dissociation-linked field scanned mass spectra of leucine enkephalin, *Int. J. Mass Spectrom. Ion Proc.*, 54, 1, 1983.
24. **Jennings, K. R. and Mason, R. S.,** Tandom mass spectrometry utilizing linked scanning of double focusing instruments, in *Tandem Mass Spectrometry*, McLafferty, F. W., Ed., Wiley-Interscience, New York, 1983, 197.
25. **Yu, T. J., Schwartz, H. A., Cohen, S. A., Vouros, P., and Karger, B. L.,** Sequence analysis of derivatized peptides by high-performance liquid chromatography-mass spectrometry, *J. Chromatogr.*, 301, 425, 1984.
26. **Hunt, D. F., Buko, A. M., Ballard, J. M., Shabanowitz, J., and Giordani, A. B.,** Sequence analysis of polypeptides by collision activated dissociation on a triple quadrupole mass spectrometer, *Biomed. Mass Spectrom.*, 8, 397, 1981.
27. **Barber, M., Bordoli, R. S., Garner, G. V., Gorden, D. B., Sedgwick, R. D., Tetler, L. W., and Tyler, A. N.,** Fast atom bombardment mass spectra of enkephalins, *Biochem. J.*, 197, 401, 1981.
28. **Gaskell, S. J., Reilly, M. H., and Porter, C. J.,** Collisionally activated decomposition of leucine enkephalin and analogues using a hybrid tandem mass spectrometer, *Rapid Commun. Mass Spectrom.*, 2, 142, 1988.

29. **Desiderio, D. M.,** FAB-MS/MS study of two neuropeptides, dynorphins 1-7 and 1-13, *Int. J. Mass Spectrom. Ion Proc.,* 74, 217, 1986.

30. **Lippstreu-Fisher, D. and Gross, M. L.,** Fast atom bombardment and tandem mass spectrometry for sequencing peptides and polyamino alcohols, *Anal. Chem.,* 57, 1174, 1985.

31. **Grotemeyer, J. and Schlag, E. W.,** Laser desorption/laser ionization mass spectrometry of biomolecules, *Biomed. Environ. Mass Spectrom.,* 12, 191, 1985.

32. **Tang, X., Ens, W., Standing, K. G., and Westmore, J. B.,** Daughter ion mass spectra from cationized molecules of small oligopeptides in a reflecting time-of-flight mass spectrometer, *Anal. Chem.,* 60, 1791, 1988.

33. **Gower, J. L.,** Matrix compounds for fast atom bombardment mass spectrometry, *Biomed. Mass Spectrom.,* 12, 191, 1985.

34. **DePauw, E.,** Liquid matrices for secondary ion mass spectrometry, *Mass Spectrom. Rev.,* 5, 191, 1986.

35. **Dass, C., Seshadri, R., Israel, M., and Desiderio, D. M.,** Fast atom bombardment mass spectrometric analysis of anthracyclines and anthracylinones, *Biomed. Environ. Mass Spectrom.,* 17, 37, 1988.

36. **Tolun, E., Dass, C., and Desiderio, D. M.,** Trace level measurements of enkephalin peptides at the attomole/femtomole level by FABMS, *Rapid Commun. Mass Spectrom.,* 1, 77, 1987.

37. **Rinehart, K. L.,** Fast atom bombardment mass spectrometry, *Science,* 218, 254, 1984.

38. **Fenselau, C. and Cotter, R. J.,** Chemical aspects of fast atom bombardment, *Chem. Rev.,* 87, 501, 1987.

39. **Renugopalkrishnan, V., Rapaka, R. S., Collette, T. W., Carreija, L. A., and Bhatnagar, R. S.,** Conformational states of Leu^5- and Met^5-enkephalins in solution, *Biochem. Biophys. Res. Commun.,* 126, 1029, 1985.

40. **Williams, D. H., Bradely, C., Bojesen, G., Santikaran, S., and Taylor, L. C. E.,** Fast atom bombardment mass spectrometry: a powerful technique for the study of polar molecules, *J. Am. Chem. Soc.,* 103, 5700, 1981.

41. **Roepstorff, P. and Fohlman, J.,** Proposal for a common nomenclature for sequence ions in mass spectra of peptides, *Biomed. Mass Spectrom.,* 11, 601, 1984.

42. **Biemann, K.,** Contribution of mass spectrometry to peptide and protein structure, *Biomed. Environ. Mass Spectrom.,* 16, 99, 1988.

43. **Rumpf, B. A., Allison, C. E., and Derrick, P. J.,** Mass discrimination in collisionally activated decomposition (CAD) and mass-analyzed ion-kinetic energy (MIKE) spectra, *Org. Mass Spectrom.,* 21, 295, 1986.

44. **Carr, S. A., Herlihy, W. C., and Biemann, K.,** Advances in gas chromatography-mass spectrometric protein sequencing: 1-optimization of the derivatization chemistry, *Biomed. Mass Spectrom.,* 8, 51, 1981.

45. **Leclerq, P. A. and Desiderio, D. M.,** A laboratory procedure for the acetylation and permethylation of oligopeptides on the microgram scale, *Anal. Lett.,* 4, 305, 1971.

46. **Yu, T. J., Schwartz, H. A., Giese, R. W., Karger, B. L., and Vouros, P.,** Analysis of N-acetyl-N,O,S-permethylated peptides by combined liquid chromatography-mass spectrometry, *J. Chromatogr.,* 218, 519, 1981.

47. **Desiderio, D. M., Yamada, S., Sabbatini, J. Z., Tanzer, F.,** Quantification of picomole amounts of leucine-enkephalin by means of field-desorption mass spectrometry, *Biomed. Mass Spectrom.,* 8, 565, 1981.

48. **Desiderio, D. M., Sabbatini, J. Z., and Stein, J. L.,** HPLC and field desorption mass spectrometry of hypothalamic oligopeptides, *Adv. Mass Spectrom.,* 8, 1198, 1980.

49. **Desiderio, D. M. and Sabbatini, J. Z.,** Field desorption collisional activation linked scanning mass spectrometry of underivatized oligopeptides, *Biomed. Mass Spectrom.,* 8, 10, 1981.

50. **McCrery, D. A., Ledford, E. B., Jr., and Gross, M. L.,** Laser desorption Fourier transform mass spectrometry, *Anal. Chem.,* 54, 1435, 1982.

51. **Mallis, L. M. and Russell, D. H.,** Fast atom bombardment-tandem mass spectrometry studies of organo-alkali metal ions, *Anal. Chem.,* 58, 1076, 1986.

52. **Renner, D., Spiteller, G.,** Linked scan investigation of peptide degradation initiated by liquid secondary ion mass spectrometry, *Biomed. Environ. Mass Spectrom.,* 15, 75, 1988.

53. **Grese, R. P., Cerny, R. L., and Gross, M. L.,** Metal ion-peptide interaction in the gas phase: a tandem mass spectrometry study of alkali metal cationized peptides, *J. Am. Chem. Soc.,* 111, 2835, 1989.

54. **Leary, J. A., Williams, T. D., and Bott, G.,** Strategy for sequencing peptides as mono- and dilithiated adducts using a hybrid tandem mass spectrometer, *Rapid Commun. Mass Spectrom.,* 3, 192, 1989.

55. **Gibson, B. W., Yu, Z., Aberth, W., Burlingame, A. L., and Bass, N. M.,** Revision of the blocked N-terminus of rat heart fatty acid-binding protein by liquid secondary ion mass spectrometry, *J. Biol. Chem.,* 263, 4182, 1988.

56. **Eckart, K., Schwarz, H., Chorev, M., and Gilon, C.,** Sequence determination of N-terminal and C-terminal blocked peptides containing N-alkylated amino acids and structure determination of these amino acid constituents by using fast-atom-bombardment/tandem mass spectrometry, *Eur. J. Biochem.,* 157, 209, 1986.

57. **Walz, D. A., Wilder, M. D., Snow, J. W., Dass, C., Desiderio, D. M., and Gantz, I.,** The complete amino acid sequence of porcine gastrotropin, an ileal protein which stimulates gastric acid and pepsinogen secretion, *J. Biol. Chem.,* 263, 14189, 1988.

58. **Desiderio, D. M. and Kai, M.,** Preparation of stable isotope-incorporated peptide internal standards for field desorption mass spectrometry quantification of peptides in biological tissue, *Biomed. Mass Spectrom.,* 10, 471, 1983.

Chapter 18

PEPTIDE-CHARTING APPLIED TO STUDIES OF PRECURSOR PROCESSING IN ENDOCRINE TISSUES

Douglas F. Barofsky, Gottfried J. Feistner, Kym F. Faull, and Peter Roepstorff

TABLE OF CONTENTS

I.   Introduction 348

II. General Scheme and Methods 348
     A.   Principle 348
     B.   Mass Spectrometry 350
          1.   Instrumentation 350
          2.   Molecular Weight Determination 351
     C.   Data Analysis 351

III. Applications to Endocrine Tissues 351
     A.   General 351
     B.   Bovine and Rat Pituitary 353
     C.   Catfish Pancreas 356

IV. Summary and Forecast 362

References 363

I. INTRODUCTION

Analyses of peptides are generally prompted by assays for specific biological functions or diseases. It is often desirable to survey families of peptides associated with certain tissues, or to monitor the responses of entire peptide profiles to pharmacological treatment or to environmental stimuli. The lack of methods and detectors capable of responding to a diversity of peptides makes these tasks formidable. The genetic and biosynthetic relationships between peptides derived from the same precursor can frequently be deduced from the encoding cDNA precursor sequences, but the peptides actually expressed in a given tissue must always be verified because the enzymatic processing of precursors is species- and tissue-specific, and the cleavage products are often modified further by post-translational glycosylation, acetylation, sulfation, phosphorylation, or amidation. Techniques well-suited to the analysis of individual compounds have led to the structural elucidation of numerous biologically active peptides, their precursors, and associated components. However, these same techniques become arduous if not outright impractical when they are used to determine peptide by peptide an entire profile in a very limited amount of sample. Hence, profile studies are usually restricted to a small number of compounds. For example, the use of radioimmunoassays (RIA) to identify final peptide products reveals only peptides with specific antigenic determinants.

Analyzing peptides in biological tissues by a methodology based on mass spectrometry has been made possible only recently by the introduction of particle-induced desorption-ionization methods for the analysis of intact biopolymers[1,2] and by advances in instrumentation for compounds with high molecular weights. These developments have resulted in mass-specific analyzers that are capable of detecting organic compounds, including peptides, regardless of provenance, composition, and prior knowledge of structure. Neither an RIA nor a chemical assay, which depends on a specific structural feature, could accomplish this task. Mass spectrometers provide fast analyses, can be used on mixtures, and can be coupled with high performance liquid chromatography (HPLC)[3-5] or capillary zone electrophoresis (CZE)[6-7] for off- or on-line characterization of complete profiles. For example, fast atom bombardment (FAB) mass spectrometry[8-11] and californium-252 plasma desorption mass spectrometry (^{252}Cf-PDMS)[12-14] have both been previously used to map complex proteolytic digests.

A methodology has emerged that exploits the newly developed capabilities of mass spectrometry to effect componental analyses of crude peptide or protein mixtures extracted from a given tissue. This form of analysis has been termed peptide-charting[15] and is distinct from peptide-mapping, which refers specifically to the compositional characterization of a particular peptide or protein by means of chemical or enzymatic digestion.[9,12] To date, only a few instances of peptide-charting have been reported. This chapter describes a general approach to the method and illustrates it with examples taken from the charting of peptides in pituitary[15-17] and pancreatic[18] tissues.

II. GENERAL SCHEME AND METHODS

A. PRINCIPLE

The mass of a peptide is uniquely determined by its amino acid composition. A mass spectrometric approach to peptide-charting is based on the rationale that molecular weight information is sufficiently specific to tentatively identify all peptides that are either known to be, or can be predicted to be, in a given tissue. Prediction can be made on the basis of classical precursor processing at consecutive basic amino acids or of processing at consensus sequence sites.[19-21] The mass of a known or predicted peptide can be calculated from its structure and measured with great accuracy. Any structural alteration in a peptide that changes its mass can, in principle, be detected.

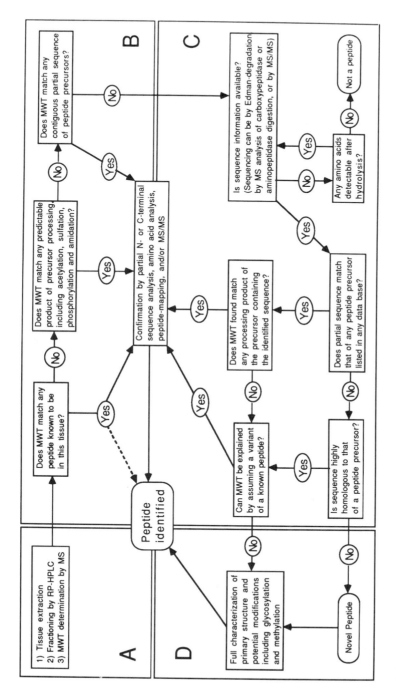

FIGURE 1.   Algorithm for peptide charting. (From Feistner, G. C., Højrup, P., Evans, C. J., Barofsky, D. F., Faull, K. F., and Roepstorff, P., *Proc. Natl. Acad. Sci. U.S.A.*, 86, 6013, 1989. With permission.)

A general algorithm for charting peptides is shown in Figure 1. The scheme takes advantage of mass spectrometry's capability for direct analysis of intact peptides in crude extracts to keep sample handling at a minimum. Mild acidic extraction followed by a single stage of reversed phase chromatography[15-17] or by an ether precipitation[18] has been found sufficient to prepare mass analyzable samples from endocrine tissues. In addition to reducing time and effort, these procedures preserve labile amino acids, e.g., Trp, Gln, Asn, and Cys, as well as labile modifications, e.g., phosphorylation, glycosylation, sulfation, and carboxyl-terminal amidation. Molecular weights, which are determined early in the scheme and are matched to available information on precursor sequences and tissue biochemistry, are used to reduce the number of peptides that have to be analyzed by more rigorous, but also more time-consuming, methods (Figure 1, boxes C and D). When classical procedures or special techniques such as mass spectrometric peptide mapping[9,12] and tandem mass spectrometry (MS/MS)[22,23] are required for structure confirmation, the molecular weights obtained by peptide-charting are used to aid in selecting the best methods. This strategy also helps to prevent failures, such as might occur in an attempt to sequence an α-N-blocked peptide by Edman degradation.

B. MASS SPECTROMETRY

1. Instrumentation

^{252}Cf-PD time-of-flight (TOF) mass spectrometry (see Chapters 1 to 4) and FAB mass spectrometry on a double-focusing magnetic sector instrument (see Chapters 6 and 7) have both been employed in peptide-charting. The nonscanning TOF mass analyzers used in ^{252}Cf-PDMS are integrating devices that detect nearly all ions produced over a wide mass range,[24-26] whereas the scanning sector analyzers used in FAB mass spectrometry are differential devices that utilize only a very small fraction of the total ion current for mass analysis.[27,28] Not surprisingly therefore, Feistner et al.,[15] in a direct comparison, experienced generally five to ten times greater detection sensitivity in the ^{252}Cf-PD mode than in the FAB mode. The absolute amounts of the peptides used were estimated from sequencing yields to be in the range of 50 to 600 pmol for ^{252}Cf-PD-analyses and 0.2 to 2 nmol for FAB analyses.

In (predictable) contrast, more accurate molecular weight data are obtained on sector instruments in the FAB mode. Mass accuracy has a direct bearing on the effectiveness of algorithms designed to match measured masses with the sequences of possible metabolic processing products of known precursors. Specifically, the efficiency of any such routine is inversely proportional to the number of precursors that must be considered and to the experimental error in a given mass measurement. The narrower the search window can be kept, the smaller the number of possible matches and, thus, the smaller the ambiguity in identifying a peptide. Because it is likely that all tissues contain multiple precursors, it is important that mass determinations be as accurate as possible. Feistner et al.[15,16] reported mass errors of ±1 U at R = 500 and ±0.2 U at R = 3000 for FAB mass measurements on a VG-ZAB-SE; in comparison, their ^{252}Cf-PD measurements, even below mass 3000, were only reliable to within ±3 U. Andrews et al.,[18] who performed only ^{252}Cf-PD-TOF analyses, reported a mass error of approximately ±2 U for sharp peaks and a somewhat greater error for diffuse peaks; data that are in good agreement with Feistner and co-workers.

At present, ^{252}Cf-PD instruments are less expensive and simpler to operate than magnetic-sector instruments operating in the FAB mode; consequently, Andrews and co-workers[18] have suggested that a ^{252}Cf-PD mass spectrometer would be an effective addition to any biochemistry or endocrinology facility. New versions of TOF and sector mass spectrometers are constantly being developed and becoming available commercially. For example, greater mass specificity in PD mass spectrometers is being pursued through the use of ion mirrors,[29] and increased detection sensitivity in sector instruments is being investigated through the use of liquid metal ion probes to promote more efficient sample desorption[30] and array detectors for the simultaneous detection of ions.[31] Other new developments, such as Fourier-transform mass spectrometry

(FTMS) with secondary ion or ^{252}Cf-PD,[32,33] multiphoton-ionization,[34] and matrix-assisted laser desorption (LD)[35] might also lead to improved methods for mass spectrometric charting of peptides.

2. Molecular Weight Determination

PD and FAB mass spectra provide mainly molecular weight information (see Chapters 1 to 4, 6, and 7). Frequently, this information is in the redundant form of multiple molecular ions due to cationization with hydrogen, sodium, and potassium. Figure 2 shows an expanded view of a ^{252}Cf-PD spectrum of a catfish pancreatic extract in the region of m/z 1600; clearly evident in the spectrum are peaks at m/z 1638, 1660, and 1678 that have been interpreted, respectively, as the (M+H)$^+$, (M+Na)$^+$, and (M+K)$^+$ molecular ion species of somatostatin (SST)-14.[18]

The binding of more than one cation to peptides can complicate further data interpretation in peptide-charting studies. Figure 3 shows a partial FAB overview spectrum of a 1-mL HPLC-fraction of a bovine pituitary extract that exhibits peaks at m/z 5436.6, 2718.5, 1813.0, 3039.3, and 1520.9; these data are compatible, respectively, with the chemical masses of the (M-H+Na+K)$^+$, (M+Na+K)$^{2+}$, and (M+H+Na+K)$^{3+}$ ions of the N-terminus of proopiomelanocortin (pOMC) and the (M+H)$^+$ and (M+2H)$^{2+}$ ions of Ac-β-endorphin(1 to 27).[15] The protonated molecular ion of the N-terminus of pOMC was observed only in one of several batches. It is especially interesting to note that, although the 2 peptides are desorbed out of the same matrix, the N-terminus is cationized preferentially by sodium and potassium, whereas Ac-β-endorphin (1 to 27) is ionized predominantly by proton attachment.

With FAB, interfering cations can be reduced only by fastidious wet-chemical procedures; ^{252}Cf-PD offers the additional possibility of removing the interfering ions by washing samples on a nitrocellulose matrix.[36,37] Unfortunately, this procedure also poses the risk of losing small peptides, as can be seen in Figure 4.

C. DATA ANALYSIS

Data analysis requires a computer with software that can calculate the molecular weights of known or predicted peptides and compare the computed values with experimental data or, alternatively, that can search for a match between the experimentally determined mass of an unknown ion and any contiguous amino acid sequence within appropriate metabolic precursors. At least three such programs have been described.[38-40]

Of course, a molecular weight and a computer match do not unequivocally identify a peptide. A tentative identification obtained in this manner is strengthened when it is supported by a sound biochemical basis for the existence of the predicted peptide. Additional supporting evidence can be obtained by peptide mapping and, when required for absolute confirmation, by amino acid analysis and amino acid sequence analysis performed by standard techniques.

III. APPLICATIONS TO ENDOCRINE TISSUES

A. GENERAL

To date, peptide charting has been attempted only on bovine pituitary,[15,17] rat pituitary,[16] and catfish pancreas.[18] The endocrine tissues used in the different investigations were chosen because they are known to contain peptides from several different protein precursors with a variety of post-translational modifications. The success that has been achieved in charting the peptides in these tissues is summarized in Figure 5, which reveals the scope and power of the technique.

A mild acidic extraction is apparently sufficient to extract peptides from endocrine tissues. More vigorous extraction conditions would probably yield degradation products from both peptides and structural biopolymers. Feistner et al.[15,16] used 1% aqueous trifluoroacetic acid (TFA) to carry out their extractions on pituitary tissues. Andrews et al.[18,41] prepared an 82%

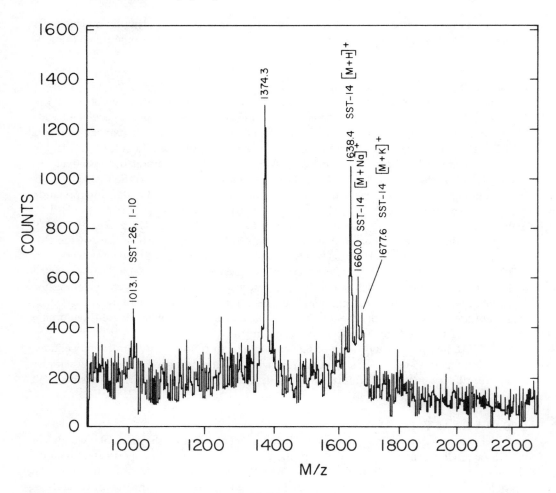

FIGURE 2.   [252]Cf-PD spectrum of a catfish pancreatic extract expanded around m/z 1600 to show the mass peaks that are compatible with the molecular ions (M+H)⁺, (M+Na)⁺, and (M+K)⁺ of SST-14. (From Andrews, P. C., Alai, M., and Cotter, R. J., *Anal. Biochem.*, 174, 23, 1988. With permission.)

ethanol/2% sulfuric acid extract from which the peptides were precipitated with diethyl ether. Both groups desalted their extracted mixtures of peptides and proteins by adsorbing and washing them on RP packing material. Feistner et al. achieved desalting in the course of performing HPLC on the extracts; Andrews and co-workers accomplished it in batches on Sep Pak cartridges. The peptides and proteins were eluted in both cases with 0.1% TFA/acetonitrile.

The pancreas contains relatively few peptides (Figure 5); hence, Andrews et al. were able to perform [252]Cf-PD mass spectrometry directly on an entire peptide extract without chromatographic separation.[18] Because the pituitary has a relatively large number of physiologically active peptides (Figure 5), Feistner et al.[15,16] carried out a single stage of gradient RP HPLC on their extracts prior to mass analysis; they used a simple linear gradient of 0 to 50% acetonitrile over 50 min and collected 1 ml fractions for subsequent mass analysis. A typical chromatogram of one of the fractions taken from a bovine pituitary extract is shown in Figure 6A. Under essentially the same conditions used by Feistner and co-workers, Andrews et al. carried out HPLC to confirm the presence of specific pancreatic peptides by comparing their retention times with those of authentic standards (Figure 6B). In only a few instances did either group of researchers isolate individual peptides for follow-up studies.

For [252]Cf-PD mass spectrometry, Andrews et al. dissolved their samples in acetic acid/

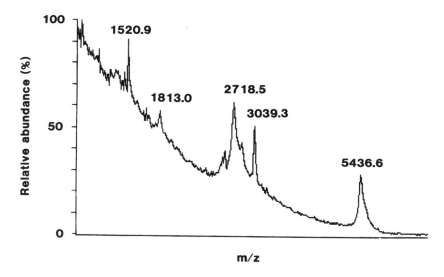

FIGURE 3. Partial FAB overview spectrum of 1-ml HPLC-fraction of a bovine pituitary extract exhibiting mass peaks that are compatible with the chemical masses of the molecular ions (M-H+Na+K)⁺ (m/z 5436.6), (M+H)⁺ (m/z 3039.3), (M+Na+K)²⁺ (m/z 2718.5), (M+H+Na+K)³⁺ (m/z 1813.0), and (M+2H)²⁺ (m/z 1520.9) of the N-terminus of pOMC. (From Feistner, G. C., Højrup, P., Evans, C. J., Barofsky, D. F., Faull, K. F., and Roepstorff, P., *Proc. Natl. Acad. Sci. U.S.A.*, 86, 6013, 1989. With permission.)

glutathione[42] and electrosprayed them onto aluminized mylar targets. Feistner et al. dissolved their samples in either 0.1% TFA or 0.1% TFA/12.5% acetonitrile and applied them to nitrocellulose-coated aluminized mylar targets[36] by spin-drying.[37]

To produce low-resolution (R = 500) FAB overview spectra, Feistner et al.[15-17] used an *m*-nitrobenzylalcohol matrix and scanned exponentially from m/z 12,000 to 1000 at a rate of 20 s/decade. The *m*-nitrobenzylalcohol was preferred for this broad mass range because of its apparent advantages over other liquid matrices for very large peptides and small proteins.[43] For peptides having molecular weights ≤4000 U, a thioglycerol matrix was employed, and unit resolution (R = 3000) mass spectra were recorded by scanning exponentially from m/z 3700 to 650 at a rate of 80 s/decade.

B. BOVINE AND RAT PITUITARY

The original overview spectra of the bovine pituitary extracts revealed 50 to 60 significant ions, about half of which have since been identified as molecular ions of peptides.[15,17] The molecular weight data alone (Table 1, columns 5 and 7) immediately pinpointed 14 of the reported processing products of propressophysin (pPP), prooxyphysin (pOP), and pOMC[44,45] (marked * in Table 1) and, with the exception of the glycopeptides Lys-γ_3-MSH and CPP (C-terminal glycopeptide of pPP, also coined copectin[46]), ultimately accounted for the entire lengths of pOMC, pPP, and pOP (Figure 5). With few exceptions, the same peptides were observed with both the ²⁵²Cf-PD and FAB methods. In some cases, the superior mass accuracy of FAB mass spectrometry was needed to distinguish unambiguously between peptides that have nearly the same mass, such as corticotropin-like intermediate peptide (CLIP) and the so-called acidic joining peptide (AJP).

Because molecular ions of glycopeptides are not always observed in mass spectrometry[47] and because the carbohydrate structure of the posterior/intermediate pituitary glycopeptides are not known, preliminary identification of Lys-γ_3-MSH and CPP(1-N) on the basis of their molecular weights was not possible. They were ultimately identified with the aid of Edman degradation (Table 1) performed on isolated compounds. The identification of CPP(1-N) was particularly

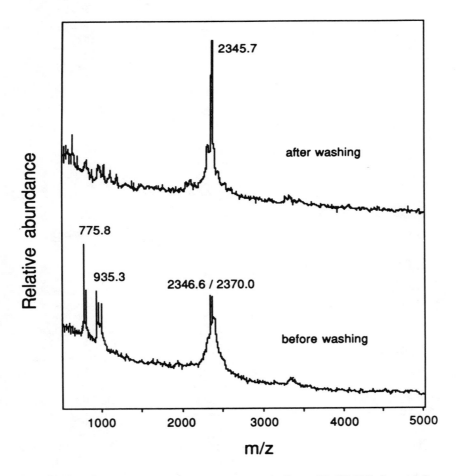

FIGURE 4.   [252]Cf-PD mass spectrometric map of a tryptic digest of Ile-VP-NP before and after washing the loaded nitrocellulose target.

interesting because the HPLC fractions in which it eluted yielded some ions (possibly fragments) from which the program for matching molecular weights proposed a possible (but what proved to be a false) candidate for this peptide.

The largest molecules observed in the peptide-charting of bovine pituitary were vasopressin-neurophysin (VP-NP) and oxytocin-neurophysin (OT-NP), both with a mol wt of almost 10,000 U. The study of these small proteins was expanded because molecular weight determinations for equivalent samples from duplicate chromatographic separations were inconsistent and because many neuronal peptides coexist in multiple forms due to partial completion of proteolytic and other post-translational modification reactions.[17] For complete characterization, the crude neurophysin fraction was resolved by RP HPLC, and the fractions corresponding to single HPLC peaks were analyzed by FAB mass spectrometry and by peptide mapping, using both [252]Cf-PD and FAB mass spectrometry. The results showed the mass variations observed in the peptide charting of the neurophysins to be largely due to sample heterogeneity, in particular, to Ile/Val microheterogeneity in VP-NP, to Gln/Leu microheterogeneity in OT-NP, and probably to C-terminal truncation.[48,49] The mass spectrometric procedure is significantly faster than the common practice[50] of identifying the variants by amino acid analysis, carboxypeptidase digestion, gel electrophoresis, and Edman degradation.

The capability for discovering novel peptides via peptide charting was demonstrated in the cases of AJP(4-24) and CPP(22-39) (Table 1). These peptides were the most favorable matches

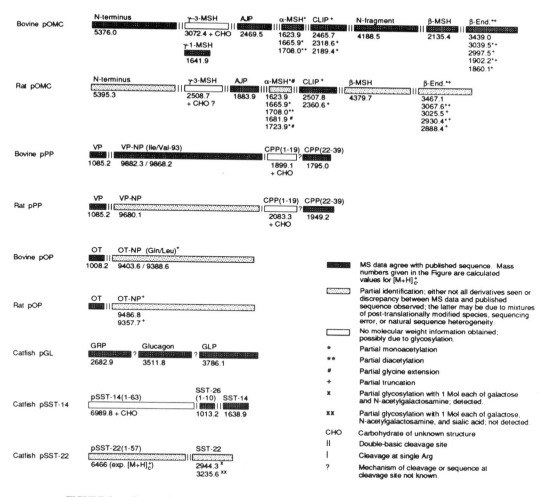

FIGURE 5.   Extent of mass spectrometric peptide charting of pituitary and pancreatic tissues.

for ion peaks observed at m/z 2081 and m/z 1975, respectively. All other peptides suggested by the matching program were judged to be poor candidates either because they contained dibasic cleavage sites not expected to survive in the intermediate pituitary or because their processing would have required the unlikely cleavage of an X-Pro or a Pro-X bond. The amino acid sequences of AJP(4-24) and CPP(22-39) were confirmed by Edman degradation. Feistner et al.[15] argue that the CPP fragments and the truncated form of AJP [AJP(4-24)] found in the peptide charting studies of bovine pituitary are likely to be natural processing products rather than artifacts of the isolation procedure; however, further studies are required to confirm this hypothesis. Computer analysis also suggested C-terminally truncated CLIPs as candidates for peptides found in two other HPLC fractions of bovine extract (Table 1); CLIP(1-21) had been isolated previously from the rat.[51,52]

One rat pituitary weighs only 10 to 20 mg; nevertheless, the sensitivity of [252]Cf-PD mass spectrometry is high enough to perform peptide charting on an extract from the pituitary of a single rat.[16] For example, the principle mass peaks in the spectrum of one crude fraction, Figure 7A, were identified tentatively by computer matching as Ac-α-MSH (M_r 1706.9) and β-MSH (M_r 4378.7). This latter assignment was confirmed by 11 cycles of Edman degradation, which, because α-MSH is N-blocked, showed only the N-terminal sequence of β-MSH. The spectrum of another crude fraction, Figure 7B, indicated the presence of a truncated OT-NP minus the C-

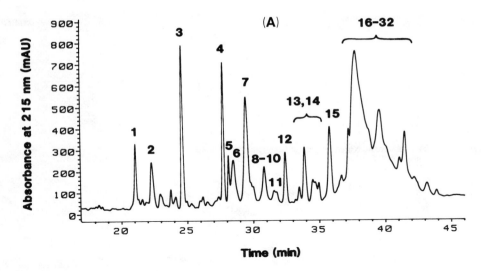

FIGURE 6.   (A) Sequential RP HPLC chromatogram of the initial peptide charting of bovine posterior/
intermediate pituitary peptides (linear gradient of 0—50% acetonitrile in 0.1% aqueous TFA over 50 min;
an extract aliquot equivalent to 0.5 pituitary was used). (From Feistner, G. C., Højrup, P., Evans, C. J.,
Barofsky, D. F., Faull, K. F., and Roepstorff, P., *Proc. Natl. Acad. Sci. U.S.A.*, 86, 6013, 1989. With
permission.) (B) Chromatographic profile of catfish pancreatic extract on a VYDAC pH-stable C-8 RP
HPLC column (4.6 mm × 25 cm). The peptides were eluted over 60 min with a linear gradient of 0—50%
acetonitrile. Both solvents contained 0.1% TFA. The flow rate was 0.7 ml min[-1] and the absorbance was
monitored at 225 nm. The peaks corresponding to SST-22a (a), SST-22b (b), SST-14 (c), insulin (d), GLP
(e), and glucagon (f) are indicated. The elution positions were identified by chromatography of the
previously purified peptides[41, 53, 54] under the same conditions. (From Andrews, P. C., Alai, M., and Cotter,
R. J., *Anal. Biochem.*, 174, 23, 1988. With permission.)

terminal Glu (M_r 9356.5, calculated from published sequence[47]) as judged by its experimentally
determined molecular weight of 9356.8 U and an HPLC retention time similar to that of bovine
neurophysins. In view of the results on bovine neurophysins,[17] this preliminary assignment
requires confirmation by mass spectrometric analysis of individual rat neurophysins isolated by
HPLC. Progress in charting the peptides in rat pituitary by mass spectrometry is summarized in
Figure 5.

C. CATFISH PANCREAS

The catfish pancreas is known to contain the hormones SST-14, SST-22a, SST-22c,
glucagon, glucagon-like peptide (GLP), and insulin, which derive from the precursors prosoma-
tostatin-14, prosomatostatin-22, proglucagon (pGlucagon), and proinsulin (Figure 5, Table 2).
[252]Cf-PD mass spectra of crude pancreatic extracts exhibit molecular ion mass peaks corresponding to all of the known hormones, except SST-22c, (Figure 8, Table 2). The presence of
the hormones in the extracts was confirmed by comparing their HPLC retention times (Figure
5B) with those of the authentic standards.[41,53,54] The sequence of catfish insulin is unknown;
however, the mass peak at m/z 5500 is consistent with the mass observed by gel permeation chromatography (Table 2).

Of the remaining 3 molecular ions observed in the spectrum for the crude pancreatic extract
(Figure 8), that at m/z 1014 could be explained using the mass analysis program, but those at
m/z 1374 and 2729 could not. The mass peak at m/z 1014 corresponds to a (M+H)[+] ion for catfish
SST-26(1-10), an expected processing product of pSST-14. The presence of SST-26(1-10) in
catfish pancreatic extract was confirmed by [252]Cf-PD mass spectrometry after performance of
one cycle of Edman degradation on the isolated peptide.[18] Of the 3 products expected from
pSST-14, viz. pSST-14(1-63), SST-26(1-10), and SST-14, only the latter had been isolated

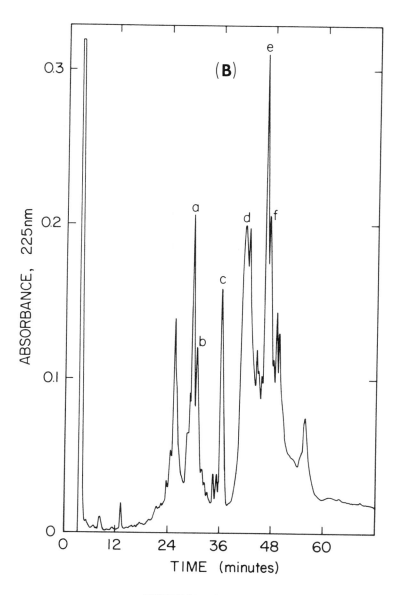

FIGURE 6 (continued).

previously from catfish pancreas. Thus, the identification of SST-26(1-10) is an example of a predictable processing product, which had not been isolated previously, being discovered by peptide charting. The peptide responsible for the molecular ion at m/z 1374 is still unknown. Andrews et al.[18] argue on biological grounds that neither of the pSST are likely precursors; they cannot rule out a peptide derived from pGlucagon or another precursor. The molecular ion at m/z 2729 was found to be due to glicentin-related peptide (GRP) from the N-terminus of pGlucagon by isolating the peptide and determining its amino acid composition and amino acid sequence. The confirmation of catfish GRP completed the identification of all the major products of pGlucagon processing in catfish pancreas (Figure 5). The complete amino acid sequence of catfish pGlucagon is not known, however, and thus, the identification of catfish GRP is an excellent example of how mass spectrometric peptide charting can be used to indicate the presence of truly unknown peptides.

TABLE 1

Summary of Expected and Observed PD and FAB Mass Spectrometric Data for Bovine Posterior/Intermediate Pituitary Peptides[15]

Fr.#	Peptide	Source	Calcd. (M+H)^+_c	PD-MS R=500	Δm	FABMS R=500	Δm	FABMS R=3000	Δm	Calcd. (M+H)^+_m	Edman degradation[a]	Ref.
22	AJP(4-24)	pOMC(83-103)	2082.1	2081.0	-1.1	2081.2	-0.9	2081.1	+0.1	2081.0	VAVGEGPGRGDDAETGPRED	15
23	AJP(1-24)	pOMC(80-103)	2469.5	2468.5	-1.0	2468.4	-1.1	2468.3	+0.2	2468.1	EEEVAVGEGPGRGDDAETGPRED	55
25	Vasopressin[b]	pPP(1-9)NH$_2$	1085.2	1083.7	-1.5	1084.7	-0.5	1084.5	+0.1	1084.4		45
28	CPP(22-39)	pPP(130-147)	1795.0	1792.8	-2.2	1794.8	-0.2	n.m.		1793.9	VOLAGAPEPAEPAOPGVY	15
29	Oxytocin[b]	pOP(1-9)NH$_2$	1008.2	1006.2	-2.0	1007.6	-0.6	1007.4	+/-0	1007.4		45
	K-γ$_3$-MSH[c]	pOMC(50-77)-CHO	s.t.								KYVMGHFRWDRFGRRNGSSSGVGGAAQ	56
30	β-MSH[b]	pOMC(189-206)	2135.4	2132.9	-2.5	2134.9	-0.5	2134.1	+0.1	2134.0	DSGPYKMEH...	57
31	CLIP(1-21)	pOMC(123-143)	2318.6	2318.6	+/-0	2318.2	-0.4	2317.3	+0.2	2317.1		44
	β-End(1-16)[b]	pOMC(209-224)	1747.0	n.o.		1746.8	-0.2	1745.8	+/-0	1745.8		44
	Des-Ac-α-MSH[b]	pOMC(106-118)NH$_2$	1623.9	1622.9	-1.0	1623.5	-0.4	1622.9	+0.1	1622.8		44
32	CLIP(1-20)	pOMC(123-142)	2189.4	2189.0	-0.4	n.o.		2188.2	+0.1	2188.1		44
33	α-MSH[b]	Ac-pOMC(106-118)NH$_2$	1665.9	1664.6	-1.3	1665.3	-0.6	1664.9	+0.1	1664.8		44
34	K-γ$_1$-MSH[b]	pOMC(50-61)NH$_2$	1641.9	1640.8	-1.1	1641.5	-0.4	1640.6	-0.2	1640.8		58
35	Ac-α-MSH[b]	Ac$_2$-pOMC(106-118)NH$_2$	1708.0	1706.4	-1.6	1707.8	-0.2	1706.6	-0.3	1706.9		44
36	CLIP	pOMC(123-144)	2465.7	2467.4	+1.7	2465.4	-0.3	2464.4	+0.2	2464.2	RPVKVYPNGAEDES...	44, 57
	β-End(1-17)[b]	pOMC(209-225)	1860.1	1860.6	+0.5	1860.0	-0.1	n.o.		1858.9	YGGFM...	44
37	CPP(1-19)[c]	pPP(109-127)	s.t.								ANDRSNATLLDGPSGALLL	15
38	N-fragment[b]	pOMC(147-186)	4188.5	4189.6	+1.1	4187.7	-0.8	n.m.		4186.0	ELTGERLEQARGP...	57
	Ac-β-End(1-17)[b]	Ac-pOMC(209-225)	1902.2	1901.3	-0.9	1902.1	-0.1	1901.0	+/-0	1901.0		44
39	VP-NP[d]	pPP(13-107)	s.t.	9846.4	s.t.	9903	s.t.	n.m.			AMSDLELRQCLPC...	45, 48, 49
40	OT-NP[d]	pOP(13-105)	s.t.	9383.3	s.t.	9414	s.t.	n.m.			AVLDLDVRTCLPCG...	45, 48, 49
	β-End(1-31)[b]	pOMC(209-239)	3439.0	n.o.		3439.8	+0.8	n.m.		3436.8	YGGFM...	44, 57
41	β-End(1-27)[b]	pOMC(209-235)	2997.5	n.o.		2996.8	-0.7	n.m.		2995.6	YGGEMTSE...	44
	N-terminus[e]	pOMC(1-49)	5375.9	5382.9	+7.0	5377.0	+1.1	n.m.			WCLESSQCQDLTTE...	59
42	Ac-β-End(1-27)[b]	Ac-pOMC(209-235)	3039.5	3044.3	+4.8	3039.2	-0.3	3037.6	-0.3	3037.6		44

Note: $(M+H)^+_c$ and $(M+H)^+_m$, chemical and monoisotopic masses, respectively, of protonated molecular ion. n.m., not measured; n.o., not observed; s.t., see text.

a   Sequence shown to the extent examined, unambiguous assignments underlined.

b   Tentative assignment immediately following mol wt determination.

c   Only observed in the glycosylated form.

d   Mixture of closely related species.

e   Extensive alkali attachment observed.

From Feistner, G. C., Højrup, P., Evans, C. J., Barofsky, D. F., Faull, K. F., and Roepstorff, P., *Proc. Natl. Acad. Sci. U.S.A.*, 86, 6013, 1989.

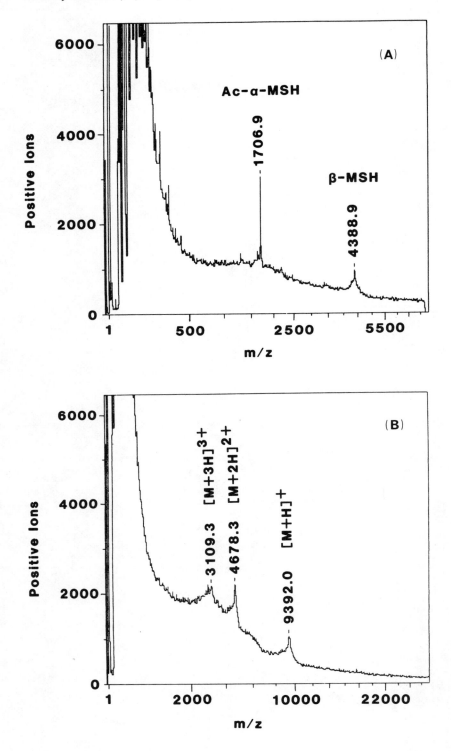

FIGURE 7. ^{252}Cf-PD mass spectra of (A) fraction 35 and (B) fraction 41 of a crude acid extract of a single rat pituitary.

TABLE 2
Summary of Expected and Observed PD Mass Spectrometric Data for Catfish Pancreatic Peptides [18]

Peptide	Source	Calculated $(M+H)^+_c$	PD-MS $(M+H)^+$	Ref.
	pSST-14(1-63)[a]	6989.8	n.o.	
SST-26(1-10)	pSST-14(65-74)	1013.2	1014	18
SST-14	pSST-14(77-90)	1638.9	1639	60
	pSST-22(1-57)	6491.4	6466	18
SST-22a	pSST-22(60-81)[b]	2944.3	2944	54, 61
SST-22c	pSST-22(60-81)[c]	3235.6	n.o.	54, 61
GRP	pGlucagon(N-term.)[d]	2682.9	2729[e]	18
Glucagon	pGlucagon(core)[d]	3511.8	3512	41
GLP	pGlucagon(N-term.)[d]	3786.1	3786	41
Insulin[d]	pInsulin[d]	—	5550[f]	18

Note: $(M+H)^+_c$, Chemical mass of protonated molecule. n.o., Not observed.

[a] Probably glycosylated.
[b] Glycosylated with galactose and *N*-acetyl-galactosamine.
[c] Glycosylated with galactose, *N*-acetyl-galactosamine, and sialic acid.
[d] The sequences for catfish proinsulin, insulin, and proglucagon are not known.
[e] Disodium adduct.
[f] Assumed to represent the unresolved molecular ion cluster of two insulin variants plus several forms of the C-peptide of insulin.[38]

From Andrews, P.C., Alai, M., and Cotter, R.J., *Anal. Biochem.*, 174, 23, 1988. With permission.

The ion at m/z 2729 must be explained as the disodium adduct of GRP, not as the $[M+H]^+$ ion, which is the predominant molecular ion for all other peptides that desorb from a nitrocellulose matrix. Andrews and co-workers[18] suggest that this strong sodium binding capacity may be due to the very acidic character of GRP; the molecule has seven carboxylic groups that can act as cation exchange sites. The same argument might be invoked to explain the strong affinity shown by the N-terminus of pOMC, which contains eight acidic amino acid residues, for sodium and potassium (see Section II.B.2.). Glucagon, which has four basic amino acid residues as well as four acidic amino acid residues and, as a consequence, has the ability to form zwitterions, does not show this tendency. In contrast to glucagon, catfish GRP and bovine pOMC(1-49) have only two and three basic residues, respectively.

As did Feistner et al., Andrews and co-workers experienced difficulty with glycopeptides. Although pSST-14(1-63) is a predicted processing product of pSST-14, no molecular ion mass peaks corresponding to this apparently glycosylated peptide can be seen in Figure 8. SST-22, another pancreatic glycopeptide, has several different glycosylated forms. The form designated SST-22c makes up approximately one third of HPLC-fraction SST-22b (Figure 6B) and contains 1 mol each of galactose, *N*-acetylgalactosamine, and sialic acid. No molecular ion mass peak for SST-22c is present in the PD mass spectrum of crude pancreatic extract (Figure 8), but a large molecular ion mass peak (m/z 2944) is present for SST-22a, which is the desialylated counterpart to SST-22c and the major form of SST-22 in catfish pancreas. It remains to be investigated whether the absence of a molecular ion for SST-22c is due to a suppression effect or to desialylation during the electrospray process.[18]

A molecular ion mass peak attributed to pSST-22(1-57), a processing product of pSST-22 that is not glycosylated, is evident in Figure 8. The identity of this peptide was confirmed by performing FAB mass spectrometry and partial N-terminal sequencing on the purified peptide.

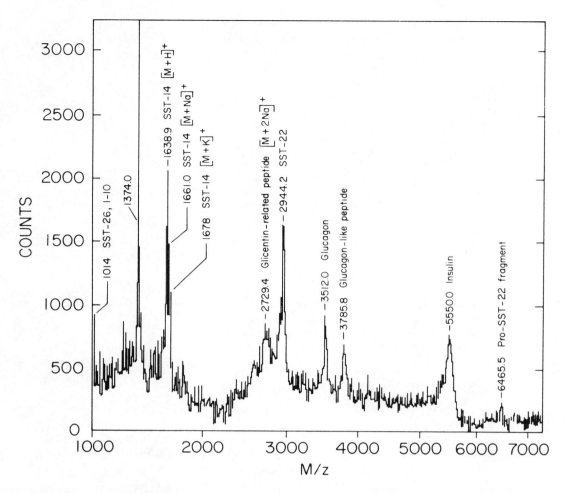

FIGURE 8. ^{252}Cf-PD mass spectrum of a crude pancreatic extract. The major mass peaks are indicated with their assigned structures. (From Andrews, P. C., Alai, M., and Cotter, R. J., *Anal. Biochem.*, 174, 23, 1988. With permission.)

The experimental molecular weight does not agree with that calculated from the published cDNA sequence of pSST-22. Andrews and co-workers[18] suggest that this discrepancy may be due to either a sequencing error or to sequence differences between populations. In any event, this example demonstrates the faculty of peptide charting for revealing possible errors in predicted peptide structures.

IV. SUMMARY AND FORECAST

Apparently, relatively few breakdown products that might interfere with mass spectrometric peptide charting are generated from other proteins in endocrine tissues under the mild acidic extraction conditions that have been used to date. Suppression of ionization of certain peptides within a mixture is observed, but not to a degree that impairs the utility of mass spectrometric peptide charting. If suppression of ionization were a problem with a particular peptide mixture, the effect could, of course, always be reduced by increased chromatographic separation. Surface effects that suppress ionization, either absolutely or relatively, might become less important if HPLC and mass spectrometry were coupled directly,[4,5] because of the transient nature the target surface would have in that arrangement.

Qualitative results obtained with ^{252}Cf-PD and FAB are similar.[15] As expected, the molecular specificity is higher on a sector instrument, whereas detection sensitivity, which is especially critical for the measurement of large peptides and for the charting of peptides in small tissue samples, is greater on a TOF instrument. Overall, both mass spectrometric techniques complement each other.

At present, the dynamic range of mass spectrometric charting is limited to one or two orders of magnitude. The abundance of molecular ions for peptides present in trace amounts is too low to detect. These shortcomings might be mitigated in the near future by emerging advances in direct couplings to HPLC and CZE, ionization techniques, and mass analyzers.

Although only a limited number of studies have been reported to date, they clearly forecast broad applicability for the technique. Its capability for revealing new, biologically active peptides is established. Feistner et al.[15] have suggested that mass spectrometric peptide charting could be used to study the regulation of whole peptide profiles in discrete tissues; Andrews et al.[18] have suggested that it could be the basis for a method to compare peptides present in homologous endocrine tissues between species or to compare peptide hormone structure in different tissues of the same animal. These potential applications are exciting possibilities.

REFERENCES

1. **Gaskell, S. J., Ed.,** *Mass Spectrometry in Biomedical Research*, Part II, John Wiley & Sons, New York, 1986.
2. **McNeal, C. J., Ed.,** *Mass Spectrometry in the Analysis of Large Molecules*, John Wiley & Sons, New York, 1986.
3. **Fridland, G. H. and Desiderio, D. M.,** Profiling of neuropeptides using gradient reversed-phase high-performance liquid chromatography with novel detection methodologies, *J. Chromatogr.*, 379, 251, 1986.
4. **Caprioli, R. M., Fan, T., and Cottrell, J. S.,** Continuous-flow sample probe for fast atom bombardment mass spectrometry, *Anal. Chem.*, 58, 2949, 1986.
5. **Caprioli, R. M.,** Analysis of biochemical reactions with molecular specificity using fast atom bombardment mass spectrometry, *Biochemistry*, 27, 513, 1988.
6. **Minard, R. D., Chin-Fatt, D., Curry, P., Jr., and Ewing, A. G.,** Capillary electrophoresis/flow FAB MS, in *Proc. 36th ASMS Conf. on Mass Spectrom. Allied Topics*, American Society for Mass Spectrometry, San Francisco, 1988, 950.
7. **Moseley, M. A., Deterding, L. J., Tomer, K. B., and Jorgenson, J. W.,** Capillary-zone electrophoresis/fast-atom bombardment mass spectrometry: design of an on-line coaxial continuous-flow interface, *Rapid Commun. Mass Spectrom.*, 3, 87, 1989.
8. **Gibson, B. W. and Biemann, K.,** Strategy for the mass spectrometric verification and correction of the primary structures of proteins deduced from their DNA sequences, *Proc. Natl. Acad. Sci. U.S.A.*, 81, 1956, 1984.
9. **Morris, H. R., Panico, M., and Taylor, G. W.,** FAB-mapping of recombinant-DNA protein products, *Biochem. Biophys. Res. Commun.*, 117, 299, 1983.
10. **Andrews, P. C. and Dixon, J. E.,** Isolation of products and intermediates of pancreatic prosomatostatin processing: use of fast atom bombardment mass spectrometry as an aid in analysis of prohormone processing, *Biochemistry*, 26, 4853, 1987.
11. **Beckner, C. F. and Caprioli, R. M.,** Verification of the DNA predicted amino acid sequence of bacteriophage P22 tail protein by mass spectrometry, *Biomed. Mass Spectrom.*, 12, 393, 1985.
12. **Tsarbopoulos, A., Becker, G. W., Occolowitz, J. L., and Jardine, I.,** Peptide and protein mapping by ^{252}Cf-plasma desorption mass spectrometry, *Anal. Biochem.*, 171, 113, 1988.
13. **Roepstorff, P., Nielsen, P. F., Klarskov, K., and Højrup, P.,** Optimization and use of nitrocellulose as a matrix for peptide and protein analysis by plasma desorption mass spectrometry, in *The Analysis of Peptides and Proteins by Mass Spectrometry*, McNeal, C. J., Ed., John Wiley & Sons, New York, 1988, 55.
14. **Nielsen, P. F., and Roepstorff, P.,** Suppression effects in peptide mapping by plasma desorption mass spectrometry, *Biomed. Environ. Mass Spectrom.*, 18, 131, 1989.
15. **Feistner, G. J., Højrup, P., Evans, C. J., Barofsky, D. F., Faull, K. F., and Roepstorff, P.,** Mass spectrometric charting of bovine posterior/intermediate pituitary peptides, *Proc. Natl. Acad. Sci. U.S.A.*, 86, 6013, 1989.
16. **Feistner, G. J., Roepstorff, P., Højrup, P., Faull, K. F., Evans, C. J., and Barofsky, D. F.,** PD-MS and FAB-MS for the charting of neuropeptides, in *Advances in Mass Spectrometry, Vol. IIB*, Longevialle, P., Ed., Heyden & Son, London, 1989, 241.

17. **Feistner, G. J., Højrup, P., Evans, C. J., Barofsky, D. F., Roepstorff, P., and Faull, K. F.,** PD/FAB-mapping of bovine neurophysins, in *Proc. 37th ASMS Conf. on Mass Spectrom. Allied Topics*, American Society for Mass Spectrometry, Miami Beach, 1989, 909.

18. **Andrews, P. C., Alai, M., and Cotter, R. J.,** The use of plasma desorption time-of-flight mass spectrometry to screen for products of prohormone processing in crude tissue extracts, *Anal. Biochem.*, 174, 23, 1988.

19. **Schwartz, T. W.,** The processing of peptide precursors: "proline-directed arginyl cleavage" and other monobasic processing mechanisms, *FEBS Lett.*, 200, 1, 1986.

20. **Benoit, R., Ling, N., and Esch, F.,** A new prosomatostatin-derived peptide reveals a pattern for prohormone cleavage at monobasic sites, *Science*, 238, 1126, 1987.

21. **Gluschankof, P. and Cohen, P.,** Proteolytic enzymes in the post-translational processing of polypeptide hormone precursors, *Neurochem. Res.*, 12, 951, 1987.

22. **McLafferty, F. W., Ed.,** *Tandem Mass Spectrometry*, John Wiley & Sons, New York, 1983.

23. **Biemann, K. and Scoble, H. A.,** Characterization by tandem mass spectrometry of structural modifications in proteins, *Science*, 237, 992, 1987.

24. **Torgerson, D. F., Skowronski, R. P., and Macfarlane, R. D.,** New approach to the mass spectroscopy of non-volatile compounds, *Biochem. Biophys. Res. Commun.*, 60, 616, 1974.

25. **Sundqvist, B. and Macfarlane, R. D.,** ^{252}Cf-plasma desorption mass spectrometry, *Mass Spectrom. Rev.*, 4, 421, 1985.

26. **Cotter, R. J.,** Plasma desorption mass spectrometry: coming of age, *Anal. Chem.*, 60, 781A, 1988.

27. **Barber, M. N., Bordoli, R. S., Sedgwick, R. D., and Tyler, A. N.,** Fast atom bombardment of solids (F.A.B.): a new ion source for mass spectrometry, *J. Chem. Soc. Chem. Commun.*, p. 325, 1981.

28. **Biemann, K. and Martin, S. A.,** Mass spectrometric determination of the amino acid sequence of peptides and proteins, *Mass Spectrom. Rev.*, 6, 1, 1987.

29. **Tang, X., Ens, W., Standing, K. G., and Westmore, J. B.,** Daughter ion mass spectra from cationized molecules of small oligopeptides in a reflecting time-of-flight mass spectrometer, *Anal. Chem.*, 60, 1791, 1988.

30. **Jiang, L. F., Barofsky, E., and Barofsky, D. F.,** Liquid assisted SIMS with a liquid metal ion microprobe, in *Secondary Ion Mass Spectrometry SIMS VI*, Benninghoven, A., Huber, A. M., and Werner, H. W., Eds., John Wiley & Sons, New York, 1988, 683.

31. **Cottrell, J. S. and Evans, S.,** Characteristics of a multichannel electrooptical detection system and its application to the analysis of large molecules by fast atom bombardment mass spectrometry, *Anal. Chem.*, 59, 1990, 1987.

32. **Hunt, D. F., Shabanowitz, J., Yates, J. R., III, Zhu, N.-Z., Russell, D. H., and Castro, M. E.,** Tandem quadrupole fourier-transform mass spectrometry of oligopeptides and small proteins, *Proc. Natl. Acad. Sci. U.S.A.*, 84, 620, 1987.

33. **Asmter, I. J., Loo, J. A., Furlong, J. J. P., and McLafferty, F. W.,** Cesium ion desorption ionization with Fourier transform mass spectrometry, *Anal. Chem.*, 59, 313, 1987.

34. **Grotemeyer, J. and Schlag, E. W.,** Multiphoton-ionization-mass spectrometry (MUPI-MS), *Angew. Chem. Int. Ed. Engl.*, 27, 447, 1988.

35. **Karas, M., Bachmann, D., Bahr, U., and Hillenkamp, F.,** Matrix-assisted ultraviolet laser desorption of non-volatile compounds, *Int. J. Mass Spectrom. Ion Proc.*, 78, 53, 1987.

36. **Jonsson, G. P., Hedin, A. B., Håkansson, P. L., Sundqvist, B. U. R., Säve, B. G., Nielsen, P. F., Roepstorff, P., Johansson, K. E., Kamensky, I., and Lindberg, M. S. L.,** Plasma desorption mass spectrometry of peptides and proteins adsorbed on nitrocellulose, *Anal. Chem.*, 58, 1084, 1986.

37. **Nielsen, P. F., Klarskov, K., Højrup, P., and Roepstorff, P.,** Optimization of sample preparation for plasma desorption mass spectrometry of peptides and proteins using a nitrocellulose matrix, *Biomed. Environ. Mass Spectrom.*, 17, 355, 1988.

38. **Andrews, P. C.,** personal communication (PROCOMP is available on request from P. C. Andrews, Department of Biological Chemistry, University of Michigan, Ann Arbor, MI 48109-0674).

39. **Lee, T. D. and Vemuri, S.,** MACPROMASS — a computer aid for the analysis of mass spectral data of peptides and proteins, in *Proc. 37th ASMS Conf. on Mass Spectrom. Allied Topics*, American Society for Mass Spectrometry, Miami Beach, 1989, 352.

40. **Højrup, P.,** General protein mass analysis (GPMA), a convenient program in studies of proteins by mass spectrometry, in *Ion Formation from Organic Solids IFOS V*, Hedin, A., Sundqvist, B. U. R., and Benninghoven, A., Eds., Springer-Verlag, Berlin, 1990, 61.

41. **Andrews, P. C. and Ronner, P.,** Isolation and structures of glucagon and glucagon-like peptide from catfish pancreas, *J. Biol. Chem.*, 260, 3910, 1985.

42. **Alai, M., Demirev, P., Fenselau, C., and Cotter, R. J.,** Glutathionine as a matrix for plasma desorption mass spectrometry of large peptides, *Anal. Chem.*, 58, 1303, 1986.

43. **Barber, M. and Green, B. N.,** The analysis of small proteins in the molecular weight range of 10-24 kDa by magnetic sector mass spectrometry, *Rapid Commun. Mass Spectrom.*, 1, 80, 1987.

44. **Smith, A. I. and Funder, J. W.,** Proopiomelanocortin processing in the pituitary, central nervous system, and peripheral tissues, *Endocrinol. Rev.,* 9, 159, 1988.

45. **Richter, D.,** The neurohypophysial hormones vasopressin and oxytocin — gene structure, biosynthesis, and processing, in *The Posterior Pituitary: Hormone Secretion in Health and Disease,* Baylis, P. H. and Padfield, P. L., Eds., Marcel Dekker, New York, 1985, 37.

46. **Smyth, D. G. and Massey, D. E.,** A new glycopeptide in pig, ox and sheep pituitary, *Biochem. Biophys. Res. Commun.,* 87, 1006, 1979.

47. **Jardine, I.,** Plasma desorption mass spectrometric analysis of glycoproteins, in *The Analysis of Peptides and Proteins,* McNeal, C. J., Ed., John Wiley & Sons, New York, 1988, 41.

48. **Burman, S., Breslow, E., Chait, B. T., and Chaudhary, T.,** Partial assignment of disulfide pairs in neurophysins, *Biochem. Biophys. Res. Commun.,* 148, 827, 1987.

49. **Chauvet, M.-T., Chauvet, J., and Acher, R.,** The neurohypophysial hormone-binding proteins: complete amino-acid sequence of ovine and bovine MSEL-neurophysins, *Eur. J. Biochem.,* 69, 475, 1976.

50. **Newcomb, R. W. and Nordmann, J. J.,** Quantitative HPLC analysis of rat neurophysin processing, *Neurochem. Int.,* 11, 229, 1987.

51. **Bennett, H. P. J., Browne, C. A., and Solomon, S.,** Characterization of eight forms of corticotropin-like intermediary lobe peptide from the rat intermediary pituitary, *J. Biol. Chem.,* 257, 10096, 1982.

52. **McDermott, J. R., Biggins, J. A., Smith, A. I., Gibson, A. M., Keith, A. B., and Edwardson, J. A.,** Removal of arg[1] and phe[22] from CLIP (ACTH$_{18-39}$) by rodent pituitary and blood peptidases, *Peptides,* 9, 757, 1988.

53. **Andrews, P. C. and Dixon, J. E.,** Isolation and structure of a peptide hormone predicted from a mRNA sequence. A second somatostatin from the catfish pancreas, *J. Biol. Chem.,* 256, 8267, 1981.

54. **Andrews, P. C., Pubols, M. H., Hermodson, M. A., Sheares, B. T., and Dixon, J. E.,** Structure of the 22-residue somatostatin from catfish. An O-glycosylated peptide having multiple forms, *J. Biol. Chem.,* 259, 13267, 1984.

55. **James, S. and Bennett, H. P. J.,** Use of reversed-phase and ion-exchange batch extraction in the purification of bovine pituitary peptides, *J. Chromatogr.,* 326, 329, 1985.

56. **Hammond, G. L., Chung, D., and Li, C. H.,** Isolation and characterization of δ-melanotropin, a new peptide from bovine pituitary gland, *Biochem. Biophys. Res. Commun.,* 108, 118, 1982.

57. **Kawauchi, H.,** Chemistry of proopiocortin-related peptides in the salmon pituitary, *Arch. Biochem. Biophys.,* 227, 343, 1983.

58. **Böhlen, P., Esch, R., Shibasaki, T., Baird, A., Ling, N., and Guillemin, R.,** Isolation and characterization of a γ1-melanotropin-like peptide from bovine neurointermediate pituitary, *FEBS Lett.,* 128, 67, 1981.

59. **Bennett, H. P. J.,** Isolation and characterization of the 1 to 49 amino-terminal sequence of pro-opiomelanocortin from bovine posterior pituitaries, *Biochem. Biophys. Res. Commun.,* 125, 229, 1984.

60. **Minth, C. D., Taylor, W. L., Magazin, M., Tavianini, M. A., Collier, K., Weith, H. L., and Dixon, J. E.,** The structure of cloned DNA complementary to catfish pancreatic somatostatin-14 messenger RNA, *J. Biol. Chem.,* 257, 10372, 1982.

61. **Magazin, M., Minth, C. D., Funckes, C. L., Deschenes, R., Tavianini, M. A., and Dixon, J. E.,** Sequence of a cDNA encoding pancreatic preprosomatostatin-22, *Proc. Natl. Acad. Sci. U.S.A.,* 79, 5152, 1982.

Chapter 19

MASS SPECTROMETRY OF BIOLOGICALLY IMPORTANT NEUROPEPTIDES

Dominic M. Desiderio

TABLE OF CONTENTS

I.   Introduction 368
    A.   Objective 368
    B.   Qualitative Analytical Mass Spectrometry 369
    C.   Quantitative Analytical Mass Spectrometry 369
    D.   Need for Neuropeptide Sequence Data in the Neurosciences 369
    E.   This Research Program 370
    F.   Hypothetical Example 370
    G.   Experimental Scheme 372

II.   Analytical Methodology 372
    A.   Tissue Acquisition 372
    B.   Acid Precipitation of Proteins 373
    C.   Sep-Pak Chromatography 374
    D.   Reversed-Phase High Performance Liquid Chromatography 374
    E.   Opioid Radioreceptor Assay 376
    F.   Immune System Methods 377
    G.   Nucleic Acid Methods 379
    H.   Post-Translational Modifications 379
    I.   Comparative Advantages of Each Analytical Method 379

III. Mass Spectrometry 380
    A.   Collision Activated Decomposition 380
    B.   Linked-Field Scans 380
    C.   Tandem Mass Spectrometry (MS/MS) Methodology 381
    D.   Internal Standards 381
    E.   Quantification of Peptides 383

IV.   Applications 383
    A.   Trigeminal System 383
        1.   Maxillary and Mandibular Branches 383
        2.   Ophthalmic Branch 385
    B.   Lumbar Cerebrospinal Fluid 387
        1.   Low Back Pain Patients 387
        2.   Senile Dementia of the Alzheimer's Type 389
    C.   Human Post-Mortem Pituitary Tissue 389
        1.   Tumor Profiles 389
        2.   Control and Drug-Overdose Pituitaries 390
        3.   β-Endorphin 391
    D.   Summary of Applications Section 393

V.   Other Peptide Systems Studied 393

A.   Tuftsin 393
B.   Ileum Protein Fragment 393
C.   FAB Matrix Studies 394
D.   Synthetic Peptide Studies 394
E.   Tandem MS Study of Two Neuropeptides, Dynorphin$_{1-7}$ and
      Dynorphin$_{1-13}$ 394
F.   Negative FABMS Analysis of Leucine Enkephalin 394

Acknowledgments 394

Abbreviations 395

References 397

I. INTRODUCTION

A. OBJECTIVE

This chapter describes the use of mass spectrometry (MS) as a key component in a multidimensional study of human neuropeptide pathways to quantify endogenous neuropeptides in extracts of human fluids and tissues.[1] These MS data are needed in the neuroscience studies of today because most of the other widely used analytical methods, except for direct amino acid sequencing methods, cannot provide the amino acid sequence of an endogenous peptide during quantification. MS provides these needed sequence data, which are critical because the molecular basis of many human diseases remains unknown and also because neuropeptide pathways are being studied in our laboratory in several human populations including controls, patients experiencing pain or stress,[2] and patients with pituitary tumor formation.

This chapter includes extensive experimental information on several other analytical methods that are used prior to MS analysis of a peptide. Although the chapter is lengthened somewhat by this important information, it is essential to discuss in one place all of the pertinent experimental details involved in analyzing human tissue and fluids (chromatography, radioreceptor assay [RRA], and radioimmunoassay [RIA]) because ultimately the quality of any MS data hinges upon the integrity of all of experimental steps preceding MS. For example, we consider any peptide that interacts with a receptor to be a prime candidate for extensive qualitative and quantitative analysis by MS. Thus, we present a total, integrated picture of tandem mass spectrometry (MS/MS) of biologic peptides. Extended studies that focus on only the MS of synthetic peptides are presented elsewhere (see Chapter 17).

The analytical scheme discussed in this chapter is based on experience gained over the past several decades by many researchers in our own and other laboratories. Electron ionization (EI) and chemical ionization (CI) MS methods have been used for many years to study chemically derivatized peptides to determine their molecular ion and amino acid sequence-determining fragment ions.[3] Introduction of "softer" ionization techniques such as field ionization (FI), field desorption (FD),[4] and fast atom bombardment (FAB)[5] MS facilitated production of molecular ions and, more recently, ^{252}Cf-plasma desorption (PD),[6] electrospray, and laser desorption (LD) (see Lubman chapter) MS techniques produced molecular ions of much larger peptides. Mass spectrometric methods to analyze peptides have been reviewed.[3]

This laboratory began analyzing biologically important neuropeptides during the amino acid sequence determination of the thyrotropin-releasing hormone (TRH),[7] providing the initial impetus for the current rapid expansion in the field of brain peptide neuroendocrinology.[8]

Following that discovery, we contributed to the development of acetylation and permethylation derivatization chemistry of synthetic oligopeptides in order to confer thermal stability to peptides.[9] The introduction of FI[4], FD, and FAB[5] desorption techniques rapidly led to our use of these methods in quantifying synthetic peptides using $(M+H)^+$ ions,[10-16] synthesizing stable isotope-incorporated peptide internal standards,[17,18] and quantifying physiological peptides in extracts of human tissues and fluids.[1,14-24]

B. QUALITATIVE ANALYTICAL MASS SPECTROMETRY

In order to place the objectives of this chapter into proper perspective, it is important to distinguish between two general stages of research involved in analyzing biologically active compounds. For example, once a physiologically active peptide has been discovered, its amino acid sequence must be determined. TRH is a case in point. That tripeptide was extracted from millions of sheep hypothalami for the purpose of accumulating a sufficient amount of material for chromatography, amino acid analysis, and high resolution mass spectrometry in order to establish the amino acid sequence as pyroglutamyl-histidyl-proline-amide.[7] With modern MS techniques, approximately nanomoles[25] to picomoles[26] of a biologically active peptide suffice to determine the amino acid sequence of a peptide, and if a collisionally activated or unimolecular decomposition fragmentation pattern is favorable, then linked-field MS can be used for amino acid sequence determination.[27]

Once the amino acid sequence of a biologic peptide is known, it is sometimes important to qualitatively corroborate the presence of that peptide in a biologic extract by means of its protonated molecular ion (or even total or partial amino acid sequence data). For example (see discussion later in chapter), detection of a protonated molecular ion of a peptide in human placenta tissue extracts is consistent with the presence of the neuropeptide $dynorphin_{1-8}$.[28,29]

C. QUANTITATIVE ANALYTICAL MASS SPECTROMETRY

It is also true that once the amino acid sequence of a biologically important peptide has been determined, the critical combination of high molecular specificity and high detection sensitivity, both of which are now available with current MS and MS/MS techniques, can accurately quantify that peptide in a biological extract.[1,24,30,31] MS has been used for many years to quantify a variety of biologically important compounds, such as steroids, cannabinoids,[32] drugs, prostaglandins,[33,34] fatty acids, and others, but until recently, it has not been used to quantify peptides. Several books discussing the various pertinent aspects of quantitative analytical MS are available.[35-37]

However, because peptides, saccharides, and nucleotides consist of various monomer units, the corresponding sequence information of their monomers must also be guaranteed during quantification to avoid any of the ambiguity inherent in most other analytical methods. The requirement for sequence information for these three classes of compounds adds an additional level of experimental difficulty (when compared to quantifying other classes of compounds) to an analytical measurement. Nevertheless, once that amino acid sequence is known, the molecular ion plus amino acid sequence-determining fragment ion information combine effectively to provide unambiguous molecular specificity and detection sensitivity levels for peptides at the picomole, femtomole, and, in favorable cases, the attomole level;[38] this level of detection sensitivity is sufficient for analyzing peptides in extracts of human tissues and fluids. Indeed, in many cases, molecular specificity, even at the expense of detection sensitivity, is a more important and readily accepted experimental datum.

In any event, it is clear that the two important experimental parameters, sensitivity and specificity, are mutually exclusive.

D. NEED FOR NEUROPEPTIDE SEQUENCE DATA IN THE NEUROSCIENCES

This chapter focuses on research involving neuroactive peptides. Because the field of neurosciences is one of the most rapidly expanding areas in science today,[8] there is an urgent need to establish the amino acid sequence of a post-translationally modified neuroactive peptide acting at a synapse, within a cell body, or in an axon, because the pertinent basic molecular mechanisms involved in neuronal and axonal chemistry are not known in many cases. For example, the amino acid sequence of a peptide is required during peptide metabolism studies to establish efficacious clinical and pharmacological intervention and treatment and to provide a firm, rigorous basis for molecular neuroscience research. Furthermore, knowledge of the amino acid sequence of a neuropeptide permits accurate conclusions concerning metabolism from its precursor, enzymes required for its metabolism, and endogenous concentrations measured in control vs. pathophysiological states.

The importance of these statements is realized when one understands that the molecular basis underpinning many human pathophysiological conditions such as pain, pituitary tumor formation, epilepsy, Tourette's syndrome, drug abuse, sudden infant death syndrome, senile dementia of the Alzheimer's type (SDAT), amyelotrophic lateral sclerosis, and stress remain unknown, even in the face of intensive and extensive research activity in these and other areas. Furthermore, new fields of neuroscience combine information from individual subspecialties into new fields such as psychoneuroendocrinoimmunology[39] to reflect the growing awareness of the cross communication that exists among several different biological subsystems that had been considered to be mutually isolated, but in which it is now known that peptides represent some of the communicating molecules. Such concepts are compatible with the viewpoint that neuropeptides contain a much higher level of information than classical amino acid neurotransmitters.[1]

E. THIS RESEARCH PROGRAM

Research discussed in this chapter describes approaches to test the hypothesis that several different human neuropeptide families contribute to human homeostasis and that derangements in those systems are involved in various pathophysiological and stress[40] conditions. In particular, this chapter focuses on opioid and tachykinin peptide families. Certainly, other neuropeptides are important,[41] and it is known that amino acid and peptide neurotransmitter systems are co-localized and apparently function as mutual neuromodulators.

There are three precursors for opioid peptides: proenkephalin A,[42] proenkephalin B,[43] and proopiomelanocortin (pOMC).[44] The proenkephalin A family contains four molecules of methionine enkephalin (ME=YGGFM), and one each of leucine enkephalin (LE=YGGFL), ME-RGL, and ME-RF. Proenkephalin B contains several different dynorphins, and the pOMC family contains ACTH, β-endorphin, α-MSH, corticotropin-like intermediate peptide (CLIP), and LPH.[45] cDNA sequencing methods demonstrated the amino acid sequences of these precursors,[46] but not, of course, of the final post-translationally modified neuropeptides.

Apparently, cellular and neuronal interactions occur among these peptide families, among individual peptides in these families, and between neuropeptides and the classical amino acid transmitter systems.[47] Our hypothesis states that a defect in any step involving preprohormone production, precursor metabolism, presynaptic peptide synthesis, receptor interaction, and postreceptor interactions could precipitate an observed clinical pathophysiology. Furthermore, the effects of stress in the human apparently cause these peptides to metabolize differentially.[40,48,49] Consequently, only the amino acid sequence and quantitative aspects of metabolic interactions among these neuropeptide families accurately define a control situation and provide the comparative analytical data needed to recognize a pathophysiology.

F. HYPOTHETICAL EXAMPLE

For example, the scheme in Figure 1 proposes a rationale that we hypothesize explains the derangements in neuropeptide processing that might occur in human controls, drug abusers, and drug addicts. This hypothesis is being tested experimentally in our laboratory (see discussion to

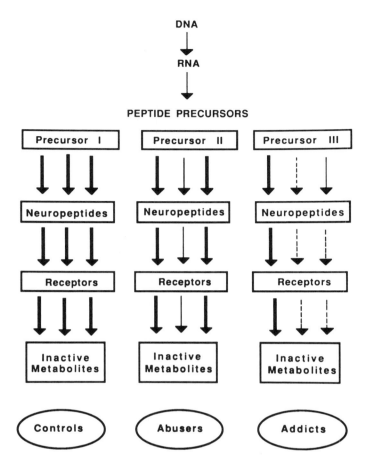

FIGURE 1. Hypothetical scheme to rationalize the several different neuropeptide pathways endogenous to controls, drug abusers, and addicts. Thick arrows represent well-developed (qualitatively, quantitatively) metabolic pathways; thin arrows represent quantitatively decreased pathways; and dotted arrows represent metabolic pathways that are absent or severely compromised.

follow). Appropriate similar schemes can be readily drawn to rationalize the other human clinical pathophysiologies listed previously.

This scheme demonstrates that post-translationally modified neuropeptides are produced from their corresponding precursor molecule. Three different types of arrows are used to represent schematically the dozens of potentially important neuropeptides that might play a role in drug abuse. A thick arrow represents a pathway that is fully functioning in a qualitative and quantitative sense. A thin arrow represents a pathway that might be different quantitatively, perhaps due to a decreased level of available enzymes, for example. A dotted line arrow represents an absent, or severely compromised pathway, such as might be caused by the absence of a key enzyme, for example.

We hypothesize that an endogenous neuropeptide system of a normal, or control, individual will not be adversely affected by the ingestion of an exogenous drug that will interact with the receptor of a particular neuropeptide. However, it is possible, of course, that a massive ingestion of a highly purified drug will overwhelm even a well-developed neuropeptide system and cause death via cardiac arrest or respiratory depression. In the latter case, for example, a drug could interact with opioid receptors in the nucleus of the solitary tract, nucleus commissuralis, and nucleus ambiguous.[50]

We hypothesize that in a drug abuser, at least one of the several neuropeptide pathways, although basically intact, functions in a quantitatively decreased manner, and that drug ingestion may or may not adversely effect, or shut down, that particular pathway. In this case, perhaps an occasional drug user can operate in such a fashion and avoid addiction.

However, we also hypothesize that an addict has a severely compromised, or absent, pathway normally required for homeostasis, and that the addict truly craves an exogenous drug, which will occupy a critical target receptor and precipitate second messenger metabolism. Of course, an exogenous drug is not equivalent in all molecular aspects to the endogenous peptide, and addiction results.

This example is only one of the representative rationales that forms the basis for the hypothesis underpinning our clinical measurements. In other words, a fully functioning complement of neuropeptide processing is necessary for homeostasis, or "well-being". A metabolic defect in any one (or more) pathway leads to a disease state. Precedence for this type of a hypothesis is found in the multiple metabolic defects that occur in the intermediate metabolism of amino acids.

Therefore, in this chapter we discuss analytical data obtained for quantitative analytical measurement of peptides in several different human systems: the trigeminal system and the effects of orthodontic stress on that system;[40,48] human pituitary tumor formation[51] of endocrinologically silent tumors; and low back pain.[52]

G. EXPERIMENTAL SCHEME

A distinct logic in this chapter leads directly to a two-stage experimental process. First, tissue is obtained rapidly and placed immediately into liquid nitrogen in a clinic, a peptide-rich fraction is analyzed by reversed phase high performance liquid chromatography (RP HPLC), the presence of biologically important peptides is established by RRA and RIA, and HPLC fractions that contain appropriate receptoractivity and immunoreactivity are funneled to subsequent MS analysis.[53]

Figure 2 contains the flow chart representing the experimental purification, screening, and identification steps in this process.

Second, this analytical scheme uses MS and MS/MS analysis of peptides. First FDMS[10,12,14,15,17,18] was used, and now FAB,[38,54] to produce a molecular ion of an endogenous peptide. The mass spectrometric behavior of many synthetic peptides is studied in great detail[55-57] (see also chapter by Dass), and peptides in biologic extracts are analyzed in a qualitative and quantitative manner.

Therefore, in this chapter, the quantitative analytical scheme is discussed in greater experimental detail (Figure 2) to test our working hypothesis (Figure 1).

II. ANALYTICAL METHODOLOGY

This research program probes several different human neuropeptide systems and combines a wide variety of analytical methods, which are only briefly explained in this section.

A. TISSUE ACQUISITION

This entire analytical process is aimed at accurately quantifying several different types of endogenous peptides. Therefore, the first important step in this process is to carefully acquire several different human tissues and fluids, including brain, pituitary, and trigeminal tissue (teeth, tooth pulp, calcified tissue) and cerebrospinal (CSF) and intracerebroventricular fluid. These tissues and fluids are located proximal to the brain, and are used in these studies to avoid extensive metabolism of neuropeptides that occur in blood and urine, for example.

Liquid nitrogen is used to ensure rapid temperature lowering of biological samples.[58] Personnel in an operating suite (orthopedic clinic, dental clinic, neurosurgical suite) acquire

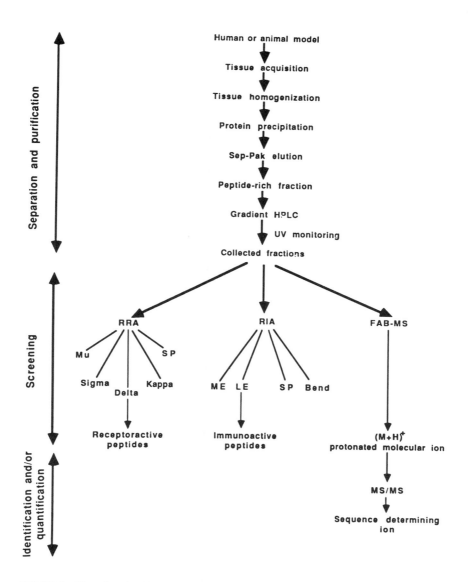

FIGURE 2.   Flow chart showing purification and separation steps leading to screening, identification, and quantification of neuropeptides with maximum molecular specificity. In some studies, FABMS is used to screen all fractions for the presence of (M + H)⁺ ions. In other studies, FABMS produces an (M + H)⁺ to corroborate the molecular weight of a peptide or B/E-SIM techniques are used to produce amino acid sequence-determining ions, which can be used either for positive identification or quantification of a peptide. (From Desiderio, D. M. and Fridland, G., *Mass Spectrometry in Biomedical Research*, Gaskell, S., Ed., John Wiley & Sons, New York, 1986, 443. With permission.)

tissue and fluid within seconds of its surgical removal. Although enzyme inhibitors minimize peptide degradation,[59] we avoid using inhibitors whenever possible because they could unnecessarily complicate the subsequent analytical methods.

B. ACID PRECIPITATION OF PROTEINS

Tissue is homogenized in acetic acid or acidified acetone to precipitate larger proteins such as opioid peptide precursors. Centrifugation removes precipitated proteins, and smaller neuropeptides remain in solution. However, when we study opioid precursors, homogenization performed in aqueous solution at neutral pH maintains those precursors in solution.

C. SEP-PAK CHROMATOGRAPHY

Supernatant from a tissue homogenate is subjected to further purification for the purpose of preferentially separating the oligopeptides from several other cellular constituents such as salts, proteins, and other water-soluble solutes. Mini-columns constructed of RP packing material are very efficient for this prepurification step. These mini-columns are washed first to extend the hydrocarbon chains and to optimize the hydrophobic surface area.

When a neuropeptide forms an ion-pair with a buffer, then that peptide:buffer ion-pair effectively becomes more hydrophobic than the peptide alone, due to efficient ion-pairing. A volatile buffer is used in these studies to minimize interferences in subsequent RRA, RIA, and MS detection methods. Carboxylic acids with various lengths of perfluorinated carbon chains have been used as the buffer to effectively regulate that important relative hydrophobicity parameter.[60] In our studies, an RP Sep-Pak is loaded with the peptide:buffer ion-pair solution, a hydrophobic peptide:buffer ion-pair interacts with the RP hydrophobic surface to preferentially retain that ion-pair, and an aqueous wash removes other water-soluble solutes. Peptide:buffer ion-pairs are removed from the hydrophobic surface with a bolus containing a high percentage of an organic modifier such as acetonitrile to yield a peptide-rich fraction.

Certain experimental parameters, which may at first seem trivial, must be analyzed carefully. For example, we studied in detail the rate of sample deposition of substance P (SP) onto a Sep-Pak, and the rate of elution from the Sep-Pak of the SP:buffer ion-pair.[61] The highest recovery (picomole-femtomole) of physiological amounts of radiolabeled SP and ME was obtained at the slowest flow rate (75 μl min^{-1}) for sample application and elution. Of course, if a detector such as ultraviolet (UV) (low detection sensitivity) is used, then nanomoles of peptide are required and the subtle differences noted in SP recovery are not measured at these application and elution rates.

These data demonstrate that femtomole-picomole equivalents of active sites for neuropeptides are present on commercial Sep-Pak columns. Such experimental data are crucial for studies involving CSF, which contains picomole amounts of peptides.

D. REVERSED-PHASE HIGH PERFORMANCE LIQUID CHROMATOGRAPHY

RP HPLC is an effective method of separating biologic neuropeptides contained in a peptide-enriched fraction obtained from a Sep-Pak. Silica gels of a known particle diameter and pore size are derivatized chemically, for example with C_{18} groups, to produce a hydrophobic surface.[62] Synthetic hydrocarbon polymer HPLC columns have also been used to separate biologic peptides. We use both types of columns in our biologic neuropeptide studies. For example, in a bovine cornea study,[63-65] following Sep-Pak preparation of a peptide-enriched fraction, RP HPLC gradient chromatography on a C_{18} silica analytical column separates peptides over a wide range of peptide lengths and relative hydrophobicities. Siliconaceous material that almost always "bleeds" from an HPLC column interferes with FABMS picomole detection sensitivity,[38] and a second isocratic separation step using a synthetic hydrocarbon polymer column diminishes that column bleed and purifies that gradient silica fraction.

In any event, a first-step RP HPLC gradient elution, such as that shown for the example of a mixture of synthetic peptides in Figure 3, elutes peptides covering a wide range of relative hydrophobicity and detects the presence of a wide range of peptide sizes and of several families of opioid peptides. Peptides of a few amino acids —up through β-endorphin, which contains 31 amino acids and beyond to precursors — are separated in one gradient.[63,66-67]

However, our experience over the past several years indicates that an HPLC peak (gradient or isocratic) of a biologic extract is hardly ever "pure" to an MS detector, and that a second HPLC separation is usually required. For example, a UV detector monitors only chromophore-containing compounds, whereas MS detects the presence of almost all organic compounds. Even if an HPLC peak contains only one peptide, organic compound buffers and column bleed can also be eluted. Although those "bleeds" may not absorb UV, both of the latter possibilities

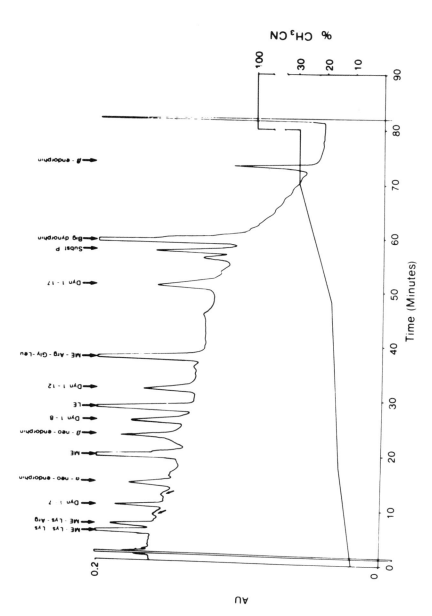

FIGURE 3.   Gradient RP HPLC detection of synthetic peptides. Varian 5000 liquid chromatograph with 15 cm μBondapak column (Waters Associates). Flow rate: 1.5 ml/min; 0.2 AUFS; UV monitoring: 200 nm; aqueous buffer: 0.04 M formic acid with triethylamine, pH 3.15 (TEAF); organic modifier: acetonitrile (CH_3CN); gradient profile is indicated as the solid line from 10 to 100% CH_3CN, indicated on the right hand axis. Arrows indicate some of the buffer peaks caused by the organic modifier step increases (see text). (From Fridland, G. and Desiderio, D. M., *J. Chromatogr.*, 379, 251, 1986. With permission from Elsevier Science Publishers, Physical Sciences and Engineering Division.)

interfere with FAB desorption processes, which are surface-sensitive phenomena. Generally, to initially detect any biologically active neuropeptides in a gradient elution, we collect 1-min fractions across either a 90- or a 120-min gradient elution. We define "biologically active" in this study to mean any endogenous peptide that interacts with a receptor to displace an appropriate radiolabeled ligand (discussed later). Such receptor activity is also a strong indicator (although not proof) of bioactivity.

It is always important to establish that, first, very low levels of biologically important peptides have not been lost onto any active sites of an HPLC analytical column, and second, that once biologic extracts have been purified by HPLC, the HPLC column has been cleaned sufficiently — meaning, for example, that no receptor activity or immunoreactivity is eluted after column cleaning. These two important experimental points are obviously mutually exclusive. For example, we studied human CSF, which contains picomole amounts of opioid peptides.[67] An HPLC column must be purified rigorously of any synthetic peptide standards, and in most cases, we do not even utilize any synthetic peptide standards if we will study biologic peptides with that column.[63-64] Rigorous column cleaning means multiple washings with long, shallow gradients of organic modifiers.[67] A low rate of change in the percentage of organic modifier for elution and cleaning is an important experimental feature, because it provides sufficient time to establish an equilibrium of a peptide between solution and active sites. On a molecular basis, most, if not all, peptide molecules have been removed from the active sites of the column; that process is the objective of cleaning a column. However, when one injects a biologic extract such as CSF onto that cleaned column, a certain quantity (in some cases, all) of the endogenous peptide content will be lost onto the column during reoccupation of those active sites. An equivalent statement was made above for active sites for peptides on Sep-Paks, where we noted femtomole-picomole equivalents of active sites for SP peptides.[61] Pituitary tissue contains nanomoles of peptides, and thus this "reoccupation" problem is less severe for that type of tissue compared to CSF, where only picomoles of endogenous peptides are present. Therefore, one should not obtain an HPLC immediately following column cleaning, but should wait until active sites are reoccupied, generally by one or a few injections of the same biologic extract that is under study.[65]

E. OPIOID RADIORECEPTOR ASSAY

An important factor in this type of research program is the post-HPLC detector. A UV detector set to 200 nm to detect peptide amide bond UV absorption is the most-used HPLC detector. We have demonstrated UV detection sensitivity for picomoles of somatostatin (SST).[11] However, for the objectives described here, UV detection, although utilized on each gradient and isocratic elution, is too nonselective and insensitive (generally low nanomole) to be of any practical use in the study of physiological levels of peptides in human fluids and tissues. Therefore, we also use RRA and RIA to detect the presence of receptor active and immunoreactive peptides, respectively, that are collected in individual fractions following gradient separation. Metabolic profiles are very helpful in these types of studies because they readily establish the relationships among peptide families and indicate which samples must be funneled to subsequent MS detection.

RRA of opioid peptides is based upon the fact that there are several different receptor types and that they interact "preferentially" with different types of opioid peptides.[68] For example, μ–, κ–, σ–, δ–, and ϵ-receptors are hypothesized to exist and to interact primarily with morphine, dynorphins, a synthetic peptide, enkephalins, and endorphin, respectively.[69] Of course, no given peptide interacts exclusively with only one receptor, but rather with several different types of receptor and with varying degrees of sensitivity and molecular specificity.

RRA is a very important aspect in neuroscience research, and MS plays a significant role in establishing which peptide interacts with which receptor. By effectively using a peptide:receptor interaction, the presence of receptor activity (and probably biologic activity) of peptides in a

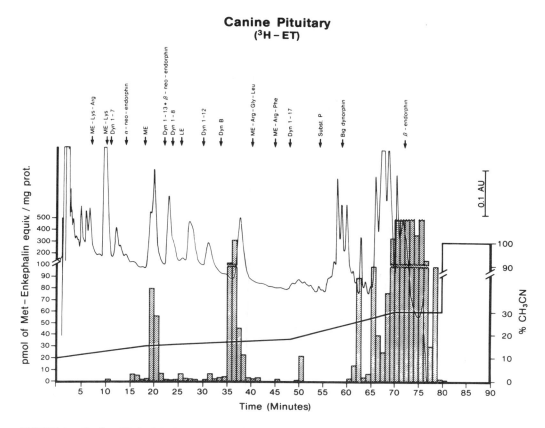

FIGURE 4. Gradient RP HPLC chromatogram and RRA of the peptide-rich fraction from pituitary tissue using ³H-etorphine. Abscissa: pmol of receptor-assayable peptide expressed as picomole ME "equivalents" per milligram protein. Ordinate: the number correlating to the 1-min fraction collected. Right axis: percentage of CH_3CN. The connected straight-line segments show the gradient. Insert shows 0.1 AUFS UV absorption at 200 nm. The UV HPLC chromatogram is shown as the solid-line profile. The arrows along the top of this curve denote the retention times of commercially available synthetic peptides under these HPLC conditions. The hatched boxes relate to the amount of receptor activity measured in that fraction. (From Takeshita, H., Desiderio, D. M., and Fridland, G., *Biomed. Chromatogr.*, 1, 126, 1986. With permission.)

gradient HPLC separation is detected.[70] However, molecular specificity of a receptor assay is not of concern in our research program because that molecular specificity will be substantiated in later steps (see Figure 2) with MS.[1] Therefore, an opioid RRA is used to screen for opioid receptor activity in each fraction collected in a gradient HPLC. Generally, picomoles of receptor activity are obtained with RRA, and tritiated etorphine is used as the ligand to interact with several different opioid receptor types of varying degrees of specificity and sensitivity.

For example, Figure 4 contains a "metabolic profile" of opioid receptor active peaks extracted from one canine pituitary. As others have shown,[71,72] the pituitary contains a greater proportion of precursor molecules, which are represented in Figure 4 by a number of later-eluting RP HPLC fractions (see also Figures 5 and 12).

In some cases, molecular specificity is increased somewhat through the use of other radiolabeled ligands, such as tritiated methionine enkephalin, which targets those peptides that interact preferentially with the δ-receptor,[73,74] or tritiated ethylketoazocine, which targets κ-receptor dynorphin peptides.[29]

F. IMMUNE SYSTEM METHODS

Analytical methods based on immunological interactions are the most widely-used peptide detection systems. In principle, these methods, which are based upon an antibody-antigen

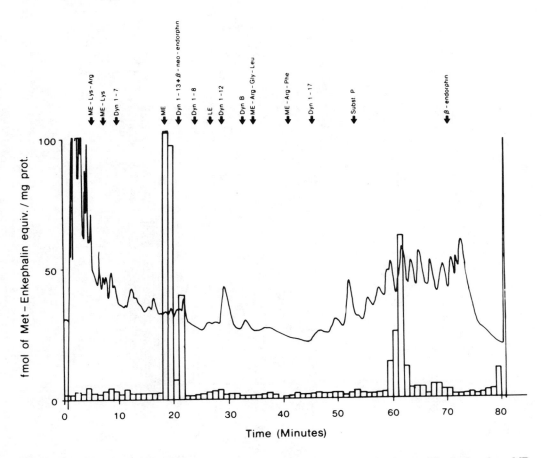

FIGURE 5. HPLC separation of the peptide-rich fraction from a pituitary chromophobe adenoma. The abscissa shows ME-LI; the ordinate shows fraction number, and the right axis the percent of CH$_3$CN. The gradient profiles is shown as the connected straight lines from 10 to 30% CH$_3$CN, followed by a 100% CH$_3$CN column wash. (From Fridland, G. H. and Desiderio, D. M., *Life Sci.*, 41, 809, 1987. ©1987 Pergamon Press. With permission.)

interaction, display a high level of specificity and detection sensitivity.[75] We certainly agree with the theory of detection sensitivity, because femtomoles of target peptide per analysis tube are generally detected. However, the claim to molecular specificity remains open to question and experimental verification; it must be kept in mind that RIA analysis does not determine an amino acid sequence, but only an antigen-antibody interaction and the radioactivity measured from a displaced antigen. Furthermore, in many cases, an antibody has been raised to a synthetic peptide:protein immunogenic complex, because a peptide with a molecular weight of less than approximately 4000 Da does not possess a sufficient level of inherent immunogenicity.

For example, in one representative study, the pentapeptide methionine enkephalin (ME) was conjugated chemically via its carboxyl terminus to a large carrier protein, thyroglobulin, (TG), to synthesize a ME:TG (30:1 mole ratio) complex.[76] That synthetic immunogenic complex was injected into an animal, in which antibodies were raised to that complex but not necessarily (and probably not) to the specific pentapeptide sequence YGGFM. Even though many commercial RIA kits list the cross-reactivity of its antibody with numerous other synthetic peptides, that information cannot guarantee that, in an unfractionated biological sample, the antibody will react only and completely with its putative target; nor, more importantly, does it ensure that, in an HPLC-purified peak, the peptide with which it does interact possesses the appropriate amino acid sequence.[77] Thus, a specific amino acid sequence cannot (ever) be guaranteed measurable by RIA.

Nonetheless, many advances in the neurosciences have been made through the use of immune-based systems.[78] For example, several immune-based systems have advanced a number of diverse fields, such as immunohistochemistry, immunofluorescence, enzyme-linked immunoassays (ELISAs), and RIAs. Reviews of these chemical neuroanatomical systems[79] indicate that immune-based analytical systems have been used to provide a great amount of information on the anatomical localization of peptide immunoreactivity.[80]

Commercial RIA kits are used in two detection modes in our analytical system (Figure 2). In one mode used during an initial study of a new human neuropeptide system, peptide immunoreactivity is measured in each gradient HPLC tube. For example (Figure 5), a human pituitary tumor chromophobe, which is a "neuroendocrinologically silent" (a misnomer that means simply that the exact peptide or immunoreactivity remains unknown) tumor was analyzed, and two areas of HPLC-separated methionine-enkephalin immunoreactivity[66] were detected; the data demonstrate clearly that the commercial antibody was not specific.

Once an appropriate level of confidence is established in the distribution of peptide-immunoreactivity across an HPLC gradient, the area where a synthetic peptide elutes can be focused upon in more intensive metabolic studies, and immunoreactivity can be measured at the retention time of that particular peptide. For example, we have extensively studied the content and distribution of opioid[52,81] and SP[82] peptide immunoreactivity in human lumbar CSF obtained from lower back pain patients.

Again, however, the molecular specificity of RIA must be questioned in these studies. For example, when the opioid or tachykinin content of unfractionated CSF was measured in these low back pain samples, though in some cases that value was 0 fmol ME-LI ml^{-1} CSF,[82] subsequent chromatographic separation demonstrated a distribution of several different immunoreactive peaks, some up to 75 fmol SP-LI ml^{-1} CSF (see data to follow). Apparently, different peptides interfere mutually in RIA when an unfractionated sample is analyzed and suppress completely any antibody-antigen interaction.

G. NUCLEIC ACID METHODS

In this section, only briefly mentioned are those analytical methods based on measurement of levels of messenger RNA and complementary DNA, or *in situ* hybridization methodology, all of which are important to obtain a high level of sensitivity in detecting peptides.[83] However, even though spectacular advancements have been made in several areas of the neurosciences with the use of these highly sensitive and relatively specific nucleic acid methods, clearly they cannot provide information on post-translational modifications to peptides, which is, of course, the only information that unambiguously identifies a particular neuropeptide.

H. POST-TRANSLATIONAL MODIFICATIONS

After a peptide has been biosynthesized, cellular metabolic machinery is available to modify that peptide or protein via, for example, sulfation of tyrosine residues, phosphorylation of tyrosine, serine, or threonine residues, acetylation of primary amine groups, sulfoxide formation[84] of thioether-containing residues such as methionine, α-amide formation[85;] pyroglutamyl cyclization, and oxidation and reduction of disulfide bonds. These multiple post-translational modifications provide dimensional stability, form an appropriate protein topology necessary for receptor interaction, and optimize biologic activity of a neuropeptide.

I. COMPARATIVE ADVANTAGES TO EACH ANALYTICAL METHOD

Current scientific literature must be read critically because from time to time, claims are made that the amino acid sequence of a biologically active peptide has been established by using techniques such as immunoassay, in some cases, even in combination with HPLC. It is not uncommon to read statements indicating that the immunoreactivity of a target peptide has been measured and that HPLC data established the sequence of a peptide because the ir-peptide co-eluted with a synthetic peptide. Of course, those types of claims are incorrect because chromatography, receptor activity, RIA, and nucleic acid methods, whether used by themselves or in various combinations with each other, cannot establish the amino acid sequence of a post-translationally modified peptide.[77]

It is also crucial to remember that a peptide in a biologic extract is known only when its amino acid sequence has been established. If one does not know the amino acid sequence of a peptide, one knows neither the peptide name nor its precursor. Presently, only three experimental methods can determine the amino acid sequence of a peptide: X-ray crystallography, which requires a relatively large amount of crystalline material, gas phase or liquid phase sequenators, and MS/MS.

III. MASS SPECTROMETRY

FABMS effectively produces the protonated or de-protonated molecular ion of a biologically important neuropeptide.[5] Sensitivity is high, ion current is long-lasting, and various matrix materials[86] optimize FAB conditions on the probe. In some cases, the protonated molecular ion of a peptide corroborates the presence of a peptide in a biological extract. However, such a process cannot convey amino acid sequence information. For example, whereas the molecular ion observed at 981 daltons corroborated the presence of dynorphin$_{1-8}$ in human placenta,[29] that study awaits amino acid sequence confirmation.

A. COLLISION ACTIVATED DECOMPOSITION

Although FABMS produces a protonated molecular ion with a high level of ion current, the total complement of amino acid sequence-determining fragment ion information is not always available in the resulting mass spectrum because insufficient energy is available to fragment all peptide bonds. However, most commercial mass spectrometers now have a collision cell immediately following the ion source. The protonated molecular ion selected by the ion source, or by the first section of a multisector mass spectrometer, can be subjected to collision activated dissociation (CAD) processes[87] to create or to increase peptide bond fragmentation. For example, dynorphins$_{1-7}$ and $_{1-13}$ were analyzed with a four-sector ($B_1E_1E_2B_2$) mass spectrometer (data shown in Figure 13), wherein the first two sectors (B_1, E_1) selected the $(M+H)^+$ ion as the precursor ion, and a B_2/E_2 linked-field scan of that precursor ion collected amino acid sequence-determining fragment ions from that molecular ion.[57] CAD of MH$^+$ significantly increased the ion current due to amino acid sequence-determining fragment ions. Also, the collision energy can be increased by electrically floating the collision cell to increase the ion current at lower masses.

B. LINKED-FIELD SCANS

Magnetic (B) and electric (E) fields in a multisector mass spectrometer can be configured electronically in several different scanning modes to obtain linked-field scans,[88] which are powerful instrumental methods for establishing important precursor-product relationships. For example, all fragment ions derived from a selected precursor ion mass are collected with a B/E linked-field scan by scanning the E sector and the B sector while maintaining a constant B/E ratio.

Also, all ions that are precursors to a selected product ion are collected in a B^2/E scan, where the ratio of B^2 and E is kept constant while that ratio is scanned. Another useful linked-field scan is the constant neutral loss scan $(B/E)(1-E)^{0.5}$, which monitors the loss of a constant neutral fragment.[89] For example, loss of a molecule of ketene from acetyl derivatives, or in the case of opioid peptides, loss of 107 amu from the tyrosine side chain (all opioid peptides contain the N-terminal YGGF–), are instances in which a constant neutral loss linked-field scan is important analytically.

C. TANDEM MASS SPECTROMETRY (MS/MS) METHODOLOGY

Molecular specificity is the primary requirement for determining and quantifying opioid peptides in human tissue and fluid extracts.[1] From the previous brief description of other analytical methods, it is clear that detection sensitivity is not a problem with those other analytical methods, especially with RIA. However, all of those other methods lack molecular specificity, which is the hallmark of MS/MS methods.[1,3,90-92] As mentioned, in principle one sector of a mass spectrometer produces the protonated molecular ion of a peptide and, in some cases, amino acid sequence-determining fragment ions as well. In that first stage, a tandem mass spectrometer selects the molecular ion of the peptide, excludes all other ion masses, and subjects that MH+ precursor ion to a CAD process. A linked-field scan in the B/E mode in the second stage selects all fragment ions, which are amino acid sequence-determining fragment ions, of that selected molecular ion.

Focal plane detectors using multichannel systems[93] operate in a fashion similar to photoplates,[94] and could increase the detection sensitivity by a factor of 100 to 1000 in a tandem four-sector MS instrument.

It is interesting to note that the use of linked-field scan MS analysis has apparently proceeded at a relatively slow rate. One reason for this could be that, in the MS/MS mode, most MH+ ion current is not detected in a linked-field scan mode. For example, in a mass spectrum of a FAB-produced molecular ion, a high ion current of (M+H)+ with little amino acid sequence-determining fragmentation occurs. One must exclude perhaps up to 95% of that ion current in a B/E linked-field scan in order to clearly detect the amino acid sequence-determining fragment ions. However, when proceeding from the B-scan to the linked-field scan mode, although a significant loss in that signal occurs, there is an even greater reduction of noise.[91] This combination of events leads to a fortuitous increase in the analytically important signal-to-noise ratio. Noise reduction results from deflecting the main ion beam in the mass spectrometer, effectively eliminating any chemical noise from impurities and FAB probe matrix, and detecting only amino acid sequence-determining fragment ions.

With that understanding of the genesis of an increased signal-to-noise ratio in MS/MS, it is easy to realize that quantification of a biologic peptide is possible by monitoring one or a few amino acid sequence-determining fragment ions unique to that peptide and comparing that ion current to the current arising from a corresponding stable isotope-incorporated peptide internal standard.[1]

Figure 6, for example, collects the B/E linked-field scan of the (M+H)+ ion of leucine enkephalin (LE=YGGFL) (top), and of $^{18}O_2$–LE (bottom). The two ^{18}O atoms are located exclusively in the carboxy-terminus group, and thus all C-terminal ions increased 4 amu in the bottom spectrum (see masses 397, 340, 282). Endogenous LE is readily quantified by comparing the integrated ion current due to the C-terminal tripeptide fragment ion -GFM at 336 (top) with the internal standard ion at 340 (bottom).

D. INTERNAL STANDARDS

Several different types of synthetic peptide internal standards serve to quantify accurately biologic peptides. Homologs are always a possibility, and were used first in this laboratory to quantify endogenous enkephalins.[10,12] For example, ^{2}ala-leucine enkephalin was used for FD-MS quantification of peptides in several canine brain regions. That internal standard performed well, even though it was separated by several minutes on the HPLC gradient system from the methionine enkephalin and leucine enkephalin target peptides.

A second type of stable isotope-incorporated internal standard used ^{18}O, and ME and leucine enkephalin were both subjected to ^{18}O-incorporation into only the carboxy-terminus.[17,18] These internal standards also worked well (see Figure 6) until we used NH_4OH on the FAB probe tip to achieve an attomole level of sensitivity in the negative ionization mode.[38] Basic pH conditions ionize the carboxyl group and facilitate back-exchange of an ^{18}O atom from the carboxy-terminus.

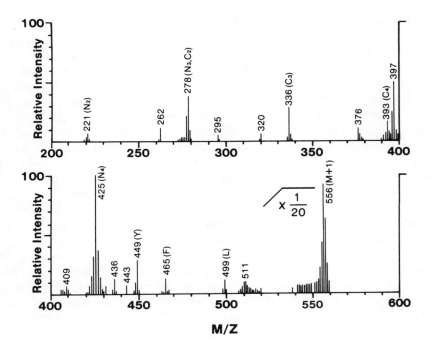

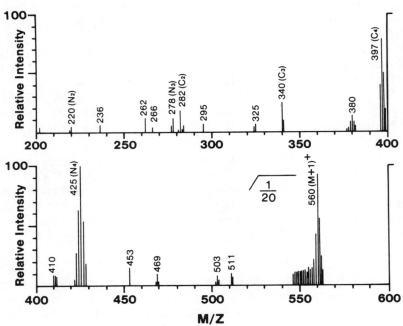

FIGURE 6.   B/E linked-field spectrum of the $(M+H)^+$ ion of LE (top) and of $^{18}O_2$-LE (bottom). (From Desiderio, D. M., *Analysis of Neuropeptides by Liquid Chromatography and Mass Spectrometry*, Elsevier, Amsterdam, 1984, 235. With permission from Elsevier Science Publishers, Physical Sciences and Engineering Division.)

Therefore, a third generation of stable isotope-incorporated peptide internal standards uses either ^{13}C- or ^2H-labeled amino acids incorporated into a synthetic peptide.[95] These internal standards are currently being evaluated.

Before an ion is used to quantify an endogenous peptide, it must satisfy seven different criteria:

1. The ion genesis is established with accurate mass, B/E, and stable isotope data.
2. A B/E linked-field spectrum of MH^+ must contain that ion.
3. The ion must be "clean", meaning that no other ion contributes to its ion current in a B/E scan.
4. The ion must shift "cleanly" following stable isotope incorporation.
5. The ion current must be relatively abundant.
6. The ion must uniquely represent the molecule being quantified.
7. Use of that ion must demonstrate appropriate detection sensitivity (femtomole, picomole) to measure endogenous amounts of peptide.

Clearly, MH^+ ions are not appropriate for these purposes.

Uniqueness of the selected fragment ion is an important consideration. For example, Tyr–Gly–Gly–Phe– is a tetrapeptide located at the N-terminus of all opioid peptides, and it produces an abundant peak at m/z 425 for both LE and ME.[55] Therefore, although perhaps that fragment ion is useful for a constant neutral screen across numerous HPLC fractions, 425 does not provide the highest level of molecular specificity for quantification of an opioid peptide compared to, for example, a carboxyl group-containing fragment ion.

E. QUANTIFICATION OF PEPTIDES

Thus, the overall process for quantifying peptides combines FAB ionization, CAD in some cases, B/E linked-field analysis, and multiple reaction monitoring of metastable transitions from endogenous and internal standard peptides.

IV. APPLICATIONS

One unique aspect of this research program is that it combines the use of several different analytical methods to determine the presence of, corroborate the protonated molecular ion of, and quantify, using an amino acid sequence-determining fragment ion, endogenous peptides in human tissues and fluids. In this section we discuss the combined HPLC, RRA, RIA, and MS analysis of neuropeptides in three human neuropeptide systems, including the trigeminal system, cerebrospinal fluid from the lumbar region, and pituitary tissue. In a brief concluding section, analysis of dynorphins in human placenta tissue is discussed.

A. TRIGEMINAL SYSTEM

The trigeminal system is comprised of three separate anatomical sections — the ophthalmic, maxillary, and mandibular branches. Because high levels of pain are experienced via this nerve system, we investigated neuropeptides contained in these three branches.

In this research effort, several separation techniques were developed, first with commercially available synthetic peptides and then with peptides extracted from canine brain, teeth, and CSF. For example, the first experiments involved a canine caudate nucleus because of its relatively large size (approximately 1 g).[14,15] Furthermore, FD-MS was used to detect protonated molecular ions, and subsequent experiments utilized FAB of human peptides.

1. Maxillary and Mandibular Branches

The demonstration of a neuropeptide by means of MS in human teeth[40] involved the use of FABMS to detect the presence of a protonated molecular ion at 574 daltons, which is a mass compatible with ME (mol wt = 573 amu).

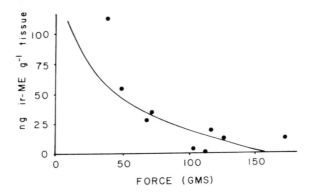

FIGURE 7. Plot of the RIA measurements of ME against the amount of force supplied by a coil-spring. Data points are for the first extracted tooth from each experimental subject. The least-squares regression line fit to these points is curvilinear because force was log-transformed for analysis. (From Walker, J. A., Tanzer, F. S., Harris, E. F., Wakelyn, C., and Desiderio, D. M., *Am. J. Orthod. Dentofac. Orthop.*, 92, 9, 1987. With permission.)

Also in that study, opioid peptides from several different patient's teeth were analyzed first by RIA and then by RRA. One set of patients, who had premolars extracted for orthodontic purposes, served as controls. A second set, also diagnosed for premolar extractions, had a coil-spring attached between their left and right maxillary premolars to supply an orthodontic force for a period of time prior to tooth extraction. HPLC, RIA, RRA, and MS were used in a series of experiments to isolate, identify, and quantify ME in extracts of the pulp tissue. These experiments showed for the first time that ME was present in human tooth pulp. Also, in the first demonstration of an experimental manipulation of a human neuropeptide system (Figure 7), orthodontic force caused a significant decrease in ME immunoreactivity in a group of experimental teeth compared to controls.[40]

Furthermore, levels of ME immunoreactivity in the first spring-attached tooth removed from each patient had a statistically significant inverse log-linear relationship to the amount of applied force. These data indicate clearly that orthodontic force mobilizes at least one neuropeptide pathway in the human tooth.[40]

Following that study and the first demonstration by MS of neuropeptides in the human tooth, we hypothesized that SP from the tachykinin neuropeptide system would behave in a metabolically similar fashion. Therefore, effects of orthodontic force on the concentration of two endogenous neuropeptides, immunoreactive ME (ir-ME) and immunoreactive SP (ir-SP), were evaluated in 20 patients from whom premolars were extracted.[48] Experimental patients in this study had a force ranging from 140 to 600 g exerted on their maxillary premolars. Data showed that ir-SP concentrations decreased in that short time when the first vs. the second tooth was extracted, suggesting that surgical manipulation depleted endogenous stores of this peptide. Concentrations increased from the second to the fourth tooth extracted. A positive correlation existed between ir-SP and ir-ME concentrations, and that correlation was highest in the first tooth extracted from controls. Furthermore, the concentration of ir-SP correlated negatively with the magnitude of orthodontic force, and this correlation was enhanced when the amount of force was log-transformed.

The third step in this human trigeminal study extended to the calcified portion of a tooth. The question was raised: if neuropeptides occur in pulp tissue, do they also exist in the calcified portion?[96] To answer that question, four wisdom teeth were removed from a patient during the course of normal dental care, and were placed immediately into perchloric acid until calcium was removed and the tissue developed a rubbery consistency. After homogenization and peptide extraction, two portions were obtained, denoted as samples V_1 and V_2. Sample V_2 was treated first with immobilized trypsin[97] and then with carboxypeptidase B; sample V_1 remained untreated. RP HPLC-RRA metabolic profiles are shown in Figure 8.

HPLC yielded two RRA fractions. The major area of opioid receptor activity in both samples occurred in fractions 19-21, which correspond to the retention time of synthetic ME (that retention time was determined in a separate experiment to avoid any column contamination). Opioid receptor activity contained in these fractions was approximately 4 ng ME equivalents per milligram protein for V_1, but greater than 43 ng for V_2. Thus, 10 times more opioid receptor activity was found in V_2 compared to V_1.

The ME-like receptor activity in sample V_2 was analyzed qualitatively and semiquantitatively by FABMS. The $(M–H)^-$ molecular anion ion at 572 was selected, and B/E linked-field scanning in the selected reaction monitoring (SRM) mode monitored the 572 to 465 metastable transition. Quantitation using an external standard demonstrated the presence of 4 ng ME mg^{-1} protein. In this case, quantitation indicated YGGFM.

These data demonstrate that 10% of the ME remains free as the pentapeptide in the calcified portion of human teeth as demonstrated by MS and MS/MS, and that 90% remains embedded within precursors, as demonstrated by enzymolysis and RRA.

Once ME and SP were demonstrated to be present in both human tooth pulp and in calcified tissue, it became clear that the human trigeminal nerve system contains proenkephalin A and tachykinin neuropeptide precursors. In an effort to determine whether other neuropeptide families were also present, human tooth pulps were analyzed for the presence of immunoreactivity corresponding to β-endorphin, which derives metabolically from proopiomelanocortin (POMC), and the octapeptide ME–Arg–Gly–Leu, which derives from the separate neuropeptide system proenkephalin A.[49] For the first time, β-endorphin-like immunoreactivity (BE-LI) was measured in human tooth pulp, and HPLC-RRA profiles were constructed. Following acute mechanical stress, a monotonic decrease in BE-LI concentrations was evident according to the extraction order of the four bicuspids.

In summary, these data demonstrate for the first time that opioid and tachykinin neuropeptides exist in human tooth pulp and opioid peptides in calcified tissue, and that an experimental manipulation such as orthodontic stress mobilizes these neuropeptide pathways and decreases the content of these peptides in pulp tissues. MS demonstrated the presence of the protonated molecular ion of ME in the tooth pulp, and MS/MS data substantiated and quantified the amount of ME in the calcified portion.

2. Ophthalmic Branch

Neuropeptides detected with RIA were purified from bovine cornea extracts by gradient followed by isocratic RP HPLC.[63,64] Neuropeptides studied included ME, SP, β-endorphin, and α-melanocyte stimulating hormone. Ir-ME and ir-SP were purified from canine cornea extracts by gradient HPLC and characterized by RIA. In an anatomic study of the bovine cornea, the cornea was separated into an epithelium- and a stroma-enriched portion. Following gradient HPLC, RIA demonstrated that all ME-like immunoreactivity (ME-LI) was located within the cornea epithelium, whereas SP-like immunoreactivity (SP-LI) was distributed between the stroma and epithelium in an approximate 2:1 ratio.[63,64] In a separate study, after partial isolation from bovine cornea, intact native ME and ME generated proteolytically from a larger precursor peptide were both identified and quantified by negative ion FABMS.[65]

In summary, a study of the maxillary, mandibular, and ophthalmic branches of animal and human trigeminal systems demonstrated the presence of neuropeptides from several different neuropeptide systems (opioid, tachykinin). MS played a significant role in these studies and increased our confidence in the amino acid sequence of selected peptides under consideration.

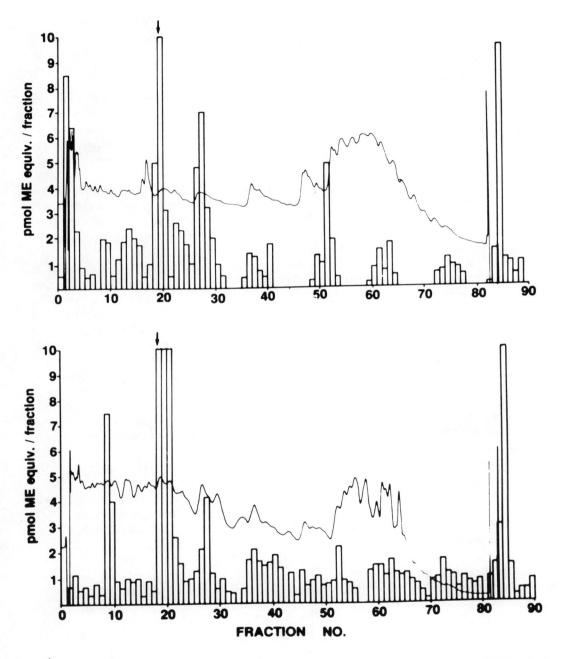

FIGURE 8.   RP HPLC-RRA data obtained from the peptide-rich fraction obtained from four fresh, decalcified, depulped human teeth. The data from untreated V_1 (top) and enzyme-treated V_2 (bottom) are shown. Proteins were precipitated. The nonlinear RP HPLC gradient was 10-15% CH3CN (0-18 min), 15-18% (18-48 min), 18-30% (48-72 min), 30% (72-80 min), 30-100% (80 min), and 100% (80-90 min); buffer was triethylamine:formate; and flow rate was 1.5 ml min⁻¹. The arrow indicates the retention time of synthetic ME, determined in a separate HPLC experiment. The receptoractivity in each one of the 90 collected fractions was determined using a canine limbic system P_2 receptor preparation with ³H-etorphine. Fractions 19-21 of V_2 were combined and analyzed by FABMS. (From Tanzer, F. S., Tolun, E., Fridland, G. H., Dass, C., Killmar, J., Tinsley, P. W., and Desiderio, D. M., *Int. J. Pept. Protein Res.*, 32, 117, 1988. With permission.)

B. LUMBAR CEREBROSPINAL FLUID

MS was useful in analyzing and quantifying neuropeptides in pituitary and most brain tissues because of the relatively high level of peptide content in those tissues.[70] However, CSF has long been considered only a waste port of disposal, metabolically not very active, and containing only low amounts of most CSF constituents.[98] Nevertheless, it has recently been demonstrated that CSF (intracerebroventricular and lumbar) is active metabolically and transports metabolically important compounds,[99] and that MS can identify neuropeptides in human CSF.[100]

When one considers the neuroanatomical relationship of CSF to internal brain regions, it is tempting to speculate that qualitative and quantitative analysis of peptides in CSF is an excellent indicator of control levels of CSF peptides and of levels that are markedly different during clinical pathophysiological conditions.[41] Toward that end, we analyzed the CSF content of immunoreactive and receptroactive opioid and tachykinin peptides in several different clinical conditions, specifically in low back pain patients[81] and in patients diagnosed with SDAT.[101]

1. Low Back Pain Patients

Opioid peptide content in lumbar CSF samples obtained from low back pain patients was measured with an opioid RRA.[52,81,100] Total (i.e., chromatographically unfractionated) opioid peptide receptor activity was measured, and desalted CSF was eluted from a Sep-Pak and subjected to RRA with tritiated etorphine.[81] Three clinical groups were found to have endogenous levels of 2.5, 4.5, and 6.4 pmol, respectively, of ME equivalents per milligram CSF, respectively. Although the differences in these three sets of data are not significant statistically, endogenous total opioid receptor activity tended to correlate with the amount of drug (lidocaine) needed to relieve a patient's perception of pain.

Of 54 patients studied for low back pain, the lumbar CSF sample of one had an atypical metabolic profile of opioid receptor activity.[52] To test the hypothesis that neuropeptides play a role in human low back pain, the amount of endogenous opioid receptor activity in a control and in that atypical patient was compared. Two lumbar puncture samples, one before, and another after clinical evaluation, were obtained from these two patients. Total opioid receptor activity was measured in each sample prior to HPLC separation, and opioid receptor activity was measured in each individual fraction after samples were subjected to HPLC. These latter data represent a metabolic profile of opioid receptor activity in human lumbar CSF. Analytical data demonstrated clearly that the atypical patient had opioid receptor activity measurements (total and profile) that differed in a qualitative and quantitative sense from the other 53 patients.

The opioid study was expanded to also incorporate tachykinin analysis. SP-LI was measured in lumbar CSF samples from low back pain patients.[82] Before the RIA measurement was obtained for the total SP-LI, all CSF samples were eluted through an octadecylsilyl disposable cartridge. In a separate procedure nontreated CSF was analyzed with a combination of RP HPLC and RIA. Qualitative differences were observed among these HPLC-RIA metabolic profiles, and immunoreactivity was found (see Figure 9) in fractions corresponding to calibrated retention times of synthetic SP sulfoxide, SP, and the C-terminal hexapeptide SP_{5-11}. The chemical instability of SP in Tris buffer was noted.[102]

Significant differences were observed in that short time (minutes) between the first and second SP-LI measurements of nonphysiological responders. Of 75 individual pairs of first and second total SP-LI measurements analyzed, 25 were below the threshold level for both the first and the second samples (see Figure 10). In the remaining 50 pairs, either one or the other measurement was equal to 0, or both measurements were above 0 fmol SPLI per milligram of CSF.

Thus, opioid neuropeptide and tachykinin content in human lumbar CSF apparently plays a significant role in lower back pain. Other workers have also correlated neuropeptide content with psychiatric disabilities.[103] Clearly, MS data are now needed for these low back pain lumbar CSF studies to confirm the amino acid sequence of the peptide indicated by immunoreactivity.[100]

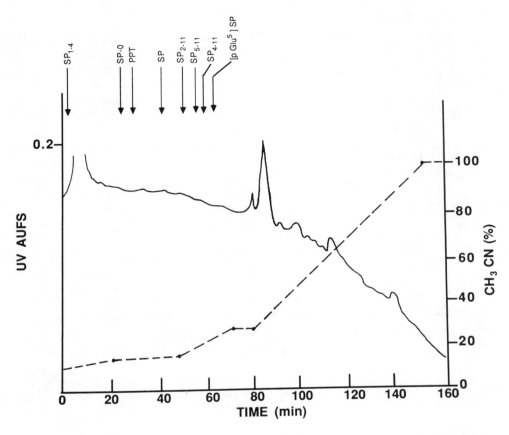

FIGURE 9.   RP HPLC gradient separation and SP-LI measured in lumbar CSF samples from lower back pain patients. (From Higa, T., Wood, G., and Desiderio, D. M., *Int. J. Pept. Protein Res.*, 33, 446, 1989. With permission.)

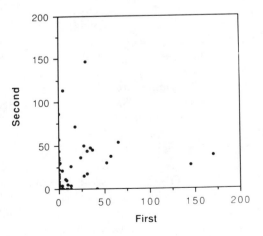

FIGURE 10.   Plot of SP-LI measurements made in first and in second lumbar CSF samples. (From Higa, T., Wood, G., and Desiderio, D. M., *Int. J. Pept. Protein Res.*, 33, 446, 1989. With permission.)

2. Senile Dementia of the Alzheimer's Type

In this study, seven patients (ages 65 to 77) were diagnosed as having SDAT based on DSM III criteria. Other causes of neurological dysfunction were excluded by laboratory examination, electroencephalogram (EEG), CSF examination, and computed tomography (CT) scanning.[101] In order to select a uniform group of patients in whom to provide an early diagnosis and also to avoid pathologies unique to the terminal stages of SDAT, this study included only those patients with classical symptoms of Alzheimer's that were present for more than 1 but less than 2 years (70 ± 5 years, range 65 to 77 years, n = 7). Age-matched controls (65 ± 3 years, range 61 to 68 years, n = 6) free of any symptoms of dementia were selected.

RP HPLC separation followed by RRA detection of peptides demonstrated that most opioid receptor activity eluted within fractions 6 to 20. The amount of opioid receptor activity measured in a second RIA was summed arithmetically, and a statistical comparison was made between the SDAT and control data. An average of 383 ± 170 (SD) pmol ME equivalents per milligram CSF (n = 7) and 89.1 ± 46.3 (n = 4) was found for SDAT and control patients, respectively. A t-test value for that sample comparison was statistically significant at the 99.99% level of confidence. Again, sequence data from MS/MS are needed.

C. HUMAN POST-MORTEM PITUITARY TISSUE

In contradistinction to human CSF, the tissue richest in opioid peptide content is the pituitary. This fact has been established in many laboratories,[71] and may be demonstrated appropriately by research utilizing HPLC-RRA conducted on the metabolic profiling of opioid peptides in canine pituitary and selected brain regions.[70] The highest amount of peptide receptor activity found is in pituitary tissue; a tenfold smaller amount occurs in the hypothalamus and caudate nucleus (Figure 11A), and still lower (30 times) amounts in the midbrain (Figure 11B), amygdala, thalamus, pons-medulla, and hippocampus. For example, Figure 11 contains RP HPLC-RRA metabolic data of canine caudate nucleus (11A) and mid-brain (11B) tissue extracts. A quantitative decrease in opioid receptor activity vs. the canine pituitary (Figure 4) is observed, as is a significant shift in the amount and type of metabolic processing.

That decrease in peptide concentration most probably reflects three separate metabolic factors that operate in those tissues: differential tissue-specific processing patterns[104] of the large peptide precursors, especially moving distally from the pituitary; specific anatomic distributions of three opioid peptide systems; and the particular receptor preparation (canine limbic system) and radioligand (tritiated etorphine) used in these studies.

1. Tumor Profiles

Opioid receptor activity in a human chromophobe ademona obtained from neurosurgery was profiled,[66] and all HPLC fractions were assayed for ir-ME content using a commercial RIA kit (Figure 5).

Fractions eluted near the retention time of synthetic ME (fraction 20) contained opioid immuno- and receptor activities. In addition, an unidentified ME-like immunoreactive peak that also demonstrated receptor activity was found in a later-eluting, more hydrophobic area (fraction 62). These analytical data demonstrate the limited structure information and molecular specificity provided by RIA and receptorassay measurements. As discussed later in this chapter, that limited molecular specificity is the reason for obtaining mass spectrometric data.

In a second study, a metabolic profile of endogenous opioid receptor active peptides was obtained for a peptide-rich fraction obtained from a homogenate of an ACTH-secreting human pituitary tumor.[51] The patient was 51 years old, with symptoms and signs of Cushing's syndrome, which was confirmed by a dexamethasone suppression test and by the presence of a 1 cm contrast-enhancing mass in an enlarged sella turcica. The HPLC-RRA metabolic profile demonstrated a preponderance of more hydrophobic (probably longer) peptides eluting in fractions 60 to 90 in the peptide-rich fraction of this ACTH-secreting tumor compared to fewer

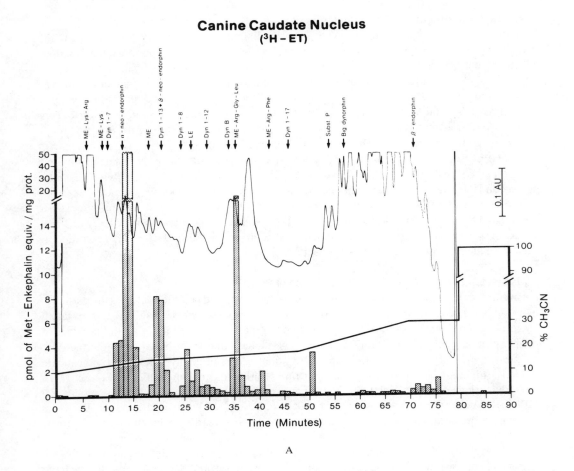

A

FIGURE 11.   RP HPLC gradient separation and radioreceptor activity metabolic profiles of canine caudate nucleus (A) and mid-brain (B) tissue. Ligand was ^{3}H-etorphine. See legend to Figure 4 for further experimental details. (From Takeshita, H., Desiderio, D. M., and Fridland, G., *Biomed. Chromatogr.*, 1, 126, 1986. With permission.)

earlier-eluting peptides. Such a distribution may indicate a decrease in metabolic processing of larger peptides. In any event, these data demonstrate that neuroendocrinologically silent tumors most probably involve derangements in opioid peptide pathways (see hypothesis in Figure 1).[105]

2. Control and Drug-Overdose Pituitaries

The two previous studies prompted an evaluation aimed at defining the opioid receptor activity content in "control" human pituitaries.[106] Toward that end, comprehensive metabolic profiles of endogenous opioid peptides were established for the human pituitary, wherein 16 pituitaries obtained post-mortem were analyzed individually by HPLC-RRA with tritiated etorphine as the ligand. These 16 assay profiles (Figure 12A) were sufficiently consistent for their average to serve as a comparative basis for other studies. Receptor activity was found mainly in 4 major fractions centering around HPLC fractions 19 to 22, 34 to 37, 73 to 78, and 84.

Not included in the data collected in Figure 12 is one pituitary from a young male who died of a drug overdose (trace of alcohol, pentazocine, tripelennamine); that corresponding HPLC-opioid radioreceptor active profile (Figure 12B) differed remarkably from the other 16 samples. For example, two areas in that HPLC-RRA demonstrated an extraordinarily high level of

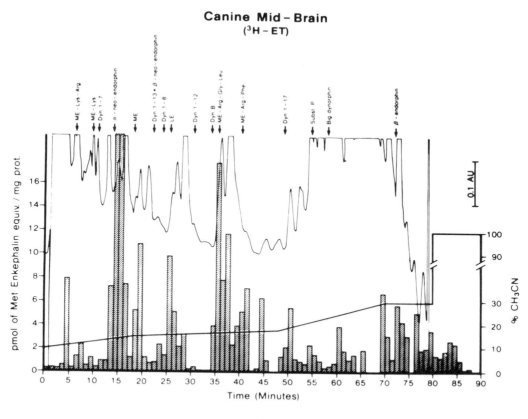

FIGURE 11B.

receptor activity centered on fractions 31 to 37 and 73 to 80. Recently, three other drug overdose cases (data not shown) had very different profiles, compared to those in Figures 12A and B.

It is interesting to speculate on the presence of such a high level of receptor activity in the two HPLC peaks in Figure 12B. It may be that either the drug or an opioid receptor active metabolite could have migrated to the pituitary in a quantity sufficiently long-lasting so that, when it was extracted from the post-mortem pituitary tissue, it was detected in the assay. Of course, only that one area of the one chromatogram where the drug elutes would be so affected — not both areas. It is also possible that the drug or an opioid-active metabolite could have stimulated metabolic production of an opioid peptide from its precursor or several products therefrom. The latter possibility is not incompatible with the hypothesis presented in Figure 1. More tissue is under study so that these important questions about human drug overdose may be answered.

3. β-Endorphin

β-Endorphin, which is a product from the POMC precursor, is also a very significant neuropeptide in the pituitary. Other workers have shown that post-translational metabolic processing of β-endorphin is important and includes sulfation, acetylation, truncation, and arylsulfation.

We extended our mass spectrometric studies to include elucidation and quantification of β-endorphin in one human pituitary by FAB.[107] The instruments used in our studies are Finnigan MAT 731 and VG 7070EH-F mass spectrometers, which have a sensitivity limitation at higher mass regions. β-Endorphin (mol wt 3463 Da) exceeds that instrumental mass range, and it was necessary to treat β-endorphin with trypsin to produce a pentapeptide of sequence NAIIK to

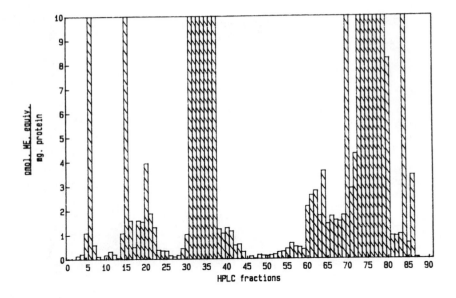

A

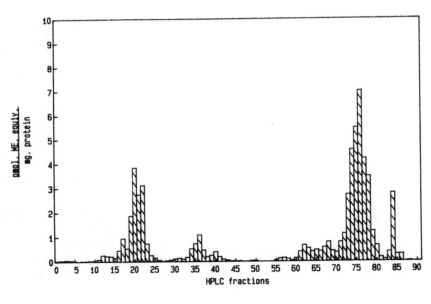

B

FIGURE 12.   (A) RP HPLC chromatographic separation of the peptide-rich fraction obtained from 16 human post-mortem pituitary tissues. The detector is an RRA using a receptor-rich preparation from a canine limbic system, and tritiated etorphine as displaced ligand. Etorphine interacts with several different opioid receptor types, and was displaced from those opioid receptors by opioid peptides eluting in specific HPLC fractions. Collected are 90 1-min HPLC fractions. The left-hand axis denotes picomoles of ME equivalents because ME was used for the calibration curve. The hatched peaks indicate the amount of ME-like receptor activity found within that fraction. (B) RP HPLC chromatographic separation of a peptide-rich fraction obtained from human post-mortem pituitary tissue. (From Desiderio, D. M., Fridland, G. H., Francisco, J. T., Sacks, H., Robertson, J. T., Cezayirli, R. C., Killmar, J., and Lahren, C., *Clin. Chem.*, 34, 1104, 1988. With permission.)

represent β-endorphin. The mass spectrometric response of this tryptic fragment β-endorphin$_{20-24}$ was used to quantitate endogenous β-endorphin. The detection limit for synthetic β-endorphin in this analytical method is 86 fmol, and human pituitary extracts contain 1.5 pmol of β-endorphin per milligram protein.[107]

D. SUMMARY OF APPLICATIONS SECTION

Research demonstrated in this applications section shows that it is important to take advantage of the analytical capabilities of several different detectors for HPLC because each method has a different level of molecular specificity. The objective of this research program is to optimize molecular specificity of peptide quantification, and whenever possible, mass spectrometry is used for that purpose. However, we have clearly shown that RP HPLC followed by receptor assay or RIA is an effective first screen to identify those peaks that must be studied in greater detail by MS. In addition, MS has sufficient detection sensitivity, and certainly the molecular specificity needed to study peptides extracted from human tissues, pituitary, brain, and CSF. Also, when the molecule is too large for a particular mass spectrometer, effective enzymolysis cleaves a peptide to an appropriate mass range. However, because only small quantities of physiological peptides are extracted from precious human samples, it is best to avoid any chemical steps whenever possible. Recently introduced four-sector tandem MS instruments with multichannel focal plane detectors[93] have sufficient molecular specificity, mass range, and detection sensitivity to overcome these limitations.

Finally, in a shorter study, it has been shown that human placental villous tissue contains opioid receptors and peptides.[28,29] Opioid peptides extracted from villous tissue were fractionated using RP HPLC and detected with an RRA. The presence of dynorphin$_{1-8}$ in human placenta was corroborated by the MS production of the MH$^+$ ion at 981 amu with FAB. These MH$^+$ data must be confirmed with sequence data.

V. OTHER PEPTIDE SYSTEMS STUDIED

Although the research effort described in this chapter focuses on several human neuropeptide systems, especially opioid and tachykinin systems as they relate to stress, and on the use of MS in combination with other analytical methods to study these peptides, other related peptide studies have also been conducted in this laboratory. In addition to the other studies and the analysis of biologically important peptides listed in the following sections, the mass spectrometric behavior of synthetic peptides was studied,[55] and the chemistry on the FAB probe tip was optimized to achieve attomole levels of detection sensitivity with sequence information.[38]

A. TUFTSIN

We described a method to identify and quantify unambiguously the phagocyte-stimulating tetrapeptide tuftsin from trypsinized human serum.[23] The polar tetrapeptide tuftsin has the sequence Thr-Lys-Pro-Arg. Serum tuftsin was separated using HPLC and pentafluorobutyric acid, which is a volatile buffer that effectively increases the hydrophobicity of the tetrapeptide:buffer ion-pair (see previous discussion). MS/MS analysis of the methyl ester of tuftsin demonstrates that tuftsin occurs in the human at the 800 pmol ml^{-1} serum level. This type of analytical system is important because, until now, no reliable analytical method has been available to quantify serum tuftsin.

B. ILEUM PROTEIN FRAGMENT

The N-terminus pentapeptide from porcine gastrotropin, an ileal protein that stimulates secretion of gastric acid and pepsinogin,[27] was sequenced with FAB MS/MS. One tryptic peptide failed to yield any sequence information, and was analyzed by FABMS. A protonated molecular ion at 565 defined the molecular weight of the peptide to 564. Based on amino acid analysis (Ala,

Phe, Thr, Gly, Lys), the calculated molecular weight of the pentapeptide equaled 522 Da, which indicated that the amino terminus was acetylated. Because it was a tryptic fragment, Lys was at the C-terminus. Deducing the sequence of the remaining amino acids from the B/E linked-field scan and CAD mass spectrum was straightforward.

C. FAB MATRIX STUDIES

Irradiation by high energy xenon atoms of peptides dissolved in glycerol, ethylene glycol, 1,4-butanediol, or 1,5-pentanediol formed several adducts having masses greater than the protonated molecular ion of the peptide.[56] Corresponding thiols demonstrated poor affinity for adduct formation, which was facilitated in dilute basic pH solutions but reduced drastically in acidic solutions. MS/MS spectra indicated that the site of attachment of glycerol to the peptide was the most basic site of the molecule, which is the free N-terminus group or the primary amine group of a basic (Lys, Arg) amino acid residue. Evidence for formation of adducts with up to three added molecules of glycerol was revealed.

D. SYNTHETIC PEPTIDE STUDIES

For all biologic peptides studied, it is imperative to have a concurrent parallel study on the mass spectrometric behavior of synthetic peptides in order to rationalize the genesis of fragment ions to detect with confidence those peptide ions in a biologic extract. Toward that end, positive and negative FAB-CAD-B/E linked-field scan spectra were obtained for leucine enkephalin,[108] SP,[109] and opioid peptides containing enkephalins, dynorphin A fragments 1 to 7 and 1 to 10, and α- and β-neoendorphins.[55] Other synthetic peptides studied included the enkephalin sequence, proctolin, β-casomorphin, and formylated nLLFnLLYK. Details of these basic mass spectrometric fragmentation behaviors are discussed in another chapter (see Chapter 17).

E. TANDEM MS STUDY OF TWO NEUROPEPTIDES, DYNORPHIN$_{1-7}$ AND DYNORPHIN$_{1-13}$

A four-sector $B_1E_1E_2B_2$ mass spectrometer was used to study the mass spectrometric behavior of two dynorphins, the heptapeptide and tridecapeptide, dynorphin$_{1-7}$ and dynorphin$_{1-13}$, respectively.[57] The protonated molecular ion of dynorphin$_{1-13}$ was selected by the first two fields (B_1, E_1) and product ions arising from either unimolecular dissociation (Figure 13A) or CAD (Figure 13B) were collected by a linked-field scan (B_2/E_2) of the second 2 fields. CAD significantly increases the abundance of amino acid sequence-determining and other fragment ions.

F. NEGATIVE FABMS ANALYSIS OF LEUCINE ENKEPHALIN

It is important to establish the limit of detection sensitivity of FAB MS/MS analysis of neuropeptides. Therefore, FABMS details of the probe tip chemistry were studied,[38] analyzed, and controlled rationally by comparing two matrix materials, glycerol and dithiothreitol-dithioerythritol (DDT-DTE), manipulating the pH value, and studying the effect of appropriate additives such as glutathione.[110] ME and leucine enkephalin were both studied. Addition of glutathione to the FAB matrix increased detection sensitivity of peptides in the negative ion mode. Attomole detection sensitivity for leucine enkephalin standards was achieved in the positive ion mode. Negative ion FABMS spectra have fewer interferences from background matrix and provide femtomole sensitivity for both ME and leucine enkephalin.

ACKNOWLEDGMENTS

The author gratefully acknowledges the financial assistance of the National Institutes of Health (GM 26666, DRR 01651), and the typing assistance of Linda Rutherford, Deanna Darling, and Dianne Cubbins.

Dynorphin 1 - 13   B2/E2

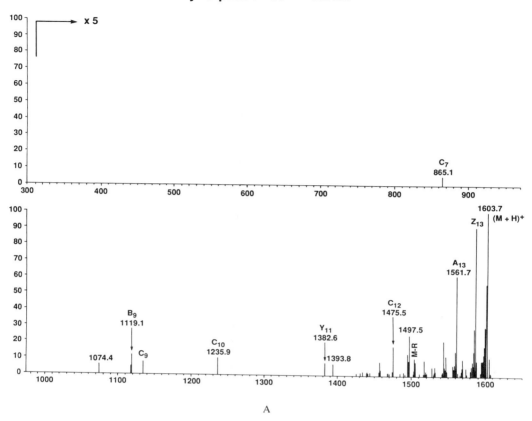

A

FIGURE 13.   (A) FABMS B$_2$/E$_2$ linked-field scan of the (M+H)$^+$ ion of dynorphin$_{1-13}$ showing unimolecular decompositions. (B) FABMS linked-field scan (B$_2$/E$_2$) scan of the collisionally activated (M+H)$^+$ ion of dynorphin 1-13. The (M+H)$^+$ ion current has saturated the ion current amplifier. (From Desiderio, D. M., *Int. J. Mass Spectrom. Ion Proc.*, 74, 217, 1986. With permission from Elsevier Science Publishers, Physical Sciences and Engineering Division.)

ABBREVIATIONS

amol	Attomole (10^{-18} moles)
αMSH	Alpha melanocyte-stimulating hormone
B/E	Linked field scan, keeping a constant B/E ratio, while scanning the magnetic (B) and electric (E) field
CAD	Collision activated dissociation
cDNA	Complementary DNA
CLIP	Corticotropin-like intermediate peptide
C18	Octadecylsyl
FAB	Fast atom bombardment
Fmol	Femtomole (10^{-15} moles)
ICV	Intracerebroventricular
Ir-	Immunoreactive-
IS	Internal standard
LD	Laser desorption
LE	Leucine enkephalin (TyrGlyGlyPheLeu=YGGFL)
-LI	Like immunoreactivity
LPH	Lipotropin hormone

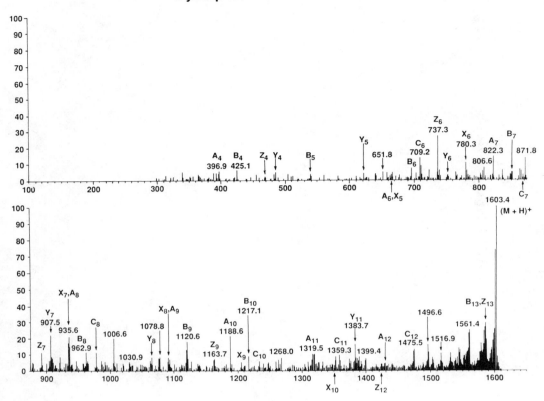

FIGURE 13B.

ME	Methionine enkephalin (TyrGlyGlyPheMet=YGGFM)
(M+H)+	Protonated molecular ion
MS/MS	Mass spectrometry/mass spectrometry
MW	Molecular weight
PD	Plasma desorption
POMC	Proopiomelanocortin
Pmol	Picomole (10^{-12} moles)
RPHPLC	Reversed-phase high performance liquid chromatography
RRA	Radioreceptorassay
RIA	Radioimmunoassay
SIM	Selected ion monitoring
SP	Substance P (RPKPQQFFGLM-NH$_2$)
TG	Thyroglobulin
TRH	Thyrotropic releasing hormone (Pyroglu His Pro NH$_2$)

REFERENCES

1. **Desiderio, D. M.,** *Analysis of Neuropeptides by Liquid Chromatography and Mass Spectrometry,* Elsevier, Amsterdam, 1984, 235.

2. **Tache, Y., Morley, J. E., and Brown, M. R. Eds.,** *Neuropeptides and Stress,* Springer-Verlag, Berlin, 1989, 345.

3. **Biemann, K. and Martin, S. A.,** Mass spectrometric determination of the amino acid sequence of peptides and proteins, *Mass Spectrom.* Rev., 6, 1, 1987.

4. **Beckey, H. D.,** Determination of the structures of organic molecules and quantitative analysis with the field ionization mass spectrometer, in *Biomedical Applications of Mass Spectrometry,* Waller, G., Ed., Wiley Interscience, New York, 1972, 795.

5. **Barber, M., Bordoli, R. S., Sedgwick, R. D., and Tyler, A. N.,** Fast atom bombardment of solids (F.A.B.): a new ion source for mass spectrometry, *J. Chem. Soc. Chem. Commun.,* 325, 1981.

6. **Macfarlane, R. D. and Torgeson, T. F.,** *Scien*ce, 191, 920, 1976.

7. **Burgus, R., Dunn, T. F., Desiderio, D. M., Ward, D. N., Vale, W., and Guillemin, R.,** Characterization of ovine hypothalamic hypophysiotropic TSH-releasing factor, *Nature,* 226, 321, 1970.

8. **Krieger, D. T., Brownstein, M. J., and Martin, J. B.,** *Brain Peptides,* Wiley Interscience, 1983, 1032.

9. **Leclercq, P. A. and Desiderio, D. M.,** A laboratory procedure for the acetylation and permethylation of oligopeptides on the microgram scale, *Anal. Lett.,* 4, 305, 1971.

10. **Desiderio, D. M., Yamada, S., Sabbatini, J. Z. and Tanzer, F.,** Quantification of picomole amounts of underivatized leu-enkephalin by means of field desorption mass spectrometry, *Biomed. Mass Spectrom.,* 8, 10, 1981.

11. **Desiderio, D. M. and Cunningham, M. D.,** Triethylamine formate buffer for HPLC-FDMS of oligopeptides, *J. Liquid Chromatogr.,* 4, 721, 1981.

12. **Desiderio, D. M., Yamada, S., Sabbatini, J. Z., Tanzer, F., Horton, J., and Trimble, J.,** High-performance liquid chromatography and field desorption mass spectrometric measurement of picomole amounts of endogenous neuropeptides in biologic tissue, *J. Chromatogr.,* 217, 437, 1981.

13. **Tanzer, F. S., Desiderio, D. M., and Yamada, S.,** HPLC isolation and FD-MS quantification of picomole amounts of met-enkephalin in canine tooth pulp, in *Peptides: Synthesis-Structure-Function,* Rich, D. H., and Gross, E., Eds., Pierce Chemical, Rockford, IL, 1981, 761.

14. **Yamada, S. and Desiderio, D. M.,** Measurement of endogenous neuropeptides in canine caudate nuclei and hypothalami with high-performance liquid chromatography and field-desorption mass spectrometry, *Anal. Biochem.,* 127, 213, 1982.

15. **Yamada, S. and Desiderio, D. M.,** Measurement of endogenous leucine enkephalin in canine thalamus by high-performance liquid chromatography and field-desorption mass spectrometry, *J. Chromatogr.,* 239, 87, 1982.

16. **Desiderio, D. M. and Yamada, S.,** FDMS measurement of picomole amounts of leucine enkephalin in canine spinal cord tissue, *Biomed. Mass Spectrom.,* 10, 358, 1983.

17. **Desiderio, D. M. and Kai, M.,** Preparation of stable isotope-incorporated peptide internal standards for field desorption mass spectrometry quantification of peptides in biologic tissue, *Biomed. Mass Spectrom.,* 10, 471, 1983.

18. **Desiderio, D. M. and Kai, M.,** Field desorption mass spectral measurement of enkephalins in canine brain with ^{18}O peptide internal standards, *Int. J. Mass Spectrom. Ion Phys.,* 48, 261, 1983.

19. **Desiderio, D. M., Katakuse, I., and Kai, M.,** Measurement of leucine enkephalin in caudate nucleus tissue with fast atom bombardment-collision activated dissociation-linked field scanning mass spectrometry, *Biomed. Mass Spectrom.,* 10, 426, 1983.

20. **Desiderio, D. M., Tanzer, F. S., Kai, M., Wakelyn, C., Fridland, G., and Trimble, J.,** Measurement with optimal molecular specificity of endogenous peptides in nerve tissue by means of FAB-CAD-B/E-B′/E′-SIM methodology, in *Peptides: Structure and Function,* Hruby, V. J. and Rich, D. H., Eds., Pierce Chemical, Rockford, IL, 1983, 715.

21. **Desiderio, D. M.,** HPLC and MS of biologically important peptides, in *Advances in Chromatography,* Vol. 22, Giddings, J. C. and Brown, P., Eds., Marcel Dekker, New York, 1983, chap. 1.

22. **Desiderio, D. M., Kai, M., Tanzer, F. S., and Wakelyn, C.,** Measurement of enkephalin peptides in various brain regions, teeth, and CSF with HPLC and mass spectrometry, *J. Chromatogr.,* 342, 245, 1984.

23. **Naim, J., Desiderio, D. M., Trimble, J., and Hinshaw, J. R.,** The identification of serum tuftsin by reverse-phase high-performance liquid chromatography and mass spectrometry, *Anal. Biochem.,* 164, 221, 1987.

24. **Desiderio, D. M., Fridland, G. H., Tanzer, F. S., Dass, C., Tinsley, P., and Killmar, J.,** The use of HPLC and mass spectrometric techniques to quantify endogenous opioid peptides, in *Proteins: Structure and Function,* L'Italien, J. J., Ed., Plenum Press, New York, 1987, 269.

25. **Hunt, D. F., Yates, J. R., Shabanowitz, J., Winston, S., and Haver, C. R.,** Protein sequence by tandem mass spectrometry, *Proc. Natl. Acad. Sci. U.S.A.,* 83, 6233, 1986.

26. **Hunt, D. F., Zhu, N.-Z., and Shabanowitz, J.,** Oligopeptide sequence analysis by collision-activated dissociation of multiply charged ions, *Rapid Commun. Mass Spectrom.*, 4, 122, 1989.

27. **Walz, D. A., Wider, M. D., Snow, J. W., Dass, C., Desiderio, D. M., and Gantz, I.,** The complete amino acid sequence of porcine gastrotropin, an ileal protein which stimulates gastric acid and pepsinogen secretions, *J. Biol. Chem.*, 263, 14189, 1988.

28. **Ahmed, M. S., Randall, L. W., Cavinato, A. G., Desiderio, D. M., Fridland, G., and Sibai, B.,** Human placental opioid peptides: correlation to the route of delivery, *Am. J. Obstet. Gynecol.*, 155, 703, 1986.

29. **Ahmed, M. S., Randall, L. W., Sibai, B., Dass, C., Fridland, G., Desiderio, D. M., and Tolun, E.,** Identification of dynorphin 1-8 in human placenta by mass spectrometry, *Life Sci.*, 40, 2067, 1987.

30. **Desiderio, D. M.,** Field desorption mass spectrometry, in *Handbook on Use of HPLC for the Separation of Amino Acids, Peptides, and Proteins,* Hancock, W. S., Ed., CRC Press, Boca Raton, FL, 1984, 197.

31. **Desiderio, D. M. and Fridland, G.,** Mass spectrometric measurement of neuropeptides, in *Mass Spectrometry in Biomedical Research,* Gaskell, S., Ed., Wiley Interscience, New York, 1986, 443.

32. **Harvey, D. J.,** Trace analysis of the cannabinoids, in *Mass Spectrometry in Biomedical Research,* Gaskell, S., Ed., Wiley Interscience, New York, 1986, 363.

33. **Watson, J. T.,** *Introduction to Mass Spectrometry,* Raven Press, New York, 1985, 64.

34. **Perry, D. L. and Desiderio, D. M.,** Utility of d_4-PGE$_2$ as an internal standard to quantify endogenous levels of PGE$_1$, PGE$_2$, 19OH PGE$_1$ and 19OH PGE$_2$ in human seminal fluid by GC-MS-SIM, *Prostaglandins*, 14, 745, 1977.

35. **Millard, B. J.,** *Quantitative Mass Spectrometry,* Heyden, London, 1978, 171.

36. **Lawson, A. M., Lim, C. K., and Richmond, W.,** *Current Developments in the Clinical Applications of HPLC, GC and MS,* Academic Press, London, 1980, 301.

37. **Baillie, T. A.,** *Stable Isotopes. Applications in Pharmacology, Toxicology, and Clinical Research,* University Park Press, Baltimore, 1978, 314.

38. **Tolun, E., Dass, C., and Desiderio, D. M.,** Trace level of enkephalin peptides at the attomole/femtomole level by FAB-MS, *Rapid Commun. Mass Spectrom.*, 1, 77, 1987.

39. **Solomon, G. F.,** Psychoneuroimmunology: interactions between central nervous system and immune system, in *Neuroimmunomodulation,* Perex-Polo, J. R., Bulloch, K., Angeletti, R. H., Hashim, G. A., and de Vellis, J., Eds., Alan R. Liss, New York, 1987, 1.

40. **Walker, J. A., Tanzer, F. S., Harris, E. F., Wakelyn, C., and Desiderio, D. M.,** The enkephalin response in human tooth pulp to orthodontic force, *Am. J. Orthod. Dentofac. Orthop.*, 92, 9, 1987.

41. **Beal, M. F. and Martin, J. B.,** Neuropeptides in neurological disease, *Ann. Neurol.*, 20, 547, 1986.

42. **Udenfriend, S. and Kilpatrick, D. L.,** Proenkephalin and the products of its processing: chemistry and biology, in *The Peptides: Analysis, Synthesis, Biology,* Udenfriend, S. and Meienhofer, J., Eds., Academic Press, New York, 1984, 25.

43. **Goldstein, A.,** Biology and chemistry of the dynorphin peptides, in *The Peptides: Analysis, Synthesis, Biology,* Udenfriend, S. and Meienhofer, J., Eds., Academic Press, New York, 1984, 96.

44. **Civelli, O., Douglass, J., and Herbert, E.,** Proopiomelanocortin: a polyprotein at the interface of the endocrine and nervous systems, in *The Peptides: Analysis, Synthesis, Biology,* Udenfriend, S. and Meienhofer, J., Eds., Academic Press, New York, 1984, 70.

45. **O'Donohue, T. and Dorsa, D. M.,** The opiomelanotropinergic neuronal and endocrine systems, *Peptides*, 3, 353, 1982.

46. **Numa, S.,** Opioid peptide precursors and their genes, in *The Peptides: Analysis, Synthesis, Biology,* Udenfriend, S. and Meienhofer, J., Eds., Academic Press, New York, 1984, 1.

47. **Shiosaka, S., Sakanaka, M., Inagaki, S., Senba, E., Hara, Y., Takatsuki, K., Takagi, H., Kawai, Y., and Tohyama, M.,** Putative neurotransmitters in the amygdaloid complex with special reference to peptidergic pathways, in *Chemical Neuroanatomy,* Emson, P., Ed., Raven Press, New York, 1983, 359.

48. **Parris, W., Tanzer, F. S., Fridland, G., Harris, E. F., Killmar, J., and Desiderio, D. M.,** Effects of orthodontic force on methionine enkephalin and substance P concentrations in human pulpal tissue, *Am. J. Orthod. Dentofac. Orthop.*, 95, 479, 1989.

49. **Robinson, Q. C., Killmar, J. T., Harris, E. F., Fridland, G., and Desiderio, D. M.,** Immunoreactive evidence of beta-endorphin and methionine enkephalin-arg-gly-leu in human tooth pulp, *Life Sci.*, 45, 987, 1989.

50. **Kelly, D. D.,** Central representations of pain and analgesic, in *Principles of Neural Science,* Kandel, E. R. and Schwartz, J. H., Eds., Elsevier, New York, 1985, 338.

51. **Desiderio, D. M., Cezayirli, R. C., Fridland, G., Robertson, J. T., and Sacks, H.,** Metabolic profiling of radioreceptorassayable opioid peptides in a human pituitary ACTH-secreting tumor, *Life Sci.*, 37, 1823, 1985.

52. **Desiderio, D. M., Liu, D., and Wood, G.,** Opioid receptoractivity in CSF from an atypical lower back pain patient, *Life Sci.*, 43, 577, 1988.

53. **Fridland, G. and Desiderio, D. M.,** Profiling of neuropeptides using gradient RP-HPLC with novel detection methodologies, *J. Chromatogr.*, 379, 251, 1986.

54. **Desiderio, D. M. and Dass, C.,** The measurement of leucine enkephalin at the femtomole level by fast atom bombardment mass spectrometry/mass spectrometry methods, *Anal. Lett.*, 19, 1963, 1986.

55. **Dass, C. and Desiderio, D. M.,** Fast atom bombardment mass spectrometry analysis of opioid peptides, *Anal. Biochem.*, 163, 52, 1987.

56. **Dass, C. and Desiderio, D. M.,** Particle beam-induced reactions between peptides and liquid matrices, *Anal. Chem.*, 60, 2723, 1988.

57. **Desiderio, D. M.,** FAB-MS/MS study of two neuropeptides, dynorphins 1-7 and 1-13, *Int. J. Mass Spectrom. Ion Proc.*, 74, 217, 1986.

58. **Jonsson, H. T., Middleditch, B. S., and Desiderio, D. M.,** Prostaglandins in human seminal fluid: two novel compounds, *Science*, 187, 1093, 1975.

59. **Roques, B. P., Fournie-Zaluski, M. C., Soroca, E., Leconte, J. M., Malfroy, B., Llorens, C., and Schwartz, J.-C.,** The enkephalin inhibitor thiorphan shows antinociceptive activity in mice, *Nature*, 228, 286, 1980.

60. **Browne, C. A., Bennett, H. P. J., and Solomon, S.,** The isolation of peptides by HPLC using predicted elution positions, in *High Performance Liquid Chromatography of Proteins and Peptides,* Hearn, M. T. W., Regnier, F. E., and Wehr, C. T., Eds., Academic Press, New York, 1982, 65.

61. **Higa, T. and Desiderio, D. M.,** Optimizing recovery of peptides from an octadecylsilyl (ODS) cartridge, *Int. J. Pept. Protein Res.*, 33, 250, 1989.

62. **Wehr, C. T.,** Commercially available columns, in *CRC Handbook of HPLC for the Separation of Amino Acids, Peptides, and Proteins*, Hancock, W. S., Ed., CRC Press, Boca Raton, FL, 1984, 31.

63. **Tinsley, P.,** Purification and Characterization of Peptides in the Cornea, Ph.D. thesis, University of Tennessee, Memphis, 1988.

64. **Tinsley, P. W., Fridland, G. H., Killmar, J. T., and Desiderio, D. M.,** Purification, characterization, and localization of neuropeptides in the cornea, *Peptides,* 9, 1373, 1989.

65. **Tinsley, P. T., Dass, C., Trimble, J. W., and Desiderio, D. M.,** Identification of methionine enkephalin in the bovine cornea by fast atom bombardment — mass spectrometry, *Exp. Eye Res.*, in press.

66. **Fridland, G. H. and Desiderio, D. M.,** Measurement of opioid peptides with combinations of reversed phase high performance liquid chromatography, *Life Sci.*, 41, 809, 1987.

67. **Liu, D. and Desiderio, D. M.,** Reversed-phase high-performance liquid chromatographic-radioreceptor assay of human cerebrospinal fluid neuropeptides, *J. Chromatogr.*, 422, 61, 1987.

68. **Simon, E. J.,** Progress in the characterization of the opioid receptor subtypes: peptides as probes. Future directions, in *Opioid Peptides: Molecular Pharmacology, Biosynthesis, and Analysis*, Rapaka, R.S. and Hawks, R.L., Eds., National Institute on Drug Abuse 70, Rockville, Maryland, 1986, 155.

69. **Paterson, S. J., Robson, L. E., and Kosterlitz, H. W.,** Classification of opioid receptors, *Br. Med. Bull.*, 39, 31, 1983.

70. **Takeshita, H., Desiderio, D. M., and Fridland, G.,** Metabolic profiling of opioid peptides in canine pituitary and selected brain regions using HPLC with a radioreceptor assay detector, *Biomed. Chromatogr.*, 1, 126, 1986.

71. **Akil, H., Ueda, Y., Lin, H. L., and Watson, S. J.,** A sensitive coupled HPLC/RIA technique for separation of endorphins: multiple forms of β-endorphin in rat pituitary intermediate vs. anterior lobe, *Neuropeptides*, 1, 429, 1981.

72. **Mains, R. E. and Eipper, B. A.,** Synthesis and secretion of corticotropins, melanotropins, and endorphins by rat intermediate pituitary cells, *J. Biol. Chem.*, 254, 7885, 1979.

73. **Liu, D. and Desiderio, D. M.,** Measurement of dynorphins with ³H-etorphine and canine limbic system receptors, *Neuropeptides*, 12, 35, 1988.

74. **Liu, D. and Desiderio, D. M.,** Radioreceptorassay of opioid peptides with ³H-methionine enkephalin and a canine limbic system receptor preparation, *Anal. Lett.*, 21, 1865, 1988.

75. **Yalow, R. S.,** Radioimmunoassay, *Annu. Rev. Biophys. Bioeng.*, 9, 327, 1980.

76. **Gros, C., Pradelles, P., Rouget, C., Bepoldin, O., Dray, F., Fournie-Zaluski, M. C., Roques, B. P., Pollard, H., Llorens-Cortes, C., and Schwartz, J. C.,** Radioimmunoassay of methionine- and leucine-enkephalins in regions of rat brain and comparison with endorphins estimated by a radioreceptorassay, *J. Neurochem.*, 31, 29, 1978.

77. **Ishizuka, J., Yanaihara, C., Yanaihara, N., Greeley, G. H., Cooper, C. W., and Thompson, J. C.,** Cross-reaction of two different somatostatin antisera with calcitonin gene-related peptide. *Proc. Soc. Exp. Biol. Med.*, 189, 1, 1988.

78. **Fuxe, K., Agnati, L. F., Harfstrand, A., Andersson, K., Mascagni, F., Zoli, M., Kalia, M., Battistini, N., Benfenati, F., Hokfelt, T., and Goldstein, M.,** Studies on peptide comodulator transmission, in *Progress in Brain Research,* Vol. 66, Emson, P. C., Rossor, M. N., and Tohyama, M., Eds., Elsevier, New York, 1986, chap. 15.

79. **Emson, P. C., Ed.,** *Chemical Neuroanatomy,* Raven Press, New York, 1983.

80. **Emson, P. C., Arregui, A., Clement-Jones, V., Sandberg, E. B., and Rossor, M.,** Regional distribution of methionine-enkephalin and substance P-like immunoreactivity in normal human brain and in Huntington's disease, *Brain Res.*, 199, 147, 1980.

81. **Desiderio, D. M., Onishi, H., Fridland, G., Wood, G., and Pagidipati, D.,** HPLC receptorassay of opioid peptides in the cerebrospinal fluid of lower back pain patients, *Biomed. Chromatogr.*, 2, 47, 1987.

82. **Higa, T., Wood, G., and Desiderio, D. M.,** Tachykinin immunoreactivity in human cerebrospinal fluid, *Int. J. Pept. Protein Res.*, 33, 446, 1989.
83. **Valentino, K. L., Eberwine, J. H., and Barchas, J. D.,** *In Situ Hybridization*, Oxford University Press, New York, 1987.
84. **Brot, N., Werth, J., Koster, D., and Weissbach, H.,** Reduction of N-acetyl methionine sulfoxide: a simple assay for peptide methionine sulfoxide reductase, *Anal. Biochem.*, 122, 291, 1982.
85. **Bradbury, A. F., Finnie, M. D. A., and Smyth, D. G.,** *Nature*, 298, 686, 1982.
86. **Gower, J. L.,** Matrix compounds for fast atom bombardment mass spectrometry, *Biomed. Mass Spectrom.*, 12, 191, 1985.
87. **Todd, P. J. and McLafferty, F. W.,** Collisionally activated dissociation of high kinetic energy ions, in *Tandem Mass Spectrometry*, McLafferty, F. W., Ed., John Wiley & Sons, New York, 1983, 149.
88. **Jennings, K. R. and Mason, R. S.,** Tandem mass spectrometry utilizing linked scanning on double focusing instruments, in *Tandem Mass Spectrometry*, McLafferty, F. W., Ed., John Wiley & Sons, New York, 1983, 197.
89. **Haddon, W.,** The constant neutral linked magnetic field-electric sector scan, *Org. Mass Spectrom.*, 15, 539, 1980.
90. **McLafferty, F. W., Ed.,** *Tandem Mass Spectrometry*, John Wiley & Sons, New York, 1983.
91. **Busch, K. L. and Cooks, R.G.,** Analytical applications of tandem mass spectrometry, in *Tandem Mass Spectrometry*, McLafferty, F. W., Ed., John Wiley & Sons, New York, 1983, 11.
92. **Busch, K. L., Glish, G. L., and McLuckey, S. A.,** *Mass Spectrometry/Mass Spectrometry*, VCH, New York, 1988.
93. **Hill, J. A., Martin, S. A., Biller, J. E., and Biemann, K.,** Use of a microchannel array detector in a four-sector tandem mass spectrometer, *Biomed. Environ. Mass Spectrom.*, 17, 147, 1988.
94. **Desiderio, D. M. and Mead, T. E.,** High resolution mass spectral photoplate data acquired and reduced with a real time remote decision making digital computer, *Anal. Chem.*, 40, 2090, 1968.
95. **Kusmierz, J. J., Sumrada, R., and Desiderio, D. M.,** *Analyt. Chem.*, in press.
96. **Tanzer, F. S., Tolun, E., Fridland, G. H., Dass, C., Killmar, J., Tinsley, P. W., and Desiderio, D. M.,** Methionine-enkephalin peptides in human teeth, *Int. J. Pept. Protein Res.*, 32, 117, 1988.
97. **May, H. E., Tanzer, F. S., Fridland, G. H., Wakelyn, C., and Desiderio, D. M.,** High performance liquid chromatography and proteolytic enzyme characterization of peptides in tooth pulp extracts, *J. Liquid Chromatogr.*, 5, 2135, 1982.
98. **Wood, J. H.,** *Neurobiology of Cerebrospinal Fluid*, Wood, J. H., Ed., Plenum Press, New York, 1983, 43.
99. **Banks, W. A., Kastin, A. J., Coy, D. H., and Angulo, E.,** Entry of DSIP peptides into dog CSF: role of physiochemical and pharmacokinetic parameters, *Brain Res. Bull.*, 17, 155, 1986.
100. **Liu, D., Dass, C., Wood, G., and Desiderio, D. M.,** Opioid and tachykinin peptides and their precursors and precursor-processing enzymes in human CSF, *J. Chromatogr.*, 500, 395, 1990.
101. **Muhlbauer, M., Metcalf, J. C., Jr., Robertson, J. T., Fridland, G., and Desiderio, D. M.,** Opioid peptides in the cerebrospinal fluid of Alzheimer patients, *Biomed. Chromatogr.*, 1, 155, 1986.
102. **Higa, T. and Desiderio, D. M.,** Chemical degradation of [³H]-substance P in solution of Tris buffer, *Anal. Biochem.*, 173, 463, 1988.
103. **Terenius, L. and Nyberg, F.,** in *Progress in Brain Research*, van Ree, J. M. and Matthysse, S., Eds., Elsevier, New York, 1986, 207.
104. **Liston, D., Patey, G., Rossier, J., Verbanck, P., and Vanderhaeghen, J. J.,** Processing of proenkephalins is tissue-specific, *Science*, 225, 734, 1984.
105. **Peterson, M. E., Orth, D. N., Halmi, N. S., Zielinski, A. C., Davis, D. R., Chavez, F. T., and Drucker, W. D.,** Plasma immunoreactive proopiomelanocortin peptides and cortisol in normal dogs and dogs with Addison's disease and Cushing's syndrome: basal concentrations, *Endocrinology*, 119, 720, 1986.
106. **Desiderio, D. M., Fridland, G. H., Francisco, J. T., Sacks, H., Robertson, J. T., Cezayirli, R. C., Killmar, J., and Lahren, C.,** Opioid peptide profile in human pituitary, *Clin. Chem.*, 34, 1104, 1988.
107. **Dass, C., Fridland, G. H., Tinsley, P. W., Killmar, J. T., and Desiderio, D. M.,** Characterization and quantitation of β-endorphin in human pituitary by fast atom bombardment mass spectrometry of trypsin-generated fragments, *Int. J. Pept. Protein Res.*, 34, 81, 1989.
108. **Desiderio, D. M. and Katakuse, I.,** Positive and negative fast-atom bombardment-collision-activated dissociation-linked-field scanned mass spectra of leucine enkephalin, *Int. J. Mass Spectrom.*, 54, 1, 1983.
109. **Desiderio, D. M. and Katakuse, I.,** Fast atom bombardment-collision activated dissociation-linked field scanning mass spectrometry of the neuropeptide substance P, *Anal. Biochem.*, 129, 425, 1983.
110. **Alai, M., Demirev, P., Fenselau, C., and Cotter, R. J.,** Glutathione as a matrix for plasma desorption mass spectrometry of large peptides, *Anal. Chem.*, 58, 1303, 1986.

Index

INDEX

A

Aberration coefficients, 231
Acceleration region, in high-pressure fluid injection, 90
Acetic anhydride, 191
Acetonitrile, 203
    HPLC/CF-FAB and, 210
    Sep-Pak chromatography and, 374
Acetylacetone, arginine derivative formation and, 319
Acetylation, 338
    amine, 379
    globin chain, 260
    STP-3 and, 150
N-Acetylgalactosamine, 361
N-Acetylglucosamine, 24
Acetyl group, 58
Acid/base chemistry, matrix, 175, 177—180
Acid hydrolysis, 179, 271, 284
Acidic joining peptide (AJP), 353
Acid precipitation, 373
Acid-stable cross linkages, 285
Acid-washing, permethylation reaction and, 317
ACTH, 328, 329, 370
ACTH CLIP, 142
Actinomycin, 135
Acylation, amino group, 317—318
Acyl-esters, 316
Acylhomoserine lactone, 322
Acylimmonium ions, 103
Acylium ions, 103, 291
Acyl-permethyl derivatives, 316
Acyl-trimethylsilyl-polyaminoalcohols, 316
Additives, matrix selection and, 175
Adduct(s), 31—32, 35—36, 394
    mass determination precision and, 68
    metal ion, 329
Adduct ion formation, 341
Adenoma, chromophobe, 389
Adenosine, 92
Adsorbate, 5
Adsorbed protein, surface concentration of, 6—7
Adsorption
    dynamics of, 6
    permethylation reaction and, 317
Adsorption/wash/desorb concept, 5
Adult hemoglobin, 259, see also Hemoglobin
AJP, see Acidic joining peptide
Alamethicin I, 44
Albumin, serum, bovine, 4
Alcohols, metal-ion cationization and, 307
Aldimine ions, 103
Aldolase, 322
Algae, blue-green, see Blue-green algae
Alkali metal cationized peptides, 307—310
Alkali metal ions, PDMS sample application and, 73
S-Alkylation, 52
Alkylation reactions, monitoring of, 77
Aluminized mylar, 352, 353

Aluminum foil, in PDMS sample preparation, 69, 71
Alzheimer's dementia, 389
α-Amide, 379
Amide, deprotonated, isomerization to, 310
Amide backbone, cleavage of, 128
Amide nitrogen atom, protonation of, 128, 129, 291
Amine groups, acetylation of, 379
Amino acid(s), see also specific amino acids
    ^{13}C-labeled, 383
    ^{2}H-labeled, 383
    residue weights of, 225—229
Amino acid analysis, 261, 351, 354
Amino acid composition, 25, 277
Amino acid mutations, digit-printing method and, see Digit-printing method
Amino acid sequence, 351, see also Cyclic peptides, amino acid sequence determination of
Amino acid substitutions, nominal mass differences and, 264
S-Aminoethylation, 244
Aminoethylation, hemoglobin β-chain, 267
S-Aminoethylcysteine, 318
Amino groups, acylation of, 317—318
Aminopeptidase M, 271
Ammonia, high-pressure injection of, 92—93
Ampholyte mixtures, 261
Amygdala, 389
Amyloidotic polyneuropathy, familial, 253
Anemia, sickle cell, 259, 261
Angiotensin, 175
Angiotensin I, 181, 215
Angiotensin II, HPLC/CF-FAB and, 212
Angiotensin III, HPLC/CF-FAB and, 212
Aniline, 98
Anion exchange, 25
Anthracene, 9
Anthroic acid, 9
Antibiotics, 44, 316
Antibody-antigen interaction, 377—378
Antibody cross-reactivity, 213
Antigenic determinants, 348
Antisubstance P immunoaffinity chromatography, 213
API, see Atmospheric pressure ionization
Apolipoprotein A, HPLC/CF-FAB and, 210
Aprotinin, 183
Aqueous carrier solution, CF-FAB, 207
Arginine derivatives, formation of, 318—319
Argon, as collision gas, 124
Argon fluoride excimer laser, 142
Aromatic group, 99
Array detectors, 122, 350
Arylsulfation, 391
Asp-N, STP-3 and, 149
Atmospheric pressure, constant-flow analysis and, CF-FAB, 209
Atmospheric pressure ionization (API), 214
Atom bombardment, 123, see also Fast atom bombardment (FAB)

Attomole, 369
Automated microsequencing, hemoglobin variants
    and, 261
Automatic Edman degradation, see Edman degradation
    tion
Avian egg white, 22, see also Hen egg-white
    lysozyme
Axon, 370
Azlactone, 191
Azo-group-containing peptides, reduction of, 180

B

Background ions, 205
Background subtraction, 66
Back pain, 372, 387—388
Basicity, 293
B/E linked-field scans, neuropeptides and, 328, 334—
    338, 340—341, 380—381
Bent-quadrupole mass filter, 140
Benzene, ionization of, 88
Beta chain, 12
Biantennary sugars, 28
Bidirectional sequencing, 340
Bile salts, metal-ion cationization and, 307
Bioactivity, cationization and, 307
Bio-Ion, 4—5
Biological potency, dimers and, 35
"Blot-based" assay methods, 6
Blue-green algae
    scytonemin A and, 297—298
    toxic cyclic peptides from, characterization of,
        152—157
Bond dissociation, 329
Bovine fetuin, glycopeptides from, 24—25, 30
Bovine growth factor, HPLC/CF-FAB and, 212
Bovine insulin
    β-chain of, 110
    sample surfactancy and, 183
Bovine pituitary, peptide charting and, 351—359
Bovine pituitary extract, 351, 353
Bovine ribonuclease, plasma desorption spectrum of,
    66, 67
Bovine serum albumin, 4
Bradykinin, 181, 213
Brain, 372
Buffer solutions, high molecular weight proteins in,
    ^{252}Cf-PDMS of, 7—8
Bull and Breese indices, 183—187
1,4-Butanediol, 341, 394
By-products, permethylation reaction and, 317

C

CAD, see Collisionally activated decomposition
Calcified tissue, 372, 384—385
Calcitonin, 193
Calcium, 311
Calibration, narrow magnetic scans and, 112
252-Californium plasma desorption mass spectrome-
    try (^{252}Cf-PDMS), 4—13, 328, 348
    gas phase ion formation and, 10

high molecular weight capabilities of, 5
mass determination of protein by, 11—13
matrix in, 5—10
    high molecular weight proteins in buffer solutions
        and, 7—8
    historical development of, 5—6
    new substrates and, 8—10
    nitrocellulose, 6
    stoichiometry of protein adsorption on nitrocellu-
        lose and, 6—7
peptide charting and, 350—356, 360—362, see also
    Peptide charting
synthetic peptide and protein analysis by, 42—61
    chemical procedures used in synthesis and, 61
    covalent structure verification in, 42—44
    disulfide bonds in, 58—61
    fully protected, 53, 55—58
    homogeneity verification in, 44—53
    limitations of, 53, 54
    unusual, 58
Calmodulin, 251, 253, 254
Camphorsulfonic acid, 179
Cancer, lung, β-2-microglobulin in, 75
Capillary chromatography
    electrokinetic, micellar, 214
    HPLC, 210, 212
Capillary zone electrophoresis (CZE), 135, 197, 348
    CF-FAB and, 207, 213—217
    mass spectrometry interfaces with, 214
Carbamylation, globin chain, 260
Carbohydrate, 25, 322
S-Carboxamidomethylcysteine, 318
γ-Carboxyglutamic acid, 322
Carboxymethylation, 77, 193
S-Carboxymethylation, 248
S-Carboxymethylcysteine, 318
Carboxypeptidase(s), 77, 79, 81, 209
Carboxypeptidase A, 277
Carboxypeptidase B, 319, 385
Carboxypeptidase digestion, 79, 81, 354
Carboxyterminal peptide identification, 322
Carrier solution
    CF-FAB, 207
    permethylation reaction and, 317
β-Casomorphin, 329, 336
Cataracts, disulfide bonds and, 276
Catfish pancreas, peptide charting and, 351, 352, 356,
    361—362
Catfish pancreatic extract, 351, 352
Cation exchange sites, 361
Cationic species, complexation of, 290
Cationization, 95, 351, see also Metal-ion cationized
    peptide decompositions
Cations, monovalent, dipeptides complexed with,
    310—311
Caudate nucleus, 389
cDNA, 379
cDNA precursor sequences, 348
Cecropin A, 44
Cell body, 370
Cellular immunity, 145

Cellulose acetate, electrophoresis and, 261
Center-of-mass collision energy, 125
Central nervous system, 328
Centroid determination, precision of, 68
Cerebrospinal fluid (CSF), 341, 372, 387—389
Cesium ion guns, 110, 180
^{252}Cf-PDMS, see 252-Californium plasma desorption
  mass spectrometry
Charge exchange, high-pressure ammonia injection
  and, 92
Charge/mass ratio, 11
Charge-remote fragmentation, 133
Charge retention, 103
Charting, see Peptide charting
Chelation, 339
Chemical derivatives, 316
Chemical ionization, neuropeptides and, 328, 329,
  338
Chemically derivatized peptides, 368
Chemical molecular weight, 111
Chemical noise, 202, 205, 312, 328
Chemotactic tetrapeptide, 320
Chicken lysozyme, 21, see also Hen egg-white
  lysozyme
Chloracetyl chloride, 191
Chlorination, performic acid and, 265
Chloroform extraction, 322
Chromatography, 368, see also Gel chromatography;
  High performance liquid chromatography
  (HPLC)
  capillary, micellar electrokinetic, 214
  CF-FAB and, 207, 209, see also High performance
    liquid chromatography (HPLC), CF-FAB and
  gas, see Gas chromatography entries
  gel, 35, 66, 74, 356
  ion-exchange, 261
  Sep-Pak, 374
  supercritical fluid, 323
Chromophobe adenoma, 379, 389
Chromosomal crossover, fusion subunits and, 259
Chromosome 16, 258
Chymotrypsin, 149, 191, 265, 271
α-Chymotrypsin, 52
CI, see Chemical ionization
CID (collision-induced dissociation), 88, 142
CID-MIKE spectra, 271
Circular dichroism, 311
^{13}C-labeled amino acids, 383
Cleavage
  of amide backbone, 128
  disulfide bonds and, 284
  homolytic, 129, 131, 133
  multistep, 142
  peptides generated by, molecular weight of, 239—
    244
  ring, 291
β-Cleavage, 103
CLIP (corticotropin-like intermediate peptide), 353,
  370
Cluster ions, 202, 232, 234
CN ion, 10

CO_2, supercritical fluids of, 89
Coaxial CF-FAB interface, 212, 214—215
Codons, 225
Collisional activation, 329
Collisional cooling, 99
Collisional interaction, 127
Collisionally activated decomposition (CAD), 124,
    133, 291, see also Fast atom bombardment
    (FAB)-B/E-CAD mass spectra
  neuropeptides and, 380
  product ion, HPLC/CF-FAB and, 212—213
Collision cells, 122, 124, 380
Collision dynamics, 124
Collision energy, 125—126
Collision gas, 124—125, 330
Collision induced dissociation (CID), 88, 142, see
    also CID-MIKE spectra
Collision regimes, 124
Complementary DNA, see cDNA entries
Complexation reactions, 177, 290, 310—311
Computer algorithms, FAB and, 196
Computer-controlled ^{252}Cf-PDMS, 5
Computer programs
  disulfide bonds in proteins and, 281, 284
  hemoglobin mass spectra and, 263
Condensed mechanisms, gas phase mechanisms
    versus, 177
Connecting peptides, 20, 82
Constant-flow analysis, CF-FAB, 207, 209—210
Constant neutral loss scan, 380—381
Contaminants, low molecular weight, 69
Continuous-flow FAB (CF-FAB), 203—218, 312, see
    also Flow-FAB analysis
  advantages of, 205—208
  applications of, 209—217
  capillary zone electrophoresis and, 213—217
  disadvantages of, 207
  liquid chromatography and, 210—213
  operation of
    memory effects in, 205
    modes of, 207, 209
    stability of, 203—205
  protein structure and, 209—210
Continuous-flow sample introduction probe, 203
Continuous injection, 90
Continuous ion beams, 312
Controlled fragmentation, 95
Convective disturbances, CZE and, 213
Conversion dynode, 140
Cooling, collisional, 99
Copper, 188
Corn, HC-toxin and, 296
Cornea, 374, 385
Correlated mass measurements, reflecting time-of-
    flight mass spectrometer in, 165—168
Corticotropin-like intermediate peptide (CLIP), 353,
    370
Coulombic energy, 175
Covalent dimers, 31—32
Covalent structure, correctness of, verification of,
    42—44

CQH (cylindrical electrical sector, quadrupole, homogeneous magnetic sector) mass spectrometer, 231
Cross contamination, 207
Cross linkages, 276, 285
Cryobaffle, in high-pressure fluid injection, 90
Cryoshield, in laser desorption/multiphoton ionization, 95
Crystallinity, matrix, 5
Crystallins, disulfide bonds and, 276
Crystallography
  ovomucoid, 24
  X-ray, 280
CSF, see Cerebrospinal fluid
CsI cluster ions, 232, 234
C_{18} silica analytical column, 374
C-terminal peptide molecule derivatizations, in FAB and LSIMS, 191
C-terminal sequence determination, 77, 209
C-terminal truncation, 354
C-terminus, deprotonated, 308
C-terminus elimination, metastable, 310
C-terminus sequence ions, 331
Cushing's syndrome, 389
Cuticle protein, insect, 74
Cyanobacterium, 152
Cyanogen bromide, 81, 322
  calmodulin and, 251, 253
  disulfide bonds and
    in peptides, 279, 280
    in proteins, 281
  peptides generated by cleavage with, molecular weight of, 239, 241—243
  STP-3 and, 149
Cycle time, narrow magnetic scans and, 111—112
Cyclic peptides, amino acid sequence determination of, 290—306
  applications of, 296—298
  synthetic peptides in method development for, 291—296
  unified approach to, 298—306
Cyclotron frequencies, 145
Cylindrical electrical sector, quadrupole, homogeneous magnetic sector (CQH) mass spectrometer, 231, see also QQHQC mass spectrometer
Cysteic acid, 279
Cysteine, 264, 318
Cysteine-containing peptides, oxidation state of, 192
Cysteinyl thiols, 276
Cytochrome c, 44
  horse, N-terminal peptide molecule derivatizations and, 190
  HPLC/CF-FAB and, 210
CZE, see Capillary zone electrophoresis

D

Damage products, 53
Dansylation analysis, of murine elongation factor, 31
Data analysis, 351
Deamination, insulin, 32

Decay event, probability of, 165
Decomposition
  collisionally activated, see Collisionally activated decomposition (CAD)
  metal-ion cationized, see Metal-ion cationized peptide decompositions
Deconvolution, 181
Degradation products, endocrine tissues and, 351
Degradation reactions, see also Edman degradation
  monitoring of, 195
  partial, 298
Degrees of freedom, 126, 329
Dehydroalanyl peptide, 318
Delay generators, 99
Dementia, Alzheimer's, 389
Deposited energy, 126
Deprotection, 50, 52
Deprotonated amide, isomerization to, 310
Deprotonated C-terminus, 308
Derivatization
  FAB and LSIMS and, 183, 190—195
  magnetic sector mass spectrometry and, 110
Desalting, 352
Desialylation, 361—362
Desorbed plume, 99
Desorption, 330, see also Field desorption (FD); Laser desorption; Plasma desorption mass spectrometry (PDMS)
  matrixless, 312—313
  particle-induced, 88, 348
  tandem mass spectrometry and, 311—313
  techniques, 18
Detection sensitivity, see Sensitivity
Deuterated glycerol, 193
Deuterated reagents, 320
Deuterated substance P, 213
Dexamethasone suppression test, 389
Diagonal electrophoresis, 276
Dichroism, circular, 311
Dicyclohexylcarbodiimide, 61
Dielectric constants, 175, 179
Diethanolamine, 178, 330
Diethylether, 319
Differential devices, 350
Diffusion constants, 177
Digit-printing method, 224—231, see also High resolution mass spectrometry
  ambiguity in, 230—231
  amino acid substitutions and, 226—229
  background of, 224—225
  sickle cell hemoglobin analysis by, 225, 229—230
Dilution effect, 202
Dimers, 31—32, 35—36, 59
Dimethylpyrimidylornithine, arginine derivative formation and, 319
Dimethylsulfinyl sodium, 317
Dimethylsulfoxide (DMSO), 58, 180
Dipeptide isomers, 92
Dipeptides
  complexed with monovalent cations, 310—311
  laser desorption/multiphoton ionization of, 99—103

Direct high-pressure liquid injection (DLI), 89, 93—95

Direct insertion probe, 320

Direct photoionization, 88

Discrimination effects, 290

Dispersion forces, 8

Dissociation
  bond, 329
  collision induced, 88, 142

Distilled water, 317

Distribution width, 12

Disulfide(s)
  mixed, insulin-glutathione, 36, 38
  rearrangements of, 276

Disulfide bond(s), 276—286, 379
  connectivities of, 284
  in peptides, detection of, 277—281
  permethylation reaction and, 317
  in proteins
    methods for location of, 276—277
    procedure for location of, 281—284
  reduction of, 277, 282
  in synthetic peptides and proteins, 58—61

Disulfide knot, 284

Disulfide-related cross linkages, 285

Disulfide/sulfhydryl transformations, importance of, in FAB and LSIMS, 192—195

Dithiothreitol, 150, 277, 281

Dithiothreitol/dithioerythritol, 180, 277, 278, 330, 341, 394

DLI, see Direct high-pressure liquid injection

DMSO, see Dimethylsulfoxide

DNA
  complementary, see cDNA entries
  recombinant, 18

DNA sequencing, 312

Dolastatin 13, 299, 304

Double-focusing mass spectrometer, 329

Drosophila, homeo box protein of, fragment condensation synthesis of, 55—58

Drug overdose, pituitary and, 390—392

Duty cycle, 90

Dynamic processes, FAB and, 202, 207

Dynamic range, 363

Dynorphin(s), 328, 334, 370, 376, 394—396

Dynorphin A, 52, 329

E

Edman degradation, 43, 79, 261, see also Degradation reactions
  arginine derivative formation and, 319
  blocked N-termini and, 320
  bovine pituitary and, 354
  calmodulin and, 253
  digit-printing method and, 224
  disulfide bonds and, 277
  neuropeptides and, 339
  oligopeptides and, 145
  pancreatic peptides and, 21, 24
  sensitivity of, 312

STP-3 and, 150—151

Egg-laying hormone, 42

Egg-white lysozyme, disulfide bonds in, 277, 279—284

Eglin C, 111

EI, see Electron ionization

Electrically floating cells, 125

Electrochemical high performance liquid chromatography, disulfide bonds and, 279, 282

Electrokinetic capillary chromatography, micellar, 214

Electronic excitation, 127

Electronic time resolution, mass determination and, 12

Electron impact ionization, 320

Electron ionization (EI), neuropeptides and, 328, 338

Electro-osmotic flow, 213—215

Electrophoresis, 177, 196
  capillary zone, see Capillary zone electrophoresis (CZE)
  diagonal, 276
  digit-printing method and, 224
  gel, see Gel electrophoresis
  hemoglobin F Izumi and, 248
  hemoglobin S and, 229
  hemoglobin variants and, 261

Electrospray, 195—197
  CZE and, 214
  endocrine tissues and, 352
  intact globin chains and, 271—272
  pneumatically assisted, 214
  sample deposition and, 69

Electrostatic analyzer, 312

Electrostatic retardation fields, 12

Eledoisin, 181

β-Elimination reaction, 318

ELISA, see Enzyme-linked immunoassay

Elongation factor, 31

Embryonic globin, 258

Embryonic hemoglobin, 259

Emitters, magnetic sector mass spectrometry and, 110

Endocrine tissues, 351—363, see also Neuropeptides; Peptide charting, endocrine tissues and

Endogenous opioids, 328, 370, see also Neuropeptides

Endo-Lys-C, 145

Endoprotease column, immobilized, 209

Endorphin, 328, 329, 376

α-Endorphin, 181

β-Endorphin, 328, 370, 391, 393

Ac-β-Endorphin, 351

Energy
  collision, 125—126
  Coulombic, 175
  deposited, 126
  internal, see Internal energy
  kinetic, see Kinetic energy
  vibrational, see Vibrational energy

Energy barrier, 310

Energy density, 7, 10

Energy losses, translational, 301

Enkephalin-related peptides, 341
Enkephalins, 320, 328, 329, 334, 376
Entry faces, non-normal, 312
Enzymatic degradation
    in situ, nitrocellulose-protein matrix and, 6
    synthetic peptides and proteins and, 61
Enzymatic hydrolysis, 50, 202, 209—210
Enzyme-catalyzed microscale chemical reaction, of
        surface-bound peptides, 46
Enzyme-linked immunoassay (ELISA), 379
Enzymic digests of proteins, 210
Enzymolysis, 385
Epidermal growth factor, molecular weight
        determination of, 75
Equilibrium-based arguments, 177
Ester groups, 322
Esterification
    N-terminal peptide molecule derivatizations and,
        190
    STP-3 and, 150
[^{3}H]Ethanolamine, 31
Ether precipitation, 350
Ethylene glycol, 178, 341, 394
Ethylketoazocine, 377
Etorphine, tritiated, 377
Even-electron molecular ions, 18
Excimer laser, argon fluoride, 142
Excitation
    of mass-analyzed ion, 329
    mechanisms of, in four-sector tandem mass
        spectrometry, 126—127
    Q-FTMS and, 145
Excitation spectrum, 88
Exopeptidase-catalyzed degradation, on-line
        monitoring of, CF-FAB and, 209
Exoprotease, 210
Exponentially scanned multichannel analyzer data,
        112
External rf pulse, 145
Extracellular proteins, three-dimensional structure of,
        276
Eye, 276, see also Ophthalmic branch of trigeminal
        system

F

FAB, see Fast atom bombardment
Familial amyloidotic polyneuropathy, 253
Fast atom bombardment (FAB), 18, 195—197, 316,
        348
    carboxyterminal peptide identification and, 322
    computer algorithms and, 196
    continuous-flow, see Continuous-flow FAB (CF-
        FAB)
    disulfide bonds and
        detection in peptides of, 277—281
        location in proteins of, 276—277
    flow-FAB, solvents for, 188—190
    hemoglobin and, 263
    magnet-scan, of neuropeptides, 328, 331—334,
        339—340

matrix selection for, 173—190, 202
    acid/base characteristics and, 177—180
    oxidizing/reducing nature and, 180—182
    sample surfactancy and, 181, 183—188
    solubility considerations in, 175—177
matrix studies with, 394
negative, of leucine enkephalin, 394
neuropeptides and, 328—338, see also Neuropep-
        tides; FAB mass spectrometry and
peptide analysis and, 202
permethylated peptides and, 320
sample derivatization for, 190—197
    C-terminal peptide molecule, 191
    degradation reactions and, 195
    disulfide/sulfhydryl transformations and, 192—
        195
    internal, 191—192
    N-terminal peptide molecule, 190—191
FAB-*B/E*-CAD mass spectra, of neuropeptides, 328,
        334—338, 340—341
FAB guns, 174, 330
FAB mapping, 195, 265
FAB probe matrix, 381
Fatty acids, metal-ion cationization and, 307
FD, see Field desorption
Fetal hemoglobin, 259
Fetuin, bovine, glycopeptides from, 24—25, 30
Field desorption (FD), 18, 316, 328
    hemoglobin and, 263
    neuropeptides and, 338
Filter pads, CF-FAB and, 203
Filter paper, digit-printing method and, 224
"Fingerprinting", 224
Fission fragment time-of-flight (TOF) mass spectrum,
        covalent structure and, 42—44
Fission track, 5
Flight time, product ion, 163
Flow-FAB analysis, 188—190, see also Continuous-
        flow FAB (CF-FAB)
Flow-injection analysis, CF-FAB, 207, 209
Flow-SIMS analysis, 190
Fluid injection, high-pressure, see High-pressure fluid
        injection
Focal place detection, 126
Focal plane detectors, 381
Formic acid, 148, 319
Formyl-Trp, 48
Forward-geometry mass spectrometer, 336
Fourier transform mass spectrometry, 350, see also
        Quadrupole Fourier transform mass
        spectrometry
Four-sector tandem mass spectrometry, 122—136
    capillary reversed-phase HPLC and, 212
    collision energy and, 125—126
    collision gas and, 124—125
    excitation and fragmentation mechanisms in, 126—
        127
    fragmentation pathways of protonated linear
        peptides and, nomenclature of, 127—132
    ionization methods in, 123—124
    neuropeptides and, 381

published spectra of peptides and, 132—135
Fractional distillation, 320
Fractional energy spreads, 125
Fragmentation(s), 114
  charge-remote, 133
  controlled, 95
  mechanisms of, in four-sector tandem mass
       spectrometry, 126—127
  metastable, 145
  neuropeptides and
    EI-CI mass spectra and, 338
    FAB mass spectrometry and, 330—338
    field desorption mass spectra and, 338
    laser desorption/laser ionization and, 338—339
    metal ion-peptide interaction and, 339
  nonergodic, 133
  of permethylated peptides, 319—320
  protonation site and, 133
  sidechain, 128
  unimolecular, 127
Fragmentation analysis, 122
Fragmentation channels, 329
Fragmentation efficiency, collision gas and, 124
Fragmentation pathways, of protonated linear
       peptides, nomenclature of, 127—132
Fragment condensation synthesis, 55—58
Fragment ions, 368
  source-produced, 299
  types of, nomenclature and, 330—331
Fragment ion series, 269
Frequency-doubled Nd:YAG pumped dye laser, in
       high-pressure fluid injection, 92
FRIT-FAB interface, 212
Fusion subunits, 259

G

Galactose, 361
Gas chromatography, 316
Gas chromatography/mass spectrometry (GC/MS),
       191, 316
  arginine derivative formation and, 318
  carboxyterminal peptide identification and, 322
  permethylated peptides and, 320, 322, 323
Gas phase ions, 4
  formation of, 10
  metastable, 5
Gas phase mechanisms
  cationization and, 307
  condensed mechanisms versus, 177
Gastrotropin, 393
GC/MS, see Gas chromatography/mass spectrometry
Gel chromatography, 35, 66, 74, 356
Gel electrophoresis, 74, 354
  isoelectric focusing and, 261
  SDS, 53, 66
Gel filtration, 319
Genes, hemoglobin abnormalities and, 258
Germanium lens, in laser desorption/multiphoton
       ionization, 99
Gibbs equation, 186
Glass sample tubes, 317

Globin, 258
Globin chains, see also Hemoglobin
  intact, analysis of, 271—272
  normal, tryptic peptides of, 232, 234—248
  reversed-phase HPLC separation of, 261, 262
GLP, see Glucagon-like peptide
Glu-C, STP-3 and, 149
Glucagon, 21, 110, 317
  in catfish pancreas, 356
  N-terminal peptide molecule derivatizations and,
    190
  tryptic digest of, 205, 212
Glucagon-like peptide (GLP), in catfish pancreas, 356
Glutathione, 71, 179, 352, 394
Glutathione-insulin transhydrogenase, 36
Glycan, 25, 28
Glycerol, 279, 330, 341, 394
  deuterated, 193
  HPLC/CF-FAB and, 210
Glycerol/thioglycerol matrix, 123
Glycinamidation, cyanogen bromide, 253
Glycolipids, 18
Glycopeptides, 322, 353
  from bovine fetuin, 24—25, 30
  high-mannose, from trypanosomes, 25, 28, 32
O-Glycosylation, 20
Glycosylation, globin chain, 260
Glycosylation heterogeneity, 20
Glycosylation sites, 22, 24—25, 28—30
Glycosylphosphatidylinositol, 28
Gramicidin S
  CF-FAB and, 205
  HPLC/CF-FAB and, 213
Growth factor, bovine, HPLC/CF-FAB and, 212
Growth hormone, 20, 212, see also Human growth
    hormone
Growth hormone releasing factor, 210
Guanidine, 319
Guanido group, 317

H

Hard ionization, 95
Hb, see Hemoglobin
Hb A, 259
Hb Barts, 259
Hb Constant Spring, 259, 260
Hb E, 260
Hb F, 259
Hb F Izumi, 248, 249
Hb F Yamaguchi, 248, 252
Hb Knossus, 260
Hb Koln, 263
Hb Lepore, 259
Hb North Chicago, 269
Hb Pasadena, 265—266, 268
Hb Providence, 248, 250, 252
Hb Q Thailand, 270
Hb Setif, 266—268
Hb Tarrant, 271
Hb Torino, 271

Hb variants, 258
  characterization of, analytical methods for, 261—
    262
  mass spectrometry of, 263—272
    background of, 263
    intact globin chain analysis in, 271—272
    isolated peptide analysis in, 269—271
    tryptic digest mixture analysis in, 263—269
HC-toxin, 296—297
Helium, as collision gas, 124, 330
Heme, 258
Hemiacetal, 304
Hemoglobin (Hb), 258—272, see also Globin entries
  abnormal, 258—261, see also Hemoglobin variants
  adult, 259
  β-chain of, tryptic fragments of, 267
  embryonic, 259
  fetal, 259
  four-sector tandem mass spectrometry and, 135, 212
  instability of, tests for, 261
  normal, 258—259
  sickle cell, digit-printing method and, 225, 229—
    230
  signal-to-noise values and, 185
  structure of, 258—261
Hemoglobinopathies, 259—261, 263
Hemolysates, 261, 272
Hen egg-white lysozyme, disulfide bonds in, 277,
    279—284
Heteroatom, 317
Heterodimers, 31
Heterogeneity, 22, 24—25, 28—31, 33—35
Heterozygosity, 260
High energy liquid secondary ion mass spectrometry,
    110, 116, 118, see also Magnetic sector mass
    spectrometry
High-mannose glycopeptide, from trypanosomes, 25,
    28, 32
High mass, desorption ionization and tandem mass
    spectrometry and, 311—312
High molecular weight capabilities, 5
High molecular weight proteins, in buffer solutions,
    ^{252}Cf-PDMS of, 7—8
High performance liquid chromatography (HPLC),
    21, 329, 348, see also Liquid chromatography
    (LC)
  amino acid sequence of cyclic peptides and, 301
  carboxyterminal peptide identification and, 322
  CF-FAB and, 207, 210—213
  electrochemical, disulfide bonds and, 279, 282
  hemoglobin variants and, 261
  reversed-phase, see Reversed-phase high perform-
    ance liquid chromatography
HPLC mapping, hemoglobin variants and, 261
High-pressure fluid injection, 89—95
  of ammonia, 92—93
  DLI or thermospray methods of, 93—95
  experimental setup for, 90—92
High resolution mass spectrometry, 231—232, see
    also Digit-printing method
  calmodulin and, 251, 253, 254

Hb F Izumi and, 248, 249
Hb F Yamaguchi and, 248, 252
Hb Providence and, 248, 250, 252
  normal globin chains and, tryptic peptides of, 232,
    234—248
  prealbumin variant and, 253, 255, 256
  synthetic globin variants and, 251
High voltage arcs, CF-FAB and, 207
Hippocampus, 389
Hirudin(s), 111, 114
HIV, see human immunodeficiency virus
^{2}H-labeled amino acids, 383
Homeo box protein, *Drosophila,* fragment condensa-
    tion synthesis of, 55—58
Homeostasis, 372
Homogeneity, of synthetic product, verification of,
    44—53
Homologs, 381
Homology, 18
Homolytic cleavage, 129, 131, 133
Homoserine lactones, STP-3 and, 149
Homozygosity, 260
Horse cytochrome *c,* N-terminal peptide molecule
    derivatizations and, 190
HPLC, see High performance liquid chromatography
Human growth hormone, 21, 81
  mixture analysis and, suppression effects in, 79
  recombinant, see Recombinant human growth
    hormone (rhGH)
Human growth hormone releasing factor, 210
Human hemoglobin, see Hemoglobin
Human immunodeficiency virus (HIV), 36
Human insulin
  disulfide-related cross linkages in, 285
  mutant, 18—20
Human monocyte chemoattractant, 145
Human pituitary, post-mortem, 389—393
Human proinsulin, see Proinsulin
Hybrid mass spectrometer, 329
Hydrazides, 319
Hydrazinolysis, arginine derivative formation and,
    319
Hydrocarbon polymer column, 374
Hydrogen bonding, 93, 316
Hydrogen peroxide, 148
Hydrolysis
  acid, see Acid hydrolysis
  enzymatic, see Enzymatic hydrolysis
  partial, blocked N-termini and, 320, 322
Hydrolysis products, 46
Hydropathic index, 188
Hydrophilicity, 183, 202
  CF-FAB and, 205, 207
  substrates for ^{252}Cf-PDMS and, 8
Hydrophobic core, 264
Hydrophobicity, 6, 21, 183, 202, 374
  CF-FAB and, 205
  molecular weight determination and, 74
  substrates for ^{252}Cf-PDMS and, 8
Hydroxylated liquid matrices, 175
Hydroxytropylium ion, 339

Hymenistatin 1, 301
Hypervariability, 22
Hypothalamus, 389

I

ICR, see ion cyclotron resonance cell
IGFs, see Insulin-like growth factors
Ileum protein, 393—394
Image magnification, high resolution mass spectrometry and, 231
Immobilized endoprotease column, 209
Immobilized trypsin, 210, 385
Immonium ions, 330, 339
Immunity, cellular, 145
Immunoaffinity chromatography, antisubstance P, 213
Immunoassay, enzyme-linked, 379
Immunofluorescence, 379
Immunogenic complex, 378
Immunogenicity, insulin and, 32
Immunohistochemistry, 379
Immunological interactions, neuropeptides and, 377—379
Immunoreactivity, 372, 375
Impulsive collision treatment, 127
Impurities, matrix selection and, 175
Incident ion energies, 125
Inhomogeneous fields, 312
Insect cuticle protein, 74
In situ hybridization, 379
Instrumentation, 350—351, see also specific type
Insulin(s), 21
    anthroic acid and, 9
    bovine
        β-chain of, 110
        sample surfactancy and, 183
    in catfish pancreas, 356
    disulfide bonds in, 276—277
    disulfide-related cross linkages in, 285
    HPLC/CF-FAB and, 210
    metastable decay of, 71
    mutant, 18—19, 20
    protein adsorption and, 5
Insulin dimerization, in long-term preparations, 32, 35—37
Insulin-glutathione mixed disulfides, nonenzymatic formation of, 36, 38
Insulin-like growth factors (IGFs)
    disulfide bonds in, 277
    molecular weight determination of, 74—75
Insulin molecular ions, lifetime of, 6
Integrating devices, 350
α-2 Interferon, in buffer solutions, ^{252}Cf-PDMS of, 7, 8
Interleukin-1, 81
Interleukin-1α, 135
Interleukin-2, mixture analysis and, suppression effects in, 79
Interleukin-3, limitations of PDMS and, 53
Intermolecular disulfide linkages, 59
Internal cyclization, 46

Internal degrees of freedom, 126
Internal derivatizations, in FAB and LSIMS, 191—192
Internal disulfide bonds, 58—61
Internal energy, 125, 126, 145, 334
Internal energy uptake, 127
Internal standards
    neuropeptides and, 381, 383
    quantitation and, 202
    stable isotope-incorporated, 381
Internal temperature, 126
Intracerebroventricular fluid, 372
Involatility, 4
Iodoacetamide, 193
Iodoacetic acid, 77
    radiolabeled, 276
    STP-3 and, 150
Iodomethane, 191
Ion(s), see also Molecular ions
    background, 205
    cluster, see Cluster ions
    fragment, see Fragment ions
    mass-analyzed, excitation of, 329
    mass determination of, 11—13
    matrix, 328
    multiply charged, 71, 196
    parent, see Parent ion entries
    precursor, collision gas and, 124
    product, see Product ion entries
    trapped, photodissociation of, 142
Ion activation, 312
Ion beam(s)
    continuous, 312
    instability of, 203
Ion bombardment, 123
Ion chromatograms, 212, 215, 217
Ion cyclotron resonance (ICR) cell, 142
Ion detection, 312
Ion-dipole interactions, 5
Ion-exchange chromatography, 261
Ion formation, energetics of, 10
Ion formation/desorption mechanism, 5—6
Ion genesis, 383
Ionic strength, "salting out" and, 175
Ionization, 312
    atmospheric pressure, 214
    chemical, 328, 338
    desorption, see Desorption
    electron, 328, 338
    electron impact, 320
    electrospray, see Electrospray
    FAB, see Fast atom bombardment
    laser, 329
    methods of, four-sector tandem mass spectrometry and, 123—124
    selective, 88
Ionization potential, R2PI and, 88
Ion mirrors, 163, 169, 350
Ion optics, 124, 231
Ionspray, 214, 271—272
Ion trajectories, 231

Ion transmission, high resolution mass spectrometry and, 231
Ion traps, 196
Iron, 258
Iron transport, 290
Irradiation, MPI and, 88
Irradiation time, FAB, fragment ion yields and, 334
Isobaric ions, 301
Isoelectric focusing, 261
Isomerization, to deprotonated amide, 310
Isopropyl alcohol, 261
Isotopically labeled peptide, 308
Isotopic doublets, 338
Isotopic envelope, 181
Isotopic pattern, 12, 160

J

Jet velocity, in high-pressure fluid injection, 92
Joule heating, capillary zone electrophoresis and, 213

K

β-Ketosulfoxide formation, 322
Kiloelectronic particle bombardment, 123
Kinetic energy, 11, 12
  conversion to vibrational energy, 140
  high-pressure ammonia injection and, 93

L

α-Lactoglobulin A, tryptic digest of, 210
Lactone groups, 322
Lanthionine, 277, 285
Laser(s)
  excimer, argon fluoride, 142
  Nd:YAG, in high-pressure fluid injection, 92
  ultraviolet, 88
Laser desorption, 18, 95—103, 195, 328
  accessible mass range and, 69
  matrix-assisted, 350—351
  multiphoton ionization and, see also Resonance enhanced multiphoton ionization (REMPI)
  experimental description of, 95—99
  neuropeptides and, 338—339
  results of, 99—103
Laser energies, soft ionization and, 88
Laser ionization, 329, see also Multiphoton ionization (MPI)
Laser photodissociation, 142, 145, 150
Laser power, benzene ionization and, 88
Laser pulses, in high-pressure fluid injection, 92
Laser source, pulsed, 90
LC, see Liquid chromatography
Lens of eye, disulfide bonds and, 276
Leucine, 329
Leucine enkephalin, 179, 328, 370
  CF-FAB and
    CZE and, 214
    HPLC and, 212, 213
  negative FAB mass spectrometry of, 394

²ala-Leucine enkephalin, 381
Lidocaine, 387
Linear peptides, 127—132, 290
Linked-field scan, 298, 328, 329, 334—338, 340—341, 380—381
Liquid chromatography (LC), 203, 209, see also High performance liquid chromatography (HPLC)
Liquid injection, high-pressure, see High-pressure fluid injection
Liquid matrix, 123, 328
Liquid metal ion probes, 350
Liquid nitrogen, 90, 372
Liquid samples, continuous introduction of, into FAB source, 203, see also Continuous-flow FAB (CF-FAB)
Liquid secondary ion mass spectrometry (LSIMS), 110, 116, 118, 195—197, see also Magnetic sector mass spectrometry
  hemoglobin and, 263
  matrix selection for, 173—188
  acid/base characteristics and, 177—180
  oxidizing/reducing nature and, 180—181
  sample surfactancy and, 181, 183—188
  solubility considerations in, 175—177
    sample derivatization for, 190—197
  C-terminal peptide molecule, 191
  degradation reactions and, 195
  disulfide/sulfhydryl transformations and, 192—195
  internal, 191—192
  N-terminal peptide molecule, 190—191
  toxic cyclic peptides and, 152—153
Locust cuticle protein, 74
"Loop trapping" multiloop injection valve system, 212
Low back pain, 372, 387—388
Low-energy conditions, tandem mass spectrometry and, 140
Low-mass discrimination, FAB-*B*/*E*-CAD and, 338
Low molecular weight contaminants, 69
LSIMS, see Liquid secondary ion mass spectrometry
Lumbar cerebrospinal fluid, 387—389
  Alzheimer's dementia and, 389
  low back pain and, 387—388
Lung cancer, β-2-microglobulin in, 75
Lys-C, STP-3 and, 149
Lysozyme, 114
  anthroic acid and, 9
  chicken, 21
  hen egg-white, disulfide bonds in, 277, 279—284
  PDMS peptide map of, 77, 78
Lysyl endopeptidase, peptides generated by cleavage with, molecular weight of, 240—242, 244

M

Macor ceramic, 98
Magnet, superconducting, 142
Magnetic repulsion principle, 97—98
Magnetic sector mass spectrometry, 110—118
  molecular weight determination with, 114, 116—118

scanning methods in
  narrow magnetic scans, 111—112
  narrow voltage scans, 112—113
  survey scans, 111
  spectrum characteristics in, 113—115
  techniques and instrumentation for, 110—111
Magnet-scan fast atom bombardment, neuropeptides
  and, 328, 331—334, 339—340
Mandelic acid, 95
Mandibular branch of trigeminal system, 384—386
Mannose, 24, 25, 28, 32
[^{3}H]Mannose, 28
Mapping
  FAB, 195
  HPLC, hemoglobin variants and, 261
  peptide, 77—79, 224
Mass accuracy, 4, 5
Mass-analyzed ion, excitation of, 329
Mass-analyzed ion-kinetic energy (MIKE), 341, see
  also CID-MIKE spectra
Mass calibration, plasma desorption spectrum and, 68
Mass determination
  narrow magnetic scans and, 112
  precision of, 58, 68—69
  of protein, by ^{252}Cf-PDMS, 11—13
Mass dispersion, high resolution mass spectrometry
  and, 231
Mass range
  accessible, 69
  high resolution mass spectrometry and, 231
Mass resolution, 5, see also High resolution mass
  spectrometry
  high, desorption ionization and mass spectrometry
    and, 312
  MS-1 and, 162
  product ion, 166—168
Mass shifts, 298
Mass-specific analyzers, 348
Mass specificity, 350
Mass spectrometry/mass spectrometry (MS/MS), see
  Tandem mass spectrometry
Matrix
  FAB, 202, see also Fast atom bombardment (FAB),
    matrix selection for
  impurities in, 5
  liquid, 123, 175, 328
  LSIMS, see Liquid secondary ion mass spectrome-
    try, matrix selection for
  m-nitrobenzylalcohol, 353
  nitrocellulose, 6, see also Nitrocellulose entries
  role in ^{252}Cf-PDMS, 5—10
  selection of, for fast atom bombardment and liquid
    secondary ion mass spectrometry, 173—190
  viscous, reduction in, 205
Matrix-assisted laser desorption, 350—351
Matrix background interference, 301
Matrix ions, 328
Matrixless desorption ionization, 312—313
Matrix modifier, 180
Maxillary branch of trigeminal system, 384—386
MCA, see Multichannel analyzer

MECC, see Micellar electrokinetic capillary
  chromatography
Meiosis, chromosomal crossover during, 259
Melittin, 45, 47
Mercaptoethanol, 77, 277, 281
Messenger RNA, 379
Metabolic radiolabeling, 31
Metal ion adducts, 329
Metal-ion cationized peptide decompositions, 306—
  311
  alkali metal and, 307—310
  determination of metal ion interaction sites and,
    tandem mass spectrometry in, 311
  dipeptides complexed with monovalent cations and,
    310—311
Metal ion probes, liquid, 350
Metal ions
  cyclic peptides and, 290
  neuropeptides and, 339
Metal-linked dimers, of *TAT III* cysteine-rich region,
  36, 38, 39
Metal-peptide complex, 307
Metal surfaces, 188
Metastable C-terminus elimination, 310
Metastable decay, 71, 291
Metastable fragmentation, 145
Metastable ions, 5, 13
Methanol, amino group acylation and, 318
Methanolysis, 298
Methionine, 329, 379
Methionine enkephalin, 179, 212, 214, 328, 370
Methionine sulfone, 255, 265
C-Methylation, 317, 320, 323
S-Methylcysteine, 318
Methyl-esterification, 82
Methyl group, permethylation reaction and, 317
Methyl iodide, 191, 317
Micellar electrokinetic capillary chromatography
  (MECC), 214
Microbore high performance liquid chromatography,
  210, 212
Microcrystalline deposits, anthroic acid and, 9
Microcystis aeruginosa, 152
β-2-Microglobulin, molecular weight determination
  of, 75, 76
Microheterogeneity, 354
Microsequencing
  hemoglobin variants and, 261
  murine elongation factor and, 31
Midbrain, 389
Migration, 52
MIKE, see Mass-analyzed ion-kinetic energy
Mini-proinsulins, 82
Mirrors, see Ion mirrors
Miscibility, 177
Mixture analysis, 79, see also Peptide mixtures
MNBA, see m-Nitrobenzylalcohol entries
Molecular disease, protein variants and, 224
Molecular ions, see also Ion entries
  doubly protonated, 5
  even-electron, 18

formation of, quenching of, 5
insulin, lifetime of, 6
protein, internal excitation of, 9
protonated, 264, 380
yield of, nitrocellulose matrix and, 71, 73
Molecular specificity, 328, 341, 362—363, 369, 381
Molecular weight, 4, 5, 351
chemical, 111
of intact peptides and proteins, determination of,
74—76
of large peptides, magnetic sector mass spectrome-
try determination of, 110—118, see also
Magnetic sector mass spectrometry
protein sequence determination and, 79
Mollusk, egg-laying hormone from, 42
Momentum transfer, 127
Monolayer films, 5
Monolayers, laser desorption and, 99
Monovalent cations, dipeptides complexed with,
310—311
Morphine, 376
Moving belt interface, 203
MPI, see Multiphoton ionization
α-MSH, 328, 370
MS/MS, see Tandem mass spectrometry
MS-1 parent ion spectrometer, 161—162
MS-2 product ion spectrometer, 162—164
Multichannel analyzer (MCA), 111, 112, 330
Multichannel plates, 312
Multiphoton ionization (MPI), 88, 95—103, 350, see
also Laser desorption
Multiply charged ions, 71, 196, 272
Multi-quadrupole mass spectrometer, 329
Multi-sector mass spectrometer, 329
Multistep cleavage, 142
Murine elongation factor, 31
Muscle regulatory protein, 311
Mutagenesis, site-directed, 82
Mutant insulin, 18—20
Mutations
amino acid, digit-printing method and, see Digit-
printing method
hemoglobin abnormalities and, 258, 259, 263
Mylar, 66, 352, 353
Myoglobin, HPLC/CF-FAB and, 210, 212
Myristyl group, 58

N

Nafion, 5
Narrow magnetic scans, 111—112
Narrow voltage scans, 112—113
Nd:YAG pumped dye laser, frequency-doubled, in
high-pressure fluid injection, 92
Negative fast atom bombardment mass spectrometry,
of leucine enkephalin, 394
Negative ion mass spectrum, 31, 132
Neoendorphin, 328, 329, 336
α-Neoendorphin, 168
Neurochemistry, 328
Neuroendocrinology, 368

Neuropeptides, 328—344, 368—396
acid precipitation and, 373
comparative advantages of analytical methods for,
379—380
disulfide bonds and, 276
experimental scheme for, 372
FAB mass spectrometry and, 329—338
FAB-*B*/*E*-CAD mass spectra, 334—338
normal magnet-scan FAB mass spectra, 331—334
types of fragment ions and, 330—331
fragmentation characteristics of, 330—339
EI-CI mass spectra and, 338
FAB mass spectrometry and, 330—338
field desorption mass spectra and, 338
laser desorption/laser ionization and, 338—339
metal ion-peptide interaction and, 339
immune system analytical methods and, 377—379
lumbar cerebrospinal fluid and, 387—389
Alzheimer's dementia and, 389
low back pain and, 387—388
mass spectrometry of, 380—383
CAD in, 380
internal standards and, 381, 383
linked-field scans in, 380—381
quantification and, 383
tandem, 381, 382, 394
nucleic acid analytical methods and, 379
post-mortem pituitary tissue and, 389—393
drug overdose and, 390—392
β-endorphin in, 391, 393
tumor profiles from, 389—390
post-translational modifications of, 379
processing derangements in, hypothetical example
of, 370—372
qualitative analytical mass spectrometry and, 369
quantitative analytical mass spectrometry and, 341,
369
radioreceptor assay and, 376—377
research program on, 370
reversed-phase HPLC and, 374—376
Sep-Pak chromatography and, 374
sequence data on, need in neurosciences for, 369—
370
sequence ion information about, 339—341
solvent-solute interactions and, 341
tissue acquisition and, 372—373
trigeminal system and, 383—386
maxillary and mandibular branches of, 384—386
ophthalmic branch of, 385
Neurophysin(s), 61, 354—356
Neutral component, retardation field and, 12
Neutral moieties, 336
Neutral plume, 95
Ninhydrin, 224
Nisins, 71—72, 135
Nitrobenzylalcohol, 180
m-Nitrobenzylalcohol (MNBA), 110
m-Nitrobenzylalcohol matrix, 353
Nitrocellulose, 9, 46, 188
aluminized mylar coated with, 353
nitro group on, 6

protein absorption on, stoichiometry of, 6—7
tryptic mapping and, 21
Nitrocellulose matrix, 5, 6, 66, 71—73, 351
Nitrocellulose/protein matrix, 6
Nitrophenyloctylether, 180
NMR studies, see Nuclear magnetic resonance studies
N_2O, supercritical fluids of, 89
Nomenclature, 195
of fragmentation pathways of protonated linear
peptides, 127—132
of fragment ions, 330—331
Noncoherent light sources, 88
Nonergodic fragmentations, 133
Non-normal entry faces, 312
Nonparabolic flow, 214, 215
Nonplanarity, 291
Nonpolar domains, 6
N-terminal peptide molecule derivatizations, in FAB
and LSIMS, 190—191
N-termini, blocked, permethylated peptides and, 320,
322
N-terminus sequence ions, 331
Nuclear magnetic resonance (NMR) studies
cyclic peptides and, 297—298
metal ion interaction and, 311
Nucleic acid, 322, 379
Nucleophilic attack, 308, 317, 339
Nucleosides, 92, 307
Nucleotide bases, 225
Nucleotides, metal-ion cationization and, 307

O

^{18}O, 381
carboxyterminal peptide identification and, 322
C-terminal peptide molecule derivatizations and,
191
Octadecylsilyl disposable cartridge, 387
Octaphosphoryl, 31
Oligonucleotides, 18
Oligopeptide sequence analysis
quadrupole Fourier transform mass spectrometer in,
142, 144, 145
TSQ-70 triple quadrupole mass spectrometer in,
140—143
Oligosaccharides, 18, 24
One-color frequencies, 88
On-line analysis, of enzymic hydrolysates, protein
structure from, 209—210
Ophthalmic branch of trigeminal system, 385
Opioid radioreceptor assay, 376—377
Opioid receptors, 328
Opioids, endogenous, 328, 370, see also Neuropep-
tides
Organic mass spectrometry, 316
Organic matrix
radiation damage to, 202
viscous, reduction in, 205
Orthodontic stress, neuropeptides and, 372, 384—385
OT-NP (oxytocin-neurophysin), 354
Ovalbumin, 69

Overlaps, 79
Ovomucoid third domains, 22, 24, 28, 29
Oxidation
of cysteinyl thiols, 276
performic acid, see Performic acid oxidation
Oxidation state, of cysteine-containing peptides, 192
Oxidative sulfitolysis, 318
Oxidizing matrices, 180
Oxytocin, 180, 181, 193, 276
Oxytocin-neurophysin (OT-NP), 354

P

Paim I, disulfide bonds in, 277
Pain, back, 372, 387—388
Pancreas, catfish, peptide charting and, 351, 352, 356,
361—362
Pancreatic extract, catfish, 351, 352
Pancreatic peptides, 20—24
Pancreatic spasmolytic polypeptide, 322
Papain, in buffer solutions, ^{252}Cf-PDMS of, 7, 9
Parent ions, 122, 123, 271
Parent ion spectrometer, 161—162
Partial degradation, 298
Partial hydrolysis, blocked N-termini and, 320, 322
Particle bombardment, see also Fast atom bombard-
ment (FAB)
kiloelectronic, 123
triple quadrupole mass spectrometry and, 140
Particle-induced desorption, 88, 348
Parvalbumin proteins, 311
PBS, see Phosphate-buffered saline
PDMS, see Plasma desorption mass spectrometry
Pentafluorobenzyl group, 192
1,5-Pentanediol, 341, 394
Pentane-2,4-dione, arginine derivative formation and,
319
Pepsin
insulin and, 285
porcine, 69
Peptide(s), 18
connecting, 82
cyclic, see Cyclic peptides
disulfide-containing, detection of, 277—281
isolated, analysis of, 269—271
laser induced multiphoton ionization of, in
supersonic beam/mass spectrometry, 88—105,
see also Laser entries; Multiphoton ionization
(MPI)
linear, see Linear peptides
PDMS of, see Plasma desorption mass spectrometry
phosphorylated, 31, 33—35
tandem mass spectra of, 132—135
Peptide adsorption, on nitrocellulose, dynamics of, 6
Peptide:buffer ion-pair, 374
Peptide charting, 348—363
data analysis and, 351
endocrine tissues and, 351—363
bovine and rat pituitary, 353—356
catfish pancreas, 356—362
mass spectrometry and, 350—351

principle of, 348—350
Peptide chromatogram, 210
Peptide concentration, fragment ion yields and, 334
Peptide mapping, 224, 351
    by PDMS, 77—79, 82
    peptide charting versus, 348
Peptide mixtures, 19—23, 26, 27, see also Mixture
        analysis
    pancreatic, 20—24
    permethylation and, 322
    tryptic mapping of, 21—22, 26, 27, 224
Peptide neuroendocrinology, 368
Peptide secondary structure, 183
Peptide sequencing, 329
Peptide solubility, 175
Peptide synthesis, see also Synthetic peptides
    chemical procedures used in, evaluation and
        optimization of, 61
    confirmation of, 18—19
Peptidoglycan, 135
Perchloric acid, 385
Perfluorocarbon sulfate cation exchange polymer, 5
Performic acid oxidation, 193, 264—265
    cysteine derivative formation and, 318
    disulfide bonds in peptides and, 279, 280
Peripheral nervous system, 328
Permethylated peptides, 316—323
    advantages and disadvantages of, 322—323
    chemistry of, 316—319
    mass spectrometry of, 319—322
    when to use, 323
Permethylation reaction, 317, 338
Phagocyte stimulation, 393
Phosphate-buffered saline (PBS), 7
Phosphodiesterase isozyme, 145
Phosphoglycoprotein, 31
Phosphorylated peptides, 31, 33—35
Phosphorylation, 379
Photoactivation, 313
Photochemistry, 181
Photodissociation, see Laser photodissociation
Photodissociation mass spectrum, 145, 147
Photoionization, direct, 88
Photon absorption, in MPI, 88
Photo-oxidation
    of poly(ethylene terephthalate), 9
    ultraviolet-induced, 6
Pituitary, 372
    bovine, peptide charting and, 351—359
    drug overdose and, 390—392
    β-endorphin in, 391, 393
    human, post-mortem, 389—393
    rat, peptide charting and, 351, 355—356, 360
Pituitary extract, bovine, 351, 353
Pituitary tumors, 389—390
Plasma desorption mass spectrometry (PDMS), 196,
        316, see also 252-Californium plasma
        desorption mass spectrometry (^{252}Cf-PDMS)
    dimers and adducts in, 31—32, 35—36
    insulin and, 32, 35—37

insulin-glutathione mixed disulfides and, 36, 38
    TAT III cysteine-rich region and, 36, 38, 39
    glycosylation site and heterogeneity assessment
        with, 22, 24—25, 28—30
    bovine fetuin and, 24—25, 30
    ovomucoid third domains and, 22, 24, 28, 29
    trypanosomes and, 25, 28, 32
    molecular weight determination in, 74—76
    peptide mapping by, 77—79
        in situ reactions in, 77, 78
        reaction monitoring in solution in, 77
        suppression effects in mixture analysis and, 79
    of peptide mixtures, 19—22, 23, 26, 27
        pancreatic peptides, 20—24
        tryptic mapping in, 21—22, 26, 27
    of peptides
        peptide conjugates and, 18—39
        proteins and, 66—83
        suppression effects and, 188
    peptide synthesis confirmation with, 18—19
    permethylated peptides and, 320
    of phosphorylated peptides, 31, 33, 34
    plasma desorption mass spectrum and, 66—69
        accessible mass range and, 69
        components of, 66—67
        mass calibration and, 68
        mass determination precision and, 68—69
    post-translational modifications and, 31
    protein biotechnology applications of, 80—82
    protein engineering applications of, 82
    protein sequence determination with, 79—80
    sample preparation for, 69—74
        nitrocellulose matrix and, 71—73
        practical application and, 73—74
    tryptic mapping by, 21—22, 26, 27
Plasma desorption mass spectrum
    accessible mass range and, 69
    components of, 66—67
    mass calibration and, 68
    mass determination precision and, 68—69
"Plug flow", 214, 215
Pneumatically assisted electrospray, 214
Poly(ethylene terephthalate), 6, 9
Polymer surfaces, protein adsorption on, 7
Polyneuropathy, amyloidotic, familial, 253
Polypropylene tubes, solvent-leached, 317
pOMC, see Proopiomelanocortin
Pons-medulla, 389
pOP, see Prooxyphysin
Porcine pepsin, 69
Porous metal frit, 203
Porphyrin, 58, 258
Post-HPLC detector, 376
Post-ionization, 122
Post-mortem pituitary tissue, 389—393
    drug overdose and, 390—392
    β-endorphin in, 391, 393
    tumor profiles from, 389—390
Post-translational modification(s), 20, 81, see also
        Disulfide bond(s)

endocrine tissues and, 351
globin chains and, 260
murine elongation factor and, 31
neuropeptides and, 370, 379
permethylation and, 323
Power meter, in laser desorption/multiphoton
    ionization, 99
pPP, see Propressophysin
Prealbumin variant, 253, 255, 256
Precipitation
    acid, 373
    ether, 350
    solute, 175
Precursor ion, collision gas and, 124
Premolars, 384
Preparative reduction reactions, monitoring of, 77
Preproenkephalin A, 328, 331
Preproenkephalin B, 328, 331
Probe(s)
    continuous-flow sample introduction, 203
    direct insertion, 320
    liquid metal ion, 350
Proctolin, 329, 334, 336
Product ion collisionally activated decomposition,
    HPLC/CF-FAB and, 212—213
Product ion mass spectrum
    hemoglobin and, 271
    reflecting time-of-flight mass spectrometer and, 160
Product ion resolution, reflecting time-of-flight mass
    spectrometer and, 166—168
Product ion spectrometer, 162—164
Proenkephalin A, 370
Proenkephalin B, 370
Proglucagon, 20, 356
Proinsulin, 18, 20, 81, 110, 356
Proopiomelanocortin (pOMC), 328, 331, 351, 353,
    361, 370
Prooxyphysin (pOP), 353
Propressophysin (pPP), 353
Prosomatostatin, 20, 356, 361
Proteases, 265
Protected peptide segments, 53
Protein(s)
    acid precipitation of, 373
    adsorbed, surface concentration of, 6—7, 9
    disulfide bonds in, see Disulfide bond(s)
    enzymic digests of, 210
    extracellular, three-dimensional structure of, 276
    gas phase ions of, formation of, 10
    high molecular weight, in buffer solutions, 7—8
    mass determination of, by ^{252}Cf-PDMS, 11—13
    peptides and, PDMS of, 66—83, see also Plasma
        desorption mass spectrometry (PDMS)
    synthetic, see Synthetic peptides, synthetic proteins
        and
Protein adsorption
    matrix and, 5
    on nitrocellulose
        dynamics of, 6
        stoichiometry of, 6—7

Proteinase K, 7, 9, 50
Proteinases, serine, 22
Protein biotechnology, PDMS in, 80—82
Protein cross linkages, disulfide-related, 285
Protein engineering, PDMS in, 82
Protein molecular ion, internal excitation of, 9
Protein sequence determination, 79—81, see also
    Edman degradation
Protein Society Symposium Test Peptide-3, see STP-
    3
Protein structure, from on-line analysis of enzymic
    hydrolysates, CF-FAB and, 209—210
Protein variants, molecular disease and, 224
Proton affinity, 79, 178
Protonated linear peptides, fragmentation pathways
    of, nomenclature of, 127—132
Protonated molecular ions, 264, 380
Protonation, 334
    amide nitrogen atom, 128, 129, 291
    site of, fragmentation and, 133
Proton-bound dimers, 31
Pseudo-genes, 259
Psychoneuroendocrinoimmunology, 370
Pulsed injection source, 90
Pulsed laser source, 90, 92, 313, see also Laser
    desorption
Pyridine, amino group acylation and, 318
Pyridinium salt, 191—193
S(4-Pyridylethyl)cysteine, 318
Pyroglutamyl, 379
Pyrolyzation, 317
Pyrylium salt, 191

Q

Q-FTMS (quadrupole Fourier transform mass
    spectrometer), 142, 144, 145
QQHQC (quadrupole, quadrupole, homogeneous
    magnetic sector, quadrupole, cylindrical
    electrical sector) mass spectrometer, 231—
    233
Quadrupole Fourier transform mass spectrometry, see
    also Tandem mass spectrometry
    methods of, 140
    oligopeptide sequence analysis with, 142, 144, 145
Quadrupole instruments, 69, 196
    CQH type mass spectrometer, 231
    multi-quadrupole mass spectrometer, 329
    Q-FTMS, 142, 144, 145
    QQHQC type mass spectrometer, 231—233
Qualitative analytical mass spectrometry, 369
Quantitation, 369, 383
    difficulties with, 202
    neuropeptide, 341
    surfactancy and, 184
Quartz window, in high-pressure fluid injection, 90
Quaternization
    with methyl iodide, 191
    prevention of, amino group acylation and, 317—
        318

Quinones, 181

R

Radiation damage, to organic matrix, 202
Radical loss, 129
Radioimmunoassay, 213, 348, 368, 379
Radiolabeled iodoacetic acid, 276
Radiolabeling, 28, 31
Radioreceptor assay, opioid, 368, 376—377
Rat pituitary, peptide charting and, 351, 355—356, 360
Real electronic state, 88
Receptoractivity, 372
Recombinant DNA, 18
Recombinant growth hormone (rGH), HPLC/CF-FAB and, 212
Recombinant hirudins, 114
Recombinant human growth hormone (rhGH), CF-FAB and
  CZE and, 215, 217
  HPLC and, 210
Redox pair, 180
Reducing matrices, 180—181
Reduction potentials, 181
Reduction reactions, 77, 180
Reflecting time-of-flight mass spectrometer, 160—168, see also Time-of-flight (TOF) method
  alkali metal cationized peptides and, 308
  correlated mass measurements with, 165—168
    efficiency of, 165—166
    product ion resolution and, 166—168
  as tandem instrument, 160—164
    parent ion spectrometer MS-1 and, 161—162
    product ion spectrometer MS-2 and, 162—164
Relaxins, 18
REMPI, see Resonance enhanced multiphoton ionization
Renin-substrate tetradecapeptide, 140
"Reoccupation" problem, 375
Reproducibility
  four-sector tandem mass spectrometry and, 127
  narrow magnetic scans and, 111
Resolution, see Mass resolution
Resonance enhanced multiphoton ionization (REMPI), 88, see also Laser desorption, multiphoton ionization and
Resonant electronic state, 90
Resonant two-photon ionization (R2PI), 88—89
Retardation fields, 12
Retention time, 385
Retro-isomers, 293, 295
Reversed-geometry mass spectrometer, 329
Reversed-phase high performance liquid chromatography, 44, 316
  arginine derivative formation and, 319
  capillary, slurry-packed, 212
  hemoglobin and, 261, 262
  microbore, 210, 212
  neuropeptides and, 374—376
rf potential, 142

rGH, see Recombinant growth hormone
rhGH, see Recombinant human growth hormone
Riboflavin-binding protein, heterogeneity of, 31, 33—35
Ribonuclease, 142
Ribonuclease A, 193
  in buffer solutions, ^{252}Cf-PDMS of, 7, 8
  disulfide bonds and, 277, 284
  plasma desorption spectrum of, 66, 67
Ribonuclease B, pyridylethylated, tryptic peptides of, 212
Ribonuclease S peptide, 209
Ring cleavage, 291
Rockefeller University Mass Spectrometric Research Resource, 42
Rovibronic states, 88
RP-HPLC, see Reversed-phase high performance liquid chromatography
R2PI (resonant two-photon ionization), 88—89

S

Saline, phosphate-buffered, 7
Salt concentration, 177
"Salting out", 175
Sample consumption, 175
Sample introduction, permethylated peptides and, 320, 321
Sample partitioning, miscibility and, 177
Sample preparation
  for FAB and LSIMS, 190—197
  for PDMS, 69—74
  permethylated peptides and, 316—317
Sample surfactancy, matrix selection and, 181, 183—188
Sarcosine, 135
Satellite peaks, 334
Schiff base, 153, 177
Schiff-base derivatives, 316
Scytonemin A, 297—298
SDS, see Sodium dodecyl sulfate entries
Secondary ion mass spectrometry (SIMS), 316, 328, 350
  flow-SIMS, 190
  liquid, see Liquid secondary ion mass spectrometry (LSIMS)
Secondary ion mass spectrometry (SIMS) guns, 174
Segment condensation method, fully protected synthetic peptides and, 53
Selected reaction monitoring, 328
Selection rules, 88
Selective ionization, 88
Self-aggregation, hydrophobic proteins and, 74
Selvedge, 187, 196, 330
Senile dementia, 389
Sensitivity, 5, 180, 369
  CF-FAB and, 205
  high, desorption ionization and mass spectrometry and, 312—313
  peptide charting and, 363
Sep-Pak chromatography, 352, 374

Sequenators, 80, 248, 380
Sequence ion information, neuropeptides and, 339—341
Serine, 379
Serine proteinases, 22
Serum albumin, bovine, 4
Shock waves, laser desorption/multiphoton ionization and, 99
Sialic acid, 361
Sickle cell anemia, 259, 261
Sickle cell hemoglobin, digit-printing method and, 225, 229—230
Sidechain fragmentations, 128
Side-products, unwanted, identification of, 44—53
Siderochromes, 290
Signal-to-noise ratio(s)
    CF-FAB and, 205
    surfactancy and, 184
    survey scans and, 111
    tandem mass spectrometry and, 335, 381
    tryptic mapping and, 21
Signal variation, CF-FAB and, 207
Silanization, 317
Silicon diffusion pump fluid, 98
Silicone grease, 317
SIMS, see Secondary ion mass spectrometry entries
Site-directed mutagenesis, 82
Size fractionation, 28
Skimmer, in high-pressure fluid injection, 90
Sodium dodecyl sulfate (SDS), permethylated peptides and, 317
Sodium dodecyl sulfate (SDS) gel electrophoresis, 53, 66
Sodium hydride, 317
Soft ionization, 88, 95
Solenoid pulser, in high-pressure fluid injection, 90
Solubility
    considerations of, in matrix selection for FAB and LSIMS, 175—177
    substrates for ^{252}Cf-PDMS and, 8
Solute-solvent cluster ions, high-pressure liquid injection and, 94
Solvation effects, 307
Solvent-leached polypropylene tubes, 317
Solvent-solute interactions, neuropeptides and, 328, 341
Somatostatin(s), 20, 181, 193, 351
    in catfish pancreas, 356
    radioreceptor assay and, 376
    trypsinolysis of, 277
Source pressure, CF-FAB and, 207
Source-produced fragment ions, 299
Spectral congestion, 88
Spectroscopic selectivity, R2PI and, 88
Spin-drying method, 71, 353
Stainless steel, 188
Staphylococcus aureus protease, 77, 82
Staphylococcus aureus V protease, Hb F Izumi and, 248
Statistical rate theories, 127
Stepwise solid-phase synthetic procedure, 42

Stoichiometry, of protein adsorption on nitrocellulose, 6—7
STP-3, 145—146, 148—152
    chemical experiments and molecular weight determinations with, 148—149
    linkage assignments of two peptide chains and, 151—152
    results of chemical and enzymatic digests with, 149—151
Stress, orthodontic, 372, 384—385
Sublimability, anthroic acid and, 9
Substance P
    carboxypeptidase Y and, 209
    CF-FAB and, 203, 205
    CZE and, 214
    liquid chromatography and, 213
    surfactancy and, 186
    TOF spectra of, 160, 161
Substrates, new, for ^{252}Cf-PDMS, 8—10
Subtractive Edman degradation, 21, see also Edman degradation
    STP-3 and, 150—151
Sugars, metal-ion cationization and, 307
Sulfation, 379
Sulfhydryl groups, 180
Sulfitolysis, oxidative, 318
S-Sulfocysteine, 318
Sulfolane, 330
Sulfoxide, 379
Sulfoxide linkage, 285, 286
Superactive insulin, 18
Superconducting magnet, 142
Supercritical fluid chromatography, 323
Supercritical fluids, high-density, 89, see also High-pressure fluid injection
Supersonic beam/mass spectrometry, laser induced multiphoton ionization of peptides in, 88—105, see also Laser entries; Multiphoton ionization (MPI)
Supersonic jet introduction, R2PI method and, 88
Suppression effects, 184, 188, 202
    CF-FAB and, 205
    in mixture analysis, 79, 82
    peptide charting and, 362
Surface activation, 313
Surface activity, 175
Surface area, effective, CF-FAB and, 207
Surface-bound peptides, enzyme-catalyzed microscale chemical reaction of, 46
Surface/bulk concentration ratios, 186
Surface phenomenon(a)
    FAB ionization as, 202
    peptide charting and, 362
Surface properties, of hydroxylic liquid matrices, 175
Surface sensitivity, 375
Surface tension, 186
Surfactancy, 175, 181, 183—188
Survey spectrum, magnetic exponential scans and, 111
Synapse, 370
Synthetic errors, 42

Synthetic globin variants, 251
Synthetic peptides, 394
  in method development for amino acid sequence
      determination of cyclic peptides, 291—296
  reversed-phase HPLC and, 374
  synthetic proteins and, ^{252}Cf-PDMS analysis of,
      42—61, see also under 252-Californium
      plasma desorption mass spectrometry (^{252}Cf-
      PDMS)

T

Tachykinin, 384
Tandem mass spectra, 132—135
  collision gas and, 124
  negative ion, 132
Tandem mass spectrometry, 46, 122, 140, 290, 329,
      see also Quadrupole Fourier transform mass
      spectrometry; Triple quadrupole mass
      spectrometry
  in amino acid sequence determination of cyclic
      peptides, 291—306, see also Cyclic peptides
  in assessment of metal-ion cationized peptide
      decompositions, 306—311, see also Metal-ion
      cationized peptide decompositions
  desorption ionization and, future prospects for,
      311—313
  four-sector, see Four-sector tandem mass spec-
      trometry
  limitations of, 140
  neuropeptides and, 328, 329, 335, 368, 380—382,
      394—396
  reflecting time-of-flight mass spectrometer in,
      160—164
TAT III cysteine-rich region, metal-linked dimers of,
      36, 38, 39
Teeth, 372, 384—385
Temperature, internal, 126
Temporal profile, in laser desorption/multiphoton
      ionization, 99
Termination peptides, 53
Tetrapeptide, chemotactic, 320
TG, see 1-Thioglycerol
TGFα, see Transforming-growth factor-alpha
Thalamus, 389
Thalassemia syndromes, 259—260
Theoretical plates, CZE and, 214
Thermal instability, 4
Thermospray, 93—95
Thioether, 379
Thioglycerol, 180, 330
α-Thioglycerol, 341
1-Thioglycerol (TG), 110, 279
Thioglycolic acid, 180
Thiol-disulfide interchange, 36
Thiols, 276, 394
Thioredoxins, 135
Third order ion optics (TRIO), 231
Threonine, 379
Thyroglobulin, 378

Thyrotropin-releasing hormone (TRH), 368
Time course digest, 79
Time digitizer(s), 12, 165
Time-of-flight (TOF) method, 11, 18, 196, 312, see
      also Reflecting time-of-flight mass spectrome-
      ter
  covalent structure and, 42—44
  high-pressure ammonia injection and, 92
  peptide charting and, 350
  plasma desorption spectrum and, 66
Tissue acquisition, neuropeptides and, 372—373
Tissue-specific processing, 389
TOF, see Time-of-flight method
p-Toluenesulfonic acid, 179
Tooth pulp, 372, 384—385
Toxic cyclic peptides, from blue-green algae,
      characterization of, 152—157
Transforming-growth factor-alpha (TGFα), 59
Transformylation, 50
Transient waveform recorder, 92
Translational energy losses, 301
TRH, see Thyrotropin-releasing hormone
Triantennary oligosaccharides, 24
Trichloroacetic acid, 179, 181
Trideuteriomethyl iodide, 317
Triethanolamine, 177, 178
Trifluorethanol, 58
Trifluoroacetic acid, 7, 58, 140, 179, 212
Trifluoroacetyl, 317
Trigeminal system, 372, 383—386
  maxillary and mandibular branches of, 384—386
  ophthalmic branch of, 385
Trimethyllysine, 132
Trimethylsilylation, 338
Trimethylsilyl derivatives, 316
TRIO, see Third order ion optics
Triphenylmethane, 317
Triple quadrupole mass spectrometry, see also
      Tandem mass spectrometry
  methods of, 140
  oligopeptide sequence analysis with, 140—143
Triple sector instrument, 297
Tritiated etorphine, 377
Troponin C, 311
Trypanosomes, high-mannose glycopeptide from, 25,
      28, 32
Trypsin, 50, 224, 263
  disulfide bonds and, 276, 281
  immobilized, 210, 385
  peptides generated by cleavage with, molecular
      weight of, 239—243
Trypsinogen, 114
Trypsinolysis, of somatostatin, 277
Tryptamine, 95
Tryptic digest, 36
  of bovine growth factor, 212
  of globin, 262—269
  of glucagon, 205, 212
  of α-lactoglobulin A, 210
  of murine elongation factor, 31

of rhGH, 215
Tryptic fragments, 21
Tryptic mapping, by PDMS, 21—22, 26, 27
Tryptic peptide(s), 224, 232, 234—248
TSQ-70 triple quadrupole mass spectrometer,
    oligopeptide sequence analysis on, 140—143
Tuftsin, 393
Tumors
    endocrinologically silent, 372, 379
    pituitary, 389—390
Two-color frequencies, 88
Tyramine, 95
Tyrosine, 379
Tyrosine-containing peptides, 336

U

Ubiquitin, 114
Ultracold molecules, 89
Ultracooling, laser desorption and, 95
Ultraviolet-induced photo-oxidation, 6
Ultraviolet lasers, 88, 313
Ultraviolet light, MPI and, 88
Unimolecular decomposition, 126
Unimolecular fragmentation, 127
Uppsala group, 5
Urea, 73, 319

V

Vapor pressure, 4, 179
Vasopressin, 181, 193, 276
[Arg⁸]-Vasopressin, disulfide bonds and, 279, 280

Vasopressin-neurophysin (VP-NP), 354
V8 protease
    peptides generated by cleavage with, molecular
        weight of, 240, 241, 243, 244
    Staphyloccocus aureus, Hb F Izumi and, 248
Vibrational energy, 126, 140
Vibrational excitation, 127
Viral proteins, 322
Viral replication, 36
Viscous organic matrix, reduction in, CF-FAB and,
    205
VP-NP (vasopressin-neurophysin), 354

W

Wavelength ionization spectra, 99
Wavelength selectivity, laser desorption and, 95
Western blotting, 66
White muscle, calcium in, 311
Wick material, 190
Wisdom teeth, 385

X

Xenobiotic modifications, four-sector tandem mass
    spectrometry and, 135
Xenon, kiloelectronic particle bombardment and, 123
Xenon atom guns, 110, 180, 330, 394
X-ray crystallography, 380
X-ray studies, ovomucoid, 24

Z

Zwitterionic peptides, 308, 316, 361